Introductory
Structural Analysis

CIVIL ENGINEERING AND

ENGINEERING MECHANICS SERIES

N. M. Newmark and W. J. Hall, editors

CHU-KIA WANG

Professor of Civil and Environmental Engineering
University of Wisconsin–Madison

CHARLES G. SALMON

Professor of Civil and Environmental Engineering
University of Wisconsin–Madison

Introductory Structural Analysis

PRENTICE-HALL, INC., *Englewood Cliffs, New Jersey 07632*

Library of Congress Cataloging in Publication Data

Wang, Chu-Kia, 1917–
  Introductory structural analysis.

  Includes bibliographical references and index.
  1. Structures, Theory of.   I. Salmon, Charles G.
II. Title.
TA645.W324 1984   624.1'71   82-20429
ISBN 0-13-501569-3

Printed in the United States of America

10 9 8 7 6 5 4 3 2 1

Editorial/production supervision and interior design by Paul Spencer
Cover design by Photo Plus Art
Manufacturing buyer: Anthony Caruso

ISBN 0-13-501569-3

Prentice-Hall International, Inc., *London*
Prentice-Hall of Australia Pty. Limited, *Sydney*
Editora Prentice-Hall do Brasil, Ltda., *Rio de Janeiro*
Prentice-Hall Canada Inc., *Toronto*
Prentice-Hall of India Private Limited, *New Delhi*
Prentice-Hall of Japan, Inc., *Tokyo*
Prentice-Hall of Southeast Asia Pte. Ltd., *Singapore*
Whitehall Books Limited, *Wellington, New Zealand*

Contents

Contents

Contents

Preface

Undergraduate students in civil engineering come to their first course in structural analysis with a background in statics, dynamics, and mechanics of materials. The essential preparations include the equilibrium equations of statics by means of free-body diagrams, together with shears, moments, and the flexure formula for beams. Students all need more practice to acquire proficiency.

Although analysis and design are really inseparable, for reason of efficiency a course in structural analysis almost invariably precedes courses in structural design. For that reason, this first course not only includes the theoretical principles of structural analysis but provides the link between pure analysis and the practical aspects of structural design. For the latter objective, much material has been included in the first chapter, and relevance to structural design has been highlighted wherever appropriate. In addition, the historical notes should help to stimulate the interest of students in regard to how and why methods of structural analysis were needed and developed.

In this text, unlike C. K. Wang's *Introductory Structural Analysis with Matrix Methods*, instead of solving the same problem by both classical and matrix methods in the same chapter, the matrix solutions are separately treated in the latter one-third of the book.

Chapter 13, on matrix operations, physically divides the chapters on classical methods from the chapters on matrix methods. Chapter 14, on general concepts of flexibility and stiffness methods, is an overview chapter which conceptionally links the two types of methods. In fact, Chapters 15 to 20 simply contain the step-by-step "how-to-do-it" procedures for solving most of the numerical examples in Chapters 1 to 12 in matrix notations. Chapter 14, then, can itself be presented in one or more lectures near the end of the semester.

There is considerable difference of opinion as to whether matrix methods can or should be included in the first course in structural analysis. While the total content that can be absorbed by a student in a typical three-credit course has its limits, the optimum choice of course content has to be a compromise between scope and proficiency. This book is presented in such a way that most students should be able to read through whatever chapters may remain unused in the classroom, when the need or opportunity arises. Of course, this is only for the student who does not choose to take additional courses in structural analysis.

The methods of determining the force and deformation response of statically determinate trusses and beams are presented in Chapters 2 to 5. To make room for matrix methods, one may consider, for instance, the omission of Sections 2.4, 2.5, 3.5, 3.6, 5.2, 5.6, and 5.7. Chapter 6 treats influence lines and moving loads; again, Sections 6.9 to 6.14 may be omitted. Chapters 7 and 8 discuss the force method of analyzing statically indeterminate trusses and beams; these are two important chapters that show how physical equilibrium and compatibility are related to the two Castigliano's theorems.

There can be debate as to whether Chapter 9, on the slope deflection method, can be bypassed in favor of Chapter 17, on the matrix displacement method. The compromise might be to assign Chapter 9 for reading only, but have homework problems taken from Chapter 17. Chapter 10, on the moment distribution method, should be considered essential in a first course; in fact, for some situations, moment distribution might be better utilized on the microcomputer than by matrix methods.

Chapter 11, on rigid frames, contains no new methods; it shows only the application of the various methods to the analysis of rigid frames. Chapter 12, on approximate methods of frame analysis, may stand alone, available for those teachers who wish to use it.

If matrix methods are included in the first course, minimum coverage should be Chapters 15, 16, and 17. Here again, there is difference of opinion as to whether the direct stiffness method, explained in Section 16.8, should be used exclusively for all types of structures, without the introduction of statics and deformation matrices as defined in Sections 16.2 and 16.3. It is true that the former is used exclusively in computer programs; however, the authors believe that students should learn the fundamentals at this stage of their education. The adaptation to large-scale problem solving on the computer can come later.

In Chapter 19, the treatment of sidesway by the matrix method, using the assumption of inextensible members, is consistent with the traditional moment distribution method. This approach seems desirable because of its practical use for design, even though in the direct stiffness method of computer analysis, axial stiffness can easily be included. Particularly, the assumption of inextensible members is useful in preliminary design of rigid frames subject primarily to bending deformation. If time is of the essence, however, Chapter 19 may be omitted, because Chapter 20 does consider both axial and bending deformation.

The use of computers has rapidly changed computational procedures, dictating in many respects the way one must learn structural analysis concepts. The authors are firm in the belief that basic principles, without being obscured in mathematical

[1]

Introduction

1.1 PURPOSE OF STRUCTURAL ANALYSIS

Structural analysis involves the determination of the forces acting *within* a structure or components of a structure. The term *force* is meant to include bending moment, shear, axial compression or tension, and torsional moment. Components of a structure are elements such as *beams* (members subject primarily to bending moment with little axial compression or tension), *columns* (members subject primarily to axial compression with little bending moment), *tension members* (members subject primarily to axial tension with relatively little bending moment), and *beam-columns* (members subject to significant magnitudes of bending moment as well as axial compression). In addition, components may refer to integral systems, such as an entire floor, a roof, a multistory wall, or even the entire structure.

Structural analysis and structural design are interlocked subjects. The structural engineer has the objective of proportioning a structure such that it can safely carry the loads to which it may be subjected. Structural analysis provides the internal forces and structural design utilizes those forces to proportion the members or systems of members. Sometimes the laws of statics are sufficient to determine the internal forces from known applied loads and structural configuration; such structures are said to be *statically determinate*. More frequently, the internal forces are dependent on the relative stiffness of members in addition to the laws of statics; such structures are said to be *statically indeterminate*. For statically indeterminate structures, the design process involves iteration; the relative sizes must be initially estimated, the structural analysis performed to determine the internal forces, the members or systems of members proportioned for the internal forces, the stiffnesses of members or systems of members

1

evaluated, and the structural analysis repeated. Without structural analysis, design is impossible.

Structural design has been defined [1] as "a mixture of art and science, combining the experienced engineer's intuitive feeling for the behavior of a structure with a sound knowledge of the principles of statics, dynamics, mechanics of materials, and structural analysis, to produce a safe economical structure which will serve its intended purpose."

For many centuries, structures were designed solely as an art. Early Roman and Greek structures were proportioned without the aid of structural analysis to determine the internal forces. When the proportioning was incorrect, the structures collapsed and for most, no historical reference remains. Some structures were overdesigned; some might be found to have been correctly-designed if reviewed according to twentieth-century accepted procedures.

Structural analysis, as it is recognized today, began about 130 years ago with the publication by Squire Whipple [2] of a rational discussion of the determination of forces in trusses. For the elastic theory relating to flexural members, Navier [3] is generally regarded as the founder of the modern theory of elastic solids, with his published memoir of 1827. This theory has essentially provided the differential equation method for the analysis of statically indeterminate beams.

Practical treatment of statically indeterminate analysis may be considered to have originated with the French civil engineer Clapeyron [4], who in 1857 published his "theorem of three moments" for the analysis of continuous beams. During the period from 1860 to 1900, structural analysis ideas and procedures became known and influenced structural design to a considerable extent. Structural configurations began changing from predominantly truss-type, involving members having primarily axial compression or tension, to frame-type, involving members having flexural as well as axial interaction between them.

The beginning student of structural engineering needs to acquire the "tools" to make the structural analysis of whatever structural layout may be appropriate for the structure to serve its intended function. The structural analysis may include establishing the loads to be carried. Certainly, the weight of the members or systems of members must be included as part of the loads. Loads in addition to the self-weight may be prescribed legally by the local building code, or in the absence of such requirement may have to be established by the structural engineer. Loads that are not fixed in position must be applied to the structure in a manner determined by the structural engineer to cause the most severe effect. Thus, structural analysis includes the study of the influence of various types and positions of loads on the internal forces within a structure, so that the structural design, which may involve iterations, may be performed.

1.2 TYPES OF STRUCTURAL FORM

Structures are load-carrying frameworks, such as bridges, buildings, ships, aircraft, transmission towers, machinery, offshore oil platforms, and dams. In accordance with the predominant structural action, structures may be identified in three general categories: (1) *framed*, where elements may consist of tension members, columns, beams, and members subject to combined bending and axial load; (2) *shell-type*, where membrane forces in the thin-walls predominate in the principal support action; and (3) *suspension-type*, where axial tension in the cables provides the main support.

Framed Structures

A framed structure may be composed of straight or curved elements, whose ends may be either simply connected (idealized as pin-connected) or rigidly connected as by welding. Typical framed structures are the truss, the beam, and the rigid frame. An ideal two-dimensional truss as shown in Fig. 1.2.1a consists of bars connected by pins at the ends. The continuous beam of Fig. 1.2.1b, although it may be actually one long member, may also be considered to be three members rigidly joined as by welding at B and C. A model of the two-dimensional rigid frame of Fig. 1.2.1c may actually be a monolithic cut from a steel plate, but in its prototype it may be a monolithic reinforced concrete frame (as the building frame of Fig. 1.2.2 or the bridge pier of Fig. 1.2.3), or a steel frame of five members with rigid (e.g., welded) joints at D, E, and F of Fig. 1.2.1c (as the welded steel rigid frame of Fig. 1.2.4).

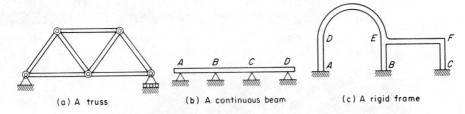

(a) A truss   (b) A continuous beam   (c) A rigid frame

Figure 1.2.1 Typical framed structures.

Figure 1.2.2 Reinforced concrete frame, showing the monolithic rigid joints at the junction of beams and columns. (Photo by C. G. Salmon)

Figure 1.2.3 Reinforced concrete bridge pier, showing monolithic rigid joints at the top of the piers. (Photo by C. G. Salmon)

Figure 1.2.4 Structural steel framed building having parallel frames containing welded rigid joints that lie in one plane, and having these rigid plane frames joined transversely by relatively flexible members and nonrigid connections. (Photo by C. G. Salmon)

Most typical buildings and bridges are framed structures. The multistory building usually consists of beams and columns, either rigidly connected (as by welding) or having simple end connections, together with diagonal bracing to provide stability. Although the multistory building is three-dimensional, when designed to have rigid joints it usually has a much greater stiffness in one direction than the other, and therefore may reasonably be treated as a series of plane frames. When trusses are used in order to carry very heavy loads across long spans as shown in Fig. 1.2.5, each is oriented and spaced to perform as a plane truss with loads primarily transmitted in one plane. However, when the framing is such that the behavior of the members in one plane substantially influences the behavior in another plane, the frame must be treated as a three-dimensional system, known as a *space frame*.

Figure 1.2.5 Plane trusses spanning between main columns at the corners of a building. (Photo courtesy of Bethlehem Steel Corporation, Bethlehem, Pa.)

Industrial buildings and special one-story buildings such as churches, schools, and arenas generally are either wholly or partly framed structures. The roof system may be a series of plane trusses, a space truss (see Fig. 1.2.6), a framed dome (see Fig. 1.2.7), or just part of a rigid flat or gabled one-story rigid frame.

Bridges are mostly framed structures, such as continuous beams (see Fig. 1.2.8), or trusses, usually continuous (see Fig. 1.2.9).

It is the primary purpose of this book to introduce the methods of analyzing framed structures: thus the title, *Introductory Structural Analysis*.

Figure 1.2.6 Space truss: a three-dimensional roof truss. Photo also shows column–girder–beam construction. (Photo courtesy of Whitehead & Kales Company, Detroit)

Figure 1.2.7 Framed dome: a steel framework covered by a roof "skin." (Photo courtesy of United States Steel Corporation, Pittsburgh)

Figure 1.2.8 Continuous beam bridge on curve, containing a reinforced concrete portion (spans at right of photo) and a structural steel portion (spans curving around at left). (Photo by C. G. Salmon)

Figure 1.2.9 Continuous truss bridge. Outerbridge Crossing, Staten Island, New York. (Photo by C. G. Salmon)

Shell-Type Structures

A shell structure is a curved thin-walled structure usually supported continuously at the edges. The shell itself serves the dual function of providing cover or containment in addition to its participation in carrying loads. One common type where the main stress is tension is the containment vessel used to store liquids (for both high and low temperatures), of which the elevated water tank is a notable example. A roof shell in the shape of a barrel, dome, or hyperbolic paraboloid is a system where compressive stress predominates. On many shell-type structures, such as roof domes, ship hulls, and airplane fuselages, a framed structure may be used in conjunction with the shell. In these cases, the "skin elements" act together with the frame to comprise an entire system.

Flat plates, such as floor slabs in buildings and deck plates on bridges and ships, are special cases of thin-walled systems without curvature. When flat elements have significant bending stiffness and have little in-plane loading on them, they are more properly included in the category of framed structures. Also, when flat elements form the "skin" of, or are attached to, an otherwise framed structure, they may more properly be considered as components (such as the flanges) of other elements (such as beams) in a framed structure.

Analysis of shell-type structures is outside the scope of this text, requiring a knowledge of theory of elasticity, and of approximate techniques such as the finite element method.

Suspension-Type Structures

In the suspension-type structure, tension cables are the major load-carrying elements. A roof may be cable supported, as shown in Fig. 1.2.10. Probably the most common structure of this type is the suspension bridge, as shown in Fig. 1.2.11. Usually, a

Figure 1.2.10 Cable-supported roof under construction, showing the cables extending radially from a ring at the center of the roof. The roofing material is to be attached to the cable system. Madison Square Garden, New York. (Photo courtesy of Bethlehem Steel Corporation, Bethlehem, Pa.)

Figure 1.2.11 Suspension bridge, showing detail of the stiffening truss. Verazzano Narrows Bridge, New York. (Photo by C. G. Salmon)

subsystem of the structure consists of a framed structure, as in the stiffening truss for the suspension bridge. Since the tension element is the most efficient way of carrying a load, increased use is being made of structures utilizing this concept.

Many unusual structures utilizing various combinations of framed, shell-type, and suspension-type structures have been built. However, the structural engineer is routinely required to be able to do structural analysis for framed structures—thus the emphasis in this book.

1.3 DEAD LOAD AND LIVE LOAD

Since structural analysis is the portion of the design process that involves the determination of internal forces caused by external loads, it is necessary at the very beginning to establish the loads to be carried by the structure. Generally, loads may be divided into two categories: dead loads and live loads. *Dead loads* are loads fixed in position and of constant magnitude over the life of the structure. *Live loads* are movable or moving loads and may vary in magnitude.

Dead Loads

Dead loads are gravity loads due to the self-weight (mass) of the structure. They may also include attachments to the structure, such as pipes, electrical conduit, air-conditioning and heating ducts, lighting fixtures, floor covering, roof covering, and suspended ceilings, which are intended to remain for the life of the structure.

Dead loads are usually known accurately, but that portion relating to the structural component being designed is not known until a design is completed. In the design process the weight (mass) of that component must be estimated, preliminary structural analysis and design made, and the member section revised if necessary.

Live Loads

Live loads are ordinary loads, including gravity loads, that vary in magnitude and location. Gravity live loads, such as human occupants, furniture, movable equipment, vehicles, and stored goods, may be called *occupancy loads*. Other live loads are snow (and rain) load (also a gravity load), wind load, earthquake, water pressure, wave forces, ice pressure, and highway and railroad vehicular traffic loads. When engineers refer to live load, they usually mean the occupancy loads. All other live loads are referred to more specifically, such as wind, snow, and highway traffic loads. These are considered separately in succeeding sections.

Occupancy live loads are those considered *movable* but not *moving*. In other words, the change in position of these loads, such as people moving about, causes a small enough dynamic effect on the structure that it may be neglected. These occupancy live loads are therefore treated as movable but static. Static and dynamic loading are compared under the topic *impact*.

Some gravity live loads may be practically permanent; others may be highly transient. The uncertain nature of the magnitude and location of live loads makes them difficult to estimate realistically.

Because of concern for adequate public safety, minimum live loads for design are prescribed by state and local building codes. These prescribed loads are generally obtained empirically and set conservatively high, based on previous experience and accepted practice. Whenever local codes do not apply or do not exist, values from a regional or national building code may be used. One such widely recognized code [5] is the *American National Standard Building Code Requirement for Minimum Design Loads in Buildings and Other Structures*, sponsored by the American National Standards Institute (ANSI), 1982. Typical occupancy live loads from this code are given in Table 1.3.1. The table gives relative magnitudes of representative loads and is not intended to include all situations.

Live load when applied to a structure should be positioned to give the most severe effect possible. Determination of where to position live loads is as important as the evaluation of the internal forces for a given positioning of loads. The use of *influence lines* to obtain the critical position of live loads is treated in Chapter 6.

In a multistory building, there is a low probability of having the prescribed loading uniformly over an entire floor, as well as over all other floors simultaneously. Most codes recognize this by allowing for some percentage reduction from full loading. For instance, except in areas to be occupied as places of public assembly, for live loads of 100 lb/ft² (4.8 kN/m²) or less, ANSI-A58.1-1982 [5] allows the design live load on any member having an *influence area* of 400 ft² (37.1 m²) to be reduced according to the equation

$$L = L_0\left(0.25 + \frac{15}{\sqrt{A_I}}\right) \tag{1.3.1}$$

where L = reduced design live load per square foot of area supported by the member

L_0 = unreduced design live load per square foot of area supported by member (as shown in a table such as Table 1.3.1)

A_I = influence area (the influence area for a column is four times the tributary area; for a beam it is two times the tributary area)

The greatest reduction will occur in the columns of a tall building, where the total area tributary to columns over 10 or 20 stories may be very large. The reduced live load L shall not, however, be less than $0.5L_0$ for members supporting one floor, nor less than $0.4L_0$ otherwise.

TABLE 1.3.1 TYPICAL UNIFORMLY DISTRIBUTED LIVE LOADS (ADAPTED FROM REF. 5)

Occupancy or use	Live load	
	lb/ft²	kN/m² [a]
1. Hotel guest rooms School classrooms Private apartments Hospital private rooms	40	2
2. Offices	50	2.5
3. Assembly halls, fixed seat Library reading rooms	60	3
4. Corridors above first floor	80	4
5. Theater aisles and lobbies Assembly halls, movable seats Office building lobbies Main floor, retail stores Dining rooms and restaurants	100	5
6. Storage warehouse, light Manufacturing, light	125	6
7. Library stack rooms	150	7
8. Storage warehouse, heavy Sidewalks, driveways, subject to trucking	250	12

[a]SI values are approximate conversions.

1.4 SNOW LOAD AND WIND LOAD

The live loads to be used for the design of roofs consist of snow load and wind load.

Snow Load

Since snow has a variable specific gravity, even if one knows the depth of snow for which design is to be made, the load per unit area of roof is at best only a guess. The recommended procedure for establishing snow load for design is to follow ANSI-A58.1-1982 [5]. This code uses a map of the United States giving isolines of ground snow corresponding to a 50-year mean recurrence interval for use in designing most permanent structures. The ground snow is then multiplied by coefficients to include the effects of wind exposure, thermal effects, and an importance factor to relate to the seriousness of the consequence of failure, thus obtaining the "unobstructed flat roof snow load." An additional multiplying factor is then applied to account for roof slope and shape of roof surface. Finally, nonuniform accumulation must be provided for on pitched or curved roofs, multiple-series roofs, multilevel roofs, and roof areas adjacent to projections on a roof level.

It is apparent that the steeper the roof, the less snow can accumulate. Also, partial snow loading should be considered, in addition to full loading, if it is believed that such loading can occur and would cause the more severe effect. Wind may also act on a structure that is carrying snow load. It is unlikely, however, that maximum snow and maximum wind loads would act simultaneously.

In general, the "unobstructed flat roof snow" load used in design varies from 30 to 40 psf (1.4 to 1.9 kN/m²) in the northern and eastern states to 20 psf (1.0 kN/m²) or less in the southern states. Flat roofs in normally warm climates should be designed for 20 psf (1.0 kN/m²) even when such an accumulation of snow may seem doubtful. This loading may be thought of as due to people gathered on such a roof. Furthermore, although wind is frequently ignored as a vertical force on a roof, it may nevertheless cause such an effect. For these reasons a 20 psf (1.0 kN/m²) minimum loading, even though it may not always be snow, is reasonable. Local codes, actual weather conditions, or ANSI-A58.1-1982 [5] should be used when designing for snow.

Wind Load

All structures are subject to wind load, but it is usually only those more than three or four stories high (other than long bridges) for which special consideration of wind is required.

On any typical building of rectangular plan and elevation, wind exerts pressure on the windward side and suction on the leeward side, as well as either uplift or downward pressure on the roof. For most ordinary situations, roof loading from vertically directed wind is neglected on the assumption that snow loading will require a greater strength than wind loading. Furthermore, the total lateral wind load, windward and leeward effect, is frequently assumed to act on the windward face of the building.

In accordance with Bernoulli's theorem for an ideal fluid striking an object, the increase in static pressure equals the decrease in dynamic pressure, or

$$q = \frac{1}{2}\rho V^2 \tag{1.4.1}$$

where q is the dynamic pressure on the object, ρ the mass density of air (specific weight $w = 0.07651$ pcf at sea level and 15°C), and V the wind velocity. In terms of velocity V in miles per hour, the dynamic pressure q (psf) would be*

$$q = \frac{1}{2}\left(\frac{0.07651}{32.2}\right)\left(\frac{5280V}{3600}\right)^2 = 0.0026V^2 \tag{1.4.2}$$

In common design procedures for usual types of buildings, the dynamic pressure q is converted into equivalent static pressure p, which may be expressed [6]

$$p = qC_eC_gC_p \tag{1.4.3}$$

where C_e is an exposure factor that varies from 1.0 (for 0 to 40-ft height) to 2.0 (for 740 to 1200-ft height); C_g is a gust factor, such as 2.0 for structural members and 2.5 for small elements including cladding; and C_p is a shape factor for the building as a whole. Excellent details of application of wind loading to structures is available in ANSI-A58.1-1982 [5] and in the National Building Code of Canada [7].

The commonly used wind pressure of 20 psf, as specified by many building codes, corresponds to a velocity of 88 miles per hour (mph) from Eq. (1.4.2). For an exposure factor C_e of 1.0, a gust factor C_g of 2.0, and a shape factor C_p of 1.3 for an airtight

*In SI units, $q = 0.63V^2$, for q in megapascals (MPa) and V in meters per second (m/sec).

building, this pressure corresponds to a 55-mph wind. For all buildings with nonplanar surfaces or plane surfaces inclined to the wind, or buildings having significant-size openings, careful examination of wind forces should be made using, for example, ANSI-A58.1-1982 [5] or the National Building Code of Canada [7]. For more extensive study of wind loads and effects, see the work of the Task Committee on Wind Forces [8].

1.5 EARTHQUAKE LOAD

An earthquake consists of horizontal and vertical ground motions, with the vertical motion usually having much the smaller magnitude. Since the horizontal motion of the ground causes the most significant effect, it is that effect which is usually thought of as earthquake load. When the ground under an object (structure) having a certain mass suddenly moves, the inertia of the mass tends to resist the movement, as shown in Fig. 1.5.1. A shear force is developed between the ground and the mass. Most building codes having earthquake provisions require the designer to consider a lateral force CW, which is usually empirically prescribed.

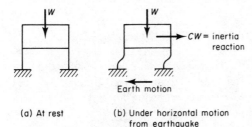

(a) At rest   (b) Under horizontal motion from earthquake

Figure 1.5.1 Force developed by earthquake.

The dynamic analysis of a structure under the action of a lateral force that varies with time, such as an earthquake, is outside the scope of this text, which treats only structural analysis for static loads. The reader is referred to Clough and Penzien [9] and Biggs [6], for example, for a treatment of structural dynamics. As the use of digital computers becomes more common, there is increased emphasis on performing an analysis for dynamic loads rather than for "equivalent" static loads. The simple "equivalent" static treatment of earthquake loads is, however, still the most prevalent method, and one of the commonly used procedures is that of the Structural Engineers Association of California (SEAOC) [10]. For the static base shear force the recommendation is

$$V = \beta CW \tag{1.5.1}$$

where V = base shear force to represent the dynamic effect of the inertia force
     W = weight of building
     $C = 0.05/\sqrt[3]{T}$ but not to exceed 0.10, the seismic coefficient equivalent to the maximum acceleration in terms of acceleration due to gravity
     T = natural period of the structure (i.e., time for one cycle of vibration)
     β = coefficient varying from 0.67 to 3.0, indicating the capacity of the members to absorb plastic deformation (low values indicate high ductility)

The total lateral load is recommended [10] to be distributed in accordance with the following:

$$F_n = \frac{W_n h_n}{\sum Wh} V \qquad (1.5.2)$$

where F_n = lateral force at the nth-floor level
$\quad W_n$ = weight at the nth-floor level
$\quad h_n$ = height above ground on the nth-floor level
$\quad \sum Wh$ = total sum of Wh for all floor levels

If the natural period T cannot be determined by rational means from technical data, it may be assumed to be

$$T = \frac{0.05H}{\sqrt{D}} \qquad (1.5.3)$$

where H = height of the building above its base
$\quad D$ = dimension of the building in the direction parallel to the applied forces (D to be in the same units as H)

The foregoing discussion based on the SEAOC is presented to show how dynamic earthquake loads may be approximated by static loads.

1.6 HIGHWAY LIVE LOADS TO REPRESENT VEHICLES

Highway vehicle loading in the United States has been standardized by the American Association of State Highway and Transportation Officials (AASHTO) [11]. The standard loadings include truck loads and lane loads that approximate a series of trucks. There are two systems, designated H and HS (M and MS for metric), which are identified by the number of axles per truck. The H system has two axles per truck, whereas the HS system has three axles per truck. Altogether there are five classes of loading: H10, H15, H20, HS15, and HS20. The loadings are shown in Fig. 1.6.1. In designing a given bridge, either one individual truck loading is applied to the entire structure, *or* the equivalent lane loading (representing a series of trucks) is

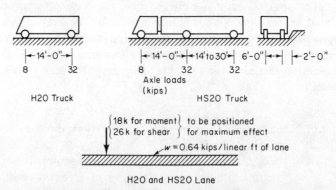

Figure 1.6.1 AASHTO-1977 Highway H20 and HS20 loadings (for H15 and HS15 use 75% of H20 and HS20, and for H10 use 50% of H20). (1 kip = 4.45 kN)

applied. When the lane loading is used, the uniform portion is distributed over as much of the span or spans as will cause the most severe effect. In addition, one concentrated load is positioned for its most severe effect. The knowledge of influence lines as treated in Chapter 6 is necessary to determine how to position these loads. The load distribution across the width of a bridge to its various supporting members is taken in accordance with semiempirical rules that depend on the type of bridge deck and supporting structure.

The single truck loading provides the effect of a heavy concentrated load and usually governs on relatively short spans. The uniform lane load is to simulate a line of traffic, and the added extra concentrated load is to account for the possibility of one extra heavy vehicle in the line of traffic. These loads have been used with no apparent difficulty since 1944, before which time a series of trucks was actually used for the design loading. On the interstate system of highways, a military loading consisting of two 24-kip (108-kN) axle loads 4 ft apart must be provided as a possible loading condition in addition to HS20.

Railroad bridges are designed to carry a similar semiempirical loading known as the Cooper E72 train, consisting of a series of concentrated loads at fixed distances apart followed by a uniform loading. This loading is prescribed by the American Railway Engineering Association (AREA) [12].

1.7 IMPACT AND DYNAMIC LOADS

The term *impact* as ordinarily used in structural design refers to the extra static load applied to approximate the dynamic effect of a suddenly applied load. In constructing a structure, the materials are added slowly; people entering a building are also considered a gradual loading. Dead loads are static loads; live loads may be either static or they may have a dynamic effect. People and furniture would be treated as static live load, but cranes and various types of machinery may also have dynamic effects.

When a force P is gradually applied to a linear elastic spring as shown in Fig. 1.7.1a, the extension X is at all times proportional to the applied force P such that

$$P = KX \qquad (1.7.1)$$

where K is the stiffness (resistance per unit length of deformation) of the spring. The work done in effecting an additional extension dX is

$$dW = P\,dX = KX\,dx$$

and the total work done in extending the spring length from L_0 to $(L_0 + X)$ is

$$W = \int dW = \int_0^x KX\,dX = \frac{1}{2}KX^2 = \frac{1}{2}PX \qquad (1.7.2)$$

Thus, the elastic energy stored inside the spring is equal to one-half of the product of the resulting force in the spring and its extension.

The behavior of a solid bar, as long as it remains elastic (i.e., stress is proportional to strain), is no different from that of a linear elastic spring, except that the stiffness K is now

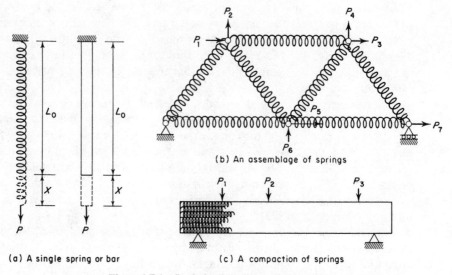

(b) An assemblage of springs

(a) A single spring or bar

(c) A compaction of springs

Figure 1.7.1 Static load on linear elastic springs.

$$K = \frac{EA}{L_0} \tag{1.7.3}$$

where E is the modulus of elasticity and A is the cross-sectional area.

If instead of applying a gradual force to the spring, a mass with weight equal to W is hung at the end as in Fig. 1.7.1a, there arises the question of whether the mass is gradually or suddenly released to act on the spring. In the former case, the load is said to be *static* and the resulting force in the spring, by use of Eq. (1.7.1), is

$$KX = W \tag{1.7.4}$$

Thus, for the static loading the force in the spring equals the weight W and the maximum deflection equals X.

When the load is *suddenly* released to act on the spring, the load is said to be *dynamic* and the maximum spring extension is $2X$, that is, twice as great as the static extension. In the latter case, the mass vibrates in simple harmonic motion with its neutral position equal to its static deflected position. In real structures, the harmonic (vibratory) motion is damped out (reduced to zero) very rapidly. Once the motion has stopped, the force remaining in the spring is the weight W.

Most structures for which structural analysis is to be made may be thought of as elastic springs or a series of springs. The truss of Fig. 1.7.1b is considered to consist of seven linear elastic springs hooked together at the five joints. The load system can be completely defined by the seven values of P. If the loads are gradually applied, the seven unknown deflections will also gradually increase from zero to their full amounts —as will the forces and deformations in the springs. Even the beam of Fig. 1.7.1c may be considered to consist of an infinite number of little springs compacted together, and its behavior under static loads is similar to that of the single bar or of a group of bars.

The structural analysis for dynamic load may be accomplished in either of two ways: (1) use a dynamic analysis on a mathematical model with actual dynamic forces (or a reasonable approximation thereof) acting on springs, where the springs themselves possess mass; or (2) use a static analysis with the dynamic forces replaced by "equivalent" static loads.

Traditionally in structural design, the latter method has been used for most situations. Gravity loads have been increased by using *impact factors*. For instance, when a single mass is suddenly applied to a spring (Fig. 1.7.2), the maximum displacement is twice the static displacement and the maximum spring force is $2W$. To account for this increased force during the time the mass is in motion, a load equal to $2W$ should be used if a static analysis were to be made. This is to add 100% of the static load to represent the dynamic effect, and is called a 100% impact factor.

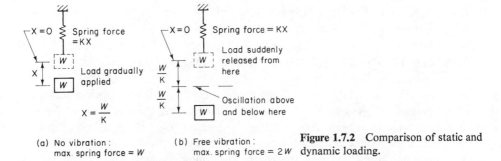

(a) No vibration:
max. spring force = W

(b) Free vibration:
max. spring force = $2W$

Figure 1.7.2 Comparison of static and dynamic loading.

Any gravity live load that can have a dynamic effect should be increased by an impact factor. Although a dynamic analysis of a structure could be made to determine these effects accurately, such a procedure (which is outside the scope of this text) is usually too complex or too costly in ordinary design. Thus empirical formulas and impact factors are usually used. In cases where the dynamic effect is small (say where impact would be less than about 20%), it is ordinarily accounted for by using a conservative (higher) value for the specified live load. The dynamic effects of persons in buildings and of slow-moving vehicles in parking garages are examples where ordinary design live load is conservative and no explicit impact factor is usually added.

For highway bridge design, however, impact is always to be considered. AASHTO-1977 [11] prescribes empirically that the impact factor expressed as a portion of live load is

$$I = \frac{50}{L + 125} \leq 0.30 \tag{1.7.5}$$

In Eq. (1.7.5), L (expressed in feet) is the length of the portion of the span that is loaded to give the maximum effect on the member. Since vehicles travel directly on the superstructure, all parts of it are subjected to vibration and must be designed to include impact. The substructure, including all portions not rigidly attached to the superstructure, such as abutments, retaining walls, and piers, are assumed to have adequate damping or be sufficiently remote from the application point of the dynamic load so that impact is not to be considered. Again, conservative static loads may account for the smaller dynamic effects.

In buildings it is principally in the design of supports for cranes and heavy machinery that impact is explicitly considered. The American Institute of Steel Construction (AISC) Specification [13] (AISC-1.3.3) states that if not otherwise specified, the impact percentage shall be:

For supports of elevators	100%
For cab operated traveling crane support girders and their connections	25%
For pendant-operated traveling crane support girders and their connections	10%
For supports of light machinery, shaft or motor driven, not less than	20%
For supports of reciprocating machinery or power-driven units, not less than	50%
For hangers supporting floors and balconies	33%

1.8 IDEALIZATION OF A STRUCTURE AND LOADING

An actual structure consists of members that have finite depth or thickness and of connecting joints that also have finite dimensions. Loads such as from a column supported by a beam shown in Fig. 1.8.1 actually act along a finite length. For structural analysis the system of members and loads is usually idealized into a series of line diagrams using for member lengths the center-to-center distance of the supporting members or elements. Loads, such as the column load, are treated as concentrated at the center of the column and shown by force arrows on the idealized structure. Of course, occupancy-type dead and live loads are shown as uniformly distributed either in part or full acting along the member.

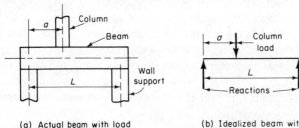

(a) Actual beam with load supported on walls

(b) Idealized beam with load and reactions

Figure 1.8.1 Comparison of an actual structure with an idealized structure.

One of the tasks of the structural designer closely allied with structural analysis is the proper idealization of the structure and its loading. When members are attached to one another, say for example a beam to a column, the attachment may be relatively flexible such that the vertical reaction of the beam is transmitted to the column but deflection of the beam causing a rotation at its ends has relatively little effect on the column. On the other hand, the attachment may be rigid, such as in monolithic construction of a concrete column with a beam. These end conditions must be idealized so that they can be mathematically defined before structural analysis can be made.

For instance, in Fig. 1.8.1 the ends of the actual beam rest on walls. The idealized assumption is that these walls act like knife-edge supports, so-called *simple supports*.

Simple supports may be schematically represented (see Fig. 1.8.2a) on a line drawing of a structure as a hinge, pin, or knife-edge, indicating that two component reactions exist but that rotation at the joint is unimpeded. Hinges in actual structures are shown in Figs. 1.8.3 to 1.8.5. A simple support may also be represented by a roller (Fig. 1.8.2c) where only one reaction is present acting in a direction perpendicular to (either toward or away from) the plane supporting the roller. Such a support in an actual structure is shown in Fig. 1.8.6. A fixed support, an ideal condition of zero

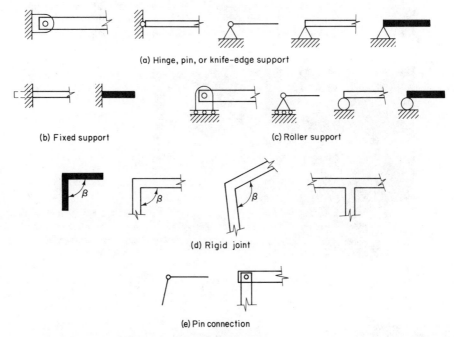

(a) Hinge, pin, or knife-edge support

(b) Fixed support

(c) Roller support

(d) Rigid joint

(e) Pin connection

Figure 1.8.2 Schematic representation of various support and connection idealizations.

Figure 1.8.3 Hinge support for a steel girder bridge. The analytical model for this support is shown in Fig. 1.8.2a. (Photo by C. G. Salmon)

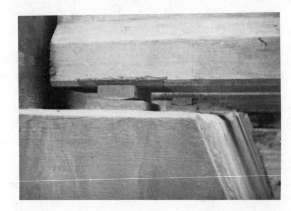

Figure 1.8.4 Hinge support for a prestressed concrete girder. The analytical model for this support is shown in Fig. 1.8.2a. (Photo by C. G. Salmon)

Figure 1.8.5 Hinge support for a steel arch. The analytical model for this support is shown in Fig. 1.8.2a. (Photo by C. G. Salmon)

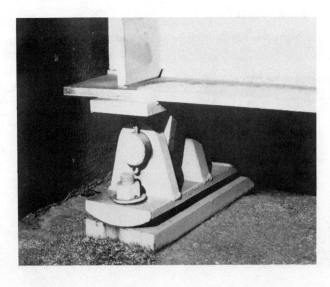

Figure 1.8.6 Roller support for a steel girder. Movement of roller along roller surface is small, as is the usual situation. The analytical model for this support is shown in Fig. 1.8.2c. (Photo by C. G. Salmon)

rotation which may be only partially attainable in actual structures, is shown modeled by Fig. 1.8.2b, where three reaction components (two mutually perpendicular forces and one moment) can develop. An actual fixed support is shown in Fig. 1.8.7.

Figure 1.8.7 Fixed support for a steel arch. The analytical model for this support is shown in Fig. 1.8.2b. (Photo by C. G. Salmon)

Rigid joints (modeled as in Fig. 1.8.2d) are joints where all internal forces (moments, shears, and axial forces) act through the joint without interruption. (See the actual rigid joints shown in Figs. 1.2.2 to 1.2.4.) Although the entire rigid joint may rotate through a certain amount, the angle β between the new directions of the member ends remains constant throughout loading and deformation of the structure. A pin connection (modeled as in Fig. 1.8.2e) between members allows force components to be carried across the joint—but not moment. (See the actual internal hinge shown in Fig. 4.1.2.) As expected, the angle between the member end directions may change freely around the pin during deformation of the structure.

Trusses are frequently idealized as a series of pin-connected members. Although in the early years of the twentieth century, trusses were actually built with pins at the joints, in the 1980s trusses are nearly always welded with the members intersecting at one point welded to the same small plate (known as a gusset plate) at the junction. The real structure involves members having end conditions that are somewhere between simple (pinned) and rigid.

Many idealizations are required as a part of structural analysis, both for the structure itself and for the loads that act on it. One should never lose sight of the fact that every structural analysis is performed on a "model" and not on the real structure.

1.9 FIRST-ORDER ANALYSIS VS. SECOND-ORDER ANALYSIS

The analysis of a framed structure, whether it be a truss, a beam, or a rigid frame, involves the determination of:

1. The deflected shape of the structure
2. The axial force, and/or the shear and bending moment at any one point in any one member on the basis of:

a. The given shape of the structure
b. The sizes and material properties of the members
c. The applied loads

To illustrate this, the possible deflected shapes of the three typical structures are shown in Fig. 1.9.1.

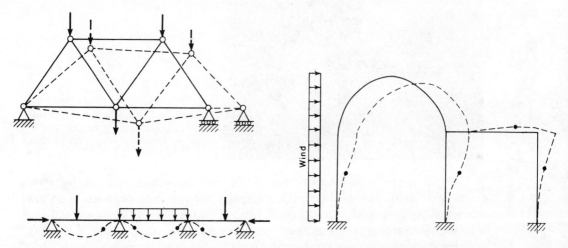

Figure 1.9.1 Possible deflected shapes of a truss, a beam, and a rigid frame.

In the case of a truss, the deflected shape is defined by the horizontal and vertical displacements of each joint, since in the idealized structure all bars are weightless and remain straight even after loads are applied. Each joint is subjected to a system of concurrent forces which include the applied loads at this joint and the pulls on it by the bars in tension, as well as the pushes on it by the bars in compression. Now there is the question of whether the directions of the pulls and pushes should be those of the solid lines in Fig. 1.9.1 or those of the dashed lines. The approach used in the former, where the equilibrium equations are based on the geometry of the original shape, belongs to *first-order* (primary) *analysis*; but that of the latter, where the equilibrium equations are based on the geometry of the deformed structure, is theoretically more accurate, is one further step ahead, and is therefore given the name *second-order analysis*. Second-order analysis is complicated because the geometry of the deformed structure depends on the axial forces, which in turn depend on the geometry. Fortunately, in ordinary cases the changes in geometry are so small that second-order analysis is not needed.

The deflected shape (known as the *elastic curve*) for the continuous beam in Fig. 1.9.1 is a function of the bending moment, which is primarily due to the transverse loads on the beam. In first-order analysis only the bending moment caused by transverse loads is used, neglecting any effect of axial compression. In second-order analysis the additional bending moment due to the product of the axial force and deflection is combined with that due to the transverse loads in producing the elastic curve. Here again, the total bending moment and the elastic curve depend on one another and may be determined only by solving a differential equation or by an iterative process.

It suffices to say at this point that only first-order analysis is treated in this introductory text. To repeat, in first-order analysis, the equilibrium equations are based on the geometry of the original, undeformed shape.

1.10 IDEALIZATION OF THE STRESS–STRAIN CURVE

In determining the response of framed structures to loads or disturbances of any kind, a stress–strain relationship is necessary. The standards used are the uniaxial tension test of a slender rod for steel, and the uniaxial compression test of a 6-in.-diameter by 12-in.-high cylinder for concrete. Whereas the stress–strain curve of steel is, for all practical purposes, linear within a substantial range, that of concrete is curved over its entire length. In addition, when concrete is reinforced with steel and subjected to the development of cracks in the tension zone, the behavior of an entire member can only be assumed statistically. In Fig. 1.10.1c an idealized stress–strain curve *ABC* is shown, which can reasonably be used in the structural analysis of the idealized structure with its idealized loading. Although the standard tests giving stress–strain curves for structural materials such as steel and concrete are made with a uniaxial state of stress, it is reasonably assumed that load-deformation behavior in actual structures conforms essentially to the idealized elastic–plastic situation. In other words, it is assumed for structural analysis that the modulus of elasticity of steel in compression (or of concrete in tension) is the same as that obtained from the standard tests. It is also assumed that the relationship between moment and curvature for beams is geometrically similar to the tension–compression stress–strain curve; that is, there is a linearly elastic part followed by a constant plastic part.

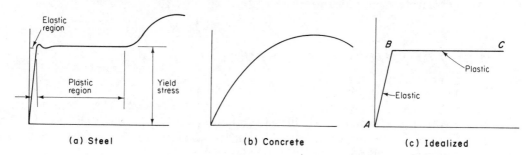

Figure 1.10.1 Typical stress–strain curves.

1.11 EQUILIBRIUM OF FORCE SYSTEMS

The basis of structural analysis is the condition of balance (equilibrium) between the applied forces and the resistance of the structure. For the structure as a whole, the applied forces are balanced by the reactions. For any joint or portion of a structure the external forces and internal forces must be in balance. Structures are in equilibrium (i.e., at rest) when the laws of statics are satisfied. The laws of static equilibrium may be stated as follows using three arbitrary orthogonal coordinates x, y, and z:

$$\sum F_x = 0; \qquad \sum F_y = 0; \qquad \sum F_z = 0$$
$$\sum M_x = 0; \qquad \sum M_y = 0; \qquad \sum M_z = 0 \tag{1.11.1}$$

Thus, the summation of all forces acting in *any* possible direction must be zero, and the summation of all moments about *any* axis must be zero. Equilibrium of real structures is a three-dimensional problem.

For analysis and design purposes, most structures can be simplified to be two-dimensional in one plane when their components and connections are actually designed and contructed so that their principal load-carrying function lies in one plane. For instance, a building may consist of a series of plane frames or trusses which are to transmit loads acting in the plane of the frames. The framing scheme perpendicular to these frames or trusses is completely different from in the primary load-carrying plane, and frequently consists only of bracing elements such as diagonal bracing. The only loads that would possibly be transmitted perpendicular to the primary planes would be wind loads, and these loads would be carried entirely by the bracing system. In other words, the load transmission in one direction may frequently be assumed to have no effect on the load-carrying function in the orthogonal direction.

For structures that may be treated as planar for structural analysis, the equations of equilibrium reduce from six to three. Summation of forces in two perpendicular directions must equal zero and summation of moments about an arbitrary point in the plane of the structure must equal zero. The equations may be written

$$\sum F_x = 0; \qquad \sum F_y = 0; \qquad \sum M_z = 0 \tag{1.11.2}$$

Although most of the general concepts treated in this text can be extended to include three-dimensional structures, specific treatment is limited to the usual category of framed structures in one plane.

1.12 STATICAL STABILITY OF A STRUCTURE

In general, when the equations of static equilibrium, Eqs. (1.11.1), are satisfied, the structure is at rest and would be said to be a *stable* structure. When the structure, or any part of it, cannot satisfy the equilibrium equations, it is said to be *unstable*. The following discussion pertains to the usual two-dimensional framed structures in the categories of trusses, beams, and rigid frames.

Members in a truss are treated as *two-force* members; that is, they may be thought of as having an axial force at each end, directed either toward or away from each other. A *stable* truss is one in which there can be no movement of parts other than that caused by elastic deformation of the members. For example, the truss of Fig. 1.12.1a is unstable because the center rectangle may change its form without developing forces in the bars, but that of Fig. 1.12.1b is stable because each triangle cannot change its form without forces arising in the bars. Thus, in general, a truss consists

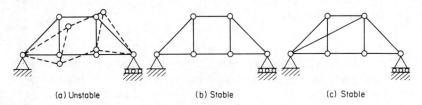

Figure 1.12.1 Stable and unstable trusses.

of an assemblage of triangles, although the truss of Fig. 1.12.1c is stable as long as a force can be applied in any direction at each pin end yet any portion of the truss when taken as a free body is in equilibrium. In Chapter 3 it will be shown that structural analysis of an unstable truss cannot be successfully completed.

Beams or continuous beams are members subjected to flexure (bending moment and shear) without axial force. First examine a simple beam, such as that shown in Fig. 1.12.2a. The beam under load W satisfies the equations of equilibrium by virtue of its vertical reactions at A and B. As the load W increases, the beam deflects and the reactions at A and B increase. This is a stable structure. The idealized structure is drawn showing a hinge (or knife-edge; see Fig. 1.8.2a) at A and a roller at B. This is done to show that any potential horizontal load which will then cause axial force in some portion of the beam is to have its reaction taken at A. When a structure is said to be stable, it means that it is stable under *any* loading. If rollers were shown at A and B, any potential horizontal component of load could not be resisted. The structure would roll to the side without inducing any internal force. Next, consider the simple beam with hinge shown in Fig. 1.12.2b. This is an unstable structure because the elements AC and BC can rotate without inducing internal forces. Even if the roller at B were changed to a hinge, there could still be an infinitesimal downward movement of point C without inducing any internal forces.

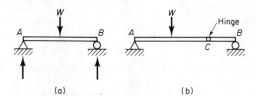

Fig. 1.12.2 Statically stable and statically unstable beams.

A rigid frame in general consists of members subject to flexure (bending moment and shear) as well as axial force. Unstable rigid frames are rare and can easily be detected, such as when the number of reactions are insufficient. However, in complex situations where there may be a mixture of pinned and rigid joints, statical instability may not be obvious until a structural analysis cannot be completed because a certain portion of the structure cannot satisfy equilibrium.

The concept of statical instability is an integral part of an understanding of elementary structural analysis; a statically unstable structure is one in which the structure as a whole, or any portion of it, cannot satisfy equilibrium under any internal force system.

In the study of mechanics of materials, the concept of *elastic stability* is treated. This is commonly referred to as *buckling*. A statically stable member subject to compression or compression combined with bending moment may become elastically unstable when the compression load becomes large enough. Even an entire rigid framework may become elastically unstable if compression forces get high enough in some or all of the members. Except for this brief section to contrast it with statical stability, elastic stability is not treated in this book.

To illustrate what is meant by elastic stability, consider a bar hinged at one end and supported on rollers at the other, such as the one shown in Fig. 1.13.1. When this bar is subjected to a compressive axial force P, it should stay straight and the roller support would be displaced downward by an amount equal to

$$X = \frac{P}{K} = \frac{PL_0}{EA} \tag{1.13.1}$$

The structural analysis is correct as long as the stress–strain relationship for the material is linear, and as long as P is below the *Euler buckling load*, given as

$$P_{cr} = \frac{\pi^2 EI}{L_0^2} \tag{1.13.2}$$

where I is the moment of inertia about axis 2-2 of the cross section. As derived in books on mechanics of materials, the Euler load is one at which the bending moment P_{cr} times the transverse deflection y would just cause the transverse deflection y. When the load P is greater than the Euler load, the one-bar structure of Fig. 1.13.1 buckles and is termed *elastically unstable*.

The structural analysis of the truss in Fig. 1.13.1b involves the determination of the deflections of all joints and the determination of the axial forces in all the bars. In the event the axial compressive force in any one bar is larger than its P_{cr} value indi-

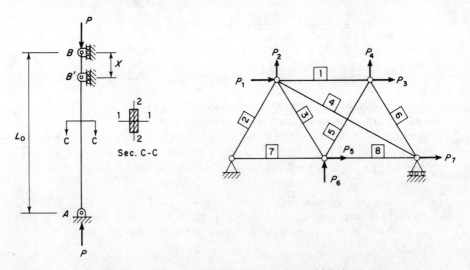

Figure 1.13.1 A one-bar structure and a multiple-bar truss.

cated by Eq. (1.13.2), the whole set of results would have no physical meaning because this particular bar would have buckled and ended its load-carrying participation when P_{cr} was reached. In this book it is assumed that such elastic instability does not occur, leaving to more advanced treatises the analysis of structures some of whose members become incapacitated due to buckling.

Elastic instability should not be confused with statical instability. For instance, if bar 2 in the truss of Fig. 1.13.1b were omitted, the structure would become incapable of keeping a general load system—an arbitrary set of seven P-values—in statical equilibrium and the structure is termed *statically unstable*. Statically unstable structures *cannot exist* and there can be no structural analysis of them.

1.14 STATICAL DETERMINACY AND INDETERMINACY

When a structure is statically stable and the laws of statics (equations of equilibrium) are sufficient to determine the internal forces in the members, the structure is said to be *statically determinate*. The simply supported beam of Fig. 1.12.2a is statically determinate; the three equations (1.11.2) are sufficient to determine the two vertical reactions and the one horizontal reaction at A if there are inclined loads on the beam. When the number of unknown reactions exceeds the number of equations of statics, the structure is said to be *statically indeterminate*. The beam of Fig. 1.14.1 has four unknown reactions, horizontal at A (if there are inclined loads on the beam) and verticals at A, B, and C. The three equations (1.11.2) are insufficient to determine the four reactions, indicative of a statically indeterminate structure.

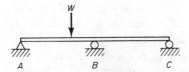

Figure 1.14.1 Statically indeterminate beam.

The complete structural analysis of a truss like the one in Fig. 1.14.2a involves the determination of the horizontal and vertical displacements of all joints and the axial forces in all bars resulting from the application of an arbitrary set of loads defined by the P-values. When the entire truss is broken apart into five free bodies of the joints, as shown in Fig. 1.14.2b, there are available two equations of statical equilibrium per joint or a total of 10 conditions. In this particular truss, the total number of unknown forces also happens to be 10: three reactions R_1 to R_3 and seven bar forces F_1 to F_7. When the number of conditions of statics is equal to the number of unknown forces, the structure is *statically determinate* provided that it is statically stable. For instance, the truss of Fig. 1.14.2c has 10 statical conditions and 10 unknown forces; however, it is statically unstable. The truss of Fig. 1.14.2d has one more bar than that of Fig. 1.14.2a; therefore, it still has only 10 statical conditions, but has 11 unknown forces. When the number of available independent conditions of statics is less than that of unknown forces, the structure is *statically indeterminate*. The number of insufficient conditions of statics is called the degree of *statical indeterminacy*. The degree of statical

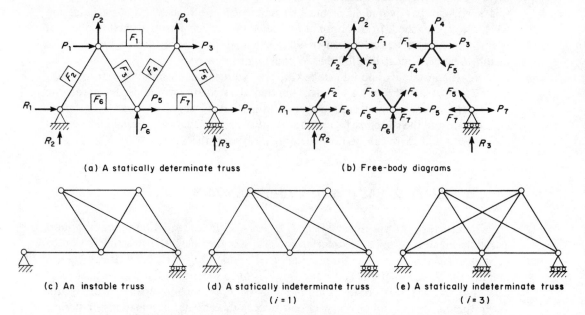

(a) A statically determinate truss

(b) Free-body diagrams

(c) An instable truss

(d) A statically indeterminate truss
(*i* = 1)

(e) A statically indeterminate truss
(*i* = 3)

Figure 1.14.2 Statically determinate and statically indeterminate trusses.

indeterminacy of the truss in Fig. 1.14.2d is one, but that of the truss in Fig. 1.14.2e is three.

Simple, cantilever, and overhanging beams as shown in Fig. 1.14.3 are statically determinate, but continuous beams in Fig. 1.14.4 are statically indeterminate to the first, second, and third degrees, respectively. Also, the rigid frames shown in Fig. 1.14.5 are statically determinate because the three unknown reactions R_1 to R_3 in each structure exactly equal the number of equations of statical equilibrium. Each of the rigid frames in Fig. 1.14.6 is statically indeterminate to the sixth degree. The one in Fig.

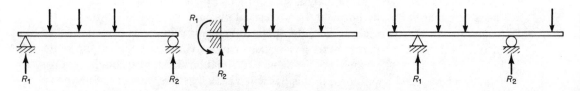

Figure 1.14.3 Statically determinate beams.

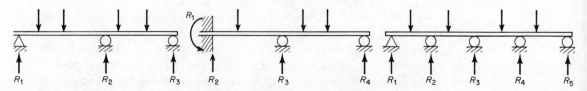

Figure 1.14.4 Statically indeterminate beams.

Introduction   Chap. 1

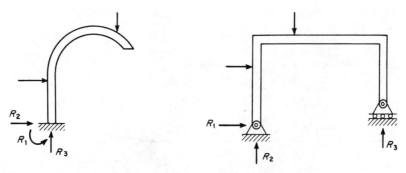

Figure 1.14.5 Statically determinate rigid frames.

1.14.6a has three extra reactions and three extra unknown internal forces when a cut is made, say in the top beam. The one in Fig. 1.14.6b has six extra reactions; had the six extra reactions been found by means other than statics, the remaining reactions and the internal forces become determinable by statics.

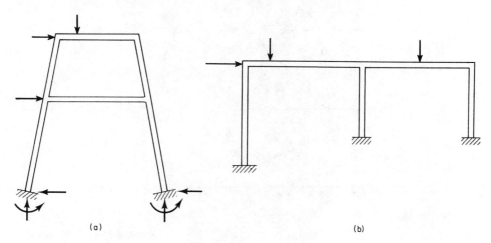

(a)   (b)

Figure 1.14.6 Statically indeterminate rigid frames.

1.15 FORCE METHOD OF STRUCTURAL ANALYSIS

There are two general philosophical approaches to structural analysis, identified according to whether forces or displacements are solved first. When reactions and internal forces are solved first, the approach is referred to as the *force method*. When displacements of the joints (pinned or rigid) are solved first, the approach is referred to as the *displacement method*.

As an illustration of the force method, examine the truss of Fig. 1.15.1a. The *P*-forces shown at the joints represent all possible external forces acting on the truss in terms of horizontal and vertical components. The internal forces in the members are

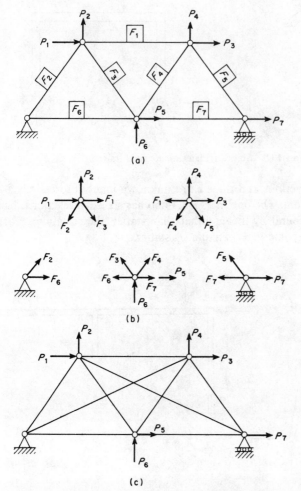

Figure 1.15.1 Truss analysis.

numbered from F_1 to F_7 and shown in boxes. The total number of unknown forces in this statically determinate truss is 10, consisting of three external reactions and seven internal forces. The free-body diagrams shown in Fig. 1.15.1b of the joints must each satisfy static equilibrium. Observing that summation of forces equal to zero in two orthogonal directions provides two equations of statical equilibrium at each joint, the five joint free bodies provide a total of 10 equations of statics. The 10 unknown forces (reactions and internal forces) then may be determined directly from the 10 equations, without knowledge of the sizes and material properties of the seven bars. The determination of joint displacements could then be determined if desired. As a rule, the *force method* is the standard way of solving statically determinate problems.

The situation is different in the case of the statically indeterminate truss of Fig. 1.15.1c. If the 12 unknown forces were to be taken as the primary unknowns, two conditions additional to the 10 statical conditions would be required. These conditions are those of compatibility of deformations. The concept involves:

1. The cutting of any two bars to create a statically determinate and statically stable truss
2. The determination of the amount of gap or overlap at the two cuts due to the applied loads
3. The determination of the axial forces in the two cut bars which, when applied in pairs at the cuts, would eliminate the gaps or overlaps

The unknown axial forces in the bars selected to be cut are to be solved first and are commonly referred to as the *redundants*. These redundant forces can be determined from the compatibility conditions as stated in item 3. When the method of structural analysis involves determining forces (redundant forces when the structure is statically indeterminante) ahead of the joint displacements, such a method is called the *force method*. In using the force method to analyze statically indeterminate structures, the elastic properties of the constituent members are required in the evaluation and then elimination of the gaps or overlaps at the cuts of the newly created statically determinate structure.

The force method applies not only to trusses but also to all types of framed structures. Further detailed treatment relating to beams and rigid frames is presented in subsequent chapters.

1.16 DISPLACEMENT METHOD OF STRUCTURAL ANALYSIS

Examine again the statically determinate truss of Fig. 1.15.1a and the statically indeterminate truss of Fig. 1.15.1c. Complete analysis of either truss, subject to an arbitrary set of forces P_1 through P_7, should include the joint displacements (seven in all in terms of horizontal and vertical components) in the directions of the forces P and the magnitudes of the internal axial force F (seven in the statically determinate truss and nine in the statically indeterminate truss of Fig. 1.15.1c).

Alternative to taking the internal forces F as the primary unknowns, as in the force method, one may take the joint displacements as the primary unknowns. Since there are five joints in each truss of Fig. 1.15.1, there are 10 total equations of static equilibrium available for use in the analysis of each truss. Note that even though additional equations may be written using the whole truss or a group of joints with connecting members between them, as a free body, these equations are only different ways of expressing the 10 independent equations for the five joints. As long as each joint is in balance, the whole or any part must also be in balance. Of the 10 equations, seven equations requiring the summation of forces in the P-directions equal to zero may be used to solve for the seven unknown displacements, regardless of whether the truss has seven or nine bars. This procedure is called the *displacement method*.

The general approach for the displacement method may be summarized as follows:

1. The change in length of each bar is expressed in terms of the changes in position of its ends.
2. The internal axial force F in each bar is expressed in terms of its change in length, assuming that stress is proportional to strain.
3. Equations of joint equilibrium are written along the directions of the unknown joint displacements.

According to this procedure, there are always as many equations of statics as unknown joint displacements because there is always one load coordinate for each displacement coordinate.

The concept of the displacement method is simple, but its application involves the processing of a number of linear transformations and the solution of simultaneous equations. With the help of matrix algebra and electronic computation, these operations now become conveniently feasible. Although a truss analysis problem has been selected to convey the fundamental principle of the displacement method, the method itself is applicable to the analysis of all types of statically determinate and indeterminate structures.

1.17 CLASSICAL METHODS, MATRIX METHODS, AND COMPUTER METHODS

Structural analysis is the cornerstone of structural engineering. A sound knowledge of structural analysis is essential if the designer is to idealize loads and apply them properly to the structure, determine the internal forces, and finally design the structure. One cannot know too much structural analysis.

The fact that the electronic digital computer is available and used extensively to perform structural analysis does not preclude the necessity for having the ability to understand and use the classical methods. Computer output providing structural analysis results must not be accepted on faith alone. The designer must always make checks on computer results. Usually, the appropriate checks may be made using classical methods on an element of the structure or on a small subassemblage of elements. For example, a large-capacity high-speed electronic computer might do the structural analysis of a 40-story frame treating it as a whole without separating it into a series of subassemblages as would have been required if the analysis were performed using a hand-held electronic calculator or slide rule. The check of results, however, would still require identifying one or more appropriate subassemblages and applying classical methods.

The study of classical methods, which are applicable for the hand-held electronic calculator or slide rule, is complementary to the study of *matrix methods*, which are applicable for the large-capacity high-speed digital computer. A good knowledge of classical methods has not been made obsolete by the digital computer.

In the chapters that follow, classical methods are presented in Chapters 2 to 12. Matrix formulations of structural analysis, in particular the matrix displacement methods, are presented in Chapters 13 to 20. The study of structural analysis may be made by studying the classical methods first or by studying the classical and matrix formulation procedures side by side.

1.18 ACCURACY OF COMPUTATION

In the *study* of structural analysis, calculations should be carried to more significant figures than would be appropriate when doing practical design. The purpose throughout this text is to emphasize concepts. To be certain that a method is properly understood, a calculation may be carried out to as many significant figures as possible, solely for the purpose of comparing the results with a similar solution using a different method. When the purpose is to learn the different methods, more "accuracy" in computations is justified; in fact, it may be required. Two methods involving the same inherent assumptions regarding idealization of the structure must give identical results to as many significant figures as can be carried.

The reader should not, however, interpret "accuracy" in the sense of comparing methods as "exactness" as far as the end result of the structural analysis is concerned. The internal axial forces, bending moments, and shears are not "exact" no matter how many significant figures are carried in the computations. As discussed in Sections 1.3 to 1.8, the loading is "modeled" either by the structural engineer or by building codes and the structural framework is also "modeled". The structural analysis, therefore, gives results that must be interpreted in context with the "models". For design, three significant figures are suggested as the *maximum* meaningful values, and even this many are recommended more for systematic control of calculations and for ease in checking than for any improved effect on the final structure.

SELECTED REFERENCES

1. Charles G. Salmon and John E. Johnson, *Steel Structures: Design and Behavior*, 2nd ed., Harper & Row, Publishers, New York, 1980, Chap. 1.

2. S. Whipple, *A Work on Bridge Building*, Utica, N.Y., 1847, 128 pp.

3. Isaac Todhunter and Karl Pearson, *A History of the Theory of Elasticity and of the Strength of Materials*, Vol. 1, Cambridge University Press, Cambridge, England, 1886 (reprinted by Dover Publications, Inc., 1960), 133–146.

4. E. P. B. Clapeyron, "Calcul d'une poutre élastique reposant librement sur des appuis inégalement espacés," *Comptes Rendus des Séances de l'Académie des Sciences*, Paris, *45* (July–December 1857), 1076–1080.

5. *American National Standard Building Code Requirement for Minimum Design Loads in Buildings and Other Structures*, American National Standards Institute, A58.1–1982.

6. John M. Biggs, *Introduction to Structural Dynamics*, McGraw-Hill Book Company, New York, 1964, Chap. 6.

7. *National Building Code of Canada*, Associate Committee on the National Building Code, National Research Council of Canada, Ottawa, 1980.

8. "Wind Forces on Structures," Task Committee on Wind Forces, Committee on Loads and Stresses, Structural Division, ASCE, Preliminary Reports, *Journal of Structural Division*, ASCE, *84*, ST4 (July 1958); and Final Report, *Transactions*, ASCE, *126* (1961), part II, 1124–1198.

9. Ray W. Clough and Joseph Penzien, *Dynamics of Structures*, McGraw-Hill Book Company, New York, 1975.

10. *Recommended Lateral Force Requirements and Commentary*, Seismology Committee, Structural Engineers Association of California, San Francisco, 1967.

11. *Standard Specifications for Highway Bridges*, 12th ed., American Association of State Highway and Transportation Officials, Washington, D.C., 1977.

12. *Specifications for Steel Railway Bridges*, American Railway Engineering Association, Chicago, 1965.

13. *Specification for the Design Fabrication and Erection of Structural Steel for Buildings*, American Institute of Steel Construction, New York, 1978.

[2]

Statically Determinate Trusses —Reactions and Member Forces

2.1 DEFINITION OF A STATICALLY DETERMINATE TRUSS

A *truss* is a framed structure consisting of a finite number of bars connected at frictionless pin joints at which external forces may be applied. A typical truss joint in the ideal situation is shown in Fig. 2.1.1a, where a pin is inserted at the center of the eyed ends of the bars. The bars are free to turn at the pin. When structural analysis is performed using the pin joint assumption, the resulting internal forces are said to be *primary forces*. The stresses* produced by these primary forces are *primary stresses*. In first-order analysis (see Section 1.9) the internal forces are assumed to act along the axes connecting the originally located joint centers at the ends of members. As the structure deforms, the directions of all internal and external forces are presumed not to change.

In reality, of course, trusses are not connected by pins. Modern trusses are usually welded at the joints. Seemingly, if welded, the structure should correctly be designed as a rigid frame. However, when the members are long and the connections relatively compact, determination of internal forces using primary analysis for a truss with pin joints is acceptable for design.

For more elaborate analysis, it is possible to consider the secondary effects. Secondary stresses in trusses are due to bending moments caused by such factors as: (1) joints are never frictionless, even if actually pin joints; (2) horizontal and inclined members are always deflected by their own weight; and (3) the centroidal axes of and

---

*The term *stress* throughout this text refers to *unit stress*, such as pounds per square inch (psi) or newtons per square meter (N/m^2). The term force, or internal force, is used to mean the total result of unit stress times area.

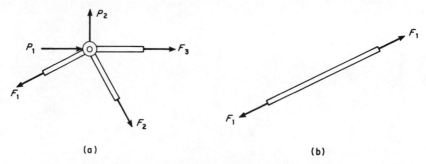

Figure 2.1.1 Typical joint and bar in a truss.

therefore the axial forces in the members meeting at a joint may not intersect at a common point. Proper alignment of members at joints and keeping joints compact would minimize secondary effects.

Thus, for purpose of practical design, one may consider external forces to be applied only at the joints, so that each bar in the truss is subjected only to a pair of equal and opposite axial forces, as in Fig. 2.1.1b. Such a truss bar is called a *two-force member*.

A statically determinate truss requires that the external reactions and the internal member forces be determinable from using only the equations of static equilibrium. For a stable truss, at least three reaction components must be acting. This could be accomplished by one hinged support and one roller support, as in Fig. 2.1.2a. A special case would be three roller supports having no more than two parallel reactions. At a hinged support, there are two unknown components of the hinge reaction; and for a roller support, there is one unknown reaction. Sometimes, although not often, the number of reaction components may be even larger than three, yet the truss is still statically determinate, such as in Fig. 2.1.2b.

Any system of external forces P acting on a two-dimensional truss may be described by the numerical magnitudes in two reference directions (horizontal and vertical are usually convenient reference directions) at all the joints, except where there exist unknown reaction components R. If NP is the total number of possible external forces that may act, and NJ is the total number of joints, then since there exists a force reaction component in each reference direction,

$$NP + NR = 2(NJ) \qquad (2.1.1)$$

The total number of available equations of static equilibrium for a two-dimensional truss is $2(NJ)$ and the total number of unknown forces, including NF internal forces and NR reactions, is $(NF + NR)$. The necessary requirement for statical determinacy is then,

$$NF + NR = 2(NJ) \qquad (2.1.2)$$

Comparing Eqs. (2.1.1) and (2.1.2) gives

$$NP = NF \qquad (2.1.3)$$

In other words, when the number of bars in a truss equals the number of possible external force components, the truss is statically determinate. Conversely, for a truss to be

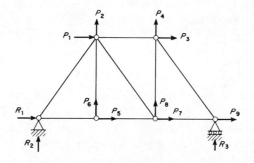

(a) $NP=9$, $NR=3$, $NF=9$; stable

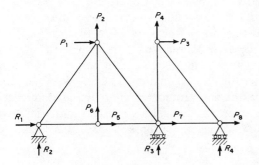

(b) $NP=8$, $NR=4$, $NF=8$; stable

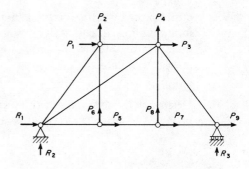

(c) $NP=9$, $NR=3$, $NF=9$; stable

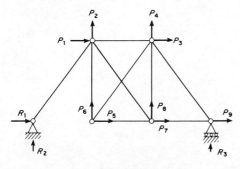

(d) $NP=9$, $NR=3$, $NF=9$; instable

Figure 2.1.2 Typical statically determinate trusses. The truss in part (d) is unstable.

statically determinate, the number of possible directions in which external forces may act is equal to the number of bars.

As illustrations, all four trusses of Fig. 2.1.2 satisfy the condition $NP = NF$, although only the first three are statically stable. The structure of Fig. 2.1.2d does not include three independently unknown reaction components because the resultant of R_1 and R_2 must be in the direction of the only bar entering the left support.

2.2 METHOD OF JOINTS

The analysis of a truss involves determination of the external reaction components and the internal axial forces for any given external-force system. The first step is to establish whether or not the truss is statically determinate. The algebraic condition $NP = NF$ as described in the preceding section is a requirement for statical determinacy but will not detect statical instability. Usually, a lack of triangular arrangement of bars will identify statical instability; however, occasionally one may not recognize the condition until encountering difficulty in completing the analysis.

The *method of joints* is frequently a simple way of analyzing a statically determinate truss. In common cases, the number of unknown reactions is three; they can first be determined by taking the entire truss as a free body. Next, the analyst will use the free-body diagrams for successive joints containing only two unknown axial forces. The two equations of static equilibrium involving summation of forces in two orthogonal directions are then used to solve for the two unknowns at the joint. The procedure continues until all internal axial forces are computed. Since the total number of joint equations of equilibrium is three more than the number of unknown axial forces, there will be three surplus equations for checking purposes after all the axial forces have been obtained.

Example 2.2.1

Determine the external reactions and the internal forces in the members in the truss of Fig. 2.2.1 using the method of joints.

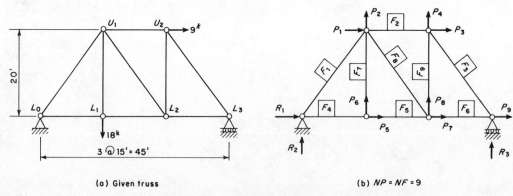

(a) Given truss   (b) $NP = NF = 9$

Figure 2.2.1 Truss for Example 2.2.1.

Solution

(a) *Establish whether or not the truss is statically determinate.* The reactions R_1 through R_3, internal forces F_1 through F_9, and the *possible* components P_1 through P_9 of external forces are numbered and marked in Fig. 2.2.1b. This establishes that $NP = NF$ and the structure is statically determinate. The systematic numbering and reference to a general force system, such as the P_1 through P_9 for this example, will be shown to be useful in integrating classical methods and matrix methods (Chapters 13 to 20).

Alternative to the use of P_1 through P_9 to establish that $NP = 9$, Eq. (2.1.2) may be applied. The number NF of internal forces (members) plus the number NR of reactions equals twice the number NJ of the joints,

$$NF + NR = 9 + 3 = 12$$
$$2(NJ) = 2(6) \quad = 12$$

which proves the truss to be statically determinate.

(b) *Solve for the reactions using the entire truss as a free body.* On Fig. 2.2.1b, $P_3 = 9$ kips, $P_6 = -18$ kips, and the other P-values are zero.

$\sum M$ about $L_0 = 0$,

$$9(20) + 18(15) - R_3(45) = 0; \qquad R_3 = 10 \text{ kips}$$

$\sum M$ about $L_3 = 0$,

$$9(20) - 18(30) + R_2(45) = 0; \qquad R_2 = 8 \text{ kips}$$

$\sum F_H = 0$,

$$R_1 = -9 \text{ kips}$$

Check by $\sum F_V = 0$: $\quad R_2 + R_3 - 18 = 0$
$$8 + 10 - 18 = 0$$

Note that reactions R_2 and R_3 were determined independently using moment equilibrium about two different points and checked by vertical force equilibrium.

(c) *Start method of joints with joint L_0* (Fig. 2.2.2). The unknown internal forces F are to be shown as "positive" on the joint free body; that is, a positive numerical value will mean that the internal force is tension.

For $\sum F_V = 0$,

$$\frac{4}{5}F_1 + 8 = 0; \qquad F_1 = -10 \text{ kips}$$

For $\sum F_H = 0$,

$$\frac{3}{5}(-10) - 9 + F_4 = 0; \qquad F_4 = +15 \text{ kips}$$

(d) *Examine joint L_1* (Fig. 2.2.3). Note that U_1 has three unknowns and cannot be solved until the number of unknowns is reduced to two.

For $\sum F_V = 0$,

$$F_7 = +18 \text{ kips}$$

For $\sum F_H = 0$,

$$F_5 = +15 \text{ kips}$$

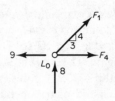

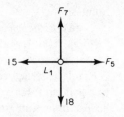

Figure 2.2.2 Joint L_0 as a free body.   **Figure 2.2.3** Joint L_1 as a free body.

(e) *Examine joint* U_1 (Fig. 2.2.4). Now there are only two unknowns at this joint.

For $\sum F_V = 0$,

$$\frac{4}{5}F_8 + 18 - \frac{4}{5}(10) = 0; \quad F_8 = -12.5 \text{ kips}$$

For $\sum F_H = 0$,

$$F_2 + \frac{3}{5}(-12.5) + \frac{3}{5}(10) = 0; \quad F_2 = +1.5 \text{ kips}$$

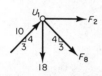

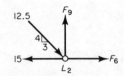

Figure 2.2.4 Joint U_1 as a free body.   **Figure 2.2.5** Joint L_2 as a free body.

(f) *Examine joint* L_2 (Fig. 2.2.5). Either joint L_2 or U_2 could be used next.

For $\sum F_V = 0$,

$$F_9 - \frac{4}{5}(12.5) = 0; \quad F_9 = +10 \text{ kips}$$

For $\sum F_H = 0$,

$$F_6 + \frac{3}{5}(12.5) - 15 = 0; \quad F_6 = +7.5 \text{ kips}$$

(g) *Examine joint* U_2 (Fig. 2.2.6).

For $\sum F_V = 0$,

$$\frac{4}{5}F_3 + 10 = 0; \quad F_3 = -12.5 \text{ kips}$$

Check for $\sum F_H = 0$,

$$+9 - 1.5 + \frac{3}{5}(-12.5) = 0 \qquad \text{OK}$$

(h) *Check by examining* L_3 (Fig. 2.2.7).

Check for $\sum F_V = 0$,

$$-\frac{4}{5}(12.5) + 10 = 0 \qquad \text{OK}$$

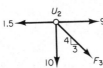
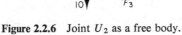

Figure 2.2.6 Joint U_2 as a free body.　　　　**Figure 2.2.7** Joint L_3 as a free body.

Check for $\sum F_H = 0$,

$$\frac{3}{5}(12.5) - 7.5 = 0 \qquad\qquad \text{OK}$$

(i) *Results of the analysis.* The results of a truss analysis are always shown on a diagram of the truss, as for this case in Fig. 2.2.8. The internal forces are shown positive (+) for tension and negative (−) for compression. In addition, arrows are given on the members showing the *actual* effect of the member on the joint. Actually, Fig. 2.2.8 is the composite free-body diagram of all the joints. An experienced analyst can usually omit diagrams in Figs. 2.2.2 to 2.2.7 and work all answers out on Fig. 2.2.8 by mentally going through the descriptions made in parts (c) to (h). Note again that the arrows drawn on the members in Fig. 2.2.8 are the *actual* directions for this set of external forces only.

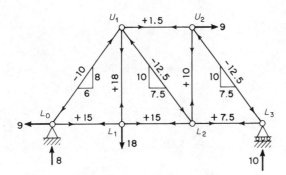

Figure 2.2.8 Solution for the truss of Example 2.2.1.

2.3 METHOD OF SECTIONS

When the method of joints leads to a situation where no subsequent joint contains only two unknown bar forces, the next approach would be to use a group of joints (called a *section of the truss*) as a free body, upon which there are no more than three unknown bar forces acting. Since the section is not subject to a concurrent force system, three unknowns can be determined using the three equations of static equilibrium. The entire section is used as the free body: hence the name *method of sections*. The method of sections may be used in lieu of the method of joints, even when the method of joints is possible. It is a convenient method for directly determining the axial force in any one particular bar in a complex truss if that truss can be split into two sections by cutting through no more than three bars.

Example 2.3.1

Determine the internal forces F_2, F_5, and F_8 for the truss of Fig. 2.2.1 directly by the method of sections.

Solution

(a) *Cut the truss through the unknown that is desired to be directly determined.* In this example, a single cut will give all three required unknown forces acting on a section. The cut divides the structure into two free bodies as given in Fig. 2.3.1.

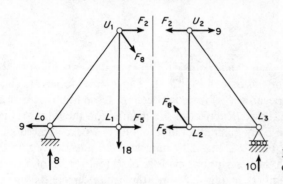

Figure 2.3.1 Method of sections for determining forces F_2, F_5, and F_8.

(b) *Use the left section as the free body.*

ΣM about $L_2 = 0$,

$$+8(30) - 18(15) + F_2(20) = 0$$
$$F_2 = +1.5 \text{ kips}$$

Note that "point" L_2 is not even on the free body, but it is the point of intersection of the lines of action of F_5 and F_8.

ΣM about $U_1 = 0$,

$$+9(20) + 8(15) - F_5(20) = 0$$
$$F_5 = +15 \text{ kips}$$

$\Sigma F_V = 0$,

$$+8 - 18 - 0.8F_8 = 0$$
$$F_8 = -12.5 \text{ kips}$$

(c) *Use the right section as the free body.*

ΣM about $L_2 = 0$,

$$-10(15) + 9(20) - F_2(20) = 0$$
$$F_2 = +1.5 \text{ kips}$$

ΣM about $U_1 = 0$,

$$-10(30) + F_5(20) = 0$$
$$F_5 = +15 \text{ kips}$$

$\Sigma F_V = 0$,

$$+10 + 0.8F_8 = 0$$
$$F_8 = -12.5 \text{ kips}$$

Example 2.3.2

Determine the internal forces in members U_2U_3, U_3U_4, L_3L_4, U_3L_3, and U_4L_4 for the truss of Fig. 2.3.2 having inclined top chord members. This truss is known as a *curved chord Pratt truss*.

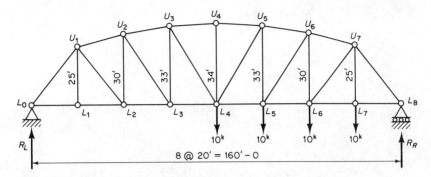

Figure 2.3.2   Curved chord Pratt truss of Example 2.3.2.

Solution

(a) *Establish whether or not the truss is statically determinate.* The truss contains 29 members, making $NF = 29$. There are 16 joints, $NJ = 16$, and there are three reaction components possible, two at the hinge at L_0 and one at the roller at L_8. Thus,

$$NF + NR = 29 + 3 = 32$$
$$2(NJ) = 2(16) \quad = 32$$

According to Eq. (2.1.2), the truss is statically determinate. Actually, the statical determinacy and stability are both obvious in this case because of the sequence of triangular arrangement.

(b) *Compute the reactions at L_0 and L_8.* Obtain these independently by using moment equilibrium, and then check summation by using vertical force equilibrium.

$\sum M$ about $L_0 = 0$,

$$10(80) + 10(100) + 10(120) + 10(140) - R_R(160) = 0; \qquad R_R = 27.5 \text{ kips}$$

$\sum M$ about $L_8 = 0$

$$10(20) + 10(40) + 10(60) + 10(80) - R_L(160) = 0; \qquad R_L = 12.5 \text{ kips}$$

Check by $\sum F_V = 0$,

$$-10 - 10 - 10 - 10 + 12.5 + 27.5 = 0$$

Note that the horizontal reaction at the hinge is zero, although usually not shown but understood.

(c) *Determine force in U_2U_3.* Cut a section through no more than three members, as shown in Fig. 2.3.3a.

$\sum M$ about $L_3 = 0$,

$$(U_2U_3)_H(33) + 12.5(60) = 0$$

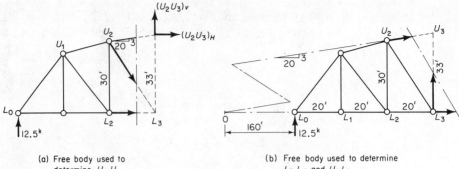

(a) Free body used to
determine U_2U_3

(b) Free body used to determine
L_3L_4 and U_3L_3

Figure 2.3.3 Free-body diagrams for truss of Example 2.3.2.

$$(U_2U_3)_H = -\frac{12.5(60)}{33}$$

$$U_2U_3 = -\frac{12.5(60)}{33}\left(\frac{20.22}{20}\right) = -22.98 \text{ kips}$$

(d) *Determine force in U_3U_4.* This time a section is cut vertically through the truss between L_3 and L_4 (see Fig. 2.3.2; the free body for this section is not given). In the same manner as for U_2U_3, U_3U_4 is determined by satisfying moment equilibrium about L_4.

$$U_3U_4 = -\frac{12.5(80)}{34}\left(\frac{20.02}{20}\right) = -29.45 \text{ kips}$$

(e) *Determine force in L_3L_4.* Using the free-body diagram of Fig. 2.3.3b, satisfying moment equilibrium about U_3 gives

$$12.5(60) - L_3L_4(33) = 0$$

$$L_3L_4 = +22.73 \text{ kips}$$

(f) *Determine force in U_3L_3.* Cut a section as shown in Fig. 2.3.3b. This force may be determined directly by locating the intersection of forces L_3L_4 and U_2U_3 at point O.

Distance from L_3 to point $O = 11(20) = 220$ ft.

$\sum M$ about point $O = 0$,

$$U_3L_3(220) - 12.5(160) = 0$$

$$U_3L_3 = -9.09 \text{ kips}$$

In this example, since U_2U_3 had already been determined, U_3L_3 could have been obtained using vertical force equilibrium for the same free body (Fig. 2.3.3b).

(g) *Determine force in U_4L_4.* A direct solution for U_4L_4 using the method of sections is not possible because member U_4L_4 cannot be cut except by cutting four members. After determining U_3U_4, however, the method of joints may be applied to joint U_4 (see Fig. 2.3.4).

$\sum F_H = 0$ and from symmetry,

$$U_4U_5 = -29.45 \text{ kips}$$

$$\Sigma F_V = 0,$$

$$+1.47 + 1.47 - U_4 L_4 = 0; \qquad U_4 L_4 = +2.94 \text{ kips}$$

(h) *Answer diagram*. The results are given on an answer diagram in Fig. 2.3.5.

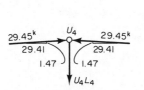

Figure 2.3.4 Joint U_4 as a free body.

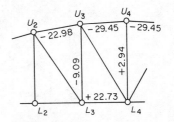

Figure 2.3.5 Answer diagram for Example 2.3.2.

2.4 METHOD OF MOMENTS AND SHEARS

In the method of moments and shears, recognition is given to the fact that a truss is really a beam without a solid web. A beam carries bending moment and shear. The bending moment may be considered to be resisted by an internal couple consisting of equal and opposite (compression and tension) internal forces acting at a distance apart. On the truss the top chord may be thought of as carrying one of those internal forces and the bottom chord the other. The shear is carried by the diagonals of the truss. On sloping chord trusses, some of the shear is carried by inclined chord members. The method of moments and shears is most appropriately applied to parallel chord trusses where none of the shear is carried by the chord members.

Example 2.4.1

Use the method of moments and shears to determine the forces in the members of the Pratt deck-type truss of Fig. 2.4.1a.

Solution

(a) *Draw the shear and bending moment diagrams*. These diagrams are as given in Fig. 2.4.1b and c. The shear is constant between panel points (locations of joints where all loads are assumed to be applied).

(b) *Forces in the chord members*. Recall that forces in chord members may be determined by cutting a section through three members, such as cutting $U_0 U_1$, $U_1 L_1$, and $L_1 L_2$ to obtain $U_0 U_1$ and $L_1 L_2$ in Fig. 2.4.2. These chord forces are obtained by taking moment equilibrium about U_1 and L_1, respectively. The summation of moments about U_1 or L_1 due to external loads is the bending moment at U_1 or L_1,

$$\text{Bending moment (BM) at } U_1 = 90(20) - 15(20) = 1500 \text{ ft-kips}$$

$$L_1 L_2 = \frac{\text{BM at } U_1}{h} = \frac{1500}{20} = +75 \text{ kips}$$

$$U_0 U_1 = -\frac{\text{BM at } L_1}{h} = -\frac{1500}{20} = -75 \text{ kips}$$

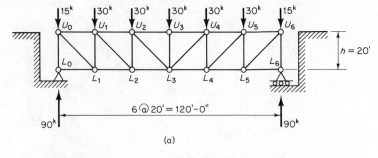

(a)

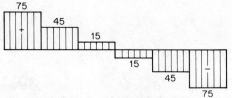

(b) Shear, V (kips)

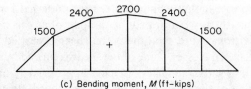

(c) Bending moment, M (ft–kips)

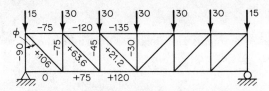

(d) Answer diagram for original loading (kips)

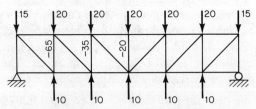

(e) Answer diagram with 2/3 load at upper panel points.
Forces not given are the same as (d) (kips)

Figure 2.4.1 Pratt truss for Example 2.4.1.

46

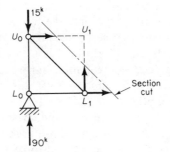

Figure 2.4.2 Cutting a section in the truss of Fig. 2.4.1.

In general,

$$\text{Chord forces} = \frac{BM}{h}$$

where the bending moment (BM) used is the value at the location that would be used as moment center in the method of sections. The sign of the internal force is determined by the sign of the bending moment. By the usual sign convention for bending moment, positive bending moment means compression in the top and tension in the bottom. For simply supported trusses or beams, the bending moment sign is positive throughout the entire span.

The forces in the other chord members are given in Fig. 2.4.1d.

(c) *Forces in diagonal members.* Visualize a section cut vertically through the truss in the first panel where the shear is 75 kips. Referring to Fig. 2.4.1d, taking the angle ϕ as that between a diagonal and vertical, the shear in the panel equals the vertical component of the force in the diagonal. Thus,

$$V = (\text{force in diagonal}) \cos \phi$$

$$\text{Force in diagonal} = \frac{V}{\cos \phi}$$

For this example, $\cos \phi = 0.707$, giving

$$\text{Force in diagonal} = \frac{V}{0.707} = 1.414V$$

To visualize the effect of shear, make the cut to divide the truss into two free bodies. Then in accordance with the shear diagram, visualize the shear displacement of the two free bodies relative to one another. With positive shear, the left free body displaces upward relative to the right one, causing on the right free body a pull on diagonals inclined upward to the left (putting them in tension), and a push on diagonals inclined upward to the right (putting them in compression). For negative shear the reverse logic can be applied.

The forces in the diagonal members are given in Fig. 2.4.1d.

(d) *Forces in vertical members.* Vertical members also carry the shear. When a section is cut to evaluate the force in a vertical by the method of sections, that section intersects the top chord in one panel and the bottom chord in the adjacent panel. When all loads are applied at the top chord as in this example, the vertical member carries a force equal to the shear in the top chord panel cut by the section. The reader is advised to cut the appropriate sections, draw the free bodies, and compare the results here with a direct solution by the method of sections.

To illustrate the effect of having some of the load applied at both top and bottom chords, refer to Fig. 2.4.1e. In this case only the forces in the vertical members are different from the original example. When a section cuts the vertical member to divide the truss into two free bodies, the 10-kip load is now acting on the left free body, whereas previously it was part of the 30-kip load and acted on the right free body. The shear carried by each vertical is now reduced by 10 kips from its value when all loads were applied to the top panel points.

The concept of the method of moments and shears is extremely important for qualitatively evaluating the behavior of a truss-like structure, whether or not the method is actually used to determine the internal forces.

2.5 GRAPHICAL METHOD OF JOINTS (MAXWELL DIAGRAM)

Identical solutions for the internal forces in a statically determinate truss by the method of joints may be obtained either algebraically or graphically. The subject of graphic statics was extensively treated in the days when the slide rule was the only computational tool. The graphical solution described here involves using a systematic notation called *Bow's* notation and a line diagram representing the forces (including reactions, external loads, and internal forces). Each line of the diagram represents the magnitude (proportional to length) and direction of a force. The line diagram is known as a *Maxwell† diagram*.

The reasoning behind the Maxwell diagram is simple; any group of forces in equilibrium, when plotted to scale successively by arrowhead lines, should form a closed polygon. The external force polygon, noted at its corners by the letters of the alphabet, represents the reactions and external loads. Each internal force polygon represents the equilibrium of the concurrent forces at each joint. Bow's notation allows the reading of external loads and reactions in a clockwise manner around the outside of the truss, and the reading of the internal forces also in a clockwise manner around each joint.

The procedure for the graphical method of joints is as follows:

1. Using Bow's notation, assign a letter to the space *outside* the truss that is separated by either a reaction or an external load, proceeding clockwise around the truss.
2. Assign a number to each triangle within a truss.
3. Begin the construction of the Maxwell diagram by starting with the external force system, drawing a line parallel to and in the direction of a known external force and of a length to represent its magnitude to scale. Proceed around the structure clockwise; the resulting force polygon should close, as the external force system is in equilibrium.
4. Then applying superposition on the external force line polygon, such as *ABCDEA* in Fig. 2.5.1, and beginning with a joint containing only two

*After Robert H. Bow [1].
†After James Clerk Maxwell [2].

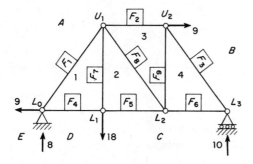

(a) Truss diagram
(Refer to Fig. 2.2.8 for algebraic solution)

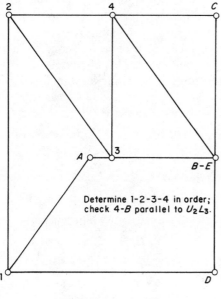

Note: Circles are added to Maxwell diagram after intersection of lines has been made, in order to separate solution points from intersection of construction lines (not shown here).

Determine 1-2-3-4 in order; check 4-B parallel to U_2L_3.

(b) Maxwell force diagram

Figure 2.5.1 Graphical method of joints for Example 2.5.1.

unknowns, construct line polygons for each joint in succession until all joints have been treated. At each joint proceed from force to force clockwise around a joint.

5. To interpret the results, at a given joint a member is designated by a letter and a number, or by two numbers, always using the sequence obtained by moving clockwise around a joint. The "letter–number" or "number–number" sequence is the same sequence used on the Maxwell diagram to establish the *directions* of the forces. The length of the line on the Maxwell diagram is proportional to the magnitude of the force.

Example 2.5.1

Determine the forces in the bars of the truss of Fig. 2.5.1a by the graphical method of joints.

Solution

    (a) *Apply Bow's notation.* The spaces between external forces and reactions are designated A-B-C-D-E-A clockwise around the truss, so the force vector A-B-C-D-E-A forms a closed polygon in Fig. 2.5.1b. The four internal triangles are numbered 1, 2, 3, and 4.

    (b) *Construct the Maxwell force diagram.* Each internal force is named by the two symbols on either side of the bar taken in the clockwise order around a joint; for instance, F_1 is in the direction of 1-A in Fig. 2.5.1b acting on the upper joint at U_1, but it is in the direction of A-1 in Fig. 2.5.1b acting on the lower joint at L_0. Points 1-2-3-4 on the force diagram of Fig. 2.5.1b are determined in succession; for instance, point 1 is the intersection of A-1 parallel to F_1 and D-1 parallel to F_4. Thus, one reads A-1-D-E-A around joint L_0 of the truss diagram and finds the closed polygon of the same name in the Maxwell diagram. The Maxwell diagram, then, is no more than the superposition of all the force polygons for the joint free bodies.

    The magnitude of each axial force may be measured from the force diagram and its direction acting on the joint (tension or compression) is determined by the indicated direction when proceeding from point to point on the Maxwell diagram according to the name of the bar obtained by reading in a clockwise pattern around a joint of the truss diagram. For this example,

$$F_1 = A\text{-}1 \text{ on } L_0 = 1\text{-}A \text{ on } U_1 = 10 \text{ kips compression}$$

$$F_2 = A\text{-}3 \text{ on } U_1 = 3\text{-}A \text{ on } U_2 = 1.5 \text{ kips tension}$$

$$F_3 = B\text{-}4 \text{ on } U_2 \text{ and } 4\text{-}B \text{ on } L_3 = 12.5 \text{ kips compression}$$

$$F_4 = 1\text{-}D \text{ on } L_0 = D\text{-}1 \text{ and } L_1 = 15 \text{ kips tension}$$

$$F_5 = 2\text{-}C \text{ on } L_1 = C\text{-}2 \text{ on } L_2 = 15 \text{ kips tension}$$

$$F_6 = 4\text{-}C \text{ on } L_2 = C\text{-}4 \text{ on } L_3 = 7.5 \text{ kips tension}$$

$$F_7 = 2\text{-}1 \text{ on } U_1 = 1\text{-}2 \text{ on } L_1 = 18 \text{ kips tension}$$

$$F_8 = 3\text{-}2 \text{ on } U_1 = 2\text{-}3 \text{ on } L_2 = 12.5 \text{ kips compression}$$

$$F_9 = 4\text{-}3 \text{ on } U_2 = 3\text{-}4 \text{ on } L_2 = 10 \text{ kips tension}$$

These forces are in agreement with the algebraic solution of Example 2.2.1 given in Fig. 2.2.8.

Example 2.5.2

Solve for the internal forces in the bars of the truss of Fig. 2.5.2a by the graphical method of joints.

Solution. Note that member L_0U_2 crosses member U_1L_1. Bow's notation involves A-3-1 at U_1 and A-1-2-D-E at L_0. The Maxwell diagram appears in Fig. 2.5.2b. The special feature relating to member L_0U_2 crossing but not attaching to member U_1L_1 is that the axial force F_8 is equal to 3-1 acting on U_1 or 2-4 acting on L_1, and F_7 is equal to 1-2 acting on L_0 or 4-3 acting on U_2.

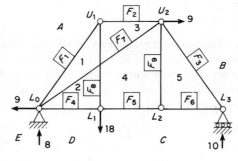

(a) Truss diagram

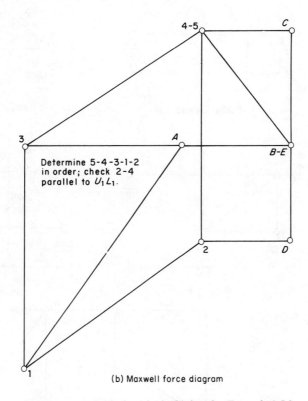

Determine 5-4-3-1-2 in order; check 2-4 parallel to $U_1 L_1$.

(b) Maxwell force diagram

Figure 2.5.2 Graphical method of joints for Example 2.5.2.

SELECTED REFERENCES

1. Robert H. Bow, *Economics of Construction in Relation to Framed Structures*, E. and F. N. Spon, London, 1873.
2. James Clerk Maxwell, "On Reciprocal Figures and Diagrams of Forces," *Philosophical Magazine* (4), *27* (1864), 294.

PROBLEMS

Each problem solution requires an answer diagram consisting of a line drawing of the truss with the axial forces (with horizontal and vertical components for inclined members) stated on the members, indicating whether tension ($+$) or compression ($-$).

2.1–2.5. Determine the reactions and the axial forces in all members of the truss in each accompanying figure using the algebraic method of joints.

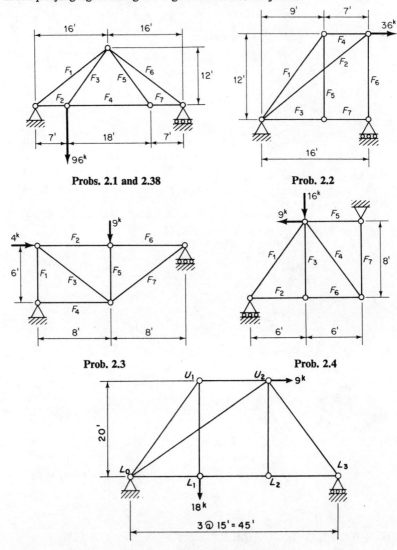

Probs. 2.1 and 2.38

Prob. 2.2

Prob. 2.3

Prob. 2.4

Probs. 2.5 and 2.39

2.6. Determine the forces in the members of the truss in the accompanying figure by the algebraic method of joints. Check member U_1L_2 by the method of sections.

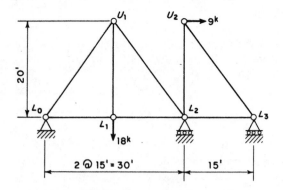

Probs. 2.6 and 2.40

2.7. Determine the forces in the members of the truss in the accompanying figure by the algebraic method of joints. Check members U_1U_2, L_1U_2, and L_1L_2 by the method of sections.

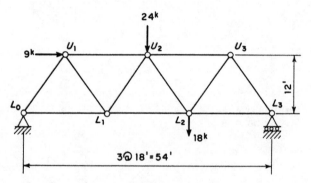

Prob. 2.7

2.8. Determine the forces in all members of the truss in the accompanying figure by the algebraic method of joints. Check members U_1U_2, L_1U_2, and L_2U_3 by the method of sections.

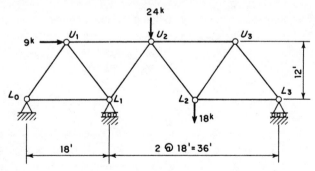

Prob. 2.8

2.9. Using the method of sections wherever possible, determine the axial forces F_1 through F_5 for the truss of the accompanying figure.

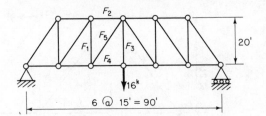

Probs. 2.9 and 2.22

2.10. Using the method of sections wherever possible, determine the axial forces F_1 through F_6 for the truss of the accompanying figure.

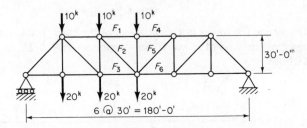

Probs. 2.10 and 2.23

2.11. Using either the method of sections or the method of joints, determine the axial forces F_1 through F_6 on the truss of the accompanying figure. Make independent solutions wherever possible.

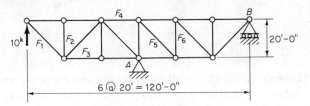

Probs. 2.11 and 2.24

2.12. For the truss of the accompanying figure, compute the reactions first; then compute axial forces F_1 through F_5 as independently as possible.

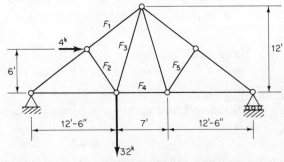

Prob. 2.12

2.13. For the truss of the accompanying figure, compute as independently as possible the axial force in all members.

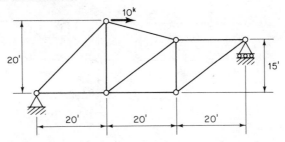

Prob. 2.13

2.14. For the truss of the accompanying figure, compute as independently as possible the axial force in bars *AE*, *BE*, *BC*, and *CE*.

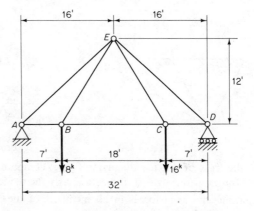

Prob. 2.14

2.15. For the truss of the accompanying figure, compute as independently as possible the axial forces F_1 through F_8. Indicate clearly the section or joint used for solution.

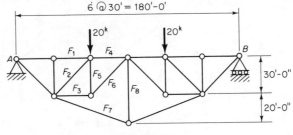

Prob. 2.15

2.16. Determine the axial forces F_1 through F_9 in the truss of the accompanying figure. Make independent solutions for the forces wherever possible. Indicate clearly the section or joint used for solution.

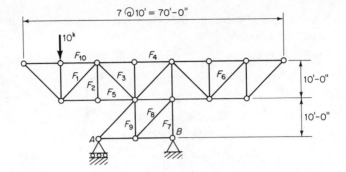

Prob. 2.16

2.17. For the truss of the accompanying figure, compute as independently as possible the axial forces F_1 through F_4. Indicate clearly the section or joint used for solution.

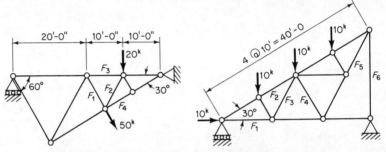

Probs. 2.17 and 2.41   **Probs. 2.18 and 2.42**

2.18. For the truss of the accompanying figure, compute as independently as possible the axial forces F_1 through F_6. Indicate clearly the section or joint used for solution.

2.19. Determine the axial forces in all members of the truss in the accompanying figure using the algebraic method of joints.

2.20. Determine independently each of the forces in the members of the truss of Prob. 2.19.

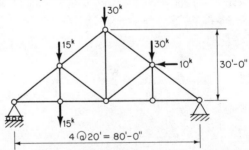

Probs. 2.19 and 2.20

2.21. For the truss of the accompanying figure, compute as independently as possible the axial forces in members $U_1 U_2$, $U_2 U_3$, $U_2 L_2$, $L_2 U_3$, $L_3 L_4$, and $U_3 L_4$ (six members in total).

2.22. Determine the axial forces in all the members of the truss of Prob. 2.9 using the method of moments and shears.

2.23. Determine the axial forces in all members of the truss of Prob. 2.10 using the method of moments and shears.

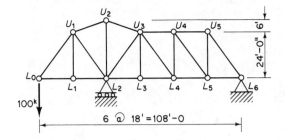

Prob. 2.21

2.24. Determine the axial forces in the six members indicated on the truss of Prob. 2.11 using the method of moments and shears.

2.25. Determine the forces F_1 through F_5 in the truss of the accompanying figure using any algebraic method.

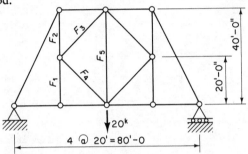

Prob. 2.25

2.26. Determine the forces F_1 through F_6 in the truss of the accompanying figure using any algebraic method.

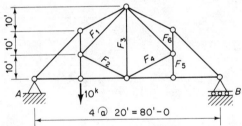

Prob. 2.26

2.27. Determine the forces in all members of the truss in the accompanying figure by any algebraic method.

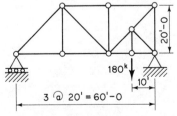

Prob. 2.27

2.28. Determine the forces F_1 through F_8 of the truss in the accompanying figure by any algebraic method.

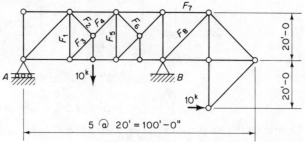

Prob. 2.28

2.29. Determine the forces F_1 through F_{10} of the truss in the accompanying figure using any algebraic method.

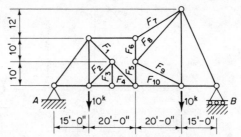

Prob. 2.29

2.30. Determine the forces F_1 through F_6 of the truss in the accompanying figure using any algebraic method.

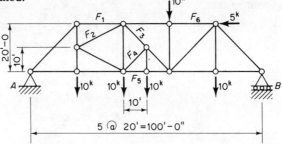

Prob. 2.30

2.31–2.36. For the truss of each accompanying figure, obtain the reactions by the algebraic

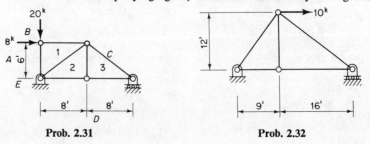

Prob. 2.31   **Prob. 2.32**

58   Statically Determinate Trusses—Reactions and Member Forces   Chap. 2

method. Then construct the Maxwell force diagram. State the scaled values of internal forces on an answer diagram. Compare with computed internal forces.

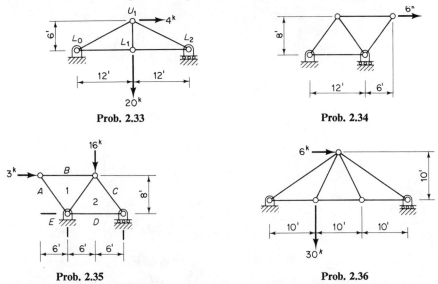

Prob. 2.33

Prob. 2.34

Prob. 2.35

Prob. 2.36

2.37. Determine the forces in all members of the truss in the accompanying figure graphically using the Maxwell diagram. Verify the forces in three members independently by algebraic solution.

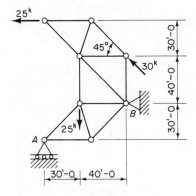

Prob. 2.37

2.38. Determine the forces in all members of the truss of Prob. 2.1 graphically using the Maxwell diagram. If Prob. 2.1 was previously assigned, show a comparison of results.

2.39. Determine the forces in all members of the truss of Prob. 2.5 graphically using the Maxwell diagram. Compare with algebraic results.

2.40. Determine graphically using the Maxwell diagram the forces in *all* members of the truss of Prob. 2.6.

2.41. Determine *all* the internal forces in the truss of Prob. 2.17 graphically using the Maxwell diagram.

2.42. Determine *all* the internal forces in the truss of Prob. 2.18 graphically using the Maxwell diagram.

2.43. Determine algebraically the forces F_1 through F_5 of the truss in the accompanying figure.

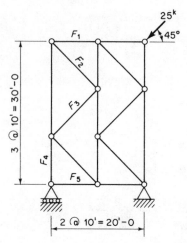

Prob. 2.43

2.44. Determine algebraically the forces F_1 through F_8 of the truss in the accompanying figure. Determine independently as many of these as possible. Note that the steel cables are capable of carrying tension only.

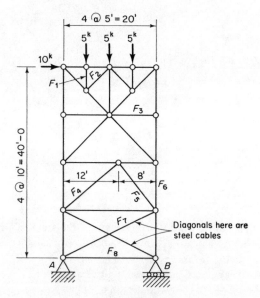

Prob. 2.44

2.45. Determine algebraically the forces in all members of the truss in the accompanying figure.

2.46. Determine graphically using the Maxwell diagram the forces in all members of the truss of Prob. 2.45. If Prob. 2.45 was previously assigned, compare graphical and algebraic results.

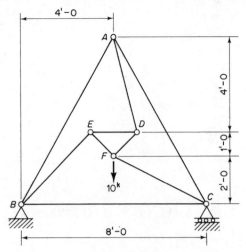

Probs. 2.45 and 2.46

[3]

Statically Determinate Trusses
—Deflections

3.1 DEFINITION OF DEFLECTIONS OR JOINT DISPLACEMENTS

The term *deflection* as used in structural analysis usually refers to a linear *displacement*. The more general term "displacement" is used to designate either the rotation of a rigid joint in a rigid frame or the slope at a point on a beam, as well as the linear movement of a point in a beam or of a joint in a truss or rigid frame. Thus, the linear movement of the joints in a truss can be called either a deflection or a joint displacement. Generally, however, except in the matrix displacement method where the displacement matrix may contain both both rotations and deflections, it is more agreeable to call a rotational displacement by the single word *slope* or *rotation*. In this way, the term "displacement" becomes synonymous with deflection.

By describing the new position of the joints of a truss in terms of the displacement components (horizontal and vertical) at each joint, the entire displaced configuration due to external loads (sometimes also due to temperature changes or fabrication tolerances) is obtained.

When the loads P_1, P_2, and P_3 are gradually applied to an unstressed, undeformed truss ABC as shown in Fig. 3.1.1a, the stressed, deformed truss will take the shape $A'B'C'$. The displacements A-A', B-B', and C-C' (usually expressed in terms of components, as for example X_1 and X_2 are the displacement components of B-B') are defined as deflections of the truss; they are functions of the shape of the initial truss, the cross-sectional areas, the modulus of elasticity, and the applied loads.

Regarding the elongations of bars, the elongation e in any bar is equal to

$$e = L' - L_0$$

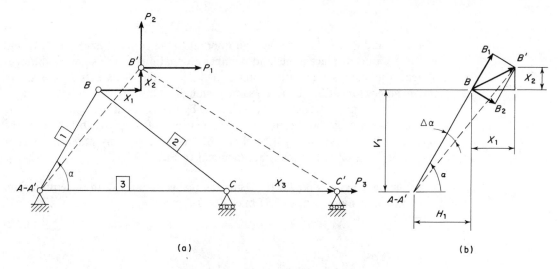

(a)   (b)

Figure 3.1.1 Definition of truss deflections.

where L' and L_0 are deformed and original lengths, respectively. For instance, the elongation e_1 of member 1 in Fig. 3.1.1 may be expressed as follows:

$$e_1 = A'B' - AB$$
$$= \sqrt{(H_1 + X_1)^2 + (V_1 + X_2)^2} - \sqrt{H_1^2 + V_1^2} \qquad (3.1.1)$$

With little error, the elongation e_1 of member 1 may be taken *approximately* as the distance BB_1, the axial component of the displacement BB'. Thus,

$$e_1 = BB_1 = X_1 \cos \alpha + X_2 \sin \alpha \qquad (3.1.2)$$

When the deflections are small relative to the original dimensions, as they generally are, there is negligible difference between the length BB_1 and distance $A'B'$-AB measured along the dashed line in Fig. 3.1.1b (i.e., the angle $\Delta\alpha$ is infinitesimal). The use of the axial component of displacement is sufficiently accurate to represent member elongation for usual situations of truss analysis. Throughout this textbook, elongations are always based on this assumption, commonly called the *first-order assumption*.

3.2 DERIVATION OF THE VIRTUAL WORK METHOD (UNIT LOAD METHOD)

In general, structural analysis procedures for determining deformations (i.e., deflections, displacements, rotations, or slopes) may be classified as either geometric or energy methods. The principle of virtual work is an energy method originating with Johann Bernoulli [1] in 1717. Proved to be true later in this section, this principle may be defined by the following two statements:

1. The total work performed by a system of external and internal forces in equilibrium in going through *any compatible displacement* in the same or different system is zero. This may be expressed mathematically as

$$\sum Q_i X_i - \sum u_i e_i = 0 \qquad (3.2.1)$$

where Q_i and u_i are the external and internal forces in one system, and X_i and e_i are the joint displacements and member elongations in the same or different system. By "compatible" it is meant that the joint displacements X_i are due to the member elongations e_i in the same system.

2. External work and internal work, each done by the forces in the *equilibrium state* of one system in going through the deformations of a compatible state of the same or different system must be equal. Mathematically, this statement is later shown as Eq. (3.2.3) but is in fact identical to Eq. (3.2.1).

Examine the truss of Fig. 3.2.1a. Using the real system for both the equilibrium and the compatible states, Eq. (3.2.1) may be written

$$P_1 X_1 + P_2 X_2 + P_3 X_3 = F_1 e_1 + F_2 e_2 + F_3 e_3 \qquad (3.2.2)$$

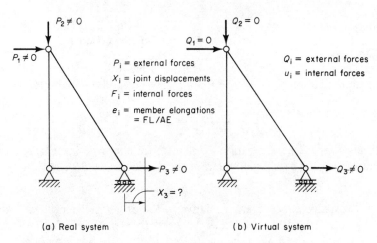

(a) Real system   (b) Virtual system

Figure 3.2.1   Real and virtual systems used in the virtual work method.

Since the displacement X_3 is desired, this energy equation is of little use because there are too many unknowns (i.e., there are three unknowns: X_1, X_2, and X_3). However, rule 1 states that Eq. (3.2.1) applies to *any compatible displacement*. Thus, one can use an imaginary set of loads (a so-called *virtual system*) acting on the actual structure. Such a virtual loading is shown in Fig. 3.2.1b. To determine the actual displacement X_3 of the truss in Fig. 3.2.1a, the virtual forces Q_i and internal forces u_i may be considered to be going through the compatible joint displacements and member elongations, respectively, of the real P-system of Fig. 3.2.1a. Thus, using rule 2,

$$\underset{\substack{\text{External} \quad \text{Joint} \\ \text{forces of} \quad \text{displacements} \\ \text{virtual} \quad \text{of real} \\ \text{system} \quad \text{system}}}{\sum Q_i X_i} \qquad = \qquad \underset{\substack{\text{Internal} \quad \text{Member} \\ \text{forces} \quad \text{elongations} \\ \text{of} \quad \text{of real} \\ \text{virtual} \quad \text{system} \\ \text{system}}}{\sum u_i e_i} \qquad (3.2.3)$$

There is distinct advantage in using a set of compatible displacements and a set of external and internal forces from different systems rather than the same system as was done with Eq. (3.2.2). Expanding Eq. (3.2.3) for the situation of Eq. 3.2.1 and letting Q_1 and Q_2 equal to zero, the left-hand summation becomes a single term. Further, using Hooke's Law, $e_i = F_i L_i / A_i E_i$. Thus,

$$Q_3 X_3 = \sum_{i=1}^{i=NF} u_i F_i \frac{L_i}{A_i E_i} \tag{3.2.4}$$

The solution for X_3 will be obtained in the simplest fashion when the force Q_3 of the virtual system is taken equal to unity: hence, the frequent reference to this method as the *unit load method*.

Thus, one may state that, according to the principle of virtual work, to determine an actual joint displacement, a unit load is applied to the structure in the direction of the desired displacement with all other loads removed (thus creating the virtual system) and the desired displacement is equal to the sum of the products of the real system member elongations times the virtual system internal forces.

To show why rules 1 and 2 hold true, examine the truss of Fig. 3.2.2. An arbitrary set of bar elongations e_1 to e_9 in a real system for the truss of Fig. 3.2.2a would cause the truss to assume the position of the dashed line relative to the original shape as defined by a set of joint displacements X_1 to X_9. The set of bar elongations are said to be compatible with the set of joint displacements; and the state in Fig. 3.2.2a is commonly called a *compatible state*. Note that the bar elongations may be due to applied loads, temperature changes, or fabrication tolerances.

By a first-order analysis based on the initial geometry of the truss, a set of bar forces u_1 to u_9 (the special designation u is used to mean internal forces due to a single external unit load) may be obtained for a single external force, such as $Q_i = 1$ kip (in direction of X_i, which would be X_6 in Fig. 3.2.2a) in the virtual system of Fig. 3.2.2b. The state in Fig. 3.2.2b is commonly called an *equilibrium state*, for which the free-body diagrams of the six joints are shown in Fig. 3.2.2c.

The work done by the forces acting on each joint of Fig. 3.2.2c, in going through the displacement X of that joint in Fig. 3.2.2a, must be zero because the resultant of the forces acting on any joint is equal to zero. The total work obtained by adding up such virtual work, done by the forces in the virtual system of Fig. 3.2.2c in going through the displacements in the real system, is still zero; or

$$W = \text{work done by } R_1 \text{ to } R_3 + \text{work done by } Q_i$$
$$+ \text{work done by nine pairs of } u_1 \text{ to } u_9 = 0 \tag{3.2.5}$$

The work done by R_1 to R_3 is zero because there are no displacements in the R-directions in the compatible state of the real system. The work done by Q_i is $(+Q_i X_i)$, as long as the positive direction of Q_i is taken in the same sense as that of X_i. The work done by the force pairs u_1 to u_9 is equal to $(-\sum ue)$ because the two forces in the same pair act toward each other (for tension) in Fig. 3.2.2c, and the relative displacement between their points of application is away from each other (for elongation) in Fig. 3.2.2a. Substituting these expressions for virtual work in Eq. (3.2.5), one obtains

$$W = 0 + Q_i X_i - \sum u_i e_i = 0$$

from which Eq. (3.2.4) is obtained.

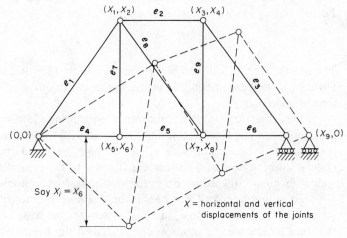

(a) Compatible state of real system

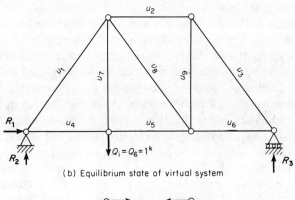

(b) Equilibrium state of virtual system

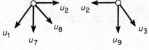

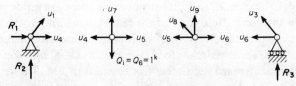

(c) Free-body diagrams of joints

Figure 3.2.2 Proof for the principle of virtual work.

66

3.3 VIRTUAL WORK METHOD (UNIT LOAD METHOD) —EXAMPLES

The use of the virtual work (unit load) method to find the deflections of the joints in a truss is illustrated by the following three examples.

Example 3.3.1

For the truss of Fig. 3.3.1, determine the vertical deflection (displacement) of joint U_2 using the virtual work (unit load) method.

Solution

(a) *Internal forces F_i for the real system.* The actual loading and internal forces are shown in Fig. 3.3.1b; note that $\sum F_x = 0$ and $\sum F_y = 0$ at every joint in this figure.

(b) *Virtual system.* Establish the virtual (imaginary) loading system by loading the actual structure with a single joint force (taken as unity for simplicity) acting in the direction of the deflection desired for the real system. The virtual system with its internal forces u_i is given in Fig. 3.3.1c. Note again $\sum F_x = 0$ and $\sum F_y = 0$ at every joint in this figure.

(c) *Utilize the virtual work concept to compute the vertical deflection of joint U_2.* Since the unit load is taken upward, the positive direction of the deflection X of joint U_2 is upward. The thought process one should keep in mind in using Eqs. (3.2.4) or (3.2.3) is that the summation of the products of the external forces on the virtual system (Fig. 3.3.1c) times their displacements X_i on the real system (Fig. 3.3.1b) equals the summation of the products of the internal forces of the virtual system times the elongations of the real system. This is best done by the following tabular approach and defining S_i as the member axial stiffness $E_i A_i / L_i$:

i	Member	F_i	$S_i = \dfrac{E_i A_i}{L_i}$	u_i	$u_i F_i / S_i$
1	$L_0 U_1$	-10	500	$+5/12$	$-8.33/1000$
2	$U_1 U_2$	$+1.5$	1000	$+1/2$	$+0.75/1000$
3	$U_2 L_3$	-12.5	500	$+5/6$	$-20.83/1000$
4	$L_0 L_1$	$+15$	1000	$-1/4$	$-3.75/1000$
5	$L_1 L_2$	$+15$	1000	$-1/4$	$-3.75/1000$
6	$L_2 L_3$	$+7.5$	1000	$-1/2$	$-3.75/1000$
7	$L_1 U_1$	$+18$	250	0	0
8	$U_1 L_2$	-12.5	500	$-5/12$	$+10.42/1000$
9	$L_2 U_2$	$+10$	250	$+1/3$	$+13.33/1000$
					$\sum u_i F_i / S_i = -15.91/1000$

$$X = -0.016 \text{ in.} \qquad \text{(negative means downward)}$$

The same result is obtained if in the thought process one thinks of computing first the elongations e_i of the real system: that is, to compute $e_i = F_i / S_i$ in the real (compatible) state (Fig. 3.3.1d); and then to obtain the summation of the products of e_i times the u_i values of the virtual (equilibrium) state.

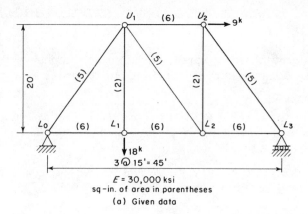

(a) Given data

$E = 30,000$ ksi
sq-in. of area in parentheses

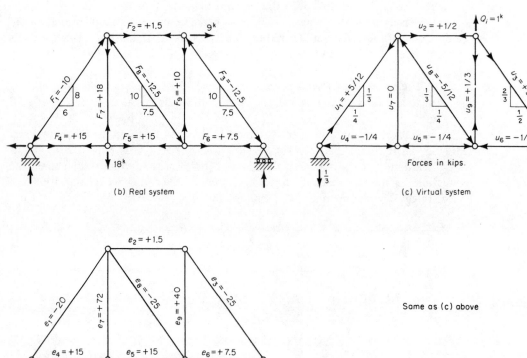

(b) Real system

(c) Virtual system

Forces in kips.

(d) Compatible state

Elongations in 10^{-3} in.

Same as (c) above

(e) Equilibrium state

Figure 3.3.1 Truss deflections by the virtual work (unit load) method.

Example 3.3.2

Determine the joint displacement X for the truss of Fig. 3.3.2 when one member is too short by an amount equivalent to the shortening due to an 80-kip compressive internal force.

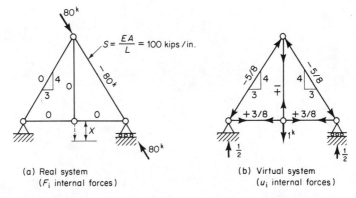

(a) Real system
(F_i internal forces)

(b) Virtual system
(u_i internal forces)

Figure 3.3.2 Truss deflection due to member made too short, for Example 3.3.2.

Solution. Referring to Fig. 3.3.2a and b and using Eq. (3.2.4),

$$X = \sum u_i F_i \frac{L_i}{A_i E} = \sum \frac{u_i F_i}{S_i}$$

In this case, there is only one term for F_i, that is, 80 kips compression. Thus,

$$X = -\frac{5}{8}(-80)\frac{1}{100} = 0.5 \text{ in.} \quad \text{(positive means downward)}$$

Example 3.3.3

Determine the relative movement apart of joints A and E of the truss shown in Fig. 3.3.3a.

Solution. In this example, the thinking is much like Example 3.3.2. Instead of applying one unit load, a pair of such loads is applied. The result will be the sum of the displacement components along the imaginary line shown in Fig. 3.3.3d, the component upward to the left of the total deflection at the upper end of the imaginary line and the component downward to the right at the lower end of this line.

The numbering system for the members is given in Fig. 3.3.3b. The forces F_i for the truss with its real loading are given on Fig. 3.3.3c. The forces u_i due to the pair of equal and opposite unit loads are shown in Fig. 3.3.3d.

The tabulation for the relative displacement X is as follows:

Member	F_i	u_i	$S_i = E_i A_i / L_i$	$F_i u_i / S_i$
1	+25/3	0	100	0
2	−20/3	+4/5	125	−128/3000
3	+25/3	−1	100	−250/3000
4	−20/3	+4/5	125	−128/3000
5	+5	+3/5	1000/3	+27/3000
6	+5	0	1000/3	0
7	0	+3/5	1000/3	0
				−479/3000

$$1 \cdot X = \sum u_i \frac{F_i}{S_i} = \frac{-479}{3000} = -0.16 \text{ in.}$$

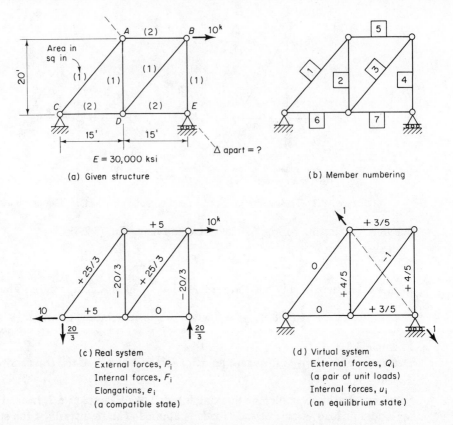

Figure 3.3.3 Relative displacement between two truss joints for Example 3.3.3.

Since the answer is negative, it means the relative movement is opposite to the direction indicated by the pair of unit loads; that is, the relative movement of joints A and E is 0.16 in. toward each other.

3.4 GEOMETRIC METHOD: USE OF THE JOINT DISPLACEMENT EQUATION

The definition of joint displacement (i.e., truss deflection) is given by Fig. 3.1.1. Assume that all the reaction components and internal forces have been obtained by a first-order analysis (the usual procedure using the undeformed geometry of the structure in writing up the equations of statical equilibrium). The joint displacement equation [2] to be developed is a more general form of Eq. (3.1.2). Referring to Fig. 3.4.1 representing a typical member AB in a truss, the elongation of the member in terms of the joint displacements of its ends may be expressed as

$$e = (X_3 - X_1) \cos \alpha + (X_4 - X_2) \sin \alpha \qquad (3.4.1)$$

Note that e as used here represents the elongation in dimensional units such as inches,

Statically Determinate Trusses—Deflections   Chap. 3

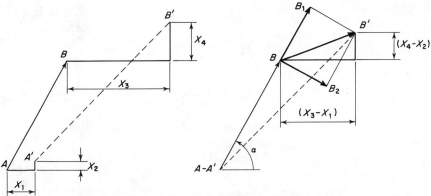

Figure 3.4.1 Elongation in terms of joint displacements.

feet, millimeters, or meters. Using member elongations as the input data, the procedure by which all joint displacements can be obtained by repeated use of Eq. (3.4.1) can best be explained by the following example.

Example 3.4.1

Determine the horizontal and vertical displacements of all joints in the truss of Example 3.3.1 by the geometric method. Note that the vertical deflection of the joint U_2 has been previously solved by the energy method (virtual work, unit load method). Refer to Fig. 3.3.1a for the given data.

Solution

(a) *Member elongations due to given loading.* From the given forces F_i (see Fig. 3.3.1b) the member elongations may be computed according to Hooke's Law,

$$e_i = F_i \frac{L_i}{E_i A_i} = \frac{F_i}{S_i}$$

where $S_i = E_i A_i / L_i$ is the axial *stiffness* of the member. These elongations e_i are presented in Fig. 3.4.2c.

(b) *Use of joint displacement equation.* The shape of the deformed truss L_0'-L_1'-L_2'-L_3'-U_1'-U_2' in Fig. 3.4.2b is dictated by the bar elongations and it should be superimposed on the original shape by placing L_0' on L_0, and L_3' on the same elevation as L_3. However, it is convenient at the beginning to place, for instance, joint L_2' over joint L_2, and bar L_2'-U_1' over bar L_2-U_1, as shown in Fig. 3.4.2b. In this orientation, L_2 and $L_2 U_1$ are called *reference point* and *reference member*, respectively. The choice of reference point and reference member is arbitrary. When displacements are desired for all truss joints, the reference point may most conveniently be taken as one that does not move, in this case L_0. If the reference point were L_0, the reference member would be either $L_0 U_1$ or $L_0 L_1$. The selection of a reference point somewhere in the middle of the truss, however, may reduce the cumulative round-off errors in hand computations for a very large truss.

The horizontal and vertical displacements of all the joints in the reference posi-

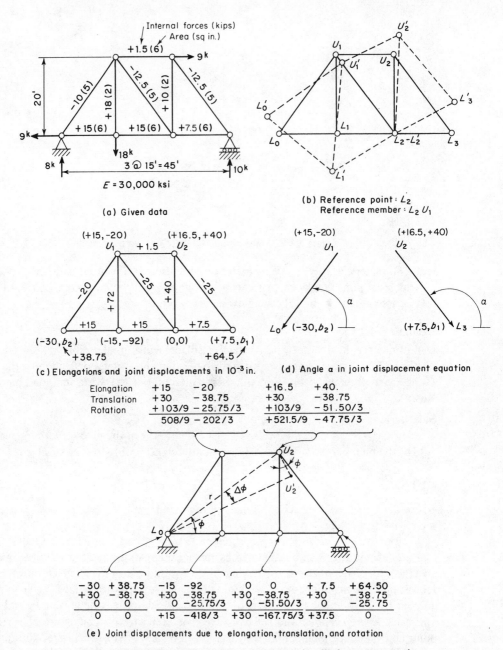

Figure 3.4.2 Truss deflections by use of the joint displacement equation.

tion shown in parentheses in Fig. 3.4.2c are partially obtained by inspection and partially by successively applying the joint displacement equation (3.4.1). In order that U_1L_2 shortens by 25 units, joint U_1 must move 25 units toward L_2, which means 15 units to the right and 20 units downward. Joint U_2 must move 1.5 units more to the right than joint U_1, or 16.5 units to the right; and it must move upward 40 units more than joint L_2. Joint L_3 moves 7.5 units to the right to accommodate the elongation of 7.5 units in L_2L_3 but its vertical displacement b_1 will as yet be unknown. Joint L_1 moves 15 units to the left and 92 units downward to accommodate an elongation of 15 units in L_1L_2 and 72 units in U_1L_1. Joint L_0 must move the sum of elongations in L_0L_1 and L_1L_2, or 30 units to the left; but its vertical displacement b_2 is unknown.

The joint displacement equation is applied to members U_2L_3 and U_1L_0, respectively, in order to obtain b_1 and b_2. For member U_2L_3 (Fig. 3.4.2d),

$$-25 = (+16.5 - 7.5)(-0.6) + (+40 - b_1)(+0.8)$$
$$b_1 = +64.5 \text{ units}$$

For member U_1L_0 (Fig. 3.4.2d),

$$-20 = [+15 - (-30)](+0.6) + (-20 - b_2)(+0.8)$$
$$b_2 = +38.75 \text{ units}$$

(c) *Truss translation and rotation from the reference position.* The deformed truss now resting in the reference position of Fig. 3.4.2b as defined by the joint displacements shown in Fig. 3.4.2c must be moved as a rigid body through a translation of 30 units to the right and 38.75 units downward so that joint L_0' may fall upon L_0, followed by a clockwise rotation around L_0 or L_0' to bring L_3' (now at $64.50 - 38.75 = 25.75$ units above L_3) down to the same level as L_3. The joint displacements due to bar elongations, due to translation, and due to rotation are shown in Fig. 3.4.2e, together with the final resulting deflections. The horizontal and vertical displacements of any joint, for instance U_2 in Fig. 3.4.2e, due to only the rotation of $\Delta\phi$ around L_0 are:

$$\text{Horizontal displacement} = U_2U_2' \sin\phi = r\,\Delta\phi \sin\phi$$
$$= (\text{vertical projection of } r)(\Delta\phi)$$
$$= (20 \text{ ft})\frac{25.75 \text{ units}}{45 \text{ ft}} = \frac{103}{9} \text{ units (rotation component)}$$

$$\text{Vertical displacement} = U_2U_2' \cos\phi = r\,\Delta\phi \cos\phi$$
$$= (\text{horizontal projection of } r)(\Delta\phi)$$
$$= (30 \text{ ft})\frac{25.75 \text{ units}}{45 \text{ ft}} = \frac{51.50}{3} \text{ units (rotation component)}$$

The computations shown above refer to the absolute quantities only; the positive or negative signs before these values for use in Fig. 3.4.2e can best be obtained by physically visualizing the effect of a clockwise or counterclockwise rotation as may be required.

(d) *Discussion.* The use of the joint displacement equation in finding truss deflections is fully displayed here, not so much to advocate the method itself as to explain the physical meaning of truss deflections. The vertical deflection of joint U_2, had it been accurately determined by the energy (virtual work, unit load) method in the tabulation of Example 3.3.1, would be $-47.75/3$ milliinches, the same as shown in Fig. 3.4.2. One can conclude that the energy and geometric methods are theoretically identical in describing a physical phenomenon on the same underlying assumptions. This is the reason why fractional numbers have been used in this example; normally, in a practical situation the usual accuracy of three or four significant figures should be used.

3.5 GEOMETRIC METHOD: WILLIOT–MOHR DIAGRAM

Just as the overall truss and each joint must be in equilibrium, the entire truss and all its joints must have compatibility between member elongations and joint displacements. In Section 2.5 it was shown that truss equilibrium can be represented graphically by a diagram known as the *Maxwell force diagram*. Similarly, the compatibility for truss deformation may be represented graphically. In fact, Fig. 3.4.2b shows the deformed shape of a truss in the reference position, only that the joint displacements are plotted on a different scale (many times larger) than that for the shape of the original truss.

In 1877, the French engineer Williot [3] in a treatise on graphic statics showed that the deformed structure can be represented graphically by using only the member elongations and member directions in the construction of the diagram, that is, assuming the original lengths of the members to be zero. This permits a large scale to be used to obtain relatively accurate deformations. Williot's method, however, was only applicable if the reference member chosen in the graphical solution did not actually rotate in the final deformed position of the truss. Otto Mohr in 1887 [4] presented a simple method for correcting the Williot diagram if the reference member rotates when the truss is deformed. Thus, the graphical method (the *Williot–Mohr method*) is identical to the geometric procedure of Section 3.4.

The use of the Williot–Mohr diagram in determining the displacements of the joints in a truss in terms of member elongations and member directions is explained in the examples that follow.

Example 3.5.1
Construct the Williot–Mohr diagram for the joint displacements of the truss of Fig. 3.4.2a.

Solution
(a) *Selection of reference point and reference member*. The elongation of each member due to its internal force is computed, as in Fig. 3.4.2c. Select a reference point, say L_2, and a reference member, say member L_2U_1. Single primes will be used to designate the displaced joints relative to the reference point and reference direction (see Fig. 3.4.2b). Thus, L'_2 coincides with L_2 (the reference point) and U_1 lies along the direction of L_2U_1.

(b) *The Williot diagram*. With the direction $L'_2U'_1$ defined by the L_2U_1 direction, start the Williot diagram at the reference point L'_2 (an arbitrary point to begin the diagram) in Fig. 3.5.1b and lay off to scale the shortening (-25 units) of the member L_2U_1. Such shortening means that if L_2 is stationary, U_1 moves to U'_1 located along the reference direction 25 units down to the right. With the deformed position [the deformed position of each joint relative to its original position is always measured from the reference point L'_2 (identical with L_2) in Fig. 3.5.1b] of L_2 and U_1 located (i.e., L'_2 and U'_1), next the third point (either L'_1 of the triangle below the refer-

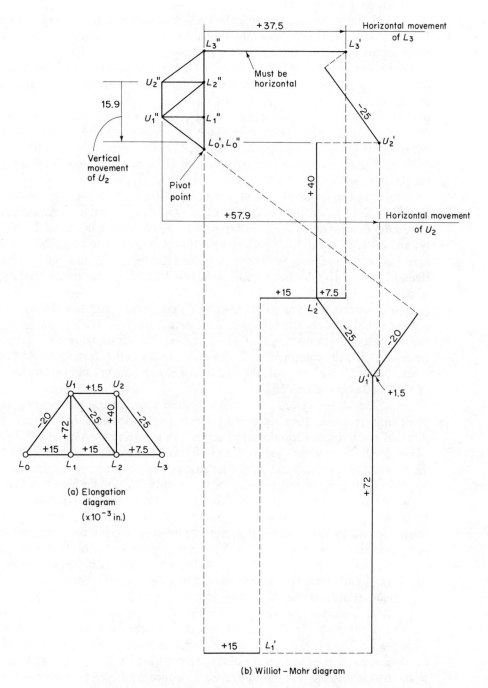

+37.5

Horizontal movement of L_3

L_3'' L_3'

U_2'' L_2''

Must be horizontal

15.9

U_1'' L_1''

-25

L_0', L_0'' U_2'

Vertical movement of U_2

Pivot point

+40

+57.9

Horizontal movement of U_2

+15   +7.5

L_2'

-25 -20

U_1' +1.5

U_1 +1.5 U_2

-20 $+72$ -25 $+40$ -25

+15   +15   +7.5

L_0 L_1 L_2 L_3

(a) Elongation diagram
($\times 10^{-3}$ in.)

+72

+15 L_1'

(b) Williot – Mohr diagram

Figure 3.5.1   Williot–Mohr diagram for Example 3.5.1.

ence member, or U_2' of the triangle above the reference member) is located by the procedure as described below.

The point L_1' should theoretically be located along an arc having a radius from L_2' that is 15 units longer than the original length of L_1L_2. It should also be located along an arc having a radius from U_1' that is 72 units longer than U_1L_1. It has been shown in Section 3.1 that it is valid to use a perpendicular to the radial line instead of an arc. Thus, point L_1' is located 15 units on a line through L_2', parallel to L_1L_2, and directed to the left because an elongation of 15 units indicates that the displaced position of L_1 is to the left of its original position when L_2 is assumed stationary. Then L_1' lies on a line perpendicular to the line extending 15 units horizontally to the left of L_2. From the known position of U_1', L_1' lies perpendicular to a line parallel to U_1L_1 and 72 units below U_1'. Lengthening of U_1L_1 means L_1 moves *down* relative to the position of U_1.

Next L_0' can be located. From stationary point L_1', L_0' moves away from, that is, to the left of, L_1' by an amount equal to 15 units. L_0' lies on a line perpendicular to the horizontal line (parallel to L_0L_1) through L_1'. From stationary point U_1', L_0' moves toward U_1' (upward to the right) along a line parallel to L_0U_1 by the amount of 20 units. L_0' lies on a line perpendicular to the line shown as -20 through U_1' (upward to right) in Fig. 3.5.1b. The intersection of the two dashed lines just located is the location of point L_0'.

In a similar fashion, points U_2' and L_3' are located and the Williot diagram is complete. Using L_2' as the origin (as the original location of all joints in the truss), the displacement components of all joints can be scaled from the diagram. For instance, L_3' has the coordinates $+7.5$ (positive means to the right) and $+64.5$ (positive means upward). Similarly, all the displacement components shown on the first line of Fig. 3.4.2e can be verified.

(c) *The Mohr diagram.* As shown in Example 3.4.1, to get the true displacements of all joints measured from their original locations in the undeformed truss, a translation and rotation correction must be applied as shown in the second and third lines of Fig. 3.4.2e. Otto Mohr showed that the desired total answers can be obtained graphically as follows. Identify the point that has zero displacement in the final position of the deformed truss, in this case L_0'. Next, identify the point where the direction of actual displacement is known; in this case L_3' must have an actual displacement that is horizontal (with zero vertical displacement). Then through L_3', draw a horizontal line. Now with L_0' as a pivot point, rotate a "miniature" truss of the same configuration and proportion as the original truss through 90°, as shown by the double-prime points in Fig. 3.5.1b. The point L_0'' of this truss should coincide with the pivot point L_0' and length $L_0''L_3''$ is dictated by the requirement that L_3'' lie on the horizontal through L_3'. The miniature truss is the Mohr diagram.

(d) *Measurement of results and proof of the Williot–Mohr method.* With the completed Williot–Mohr diagram, the true displacement of any joint can be measured from the double-prime point to the single-prime point in Fig. 3.5.1b. For instance, the displacement of joint L_3 is the distance from L_3'' toward L_3'; in this case, 37.5 units to the right of its original position. Also identified on Fig. 3.5.1b are the horizontal displacement (57.9 units to the right from U_2'' toward U_2') and the vertical displacement (15.9 units downward from U_2'' to U_2') for the joint U_2. All other displacements can be obtained similarly by measuring from the double-prime point to the single-prime point.

Now the proof for the correctness of the Williot–Mohr method can be made. Take joint U_2, for instance. First, the displacement from L_2' to U_2' in Fig. 3.5.1b is obtained due to member elongations, by holding the reference point L_2 stationary and the reference member $L_2 U_1$ without rotation. Next, the displacement of U_2 due to translation is from L_0' to the reference point L_2'. Finally, the displacement due to rotation is from U_2'' to L_0''. Observing Fig. 3.5.2, one can see that the final displacement is the vector sum of all three parts; that is, it is from the double-prime point to the single-prime point.

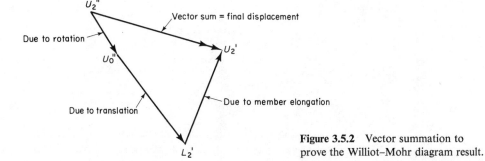

Figure 3.5.2 Vector summation to prove the Williot–Mohr diagram result.

That the displacement due to rotation is always equal to that from the double-prime point to the pivot point L_0'' or L_0' can be proved by the fact that the magnitude of this displacement is always equal to the straight-line distance of any joint from the hinge times the angle of rotation and its direction is always perpendicular to this straight line (imaginary as far as joint U_2 is concerned). This is the reason why the miniature truss is at 90° to the original truss and it is geometrically similar to the original truss because all dimensions are equal to the respective true dimensions times the angle of rotation. As to which side of $L_0'' L_3''$ the miniature truss should be drawn, the requirement is that it can be rotated 90° to a horizontal position in the plane of the paper and its appearance is the same as the original truss.

Example 3.5.2

Determine the joint displacements for the truss of Fig. 3.3.2, which contains one member too short by an amount equivalent to that produced by a compressive internal force of 80 kips. Use the Williot–Mohr diagram.

Solution

(a) *Selection of reference point and reference member.* The elongations of all members due to internal forces are computed, as shown in Fig. 3.5.3a. Next construct the Williot diagram, choosing L_0 as the reference point and $L_0 L_1$ as the reference member.

(b) *The Williot diagram.* Starting with L_0' in Fig. 3.5.3b, L_1' is coincident with L_0' because the elongation of $L_0 L_1$ is zero. Next, determine U_1' also to be coincident with L_0' and L_1' because it has no relative movement from both L_0 and L_1. Next, L_2' is located by measuring horizontally (the direction of $L_1 L_2$) an amount equal to the elongation

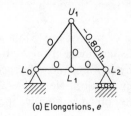

(a) Elongations, e

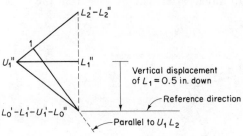

$L_2'-L_2''$

U_1''

L_1''

Vertical displacement
of $L_1 = 0.5$ in. down

Reference direction

$L_0'-L_1'-U_1'-L_0''$

Parallel to $U_1 L_2$

(b) Williot–Mohr displacement diagram

Figure 3.5.3 Displacements for Example 3.5.2.

of $L_1 L_2$, zero in this case. Then L_2 lies on a line perpendicular to the original direction of $L_1 L_2$. Also L_2 moves toward the already determined point U_1' by virtue of the shortening of the member $U_1 L_2$, that is, 0.80 in. to scale up to the left along a line through U_1' parallel to $U_1 L_2$, thus locating point 1 in Fig. 3.5.2b. Then L_2' is located along the perpendicular to the line just scaled off at its intersection with the perpendicular to the direction $L_1 L_2$ through point L_1'.

(c) *The Mohr diagram.* The Mohr correction diagram is then obtained using L_0' (also called L_0'') as the pivot point. Also noting that no vertical displacement of L_2 can occur, L_2'' must lie on a horizontal through L_2'. In this case, since $L_0''-L_2''$ must be vertical, L_2'' becomes coincident with L_2'. The Mohr diagram is then drawn as shown.

(d) *Measurement of results.* In this example there is neither horizontal nor vertical displacement of L_2. The vertical displacement of L_1 is measured from L_1'' toward L_1', giving a downward displacement of 0.5 in. This agrees with the result obtained by the virtual work (unit load) method in Example 3.3.2. Of course, the Williot–Mohr diagram can give the displacement components of every joint.

Example 3.5.3

Determine the displacements of joints L_2 and L_3 of the truss of Fig. 3.5.4a using the Williot–Mohr diagram.

Solution. First, the elongations of all members are computed and presented in Fig. 3.5.3b. The Williot diagram construction is started by taking L_0 as the reference point and $L_0 L_1$ as the reference member. The diagram is constructed as in the previous two examples and shown in Fig. 3.5.4c.

Next the Mohr rotation diagram must be constructed so that the true displacements can be directly measured. L_0 is the pivot point; because of the hinged support that joint has no displacement. The structure is now to be rotated 90° about the pivot point L_0. The scale of the rotated structure is established by noting that the dis-

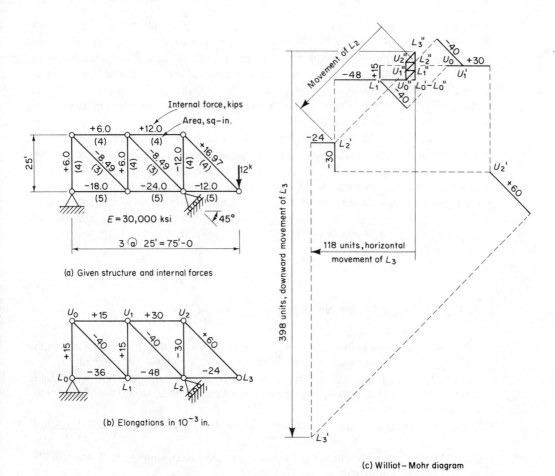

Figure 3.5.4 Displacement determination for Example 3.5.3.

placement of joint L_2 must be along a 45° line through point L_2' of the Williot diagram. The true displacement of L_2 is defined by the slope of the roller support of the truss. Thus L_2'' lies along a 45° line through L_2' parallel to the roller support and on a line vertically through L_0'. Knowing L_0'' and L_2'' locations, the entire structure is reproduced to scale, designating the joints by using double primes on the original joint designations.

The displacement of L_2 is 120 units from L_2'' to L_2' (downward to the left at 45°). The horizontal component of displacement of L_3 (from L_3'' to L_3') is 118 units to the left, and the vertical component of displacement of L_3 is 398 units downward.

3.6 COMPARISON OF METHODS

Two approaches to truss displacement have been presented: an energy method and a geometric method. The energy method, known as the *method of virtual work*, is a powerful structural analysis tool whose application will reappear numerous times throughout the remainder of this text. The usual application of the virtual work method in solving numerical problems is to use a unit load to establish the virtual system; hence, the method is commonly referred to as the *unit load method*.

The geometric method, using visual inspection and the algebra of the joint displacement equation or using graphical means with the Williot–Mohr diagram, is useful when many joint displacements are to be computed, but more important, it is useful for understanding how a truss deforms.

Both the energy and geometric methods must give identical results. Both are "exact," consistent with the basic structural analysis assumptions that the truss members are linearly elastic, that no moments are transmitted at the joints, that each member is subjected to axial force only, that the original structural configuration (i.e., first-order analysis) may be used for all geometry relationships needed to solve for internal forces and displacements, and that member elongation is equal to the axial component of the displacement of the ends of the member (this is also part of the first-order assumption).

SELECTED REFERENCES

1. P. Duhem, "Les origines de la statique," 2 vols, Paris, 1905–1906. (See also Hans Straub, *A History of Civil Engineering*, The MIT Press, Cambridge, Mass., 1964, p. 69.)

2. Chu-Kia Wang and C. L. Eckel, "Truss Deflections and Indeterminate Stresses by Joint Displacements," *Bulletin No. 4*, Engineering Experiment Station, University of Colorado, August 1945.

3. Williot, *Notations pratiques sur la statique graphique*, Publications scientifiques industrielles, 1877.

4. Otto Mohr, "Ueber Geschwindigkeitsplane und Beschleunigungsplane," *Der Civilingenieur*, **33** (1887), 631–650.

PROBLEMS

Each problem solution requires, in addition to the specific displacement to be determined, two line drawings of the truss, one containing the internal axial forces stated on the members and the other containing member elongations stated on the members. The modulus of elasticity E is to be taken as 30,000 ksi for all problems.

Problems Using Virtual Work (Unit Load) Method (Probs. 3.1 to 3.19)

3.1. For the truss of the accompanying figure, determine horizontal displacement of joint C by the virtual work (unit load) method. The elongations of the members are given in units of 10^{-3} in.

3.2. For the truss of the accompanying figure, determine the vertical deflection of joint D by the virtual work (unit load) method. All bars have the same EA.

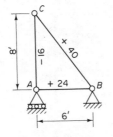

Probs. 3.1, 3.21, and 3.30　　　　**Prob. 3.2**

3.3. For the truss of the accompanying figure with the bar elongations given on the figure, obtain the horizontal deflection of joint C by the virtual work (unit load) method.

3.4. For the truss of the accompanying figure, compute the vertical and horizontal displacements of joint B by the virtual work (unit load) method.

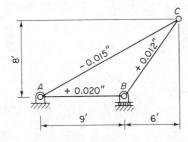

Prob. 3.3

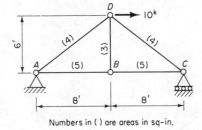

Numbers in ( ) are areas in sq-in.

Prob. 3.4

3.5. Given the elongations in 10^{-3} in. as shown on the members of the truss of the accompanying figure, compute the vertical deflection at D and the horizontal deflection at B by the virtual work (unit load) method. Be sure to state the directions of the displacements.

3.6. For the truss of the accompanying figure with the given elongations, determine the horizontal deflection of joint U_1 and the horizontal deflection of joint U_2 by the virtual work (unit load) method.

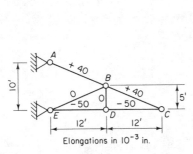

Elongations in 10^{-3} in.

Prob. 3.5

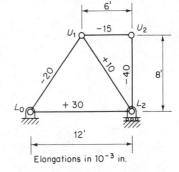

Elongations in 10^{-3} in.

Probs. 3.6, 3.24, and 3.33

3.7. For the truss of the accompanying figure with the given elongations, compute the horizontal deflection of joint U_1 by the virtual work (unit load) method.

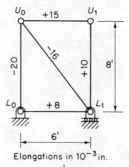

Elongations in 10^{-3} in.

Probs. 3.7, 3.25, and 3.34

3.8. For the truss of the accompanying figure, use the virtual work (unit load) method to determine the horizontal and vertical displacements of joint U_1.

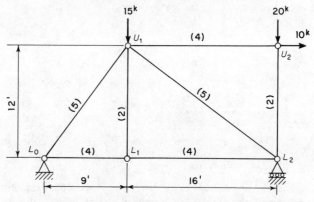

Sq-in. of area in parentheses

Probs. 3.8, 3.19, and 3.35

3.9. Determine the vertical displacement of L_1 for the truss in the accompanying figure, using the virtual work (unit load) method.

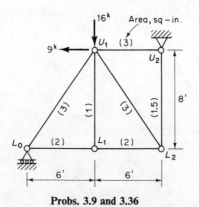

Probs. 3.9 and 3.36

3.10. For the truss of the accompanying figure, compute the horizontal deflection of joint U_2 and the horizontal deflection of joint L_0 by the virtual work (unit load) method. State the direction of each deflection.

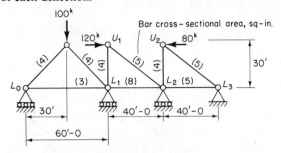

Probs. 3.10 and 3.37

3.11. For the truss of the accompanying figure, compute the horizontal and vertical deflections of joint C by the virtual work (unit load) method. State the direction of each deflection.

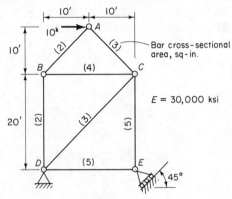

Prob. 3.11

3.12. For the truss of the accompanying figure, determine the horizontal displacement of joint U_2 and the vertical displacement of joint L_1, using the virtual work (unit load) method. (The results of Prob. 2.3 may be used.)

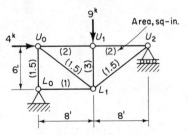

Probs. 3.12 and 3.38

3.13. For the truss of the accompanying figure, determine the vertical displacement of joint U_3 using the virtual work (unit load) method.

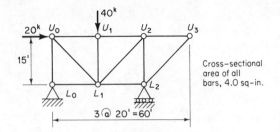

Probs. 3.13 and 3.14

3.14. For the truss of Prob. 3.13, use the virtual work (unit load) method to determine the horizontal and vertical displacements of joint U_3 caused by a 60°F drop in temperature in the bottom chord L_0 to L_2. The coefficient of expansion is 0.0000065 in./in./°F. Assume that no external loads are simultaneously acting.

3.15. For the bar elongations shown in Fig. 3.3.3b, find the horizontal displacement of joint U_2 and the vertical displacement of joint L_1 by the virtual work (unit load) method.

3.16. For the truss of the accompanying figure, determine the vertical displacement of joint U_1 and the horizontal displacement of joint U_2 using the virtual work (unit load) method. (The results of Prob. 2.6 may be used.)

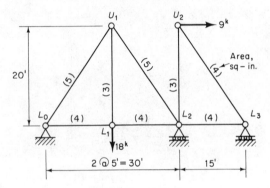

Probs. 3.16 and 3.39

3.17. Determine the vertical displacement of U_2 for the truss of the accompanying figure using the virtual work (unit load) method. (The results of Prob. 2.5 may be used.)

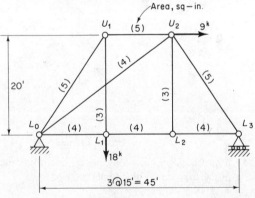

Prob. 3.17

Statically Determinate Trusses—Deflections   Chap. 3

3.18. For the truss of the accompanying figure, determine the horizontal and vertical displacements of joint U_2 using the virtual work (unit load) method. (The results of Prob. 2.7 may be used.)

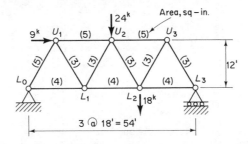

Prob. 3.18

3.19. For the truss of Prob. 3.8, use the virtual work (unit load) method to determine the relative movement of joints L_1 and U_2 measured along the direction of a line connecting the two joints.

Problems Using Geometric Methods—Joint Displacement Equation (Probs. 3.20 to 3.29)

3.20. Given the elongations of the bars as indicated in the accompanying figure, compute the horizontal and vertical deflections of joint C by use of the joint displacement equation.

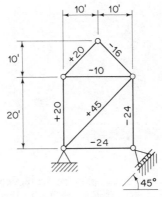

Prob. 3.20

3.21. For the truss of Prob. 3.1, obtain the horizontal and vertical displacements of joints A, B, and C by using the joint displacement equation, taking point B as the reference point and member BC as the reference member. Repeat the solution using AB instead of BC as the reference member.

3.22. For the truss of the accompanying figure with the elongations of the bars given, obtain the vertical and horizontal deflections of all joints by means of the joint displacement equation. Solve twice, using two different members as reference members.

3.23. For the truss of the accompanying figure with elongations given, determine the horizontal and vertical deflections of all joints by means of the joint displacement equation, using A as the reference point and AD as the reference member.

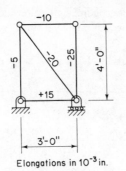

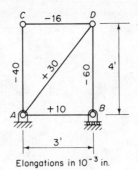

Elongations in 10^{-3} in.   Elongations in 10^{-3} in.

Probs. 3.22 and 3.31   **Probs. 3.23 and 3.32**

3.24. For the truss of Prob. 3.6, using L_0 as the reference point and L_0U_1 as the reference member, compute all joint displacements using the joint displacement equation. Verify the solution by solving the problem using the same reference point but with member L_0L_2 as the reference member.

3.25. For the truss of Prob. 3.7, using L_0 as the reference point and L_0L_1 as the reference member, determine the horizontal and vertical displacements of all joints by using the joint displacement equation. Repeat the solution using U_0L_0 as the reference member to confirm the correctness of the solution.

3.26. For the rigid body $ABCD$ of the accompanying figure, determine the horizontal and vertical displacements of A, B, C, and D if the rigid body is rotated counterclockwise around point B so that point A would have a downward displacement of 3×10^{-3} in.

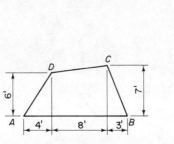

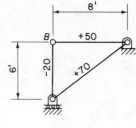

Elongations in 10^{-3} in.

Prob. 3.26   **Prob. 3.27**

3.27. For the truss of the accompanying figure, determine the horizontal and vertical deflections of joints A, B, and C by means of the joint displacement equation. (*Hint:* For this problem, it is easiest to obtain the final answers directly, without translation or rotation.)

3.28. Using the joint displacement equation, determine the joint displacements for all members of the truss of Example 3.4.1 (Fig. 3.4.2) by using U_1 as the reference point and U_1L_1 as the reference member.

3.29. Repeat Prob. 3.28 but use L_0 as the reference point and L_0U_1 as the reference member.

Problems Using Geometric Methods—Williot–Mohr Diagram (Probs. 3.30 to 3.39)

3.30. Draw a Williot–Mohr diagram for the truss member elongations of Prob. 3.1. Dimension the horizontal displacement of joint C.

3.31. Draw a Williot–Mohr diagram for the truss of Prob. 3.22. (Compare with the results of Prob. 3.22 if that problem was solved.)

3.32. Draw a Williot–Mohr diagram for the truss of Prob. 3.23. (Compare with the results of Prob. 3.23 if that problem was solved.)

3.33. Draw a Williot–Mohr diagram for the conditions of Prob. 3.24. Tabulate scaled horizontal and vertical displacements and, if Prob. 3.24 was solved, compare the results.

3.34. Draw a Williot–Mohr diagram for the conditions of Prob. 3.25. Tabulate scaled horizontal and vertical displacements, and if Prob. 3.25 was solved, compare the results.

3.35. Draw a Williot–Mohr diagram for the truss of Prob. 3.8. Dimension the horizontal and vertical displacements of joint U_1 and state the direction.

3.36. Draw a Williot–Mohr diagram for the joint displacements for the truss of Prob. 3.9. Dimension the vertical displacement of L_1 and state the direction. Tabulate all other joint displacements in terms of horizontal and vertical components.

3.37. Draw a Williot–Mohr diagram for the truss of Prob. 3.10. Dimension the vertical deflection of joint L_0 and state its direction. Tabulate all other joint displacements in terms of horizontal and vertical components.

3.38. Draw a Williot–Mohr diagram for the truss of Prob. 3.12. Dimension the horizontal displacement of joint U_2 and the vertical displacement of joint L_1, and state the direction of each. Tabulate all joint displacements in terms of horizontal and vertical components.

3.39. Draw a Williot–Mohr diagram for the truss of Prob. 3.16. Dimension the vertical displacement of joint U_1 and the horizontal displacement of joint U_2, and state the direction of each.

[4]

Statically Determinate Beams
—Shears and Moments

4.1 DEFINITION OF A STATICALLY DETERMINATE BEAM

The term *beam* is given to members subject to transverse loads only. Transverse loads may include concentrated load, uniform load, moment, or combination thereof. A *statically determinate beam* is one that can be completely analyzed by the equations of statics. Simple, cantilever, and overhanging beams as shown in Fig. 4.1.1a, b, and c are three common types of statically determinate beams. The reactions on these beams can be determined by the equilibrium of coplanar forces. Since there exist only two independent equations of statics when the whole beam is taken as a free body, each of these three types of beams contains only two unknown reactions, R_1 and R_2. However, the beam of Fig. 4.1.1d is still statically determinate, even though it has four unknown reactions, because the existence of two internal hinges provides two additional conditions of statics which require that zero moment exists at each internal hinge. An *internal hinge*, that is, a hinge occurring at a location other than at an external support, may be found in actual construction as shown in Fig. 4.1.2.

4.2 SHEARS AND BENDING MOMENTS IN BEAMS

The design of a beam requires a knowledge of the variation of the internal forces. For beams the internal forces are shears and bending moments.

Imagine the beam of Fig. 4.2.1a to be cut into two parts as indicated. Assume that the action of the left segment on the right is upward, causing an upward transverse shear force V acting on the right-hand section of Fig. 4.2.1b, together with a clockwise

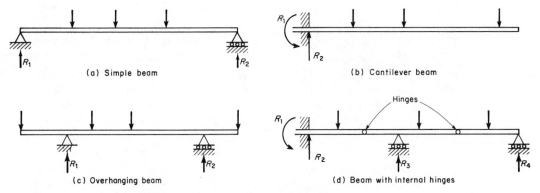

(a) Simple beam

(b) Cantilever beam

Hinges

(c) Overhanging beam

(d) Beam with internal hinges

Figure 4.1.1 Statically determinate beams.

Figure 4.1.2 Actual girder containing an internal hinge that may be modeled as shown in Fig. 4.1.1d. (Photo by C. G. Salmon)

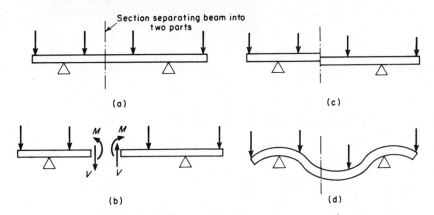

Section separating beam into two parts

(a)

(c)

(b)

(d)

Figure 4.2.1 Shear and bending moment at a section.

rotational effect identified by the bending moment M also acting on the right-hand section of Fig. 4.2.1b. The right-hand section with V and M acting at the cut location must be in equilibrium, and would be referred to as a *free body*. Since action and reaction are equal and opposite, the action of the right-hand free body on the left-hand one is indicated by V and M acting oppositely to those in the right free body. The standard sign convention refers to the situation of Fig. 4.2.1b as *positive*.

Thus, shear is positive when the left side of the beam tends to move upward when the section is cut, as shown in Fig. 4.2.1c. The bending moment at a section is positive when the net rotational effect of the left side on the right is clockwise. This means that bending moment at a section is positive if the beam is concave (causing compressive stress) at the top due to its action. That the signs in Fig. 4.2.1b, c, and d are consistent may be visualized from the condition that the right part needs to act on the left part by a downward force V to keep the left part from shifting upward relative to the right part, and that the moments M-M are produced by tensile and compressive stresses acting, respectively, on the lower and upper portions of the section.

By considering the equilibrium of the left free body in Fig. 4.2.1b, the shear V is the sum of all *upward* forces between the left end and the section; and the bending moment M is the sum of all *clockwise* moments about the section. If the right free body is used, however, the shear V is the sum of all *downward* forces between the right end and the section, but the bending moment M is the sum of all *counterclockwise* moments about the section. In the following examples, the values of V and M at a designated section of a beam are computed from using first the left free body, and then checked by using the right free body. In every case, not only the same magnitudes but also the same signs must be obtained.

Example 4.2.1

For the beam in Fig. 4.2.2, compute the reactions R_1 and R_2, and the values of V and M at a section 8 ft to the right of the left support.

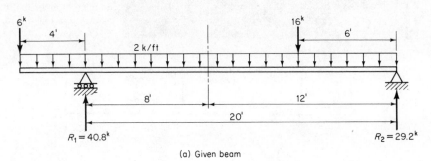

(a) Given beam

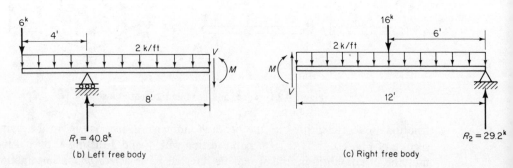

(b) Left free body   (c) Right free body

Figure 4.2.2 Beam of Example 4.2.1.

Statically Determinate Beams—Shears and Moments   Chap. 4

Solution

(a) *Reactions R_1 and R_2.* Taking moments about R_2,

$$20R_1 = 16(6) + 6(24) + 48(12); \qquad R_1 = 40.8 \text{ kips}$$

Taking moments about R_1,

$$20R_2 = 16(14) - 6(4) + 48(8); \qquad R_2 = 29.2 \text{ kips}$$

Checking by $\sum F_y = 0$,

$$40.8 + 29.2 = 6 + 16 + 48; \qquad 70 = 70$$

(b) *V and M at 8 ft to right of R_1.* Using the left free body,

$$V = +40.8 - 6 - 24 = +10.8 \text{ kips}$$

$$M = 40.8(8) - 6(12) - 24(6) = -110.4 \text{ ft-kips}$$

Checking by using the right free body,

$$V = +16 + 24 - 29.2 = +10.8 \text{ kips}$$

$$M = 29.2(12) - 16(6) - 24(6) = +110.4 \text{ ft-kips}$$

Example 4.2.2

For the beam in Fig. 4.2.3, compute the reactions R_1 to R_4, and the bending moment M at R_3.

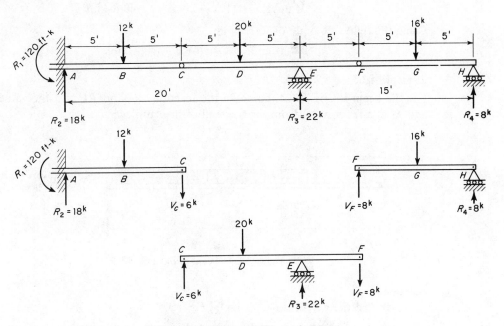

Figure 4.2.3 Beam of Example 4.2.2.

Solution

(a) *Reactions R_1 to R_4.* For a beam with internal hinges, it is best to separate the whole beam into several free bodies at the hinges and then look for a free body with only two unknown forces. In this case, taking *FGH* as a free body,

$$V_F = 8 \text{ kips}; \qquad R_4 = 8 \text{ kips}$$

Taking $CDEF$ as a free body,

$$V_C = 6 \text{ kips}; \qquad R_3 = 22 \text{ kips}$$

Taking ABC as a free body,

$$R_1 = 120 \text{ ft-kips}; \qquad R_2 = 18 \text{ kips}$$

Checking by taking the whole beam as a free body,

$$18 + 22 + 8 = 12 + 20 + 16; \qquad 48 = 48$$

$$\sum M_A = 120 + 22(20) + 8(35) - 12(5) - 20(15) - 16(30) = 840 - 840 = 0$$

(b) *Bending moment M at R_3.* The bending moment M at R_3 may be obtained by using either the whole beam or the free body $CDEF$. If the latter is available, as it usually is, the computation is much shorter. Using the left free body of the whole beam,

$$M \text{ at } R_3 = -120 + 18(20) - 12(15) - 20(5) = -40 \text{ ft-kips}$$

Using the right free body of the whole beam,

$$M \text{ at } R_3 = 8(15) - 16(10) = -40 \text{ ft-kips}$$

Using the left free body of $CDEF$,

$$M \text{ at } R_3 = 6(10) - 20(5) = -40 \text{ ft-kips}$$

Using the right free body of $CDEF$,

$$M \text{ at } R_3 = -8(5) = -40 \text{ ft-kips}$$

Example 4.2.3

For the beam in Fig. 4.2.4, compute the reactions R_1 and R_2, and the values of V and M at a section 7 ft from the free end of the cantilever.

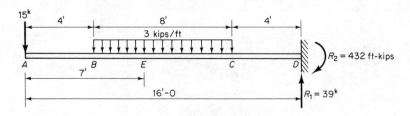

Figure 4.2.4 Beam of Example 4.2.3.

Solution

(a) *Reactions R_1 and R_2.* From $\sum F_v = 0$,

$$-15 - 3(8) + R_1 = 0; \qquad R_1 = 39 \text{ kips}$$

Taking moments about R_1,

$$R_2 = 15(16) + 3(8)(8) = 432 \text{ ft-kips}$$

(b) *V and M at 7 ft from A.* Unlike in Example 4.2.1, where separate free bodies are drawn to the left and right of the designated section, the values V and M at the point E are determined by making use of the loading diagram of the whole beam. Using the left free body,

$$V = -15 - 3(3) = -24 \text{ kips}$$
$$M = -15(7) - 3(3)(1.5) = -118.5 \text{ ft-kips}$$

Checking by using the right free body,

$$V = -39 + 3(5) = -24 \text{ kips}$$
$$M = -432 + 39(9) - 3(5)(2.5) = -118.5 \text{ ft-kips}$$

Note carefully again that when taking the right free body, an upward force causes negative shear. Also, upward forces, regardless of whether they act on the left or right free bodies, always cause positive bending moment. Finally, the moment $R_2 = 432$ ft-kips, as shown in Fig. 4.2.4, is causing compression at the bottom of the beam; therefore, it is a negative bending moment at D.

4.3 RELATIONSHIPS BETWEEN LOAD, SHEAR, AND MOMENT EQUATIONS

The variation of the shear and bending moment is the information required to design beams. This variation may be expressed mathematically by algebraic equations, and those equations may be represented by diagrams. The diagrams for shear and bending moment are the usually desired end products of the structural analysis.

The shear and bending moment diagrams can be plotted from the algebraic shear and moment equations, or more conveniently, they may be obtained without having the shear and moment equations first. The latter procedure can be demonstrated to be correct once the relationships between shear and moment equations are understood. Nevertheless, it is essential that one have the ability to write the moment equations, even though not for the purpose of obtaining moment diagrams from them, because moment equations are needed later in the virtual work (unit load) method of finding slopes and deflections in beams.

To develop the interrelationships between the loading, the shear, and the bending moment, examine the beam of Fig. 4.3.1a with an arbitrary loading. The free body of an element of the beam having an infinitesimal length dx is shown in Fig. 4.3.1b,

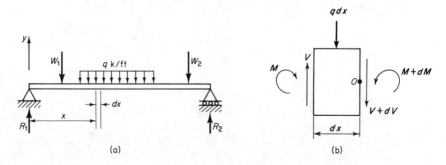

Figure 4.3.1 Beam and free-body diagram of an elemental length dx.

where all forces are in their positive directions. From $\sum F_y = 0$,

$$V - q\,dx - (V + dV) = 0$$

$$q = -\frac{dV}{dx} \qquad (4.3.1)$$

Equation (4.3.1) states that the *rate of decrease* of V at any point x equals the intensity of load at that point, *as x increases from the left toward the right of the beam*. Or the rate of change of V equals the negative of the intensity of load; thus,

$$\frac{dV}{dx} = -q \qquad (4.3.2)$$

From $\sum M_0 = 0$, taking point O as the moment center,

$$M - (M + dM) + V\,dx - q\,dx\,\frac{dx}{2} = 0$$

Neglecting the $q(dx)^2/2$ term as an infinitesimal of higher order, the equation above gives

$$\frac{dM}{dx} = +V \qquad (4.3.3)$$

Equation (4.3.3) states that the *rate of increase* of bending moment at any point x equals the shear at the section as *x increases from the left toward the right of the beam*. Combination of Eqs. (4.3.2) and (4.3.3) gives

$$\frac{d^2M}{dx^2} = -q \qquad (4.3.4)$$

Thus, from Eqs. (4.3.2) to (4.3.4) one may note that the integral of the loading (taken negative) is the shear and the integral of the shear is the bending moment. Note here that the signs in Eqs. (4.3.2) and (4.3.3) should be reversed if x increases from the right toward the left of the beam, but Eq. (4.3.4) holds true regardless of which way x is increasing.

Maximum shear occurs when $dV/dx = 0$, that is, where the loading intensity is zero. Maximum bending moment occurs where $dM/dx = 0$, that is, where the shear is zero. The latter relationship is particularly useful when drawing diagrams for V and M, as discussed in the next section.

Example 4.3.1

Obtain the equations for shear V and moment M for the beam of Fig. 4.3.2. The uniform loading due to the beam weight is neglected for this example.

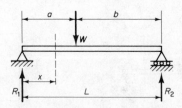

Figure 4.3.2 Simply supported beam having a single concentrated load.

Solution

(a) *Shear equation.* From a value of x infinitesimally small to a value an infinitesimal less than a, the equation for V is a constant,

$$0 < x < a, \qquad V_x = R_1 = \frac{Wb}{L}$$

When x is an infinitesimal amount larger than a and until $x = L$, the following is the equation for V_x:

$$a < x < L, \qquad V_x = R_1 - W = W\left(\frac{b}{L} - 1\right) = -\frac{Wa}{L}$$

Note that shear is not definable *at* a concentrated load. Thus, the summation of loads, taken on either free body created by cutting a section at a distance x from R_1, can only be done when the section so cut lies at least an infinitesimal distance from a concentrated load. If a section cut were attempted *at* a concentrated load or reaction, it would be undefined as to how that concentrated load or reaction would be divided between the two free bodies.

Even though x has been measured from the left end of the beam, the expression for V_x could still be obtained by using the right side of the cut as the free body. If so,

$$0 < x < a, \qquad V_x = \text{sum of downward forces to right side of cut}$$

$$= -R_2 + W = -\frac{Wa}{L} + W = W\left(-\frac{a}{L} + 1\right) = \frac{Wb}{L}$$

$$a < x < L, \qquad V_x = -R_2 = -\frac{Wa}{L}$$

(b) *Bending moment equation.* Since V and M are related by Eq. (4.3.3), when two equations are required to express shear for the entire beam, two equations will also be required to express bending moment. Cutting a section x such that $x \leq a$, the clockwise summation of moments of the forces on the left free body taken about the cut location gives

$$0 \leq x \leq a, \qquad M_x = R_1 x = \frac{Wbx}{L}$$

When x exceeds a,

$$a \leq x \leq L, \qquad M_x = R_1 x - W(x - a) = W\left(\frac{bx}{L} - x + a\right) = \frac{Wa}{L}(L - x)$$

Note here the inequality sign also has the equal sign under it because the M_x expressions are valid at $x = 0$, $x = a$, and $x = L$. A general conclusion from this example is that a concentrated load causes a discontinuity in both the shear and bending moment equations. As before in the case of V_x, the same expressions for M_x can be obtained using the right side of the cut even though x is measured from the left end. Thus,

$$0 \leq x \leq a, \qquad M_x = R_2(L - x) - W(a - x) = \frac{Wa}{L}(L - x) - W(a - x) = \frac{Wbx}{L}$$

$$a \leq x \leq L, \qquad M_x = R_2(L - x) = \frac{Wa}{L}(L - x)$$

From the bending moment equations, the reader may note that the maximum bending moment occurs when $x = a$, that is, directly under the concentrated load, for which location,

$$\text{Max. } M_x = \frac{Wab}{L}$$

Example 4.3.2

Obtain the equations for shear V and moment M for the beam of Fig. 4.3.3.

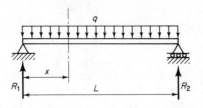

Figure 4.3.3 Simply supported beam having uniform loading.

Solution

(a) *Shear equation.* For this loading there is no discontinuity in the loading q; thus, a single equation for shear V applies from $x = 0$ to $x = L$.

$$V_x = R_1 - qx = \frac{qL}{2} - qx = q\left(\frac{L}{2} - x\right)$$

(b) *Bending moment equation.* When the shear equation has no discontinuity, the bending moment M equation also has no discontinuity.

$$M_x = R_1x - qx\left(\frac{x}{2}\right) = \frac{qLx}{2} - \frac{qx^2}{2} = \frac{q}{2}(x)(L - x)$$

This shows that for uniform loading on a simply supported beam the bending moment at any location x equals one-half the intensity q of the loading times the left segment x times the right segment $(L - x)$.

Example 4.3.3

For the beam of Example 4.2.1 (shown again in Fig. 4.3.4), obtain equations for q, V, and M at a section located a distance x to the right of support B, using x as the independent variable. First use the left free body and then use the right free body. Verify the relationships $dV/dx = -q$ and $dM/dx = V$.

Solution

(a) *Using the left free body,*

$$q_x = +2 \text{ kips/ft}$$

$$V_x = +40.8 - 6 - 2(4 + x) = +26.8 - 2x$$

$$M_x = +40.8x - 6(4 + x) - \frac{2(4 + x)^2}{2} = -40 + 26.8x - x^2$$

Note that $dM/dx = V$ and $dV/dx = -q$.

(b) *Using the right free body,*

$$q_x = +2 \text{ kips/ft}$$

$$V_x = -29.2 + 16 + 2(20 - x) = +26.8 - 2x$$

$$M_x = +29.2(20 - x) - 16(14 - x) - \frac{2(20 - x)^2}{2} = -40 + 26.8\,x - x^2$$

The equations using each free body are identical, as they must be.

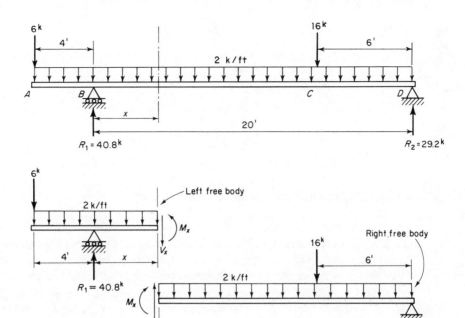

Figure 4.3.4 Beam of Example 4.3.3, showing left and right free-body diagrams that result when the beam is cut at section x.

Example 4.3.4

For the beam of Example 4.2.3 shown again in Fig. 4.3.5, obtain equations for q, V, and M at a section inside the uniform load in terms of its distance x from the fixed support as the independent variable, using first the left free body and then the right free body. Verify the relationships $dV/dx = +q$ and $dM/dx = -V$; the signs in these two equations are opposite to those in Eqs. (4.3.2) and (4.3.3) because x increases from right to left in this problem.

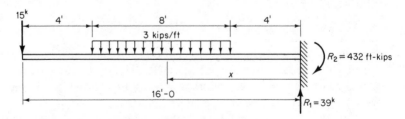

Figure 4.3.5 Beam of Example 4.3.4.

Solution

(a) *Using the left free body,*

$$q_x = +3 \text{ kips/ft}$$

$$V_x = -15 - 3(12 - x) = -51 + 3x$$

$$M_x = -15(16 - x) - \frac{3(12 - x)^2}{2} = -456 + 51x - 1.5x^2$$

(b) *Using the right free body,*

$$q_x = +3 \text{ kips/ft}$$

$$V_x = -39 + 3(x - 4) = -51 + 3x$$

$$M_x = -432 + 39x - \frac{3(x - 4)^2}{2} = -456 + 51x - 1.5x^2$$

4.4 SHEAR AND BENDING MOMENT DIAGRAMS

Since a beam must have adequate shear and moment* strength at every point along its length, it is usually desirable for the designer to have the shear and bending moment *diagrams*. Diagrams are generally preferable to equations because they show the entire relationship, discontinuities are easily accounted for, and from a practical point of view, values needed for design calculations may be scaled directly from the diagrams. Once the designer recognizes the true degree of accuracy of the loading, the structural system idealization, and the material properties, values scaled from moment and shear diagrams can be accepted as satisfactorily accurate for design purposes.

To construct the diagrams, the shear diagram is drawn first. Since the shear at a section is the sum of the upward forces between the left end and the section, the shear diagram can be constructed by starting at the left end and going up and down with the load until the right end is reached. Because the slope dV/dx of the shear curve is the load intensity q, the shear variation is horizontally straight (zero slope) over an unloaded segment, and it should slope linearly downward to the right (i.e., constant slope) over a uniformly loaded segment. Where a concentrated load is applied, there is a sudden vertical drop in the shear ordinate.

The moment diagram may be constructed by first computing its ordinates at the sections of discontinuity and at the sections of maximum positive bending moment if any (where slope $V = dM/dx$ is zero). Once these critical values are plotted, the entire diagram may be sketched by freehand or by French curve if it is to be scaled from. From Eq. (4.3.3),

$$\int dM = \int V \, dx \tag{4.4.1}$$

which means that the increase in moment between one section and an adjacent section to the right is equal to the area of the shear diagram between these two sections. By starting at the left end and successively using Eq. (4.4.1), the moments at as many sections as needed may be computed and plotted.

One should keep in mind that Eq. (4.4.1) only states that the change in moment between two sections is equal to the shear area between the two sections. Any concentrated moment that is actually applied to the beam, such as that due to a horizontal load applied at a certain distance above the beam axis, must cause a sudden rise or drop in the bending moment at the point of application of the concentrated moment.

*The term "moment" is often used to mean "bending moment."

A visual check of the relationship between the moment and shear curves should be made by noting that the slope of the moment curve at any point is equal to the shear intensity at that point. This visual check would help to determine whether the moment curve is linear (V is constant), concave at the top (negative V is decreasing in numerical magnitude or positive V is increasing in numerical magnitude), or convex at the top (positive V is decreasing in numerical magnitude, or negative V is increasing in numerical magnitude).

If the reactions are correctly computed, the shear diagram starts from a zero value at the left end, and terminates with a zero value at the right end; and the moment diagram starts with its known value (if different from zero) at the left end, and terminates with its known value (if different from zero) at the right end. In other words, the shear diagram must close and the moment diagram must close. When both diagrams close, the correctness of the shear and moment diagrams becomes certain.

Example 4.4.1

Draw the shear and bending moment diagrams for the beam of Example 4.3.1, shown in Fig. 4.4.1.

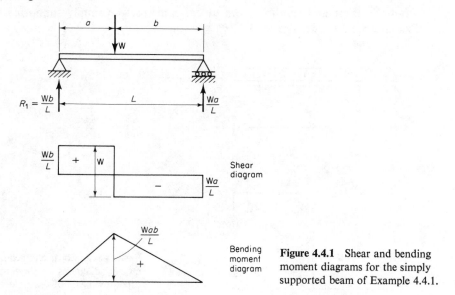

Figure 4.4.1   Shear and bending moment diagrams for the simply supported beam of Example 4.4.1.

Solution

(a) *Shear diagram.* Since the shear at any point is the sum of upward forces between the left end and that point, start at the left end and draw $R_1 = Wb/L$ upward. Over the distance a there is no downward load; thus, the slope of the shear diagram is zero (a horizontal straight line). At the concentrated load W, the shear drops by the amount W from positive Wb/L to negative Wa/L. Then with no transverse load over the distance b the slope of the shear diagram is zero again (i.e., the value remains constant). Finally, at the right end, the shear rises by the value of R_2 to zero. Thus, the shear diagram closes, as equilibrium requires that it must.

(b) *Bending moment diagram.* Starting at the left end, the bending moment is zero at the simple support. The *ordinates* on the shear diagram are constant (and positive) over the distance a; therefore, the *slope* on the bending moment diagram is constant (and positive) over the distance a. The value of bending moment at the concentrated load W equals the area under the shear diagram between the end and the load W. Thus,

$$M \text{ at } W = 0 + \int V \, dx = 0 + \frac{Wb}{L}(a) = \frac{Wab}{L}$$

This expression for bending moment can, of course, just as conveniently be obtained by the usual definition of bending moment.

From the concentrated load to the right end, the shear is constant (and negative), which means that the slope of the M diagram is constant (and negative). Thus, at the right end,

$$M \text{ at the right end} = \frac{Wab}{L} + \int V \, dx = \frac{Wab}{L} + \left(-\frac{Wa}{L}\right)b = 0$$

Thus, the bending moment diagram closes as it must.

Example 4.4.2

Draw the shear and bending moment diagrams for the simply supported beam of Example 4.3.2 shown in Fig. 4.4.2.

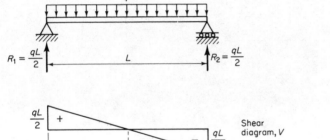

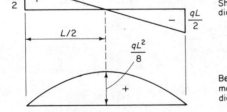

Shear diagram, V

Bending moment diagram, M

Figure 4.4.2 Shear and bending moment diagrams for the simply supported beam of Example 4.4.2.

Solution

(a) *Shear diagram.* Starting from the left, the first transverse load (the reaction R_1) is upward, $qL/2$. Then the load is constant, downward at q kips/ft, making the *slope* of the shear diagram constant at q kips/ft. At the right end, the sum of transverse forces is $qL/2$ (up) $- qL$ (down), giving $-qL/2$ as the shear at an infinitesimal distance to the left of R_2. Then R_2 is upward, bringing the shear diagram to close at zero.

(b) *Bending moment diagram.* Starting at the left end, the bending moment is zero at the simple support. The ordinates on the shear diagram equal the slope on the moment diagram. Since the shear ordinates are linearly decreasing, the slope on the moment diagram changes from a maximum positive value at the left support to zero at midspan; it then becomes increasingly negative until reaching the right support. The midspan bending moment (a maximum here since the slope is zero) equals the bending moment at R_1 plus the area of the shear diagram between R_1 and midspan. Thus, at midspan

$$M = 0 + \int V\,dx = 0 + \frac{1}{2}\left(\frac{qL}{2}\right)\frac{L}{2} = \frac{qL^2}{8}$$

The shape of this bending moment diagram is compatible with the slopes represented by the shear diagram ordinates; also, it agrees with the second-degree parabola equation developed in Example 4.3.2.

Note is made here that the location where the shear diagram passes through zero is a point of *possible* maximum moment. In this example, the zero shear point is indeed the maximum moment point. The next example shows situations where the shear diagram does not identify the maximum moment location.

Example 4.4.3

Draw the shear and bending moment diagrams for the three beams of Fig. 4.4.3.

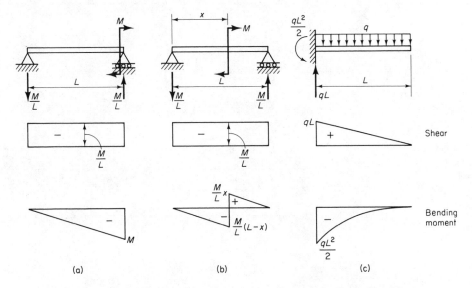

Figure 4.4.3 Shear and bending moment diagrams for Example 4.4.3.

Solution. For each of the three beams of Fig. 4.4.3, the maximum bending moment location is not apparent from examining only the shear diagram. Equations (4.3.1) to (4.3.4) are still valid; but because of the discontinuity of slope of the bending moment diagrams, setting $dM/dx = V = 0$ does not give the maximum bending moment. The cases of concentrated moment give no solution for $dM/dx = 0$, while the cantilever case gives zero shear at the free end.

Note that in Fig. 4.4.3b the *ordinate* on the shear diagram has a constant negative value of M/L. Thus, the *slope* of the bending moment diagram has a constant negative value of M/L across the entire span L, even though the magnitude of bending moment has a discontinuity at the location of concentrated bending moment. This moment is clockwise; therefore, the moment diagram rises suddenly as one travels on the moment diagram from left to right.

The shear and bending moment diagrams for all three cases are given in Fig. 4.4.3. For practical use, the shear and bending moment diagrams must be drawn along with the beam showing its loading and be aligned directly below the diagram of the structure with its loading.

Example 4.4.4

Construct the shear and bending moment diagrams for the beam of Example 4.3.3, shown again in Fig. 4.4.4a.

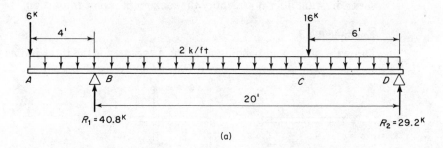

(a)

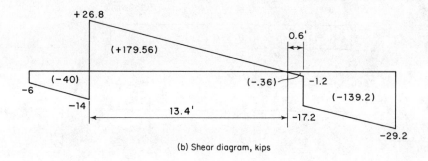

(b) Shear diagram, kips

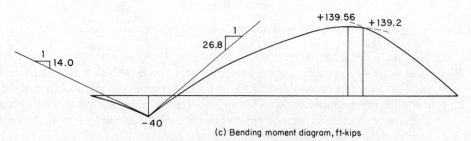

(c) Bending moment diagram, ft-kips

Figure 4.4.4 Solution for Example 4.4.4.

Solution. The shear diagram in Fig. 4.4.4b is drawn first and as the sum of transverse forces is taken from left to right, the shear diagram closes. The point of zero shear is at 26.8/2.0 = 13.4 ft from point *B*, or 1.2/2.0 = 0.6 ft from point *C*. The shear areas are computed to be −40, +179.56, −0.36, and −139.2 ft-kips. Since the change of moments between *A* and *D* is zero, the sum of the four shear areas must be zero.

The tangent to the moment curve is horizontal at the section of zero shear; thus, there is a maximum positive moment at this point. Note in Fig. 4.4.4c the slope of the moment curve changes from −14.0 to +26.8 at point *B*.

Example 4.4.5

Construct the shear and bending moment diagrams for the beam of Example 4.3.4, shown again in Fig. 4.4.5a.

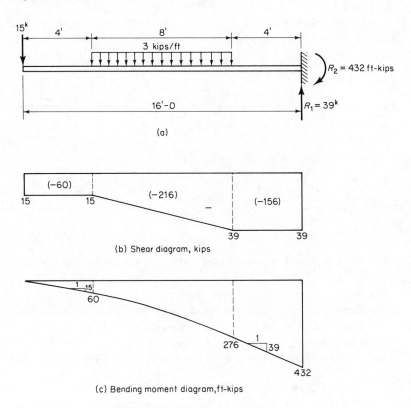

Figure 4.4.5 Solution for Example 4.4.5.

Solution

(a) *Shear diagram.* The shear diagram in Fig. 4.4.5b starts from zero at the left end, drops immediately to −15 at the left end, goes horizontally to the beginning of the uniform load, slopes down linearly to −39 at the end of the uniform load, goes horizontally again to a point an infinitesimal distance from the fixed support, and rises abruptly to zero. The shear areas are computed to be (-60), (-216), and (-156). The sum of the three shear areas is (-432), which checks with the clockwise concentrated moment acting at the right end (a clockwise concentrated moment adds to the bending moment as one travels on the moment diagram from the left to the right, but it means a drop on the bending moment ordinate if the direction of travel is from the right to the left).

(b) *Bending moment diagram.* The moment diagram starts from zero at the left end, slopes down linearly to (-60), then curves with increasing negative slope to $(-60 - 216 = -276)$, and finally again slopes down linearly to $(-276 - 156 = -432)$, an infinitesimal distance from the right end. At the right end, the bending moment ordinate suddenly rises from -432 to zero because of the reaction R_2.

Example 4.4.6

Construct the bending moment diagram for the beam of Fig. 4.4.6a containing uniform loading over part of the span.

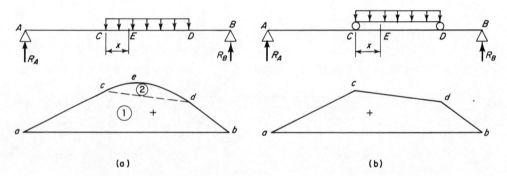

Figure 4.4.6 Construction of the bending moment diagram for Example 4.4.6.

Solution. When a uniform load does not cover the entire length of a simple beam, the moment diagram shown in Fig. 4.4.6a is the sum of area $(1) = abdca$ with straight sides; and area $(2) = cdec$, which is identical with the moment diagram of a simple beam with span equal to CD. This may be proved by comparing the moment equation for a point E in the beam of Fig. 4.4.6a with that of Fig. 4.4.6b; thus,

$$M_E \text{ (Fig. 4.4.6a)} = R_A(AC + x) - \frac{wx^2}{2}$$

$$M_E \text{ (Fig. 4.4.6b)} = R_A(AC + x) - \frac{w(CD)}{2}x$$

and

$$M_E \text{ (Fig. 4.4.6a)} - M_E \text{ (Fig. 4.4.6b)}$$

$$= \frac{w(CD)}{2}x - \frac{wx^2}{2} = \text{moment in the simple beam } CD$$

PROBLEMS

4.1–4.19. For the beam of each accompanying figure, draw the shear and bending moment diagrams to scale directly below a diagram of the beam with its loading. (For Probs. 4.5, 4.6, and 4.16 to 4.19, draw diagrams only for the horizontal portion identified by letters.) Use a French curve to show proper curvature and clearly show all breaks in curvature on the diagrams. Label the maximum values (both + and −) and values where abrupt changes occur. Mark the areas of the diagrams with + and − signs in accordance with standard sign convention. For one of the segments of the beam, write the equations for shear and bending moment.

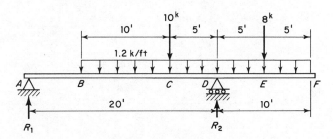

Prob. 4.1

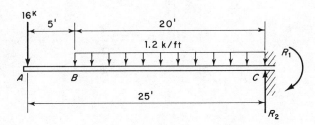

Prob. 4.2

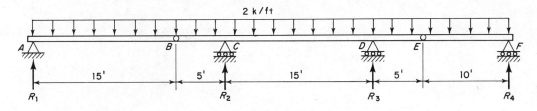

Prob. 4.3

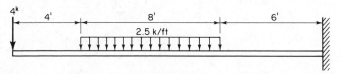

Prob. 4.4

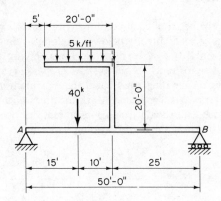

Prob. 4.5

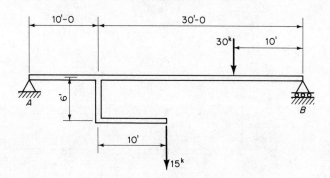

Prob. 4.6

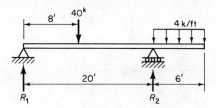

Prob. 4.7

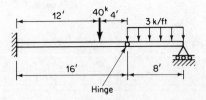

Prob. 4.8

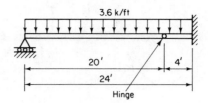

Prob. 4.9

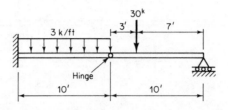

Prob. 4.10

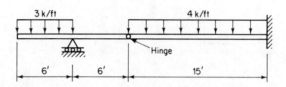

Prob. 4.11

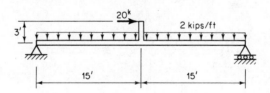

Prob. 4.12

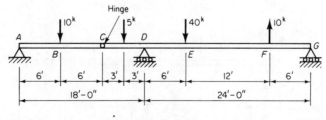

Prob. 4.13

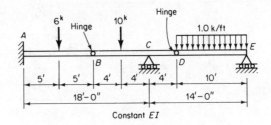

Prob. 4.14

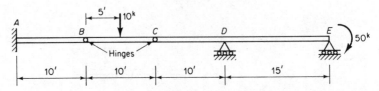

Prob. 4.15

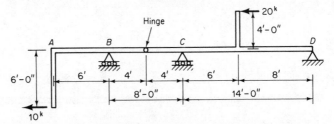

Prob. 4.16

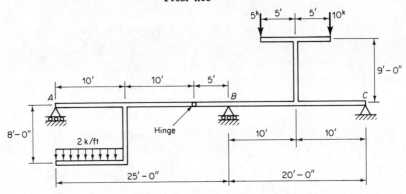

Prob. 4.17

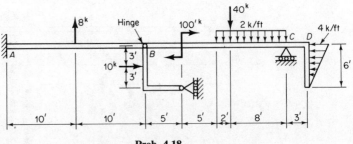

Prob. 4.18

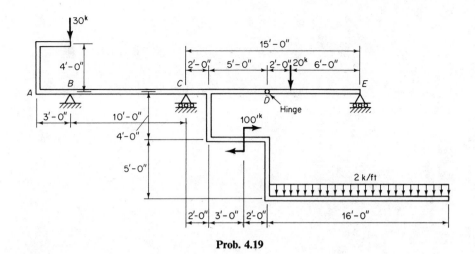

Prob. 4.19

[5]

Statically Determinate Beams —Slopes and Deflections

5.1 ELASTIC CURVES, SLOPES, AND DEFLECTIONS

The deflected shape of a flexural member is called its *elastic curve*. The elastic curve in mathematical form is the equation of the transverse deflection y as a function of the location x along the span. More than merely a mathematical exercise, knowledge of the qualitative elastic curve for a beam or frame aids the designer in evaluating the correctness of the structural analysis.

In Chapter 1, design was defined as involving "the experienced engineer's intuitive feeling for the behavior of a structure." The ability to compute slopes and deflections is a necessary structural analysis tool (1) for developing a "feeling" for the structural behavior; (2) for investigating that a structure can perform satisfactorily when in service with regard to deflection (excessive deflection may cause damage to nonstructural elements, unsightly cracks, doors prevented from being opened or closed, "bouncy" floors, or problems from ponding of water); and (3) for checking computer output of structural analysis to make certain that deformations from point to point are compatible with the external and internal forces.

Whereas the equation of the elastic curve defines the resulting deflection, it is the bending moment that causes the deflected shape. Curvature $(1/\rho)$, which is the inverse of the radius of curvature ρ, is the rate of change of slope. Thus, referring to Fig. 5.1.1,

$$\text{Curvature} = \frac{d\theta}{dx} = \frac{d(dy/dx)}{dx} = \frac{d^2y}{dx^2} \tag{5.1.1}$$

If one defines positive bending moment M as that causing elongation of the fibers

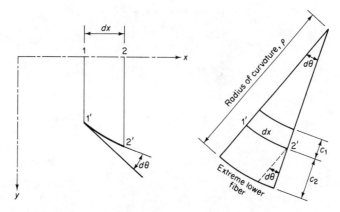

Figure 5.1.1 Differential equation of the elastic curve.

below the neutral axis, which for a downward positive y-axis means $d\theta/dx$ is *decreasing* for $+M$, then

$$d\theta = -\frac{\text{elongation of extreme lower fiber}}{c_2} \tag{5.1.2}$$

and

$$\text{elongation of extreme lower fiber} = +\frac{(\text{stress})\,dx}{E}$$
$$= +\frac{Mc_2\,dx}{EI} \tag{5.1.3}$$

Substituting Eq. (5.1.3) in Eq. (5.1.2),

$$d\theta = -\frac{M}{EI}\,dx \tag{5.1.4}$$

from which

$$\frac{d\theta}{dx} = \frac{d^2y}{dx^2} = -\frac{M}{EI} \tag{5.1.5}$$

For the axes shown in Fig. 5.1.1, the slope is positive when downward to the right. As positive bending moment has been defined for the elastic curve to be concave on the top, this condition physically requires the positive slope to decrease from left to right (like a ball rolling down inside a circular hoop from left to right). Thus, the curvature (d^2y/dx^2), or the rate of change in slope, is negative for positive bending moment. This is the reason for the negative sign in Eq. (5.1.5).

A summary of the differential relationships between deflection, slope, bending moment, shear, and load is given in Fig. 5.1.2. Note that if M is proportional to d^2y/dx^2, from Eqs. (4.3.2) and (4.3.3) the shear is proportional to the third derivative and the load intensity is proportional to the fourth derivative.

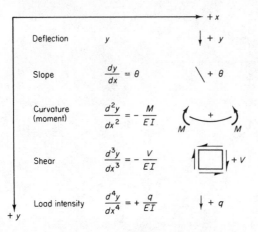

Figure 5.1.2 Relationships between deflection, slope, moment, shear, and load intensity.

Deflection	y	$\downarrow + y$
Slope	$\dfrac{dy}{dx} = \theta$	$\diagdown + \theta$
Curvature (moment)	$\dfrac{d^2y}{dx^2} = -\dfrac{M}{EI}$	
Shear	$\dfrac{d^3y}{dx^3} = -\dfrac{V}{EI}$	$+V$
Load intensity	$\dfrac{d^4y}{dx^4} = +\dfrac{q}{EI}$	$\downarrow + q$

5.2 DOUBLE-INTEGRATION METHOD

The double-integration method of determining the equation of the elastic curve is a basic classical method using the differential equation. It is given the name "double integration" because one usually starts with the bending moment M, which relates to the curvature, d^2y/dx^2. Integration twice gives the equation for deflection y. Study of Fig. 5.1.2 will show that one may start from load intensity, shear, moment, or slope and integrate the appropriate number of times to get the deflection equation.

Use of the differential equation as a basic method is occasionally useful when the complete elastic curve equations are required. This method is, however, cumbersome when discontinuities exist in the function being integrated, creating a large number of integration constants. When only the deflections or slopes at certain *chosen points* on the beam are required, the virtual work (unit load) or the moment area methods treated later in this chapter are far more convenient.

Example 5.2.1

For the beam of Fig. 5.2.1, determine the equation of the elastic curve and also the maximum deflection using the method of double integration.

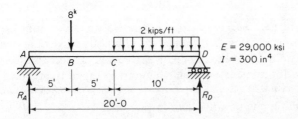

Figure 5.2.1 Beam of Example 5.2.1.

Solution

(a) *Determine the reactions.* Taking moments about end A,

$$R_D = \frac{2(10)15 + 8(5)}{20} = 17 \text{ kips}$$

Taking moments about end D,

$$R_A = \frac{2(10)5 + 8(15)}{20} = 11 \text{ kips}$$

Check that $R_A + R_D$ equals the total downward load of $8 + 2(10) = 28$ kips.

(b) *Establish the equations for bending moment.* Starting from end A, for $x \le 5$,

$$M_x = R_A x = 11x$$

For $5 \le x \le 10$,

$$M_x = R_A x - 8(x - 5) = 11x - 8x + 40 = 3x + 40$$

For $10 \le x \le 20$,

$$M_x = R_A x - 8(x - 5) - 2(x - 10)\left(\frac{x - 10}{2}\right)$$

$$= 11x - 8x + 40 - x^2 + 20x - 100$$

$$= -x^2 + 23x - 60$$

(c) *Obtain the slope equations.* Using

$$\frac{d^2 y}{dx^2} = -\frac{M_x}{EI}$$

and integrating once:
For $x \le 5$,

$$\frac{dy}{dx} = \frac{-1}{EI}\left(11\frac{x^2}{2} + C_1\right)$$

For $5 \le x \le 10$,

$$\frac{dy}{dx} = \frac{-1}{EI}\left(3\frac{x^2}{2} + 40x + C_2\right)$$

For $10 \le x \le 20$,

$$\frac{dy}{dx} = \frac{-1}{EI}\left(-\frac{x^3}{3} + 23\frac{x^2}{2} - 60x + C_3\right)$$

(d) *Obtain the deflection equations.* Integration once of the slope equations gives the following:
For $x \le 5$,

$$y = \frac{-1}{EI}\left(11\frac{x^3}{6} + C_1 x + C_4\right)$$

For $5 \le x \le 10$,

$$y = \frac{-1}{EI}\left(3\frac{x^3}{6} + 40\frac{x^2}{2} + C_2 x + C_5\right)$$

For $10 \le x \le 20$,

$$y = \frac{-1}{EI}\left(-\frac{x^4}{12} + 23\frac{x^3}{6} - 60\frac{x^2}{2} + C_3 x + C_6\right)$$

Since the deflection equations contain six constants of integration, six boundary conditions of geometry are required. These conditions are:

1. At $x = 0$, $y = 0$
2. At $x = 5$, $y_{\text{segment } AB} = y_{\text{segment } BC}$

3. At $x = 5$, $\left(\dfrac{dy}{dx}\right)_{\text{segment } AB} = \left(\dfrac{dy}{dx}\right)_{\text{segment } BC}$

4. At $x = 10$, $y_{\text{segment } BC} = y_{\text{segment } CD}$

5. At $x = 10$, $\left(\dfrac{dy}{dx}\right)_{\text{segment } BC} = \left(\dfrac{dy}{dx}\right)_{\text{segment } CD}$

6. At $x = 20$, $y = 0$

From condition 1,

$$0 = C_4 \tag{a}$$

From condition 2,

$$11\frac{(5)^3}{6} + C_1(5) = 3\frac{(5)^3}{6} + 40\frac{(5)^2}{2} + C_2(5) + C_5$$

$$5C_1 - 5C_2 - C_5 = 333.33 \tag{b}$$

From condition 3,

$$11\frac{(5)^2}{2} + C_1 = 3\frac{(5)^2}{2} + 40(5) + C_2$$

$$C_1 - C_2 = 100 \tag{c}$$

From condition 4,

$$3\frac{(10)^3}{6} + 40\frac{(10)^2}{2} + C_2(10) + C_5$$

$$= -\frac{(10)^4}{12} + 23\frac{(10)^3}{6} - 60\frac{(10)^2}{2} + C_3(10) + C_6$$

$$10C_2 - 10C_3 + C_5 - C_6 = -2500 \tag{d}$$

From condition 5,

$$3\frac{(10)^2}{2} + 40(10) + C_2 = -\frac{(10)^3}{3} + 23\frac{(10)^2}{2} - 60(10) + C_3$$

$$C_2 - C_3 = -333.33 \tag{e}$$

From condition 6,

$$0 = -\frac{(20)^4}{12} + 23\frac{(20)^3}{6} - 60\frac{(20)^2}{2} + C_3(20) + C_6$$

$$-20C_3 - C_6 = 5333.33 \tag{f}$$

Solving Eqs. (a) to (f) gives the following:

$$C_1 = -466.67; \qquad C_4 = \quad 0$$
$$C_2 = -566.67; \qquad C_5 = +166.67$$
$$C_3 = -233.33; \qquad C_6 = -666.67$$

The deflection equations are, therefore, the following:
For $x \leq 5$,

$$y = \frac{-1}{EI}(1.833x^3 - 466.67x)$$

For $5 \leq x \leq 10$,

$$y = \frac{-1}{EI}(0.5x^3 + 20x^2 - 566.67x + 166.67)$$

For $10 \leq x \leq 20$,

$$y = \frac{-1}{EI}(-0.0833x^4 + 3.833x^3 - 30x^2 - 233.33x - 666.67)$$

(e) *Determine maximum deflection.* Since most of the load is on segment CD, it is likely that maximum deflection occurs in that segment. For maximum y, set dy/dx equal to zero. For $10 \leq x \leq 20$,

$$\frac{dy}{dx} = 0 = \frac{-1}{EI}\left(-\frac{x^3}{3} + 23\frac{x^2}{2} - 60x - 233.33\right)$$

$$x^3 - 34.5x^2 + 180x + 700 = 0$$

$$x = 10.24 \text{ ft}$$

$$\text{Max. } y = \frac{-1}{EI}[-0.0833(10.24)^4 + 3.833(10.24)^3 - 30(10.24)^2$$

$$- 233.33(10.24) - 666.67]$$

$$\text{Max. } y = \frac{3000}{EI} = \frac{3000(1728)}{29,000(300)} = 0.596 \text{ in.}$$

In this problem, the maximum deflection is very close to the center of span, where the deflection is exactly 3000 kip-ft$^3/EI$.

5.3 VIRTUAL WORK (UNIT LOAD) METHOD

In Section 3.2 the virtual work (unit load) method was developed to determine the displacement in a specified direction for a particular joint in a truss. This method can be similarly used to determine the slope and deflection at a particular location along a beam.

The principle of virtual work may be restated as follows. For a given elastic structure, the external work done by *any* set of forces in equilibrium (i.e., the equilibrium state) in going through a set of compatible displacements (i.e., the compatible state) must be equal to the internal work done by the internal forces in the equilibrium state in going through the internal deformations of the compatible state. As was done for truss displacement computations in Chapter 3, the convenient procedure is to use a single unit load applied in the direction of the desired displacement, together with the reactions arising from the applied unit load, as the "set of forces in equilibrium," that is, the virtual system. The "set of compatible displacements" is the set of displacements arising from the real system containing the given loads on the structure.

Before beginning the specific development for the beam, the reader is reminded of the difference between "real work" and "virtual work." *Real work* relates to loads on a structure moving through the displacements caused by these *same* loads, in which case the work equals *one-half* of the product of the force and the displacement. *Virtual work* is the work done by forces that may be thought of as fully in place before any displacements occur, moving through the set of compatible displacements. Since the loads are constant as the displacements occur, the work equals the product of the force and the displacement (i.e., without the multiplier one-half). In fact, the word "virtual" is used to mean that the forces and displacements are unrelated.

In the case of the truss, the internal virtual work equals the sum of the products of the *u*-forces in the virtual system times the elongations of the bars in the real system. An element in the cross section of a beam may be thought of as a minute spring of length dx, corresponding to the isolated two-force member of a truss. Referring to Fig. 5.3.1, consider a slice of dimension $b\,dy$ taken at a location x along the span. The internal force is the product of the stress σ_x and the area $b\,dy$. Thus, in the virtual system having $\sigma_x = my/I$, the internal force u becomes

$$u = \sigma_x b\,dy = \frac{my}{I} b\,dy \tag{5.3.1}$$

in which m is the bending moment in the virtual system at location x along the span.

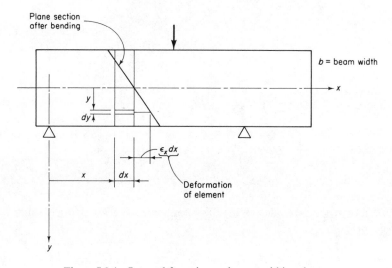

Figure 5.3.1 Internal force in an element within a beam.

In this development it is assumed that the contribution to the internal virtual work from the shear forces in the virtual system in going through the shear deformations in the real system is small (as it is except in deep, short beams). The contribution of shear deformation to the geometry of the elastic curve is not treated in this text.

Referring to Fig. 5.3.1 again, the elongation $\epsilon_x\,dx$ of an element having cross section $b\,dy$ and length dx at a location x along the span is $(\sigma_x/E)\,dx$. For the real system (Fig. 5.3.2a) having $\sigma_x = My/I$, the elongation e is

$$e = \frac{\sigma_x}{E}\,dx = \frac{My}{EI}\,dx \tag{5.3.2}$$

in which M is the bending moment in the real system at location x along the span.

Letting Fig. 5.3.2a and b represent the real and virtual systems, respectively, and applying the principle of virtual work,

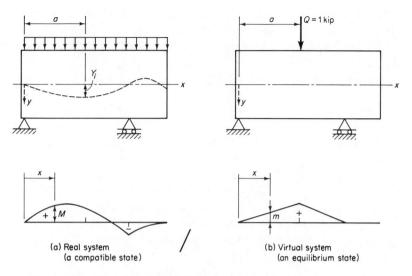

Figure 5.3.2 Virtual work (unit load) method for displacements on beams.

$$\sum Q_i Y_i = \sum ue \qquad (5.3.3)$$

| External forces of virtual system | Beam deflection of real system at location Q_i | Internal fiber forces of the virtual system | Internal fiber elongations of real system |

Substitution of Eqs. (5.3.1) and (5.3.2) into (5.3.3) gives

$$\sum Q_i Y_i = \sum \left(\frac{my}{I} b \, dy\right)\left(\frac{My}{EI} \, dx\right) \qquad (5.3.4)$$

When the virtual system contains only one load Q (and that of unit magnitude) acting in the direction of the desired displacement, the summation sign on the left side of Eq. (5.3.4) disappears. This is the practical procedure giving rise to the term *unit load method*; actually, the method is based on the principle of virtual work. Thus, taking a single unit load in the Q-system,

$$(1.0)(Y_i) = \int_0^L \frac{Mm \int_A by^2 \, dy}{EI} \, dx \qquad (5.3.5)$$

Noting that $\int_A by^2 \, dy$ is the moment of inertia I of the cross section, Eq. (5.3.5) becomes

$$Y_i = \int_0^L \frac{Mm}{EI} \, dx \qquad (5.3.6)$$

In the discussion leading up to Eq. (5.3.6) the virtual load Q_i is a force and Y_i is a linear displacement. Equation (5.3.6) equally applies when Q_i is a unit moment (which still gives rise to a moment function m in the virtual system) and Y_i then becomes the slope θ_i of the elastic curve in the real system.

In Chapter 3, a proof for an equation like Eq. (5.3.3) as it applied to trusses was presented. In that proof the truss under virtual loading was separated into a finite number of joints with the resultant of all forces acting at each joint equal to zero. Then the virtual work done by all these forces (having zero resultant) in going through the displacements or deformations in the real system is set equal to zero. In the present case, one would have to visualize that the beam under virtual loading is separated into an infinite number of small slices having cross-sectional area $b\,dy$ and length dx. The resultant of all the forces (including tension–compression and shear forces) acting on each slice is zero. Then the virtual work done by all these forces (excluding shear forces because their contribution is small except in deep, short beams) in going through the displacements and deformations in the real system is set equal to zero. Thus, the two proofs are really identical, only in the case of the beam one would not attempt to draw the infinite number of free-body diagrams for the slices of cross-sectional area $b\,dy$ and length dx.

Example 5.3.1

For the beam of Example 5.2.1 (see Fig. 5.3.3), find the deflection at midspan and the slope at A using the virtual work (unit load) method.

Solution

(a) *Determine the bending moment equations for the real system.* This is shown in Fig. 5.3.3a drawn for the equations established in Example 5.2.1, wherein x is measured from A for all segments. Alternatively, it is sometimes desirable to work with equations where each component contributing to the bending moment is kept separate; this is shown in Fig. 5.3.3b.

(b) *Establish the virtual system for midspan deflection.* Apply a unit load acting vertically downward at midspan and write the corresponding bending moment equations, as shown in Fig. 5.3.3c.

(c) *Apply the virtual work equation (5.3.6) to solve for* Δ_C. Referring to the deflection at C as Δ_C, and using M from Fig. 5.3.3a,

$$\Delta_C = \int_0^L \frac{Mm}{EI}\,dx$$

From A to B, with x measured from A,

$$\Delta_{C1} = \int_0^5 \frac{11x(0.5x)\,dx}{EI}$$

$$= \frac{1}{EI}\left[\frac{5.5x^3}{3}\right]_0^5 = \frac{229.17}{EI}$$

From B to C, with x measured from A,

$$\Delta_{C2} = \int_5^{10} \frac{(3x + 40)(0.5x)\,dx}{EI}$$

$$= \frac{1}{EI}\left[1.5\frac{x^3}{3} + 20\frac{x^2}{2}\right]_5^{10} = \frac{1187.50}{EI}$$

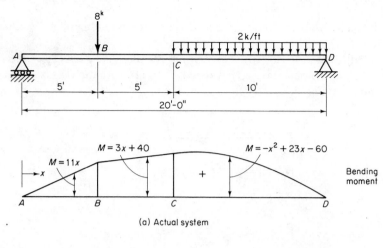

(a) Actual system

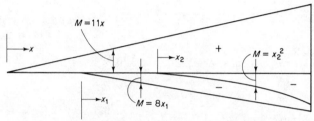

(b) Actual system; bending moment in parts

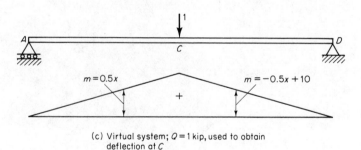

(c) Virtual system; $Q = 1$ kip, used to obtain deflection at C

Figure 5.3.3 Beam of Example 5.3.1 (virtual work method).

From C to D, with x measured from A,

$$\Delta_{C3} = \int_{10}^{20} \frac{(-x^2 + 23x - 60)(-0.5x + 10)\, dx}{EI}$$

$$\Delta_{C3} = \int_{10}^{20} \frac{(-0.5x^3 - 21.5x^2 + 260x - 600)\, dx}{EI}$$

$$= \frac{1}{EI}\left[+0.5\frac{x^4}{4} - 21.5\frac{x^3}{3} + 260\frac{x^2}{2} - 600x \right]_{10}^{20} = \frac{1583.33}{EI}$$

$$\Delta_C = \frac{1}{EI}(229.17 + 1187.50 + 1583.33) = \frac{3000}{EI}\ \text{kip-ft}^3$$

This agrees with the result obtained by the double-integration method in Example 5.2.1.

(d) *Establish the virtual system for obtaining the slope* θ_A *of the elastic curve at end A.* A unit moment is applied at end A; and the equation for m of the virtual system is shown in Fig. 5.3.4.

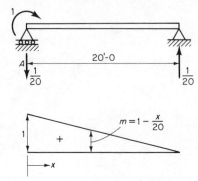

Figure 5.3.4 Virtual system used in solving for the slope θ_A at end A.

(e) *Apply the virtual work equation (5.3.6) to solve for* θ_A.

$$\theta_A = \int_0^L \frac{Mm\,dx}{EI}$$

From A to B, with x measured from A,

$$\theta_{A1} = \int_0^5 \frac{11x(1 - x/20)\,dx}{EI}$$

$$= \frac{1}{EI}\left[11\frac{x^2}{2} - \frac{11}{20}\frac{x^3}{3}\right]_0^5 = \frac{114.58}{EI}$$

From B to C, with x measured from A,

$$\theta_{A2} = \int_5^{10} \frac{(3x + 40)(1 - x/20)\,dx}{EI}$$

$$= \int_5^{10} \frac{(-3x^2/20 + x + 40)\,dx}{EI}$$

$$= \frac{1}{EI}\left[-\frac{3}{20}\frac{x^3}{3} + \frac{x^2}{2} + 40x\right]_5^{10} = \frac{193.75}{EI}$$

From C to D, with x measured from A,

$$\theta_{A3} = \int_{10}^{20} \frac{(-x^2 + 23x - 60)(1 - x/20)\,dx}{EI}$$

$$= \int_{10}^{20} \frac{1}{EI}\left(\frac{x^3}{20} - \frac{43}{20}x^2 + 26x - 60\right)dx$$

$$= \frac{1}{EI}\left[\frac{x^4}{80} - \frac{43}{60}x^3 + 13x^2 - 60x\right]_{10}^{20}$$

$$= \frac{158.33}{EI}$$

$$\theta_A = \frac{1}{EI}(114.58 + 193.75 + 158.33) = \frac{466.67}{EI} \text{ kip-ft}^2$$

The angle θ is measured from the original undeflected position, with a positive value indicating an angle of rotation in the same direction as the unit moment. This answer is identical to that obtained in Example 5.2.1.

(f) *Show an alternate solution using the bending moments of the real system in parts, as in Fig. 5.3.3b.* This arbitrary separation of the bending moment components shows all three contributions drawn with respect to point D. When using the virtual work equation (5.3.6), the expressions for M and m must have the variable x measured from the same location, but that location may be different for each of the intervals of integration.

From A to B, with x measured from A, the solution is the same as that obtained previously. Of the parts contributing to the total bending moment, only the contribution $(11x)$ due to the reaction at A is applicable over the beam segment from A to B.

From B to C, there are two components in the bending moment of the actual system.

$$\Delta'_{C2} = \int_5^{10} \frac{11x(0.5x)\, dx}{EI} \quad (x \text{ measured from } A)$$

$$\Delta''_{C2} = \int_0^5 \frac{(-8x_1)[0.5(5 + x_1)]\, dx_1}{EI} \quad (x_1 \text{ measured from } B)$$

Thus,

$$\Delta'_{C2} = \frac{1}{EI}\left[\frac{5.5}{3}x^3\right]_5^{10} = +\frac{1604.17}{EI}$$

$$\Delta''_{C2} = \frac{-1}{EI}\left[10x_1^2 + \frac{4}{3}x_1^3\right]_0^5 = -\frac{416.67}{EI}$$

$$\Delta_{C2} = \Delta'_{C2} + \Delta''_{C2} = \frac{1187.50}{EI}$$

From C to D, there are three components in the bending moment of the actual system.

$$\Delta'_{C3} = \int_{10}^{20} \frac{11x[0.5x - 1.0(x - 10)]\, dx}{EI} \quad (x \text{ measured from } A)$$

$$\Delta''_{C3} = \int_5^{15} \frac{(-8x_1)[0.5(5 + x_1) - 1.0(x_1 - 5)]\, dx}{EI} \quad (x_1 \text{ measured from } B)$$

$$\Delta'''_{C3} = \int_0^{10} \frac{(-x_2^2)[0.5(10 + x_2) - 1.0x_2]\, dx}{EI} \quad (x_2 \text{ measured from } C)$$

$$\Delta'_{C3} = \frac{1}{EI}\left[\frac{-5.5}{3}x^3 + 55x^2\right]_{10}^{20} = \frac{+3666.67}{EI}$$

$$\Delta''_{C3} = \frac{1}{EI}\left[\frac{4}{3}x_1^3 - 30x_1^2\right]_5^{15} = \frac{-1666.67}{EI}$$

$$\Delta'''_{C3} = \frac{1}{EI}\left[\frac{0.5}{4}x_2^4 - \frac{5}{3}x_2^3\right]_0^{10} = \frac{-416.67}{EI}$$

$$\Delta_{C3} = \frac{1}{EI}(+3666.67 - 1666.67 - 416.67) = \frac{1583.33}{EI}$$

In general, in using the virtual work (unit load) method, one will attempt to select a reference location and draw the moment diagrams for M and m in such a manner that the simplest mathematics will result. For instance, for portion C to D, one could measure x from D, positive toward the left. Then $M = 17x - x^2$, $m = 0.5x$, with limits of x from 0 to 10; or

$$\Delta_{C3} = \int_0^{10} \frac{(17x - x^2)(0.5x)\,dx}{EI}$$

$$= \frac{1}{EI}\left[\frac{8.5x^3}{3} - \frac{0.5x^4}{4}\right]_0^{10} = \frac{1583.33}{EI}$$

Example 5.3.2

Using the virtual work (unit load) method, compute the slope and deflection at point A of the beam of Fig. 5.3.5.

Solution

(a) *Draw the bending moment diagrams for the real system.* The diagrams and the equations are given in Fig. 5.3.5b.

(b) *Establish the virtual loadings to be used.* In this case, a unit moment is applied at A, taking counterclockwise rotation as positive by using a counterclockwise unit moment. The virtual system with its bending moment m equations is given in Fig. 5.3.5c.

For downward deflection at A, a unit load is applied downward at A as the virtual loading system, shown in Fig. 5.3.5d.

(c) *Summary of the equations to be used.*
Segment AB, origin at A, limits on x_1 from 0 to 4.

$$M = -6x_1 - x_1^2$$
$$m \text{ for } \theta_A = -1.0$$
$$m \text{ for } \Delta_A = -x_1$$

Segment BC, origin at D, limits on x from 6 to 20,

$$M = 29.2x - x^2 - 16(x - 6)$$
$$m \text{ for } \theta_A = -0.05x$$
$$m \text{ for } \Delta_A = -0.20x$$

Segment CD, origin at D, limits on x from 0 to 6,

$$M = 29.2x - x^2$$
$$m \text{ for } \theta_A = -0.05x$$
$$m \text{ for } \Delta_A = -0.20x$$

(d) *Solution for θ_A and Δ_A.* Applying Eq. (5.3.6),

$$EI\theta_A = \int (M)(m \text{ for } \Delta_A)\,dx$$

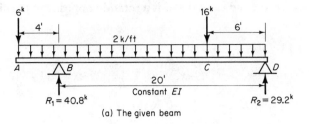

(a) The given beam

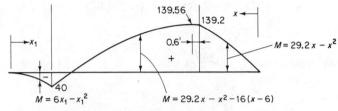

(b) Actual system bending moments

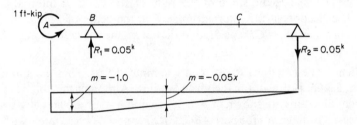

(c) Virtual system for counterclockwise θ_A

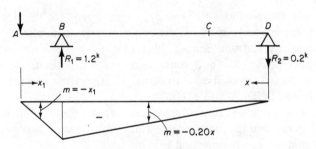

(d) Virtual system for downward Δ_A

Figure 5.3.5 Beam of Example 5.3.2—virtual work (unit load) method.

$$= \int_0^4 (-6x_1 - x_1^2)(-1.0)\, dx_1 + \int_6^{20} [29.2x - x^2 - 16(x - 6)](-0.05x)\, dx$$

$$+ \int_0^6 (29.2x - x^2)(-0.05x)\, dx$$

$$= (+69.33) + (-88.92) + (-602.28) = -621.87$$

$$\theta_A = \frac{621.87}{EI} \text{ kip-ft}^2 \quad \text{clockwise}$$

The negative result indicates that θ is a rotation opposite to the direction of the applied unit moment.

$$EI\Delta_A = \int (M)(m \text{ for } \Delta_A)\, dx$$

$$= \int_0^4 (-6x_1 - x_1^2)(-x_1)\, dx_1 + \int_6^{20} [29.2x - x^2 - 16(x - 6)](-0.20x)\, dx$$

$$+ \int_0^6 (29.2x - x^2)(-0.20x)\, dx$$

$$= (+192) + (-2409.12) + (-355.68) = -2572.8$$

$$\Delta_A = \frac{2572.8}{EI} \text{ kip-ft}^3 \quad \text{upward}$$

The negative result indicates that Δ_A is a deflection opposite to the direction of the applied unit load.

Example 5.3.3

Using the virtual work (unit load) method, compute the slope of the left elastic curve at internal hinge F, the slope of the right elastic curve at internal hinge F, and the vertical deflection of the internal hinge F in the beam of Fig. 5.3.6. The moment of inertia of segment $ABCDE$ is $5I_c$ and that of segment $EFGH$ is $2I_c$, where I_c is a reference value.

Solution. The reactions R_1 to R_4 are computed first. Then the moment diagram, as in Fig. 5.3.6b, is drawn for the given loading. Although the composite free-body and moment diagrams are shown in Fig. 5.3.6a and b, they are actually solved by drawing the free-body diagrams for part ABC, $CDEF$, and FGH separately with the corresponding moment diagrams directly under the free-body diagrams. These separate diagrams are not shown.

Using the notations θ_{FL} and θ_{FR}, respectively, for the slopes at F of the elastic curves to the left and right of F, the unit moments are applied as shown in Fig. 5.3.6d and e. Similarly, the unit load is applied at F in Fig. 5.3.6c for finding the deflection Δ_F at F. Again, in Fig. 5.3.6c, d, and e, only the composite free-body and moment diagrams are shown for each case. In actual practice, these diagrams should be separated for parts ABC, $CDEF$, and FGH.

The moment equations for M and m, the origin of reference, the value of n where $I = nI_c$, and the limits of integration for the various segments between sections of discontinuity are summarized below.

Segment	n^*	Origin	Limits	M	m for θ_{FL}	m for θ_{FR}	m for Δ_F
AB	5	A	0 to 5	$-120 + 18x$	$-1 + 0.10x$	$-0.5 + 0.05x$	$+5 - 0.50x$
BCD	5	A	5 to 15	$-120 + 18x - 12(x - 5)$	$-1 + 0.10x$	$-0.5 + 0.05x$	$+5 - 0.50x$
DE	5	A	15 to 20	$-120 + 18x - 12(x - 5)$ $- 20(x - 15)$	$-1 + 0.10x$	$-0.5 + 0.05x$	$+5 - 0.50x$
EF	2	H	10 to 15	$8x - 16(x - 5)$	$+1$	$-1 + 0.10x$	$-(x - 10)$
FG	2	H	5 to 10	$8x - 16(x - 5)$	0	$+0.10x$	0
GH	2	H	0 to 5	$8x$	0	$+0.10x$	0

$^*I = nI_c.$

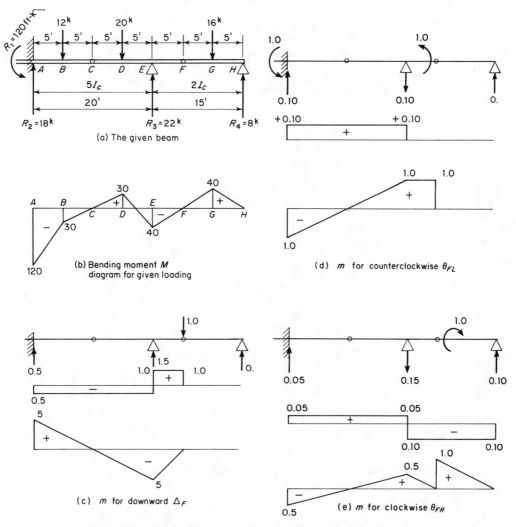

Figure 5.3.6 Beam of Example 5.3.3—virtual work (unit load) method.

Noting that the moment of inertia is not constant for the entire beam, the factor EI_c is used as a multiplier on the left side of the virtual work equation (5.3.6), and the integrals on the right side will have no EI in the denominator but the value of the n because $I = nI_c$. Thus,

$$EI_c\theta_{FL} = \frac{1}{5}\int_0^5 (-120 + 18x)(-1 + 0.10x)\,dx$$

$$+ \frac{1}{5}\int_5^{15}[-120 + 18x - 12(x - 5)](-1 + 0.10x)\,dx$$

$$+ \frac{1}{5}\int_{15}^{20}[-120 + 18x - 12(x - 5) - 20(x - 15)](-1 + 0.10x)\,dx$$

$$+ \frac{1}{2} \int_{10}^{15} [8x - 16(x - 5)](+1)\, dx$$

$$= (+60) + (+10) + \left(-\frac{20}{3}\right) + (-50) = +\frac{40}{3}$$

$$\theta_{FL} = \frac{40}{3EI_c} \text{ kip-ft}^2 \quad \text{counterclockwise}$$

$$EI_c\theta_{FR} = \frac{1}{5} \int_0^5 (-120 + 18x)(-0.5 + 0.05x)\, dx$$

$$+ \frac{1}{5} \int_5^{15} [-120 + 18x - 12(x - 5)](-0.5 + 0.05x)\, dx$$

$$+ \frac{1}{5} \int_{15}^{20} [-120 + 18x - 12(x - 5) - 20(x - 15)](-0.5 + 0.05x)\, dx$$

$$+ \frac{1}{2} \int_{10}^{15} [8x - 16(x - 5)](-1 + 0.10x)\, dx$$

$$+ \frac{1}{2} \int_5^{10} [8x - 16(x - 5)](+0.10x)\, dx$$

$$+ \frac{1}{2} \int_0^5 (8x)(+0.10x)\, dx$$

$$= (+30) + (+5) + \left(-\frac{10}{3}\right) + \left(-\frac{50}{3}\right) + \left(+\frac{100}{3}\right) + \left(+\frac{50}{3}\right) = +65$$

$$\theta_{FR} = \frac{65}{EI_c} \text{ kip-ft}^2 \quad \text{clockwise}$$

$$EI_c\Delta_F = \frac{1}{5} \int_0^5 (-120 + 18x)(+5 - 0.50x)\, dx$$

$$+ \frac{1}{5} \int_5^{15} [-120 + 18x - 12(x - 5)](+5 - 0.50x)\, dx$$

$$+ \frac{1}{5} \int_{15}^{20} [-120 + 18x - 12(x - 5) - 20(x - 15)](+5 - 0.50x)\, dx$$

$$+ \frac{1}{2} \int_{10}^{15} [8x - 16(x - 5)][-(x - 10)]\, dx$$

$$= (-300) + (-50) + \left(+\frac{100}{3}\right) + \left(+\frac{500}{3}\right) = -150$$

$$\Delta_F = \frac{150}{EI_c} \text{ kip-ft}^3 \quad \text{upward}$$

As a final note for this problem, the reader may find it a useful experience to work this problem again by treating the entire beam as three separate free bodies: *ABC*, *CDEF*, and *FGH*. Then the expression for *M* and *m* can be obtained by (1) using *A* as origin for segment *AB*, and *C* as origin for segment *CB*; (2) using *C* as origin for segments *CD* and *DE*, and *F* as origin for segment *FE*; and (3) using *F* as origin for segment *FG*, and *H* as origin for segment *HG*.

5.4 MAXWELL'S THEOREM OF RECIPROCAL DEFLECTIONS

In his development of the first systematic method for analysis of statically indeterminate structures, James Clerk Maxwell [1] in 1864 set forth the *Theorem of Reciprocal Deflections*. At that time, because of the brief treatment and lack of examples, the importance of the theorem was not appreciated. The publication in 1886 of Heinrich Müller-Breslau's classic textbook [2] expanding on his earlier paper [3] provided a much improved version of the general Maxwell–Mohr method for analyzing statically indeterminate structures. In his presentation, Müller-Breslau called attention to the importance of the reciprocal theorem, with corresponding credit to Maxwell.

The theorem states: *The displacement of any point A, caused by the application of a unit load at any other point B, is equal to the displacement at B caused by a unit load at A.* The displacement at A is measured in the direction of the unit load at A, and the displacement at B is measured in the direction of the unit load at B. The term "unit load" refers to either a force or moment. It is assumed that the structure is elastic and a first-order analysis is used.

Consider first a simple illustration using the simply supported beam of Fig. 5.4.1a. The beam is loaded with the force P_A applied vertically at A. The deflection at A is $P_A\delta_{aa}$, where δ_{aa} and δ_{ba} represent the deflections at A and B, respectively, due to a unit load applied at A. Since superposition applies, the deflection due to P_A is P_A times the deflection due to the unit load.

Next the same beam is loaded with P_B applied at B as in Fig. 5.4.1b. The deflection at A is $P_B\delta_{ab}$ and the deflection at B is $P_B\delta_{bb}$, where δ_{ab} and δ_{bb} represent the deflections at A and B, respectively, due to a unit load applied at B.

The total work done when P_A is applied first and comes to rest is

$$\frac{1}{2}P_A(P_A\delta_{aa})$$

When P_B is slowly added, with P_A already in position, point A will deflect the additional amount of $P_B\delta_{ab}$. The work done by P_A is

$$P_A(P_B\delta_{ab})$$

and the work done by P_B is

$$\frac{1}{2}P_B(P_B\delta_{bb})$$

Thus, the total work (Fig. 5.4.1c) is

$$W = \frac{1}{2}P_A^2\delta_{aa} + P_AP_B\delta_{ab} + \frac{1}{2}P_B^2\delta_{bb} \tag{5.4.1}$$

However, had the loads P_A and P_B been applied simultaneously, the contribution due to P_A is

$$\frac{1}{2}P_A(P_A\delta_{aa} + P_B\delta_{ab})$$

and the contribution due to P_B is

$$\frac{1}{2}P_B(P_A\delta_{ba} + P_B\delta_{bb})$$

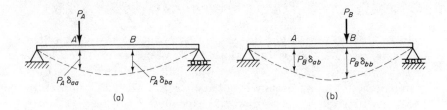

(a)   (b)

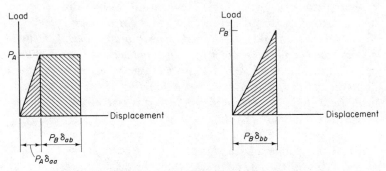

(c) Work done when P_A is applied first, and then P_B is slowly applied

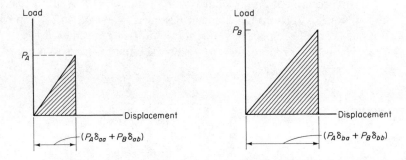

(d) Work done when P_A and P_B are applied simultaneously

Figure 5.4.1   Maxwell's theorem of reciprocal deflections.

and the total work done (Fig. 5.4.1d) is

$$W = \frac{1}{2}P_A^2\delta_{aa} + \frac{1}{2}P_AP_B\delta_{ab} + \frac{1}{2}P_AP_B\delta_{ba} + \frac{1}{2}P_B^2\delta_{bb} \qquad (5.4.2)$$

Since the sequence of loading cannot affect the total work done, Eqs. (5.4.1) and (5.4.2) may be equated. Thus,

$$P_AP_B\delta_{ab} = \frac{1}{2}P_AP_B\delta_{ab} + \frac{1}{2}P_AP_B\delta_{ba}$$

$$\frac{1}{2}P_AP_B\delta_{ab} = \frac{1}{2}P_AP_B\delta_{ba}$$

$$\delta_{ab} = \delta_{ba} \qquad (5.4.3)$$

Although Maxwell's theorem has been more popularly known as the theorem of reciprocal *deflections*, calling it the theorem of reciprocal *displacements* would generalize it. "Displacement" can be a linear deflection, a slope at a point on the elastic curve, or a deflection in any direction of a joint in a truss or a rigid frame. The important thing to remember is that the unit load and the unit displacement when referred to the same point on the structure must be consistent; that is, they must be in the same rotational direction or in the same linear direction.

In the case of the reciprocity between an angular and linear displacement, note in Fig. 5.4.1a that the unit moment applied at A and δ_{ab} at A are both in clockwise rotation, and that the unit load at B and δ_{ba} at B are both in the downward direction. The reciprocal relationship is $\delta_{ab} = \delta_{ba}$, in which δ_{ab} is in units of radians per kip, and δ_{ba} is in units of feet per ft-kip, indicating 1.0 per kip in either side of the equation. As shown by Fig. 5.4.2b, the reciprocity between the two deflections of the joints in a truss is again $\delta_{ab} = \delta_{ba}$, in which δ_{ab} and δ_{ba} are the vertical components of the total displacement at joints A and B, respectively, because both unit loads are applied in the vertical deflection.

The Theorem of Reciprocal Deflections may also be demonstrated by using the virtual work (unit load) method of determining deflections. Referring to Fig. 5.4.2a, in order to compute δ_{ba}, the actual system is represented by the left-hand part of the

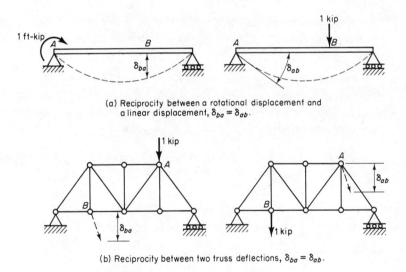

(a) Reciprocity between a rotational displacement and a linear displacement, $\delta_{ba} = \delta_{ab}$.

(b) Reciprocity between two truss deflections, $\delta_{ba} = \delta_{ab}$.

Figure 5.4.2 The theorem of reciprocal deflections.

figure and the virtual system, by the right-hand part. In order to compute δ_{ab}, the actual system is represented by the right-hand part of the figure and the virtual system, by the left-hand part. Calling m_a the bending moment on the left figure and m_b the bending moment on the right figure, and using Eq. (5.3.6),

$$\delta_{ba} = \int \frac{Mm \, dx}{EI} = \int \frac{m_a m_b \, dx}{EI}$$

and

$$\delta_{ab} = \int \frac{Mm \, dx}{EI} = \int \frac{m_b m_a \, dx}{EI}$$

Thus,

$$\delta_{ab} = \delta_{ba}$$

5.5 MOMENT AREA THEOREMS

As discussed previously, methods for displacement computations may be divided into two classes, energy methods and geometric methods. The *moment area method* is a geometric method of determining deflection or slope at chosen points on the elastic curve of a beam.

The method originates from concepts stated by Otto Mohr [4], Professor at the Technische Hochschule at Dresden, Germany in the 1860's. Mohr's approach is more closely allied with the conjugate beam method as presented in Section 5.6. Charles E. Greene of the University of Michigan presented [5, 6] in 1873 the basic moment area method (called area moment by Greene) of focusing on angles and deflections between tangents at two points on the elastic curve.

The moment area method is based on two theorems, known as Moment Area Theorems I and II:

Moment Area Theorem I. *The change in slope between any two points on a continuous elastic curve is equal to the area of the M/EI diagram between these two points.*

Moment Area Theorem II. *The deflection between point 2 on a continuous elastic curve and the tangent to the elastic curve at point 1, measured in a direction perpendicular to the original axis of the beam, is equal to the moment of the area of the M/EI diagram between points 1 and 2 about point 2.*

Referring to Fig. 5.5.1, the first moment area theorem gives the angle θ between tangents at two points on the elastic curve as

$$\theta_{12} = \int_1^2 d\theta \tag{5.5.1}$$

and using but omitting the negative sign in Eq. (5.1.4),

$$d\theta = \frac{M}{EI} dx \tag{5.1.4}$$

The negative sign in Eq. (5.1.4) has been omitted, but one should note that a positive M/EI area indicates a counterclockwise turn of the elastic curve in going from left to right. For the initial development of the moment area concept concern with a sign convention is omitted in favor of visualizing the physical effect of a positive or negative bending moment diagram as the case may be. Thus, in absolute terms the first moment area theorem is

$$\theta_{12} = \int_1^2 \frac{M}{EI} dx \tag{5.5.2}$$

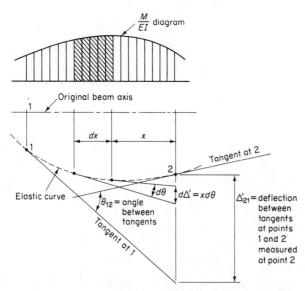

Figure 5.5.1 Moment area method.

From Fig. 5.5.1, the deflection Δ'_{21} between tangents, measured at 2, is

$$\Delta'_{21} = \int_1^2 d\Delta' \tag{5.5.3}$$

On the basis of the first-order assumption that all deflections are small, the lengths of all slant lines are equal to their projections on the original beam axis; thus,

$$d\Delta' = x \, d\theta \tag{5.5.4}$$

Then the second moment area theorem becomes

$$\Delta'_{21} = \int_1^2 x \, d\theta = \int_1^2 \frac{Mx \, dx}{EI} \tag{5.5.5}$$

As a notation standard, the subscripts of Δ' designate the two points defining the boundaries of the M/EI area being used, with the first subscript indicating the point about which the moment of the M/EI area is taken and at which Δ' is measured. In addition, the single prime sign on the Δ symbol indicates that this is the distance between tangents at two points on the elastic curve and it may not necessarily be the true deflection measured from the original beam axis.

Some areas and centroid locations that may be useful in applying the method are to be found in Fig. 5.5.3.

The following five examples show how the two moment area theorems can be used to find the slope and deflection at chosen points on an elastic curve.

Example 5.5.1

Using the moment area method, determine for the cantilever beam of Fig. 5.5.2 the slopes θ_B and θ_C and the deflections at B and C.

Solution

(a) *Compute the reactions and draw the shear and bending moment diagrams.* The shear diagram is omitted from Fig. 5.5.2, but the bending moment diagram is

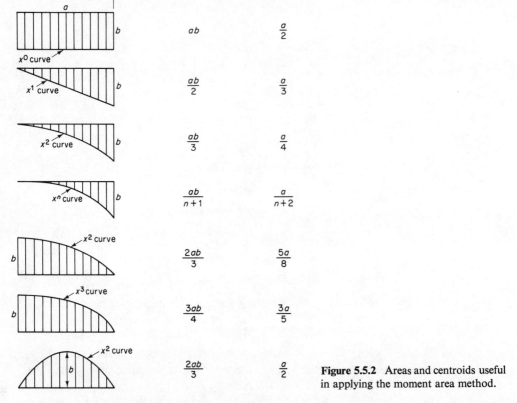

	Area	Centroid $\bar{x}$
x^0 curve	ab	$\dfrac{a}{2}$
x^1 curve	$\dfrac{ab}{2}$	$\dfrac{a}{3}$
x^2 curve	$\dfrac{ab}{3}$	$\dfrac{a}{4}$
x^n curve	$\dfrac{ab}{n+1}$	$\dfrac{a}{n+2}$
x^2 curve	$\dfrac{2ab}{3}$	$\dfrac{5a}{8}$
x^3 curve	$\dfrac{3ab}{4}$	$\dfrac{3a}{5}$
x^2 curve	$\dfrac{2ab}{3}$	$\dfrac{a}{2}$

Figure 5.5.2 Areas and centroids useful in applying the moment area method.

shown. Next, obtain the M/EI diagram. In this case with EI constant the M/EI diagram has the same shape as the moment diagram. Then, make a freehand sketch of the elastic curve by observing the sign of the bending moment along the beam. In this case, with negative bending moment over the entire length and knowing that there is zero slope at the fixed support A, the elastic curve is as shown in Fig. 5.5.3.

(b) *Obtain slopes at B and C.* Applying the moment area theorem, Eq. (5.5.2), gives the angle between tangents; thus,

$$\theta_{AB} = \int_A^B \frac{M}{EI}\,dx = \frac{M}{EI}\ \text{area on } AB$$

$$= \frac{WL}{2EI}\left(\frac{L}{2}\right) + \frac{1}{2}\left(\frac{WL}{2EI}\right)\frac{L}{2} = \frac{3WL^2}{8EI}$$

Note that only the absolute magnitude of the M/EI area is computed. Observing physically the elastic curve in Fig. 5.5.2, since $\theta_A = 0$,

$$\theta_B = \theta_{AB} = \frac{3WL^2}{8EI} \quad \text{(clockwise)}$$

The word "clockwise" is added from the appearance of the elastic curve. The slope θ_C at C equals the M/EI area on AC since the tangent at A is zero. Thus,

$$\theta_C = \theta_{AC} = \frac{M}{EI}\ \text{area on } AC = \frac{1}{2}\left(\frac{WL}{EI}\right)L = \frac{WL^2}{2EI} \quad \text{(clockwise)}$$

The directions of the slopes are apparent from the sketch of the elastic curve, without having to resort to a sign convention. If, however, one considers clockwise rotation of the elastic curve as positive (see Fig. 5.1.2), then proceeding from left to right the change in slope is obtained by *subtracting* the M/EI area (keeping in mind that the M/EI areas have signs too in accordance with Fig. 5.1.2). Thus, in this example, starting with $\theta_A = 0$, a negative area (M/EI) would be subtracted, giving a positive result for θ_B and θ_C, positive meaning clockwise. In most cases, the correct results using this geometric method should be apparent by observing the sketch of the elastic curve and a sign convention need not be used.

(c) *Obtain the deflections Δ_B and Δ_C at points B and C.* For point C, since the tangent at A is horizontal, the deflection Δ'_{CA} between tangents at A and C measured at C is the actual deflection Δ_C at C. Thus,

$$\Delta_C = \Delta'_{CA} = \text{moment of } \frac{M}{EI} \text{ area on } AC \text{ about } C$$

$$= \underbrace{\frac{WL}{EI}\left(\frac{1}{2}\right)L}_{\text{area}} \underbrace{\frac{2L}{3}}_{\text{arm}} = \frac{WL^3}{3EI} \quad \text{(downward)}$$

Similarly, the deflection Δ'_{BA} between tangents at A and B is the actual deflection Δ_B. Thus,

$$\Delta_B = \Delta'_{BA} = \text{moment of } \frac{M}{EI} \text{ area on } AB \text{ about } B$$

$$= \underbrace{\frac{WL}{2EI}\left(\frac{L}{2}\right)}_{\text{area}} \underbrace{\left(\frac{L}{4}\right)}_{\text{arm}} + \underbrace{\frac{1}{2}\left(\frac{WL}{2EI}\right)\left(\frac{L}{2}\right)}_{\text{area}} \underbrace{\left(\frac{2}{3}\right)\left(\frac{L}{2}\right)}_{\text{arm}}$$

$$= \frac{5WL^3}{48EI} \quad \text{(downward)}$$

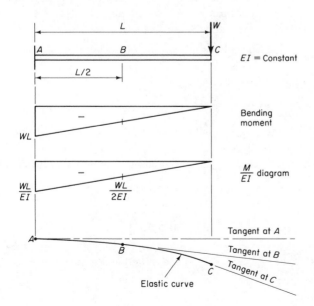

Figure 5.5.3   Cantilever beam for Example 5.5.1.

Example 5.5.2

Using the moment area method, determine the end slopes and maximum deflection for a uniformly loaded simply supported beam, as shown in Fig. 5.5.4.

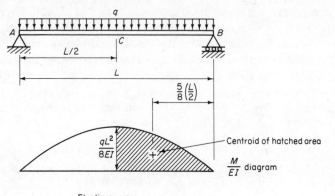

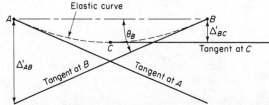

Figure 5.5.4 Simply supported uniformly loaded beam of Example 5.5.2.

Solution

   (a) *Compute reactions, draw shear and bending moment diagrams, obtain M/EI diagram, and sketch elastic curve.* The *M/EI* diagram and elastic curve are shown in Fig. 5.5.4.

   (b) *Deflection at C.* With this symmetrical loading, the tangent at midspan (point C) is horizontal. Thus, the deflection Δ'_{BC} between the tangents at B and C measured at B equals the midspan deflection Δ_C. Thus,

$$\Delta_C = \Delta'_{BC} = \text{moment of } \frac{M}{EI} \text{ area on } CB \text{ about } B$$

$$= \underbrace{\frac{2}{3}\left(\frac{qL^2}{8EI}\right)\left(\frac{L}{2}\right)}_{\text{area}} \underbrace{\left(\frac{5}{8}\right)\left(\frac{L}{2}\right)}_{\text{arm}} = \frac{5qL^4}{384EI}$$

   (c) *Slopes at A and B.* Due to symmetry the slopes to the elastic curve at points A and B are identical (although one is clockwise and one is counterclockwise). Note from Fig. 5.5.4 that θ_B can be obtained if Δ'_{AB} is computed and then divided by span L.

$$\theta_B = \frac{\Delta'_{AB}}{L}$$

$$= \frac{1}{L}\left(\text{moment of } \frac{M}{EI} \text{ area on } AB \text{ about } A\right)$$

$$= \frac{1}{L}\left[\underbrace{\frac{2}{3}\left(\frac{qL^2}{8EI}\right)L}_{\text{area}}\underbrace{\left(\frac{L}{2}\right)}_{\text{arm}}\right] = \frac{qL^3}{24EI} \quad \text{(counterclockwise)}$$

Another way of thinking is to obtain the angle between the tangents at C and B, which is the M/EI area on CB; thus,

$$\theta_B = \theta_{CB} = \frac{M}{EI} \text{ area on } CB = \frac{2}{3}\left(\frac{qL^2}{8EI}\right)\left(\frac{L}{2}\right) = \frac{qL^3}{24EI} \quad \text{(as before)}$$

Example 5.5.3

Using the moment area method, determine the deflection under the load W and the maximum deflection for the beam of Fig. 5.5.5.

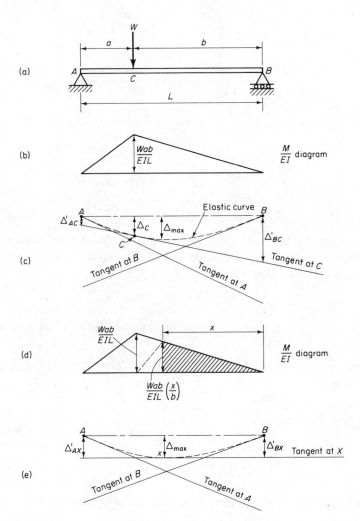

Figure 5.5.5 Beam of Example 5.5.3.

Solution

(a) *Compute reactions, draw shear and bending moment diagrams, obtain M/EI diagram, and sketch elastic curve.* The M/EI diagram and the elastic curve are shown in Fig. 5.5.5b and c.

(b) *Compute deflection Δ_C at concentrated load.* Since the slope θ_C is unknown, this solution must be done in a more general way than in Example 5.5.2. From Fig. 5.5.5c note that Δ'_{BC} and Δ'_{AC} define the slope of the elastic curve at C, and furthermore Δ_C can be obtained from Δ'_{AC} and Δ'_{BC} by straight-line proportion. Thus,

$$\Delta_C = \Delta'_{AC} + \left(\frac{\Delta'_{BC} - \Delta'_{AC}}{L}\right)a$$

$$\Delta'_{AC} = \text{moment of } \frac{M}{EI} \text{ area on } AC \text{ about } A$$

$$= \underbrace{\frac{1}{2}\left(\frac{Wab}{EIL}\right)(a)}_{\text{area}}\underbrace{\left(\frac{2a}{3}\right)}_{\text{arm}} = \frac{Wa^3b}{3EIL}$$

$$\Delta'_{BC} = \text{moment of } \frac{M}{EI} \text{ area on } CB \text{ about } B$$

$$= \underbrace{\frac{1}{2}\left(\frac{Wab}{EIL}\right)(b)}_{\text{area}}\underbrace{\left(\frac{2b}{3}\right)}_{\text{arm}} = \frac{Wab^3}{3EIL}$$

$$\Delta_C = \frac{Wa^3b}{3EIL} + \frac{Wa^2b^3}{3EIL^2} - \frac{Wa^4b}{3EIL^2}$$

$$= \frac{Wa^2b}{3EIL}\left(a + \frac{b^2}{L} - \frac{a^2}{L}\right) = \frac{Wa^2b^2}{3EIL}$$

(c) *Compute maximum deflection, $\Delta_{\max}$.* Maximum deflection will occur where slope is zero on the longer segment CB. Drawing a horizontal tangent at the point of maximum deflection, one observes

$$\Delta_{\max} = \Delta'_{AX} = \Delta'_{BX}$$

$$\Delta'_{AX} = \text{moment of } \frac{M}{EI} \text{ area between } A \text{ and } X \text{ about } A$$

$$= \frac{Wab}{EIL}\left(\frac{a}{2}\right)\left(\frac{2a}{3}\right) + \frac{Wab}{EIL}\left(\frac{b-x}{2}\right)\left(a + \frac{b-x}{3}\right)$$

$$+ \frac{Wax}{EIL}\left(\frac{b-x}{2}\right)\left[a + \frac{2(b-x)}{3}\right]$$

$$\Delta'_{BX} = \text{moment of } \frac{M}{EI} \text{ area between } X \text{ and } B \text{ about } B$$

$$= \frac{Wab}{EIL}\left(\frac{x}{b}\right)\left(\frac{x}{2}\right)\left(\frac{2x}{3}\right)$$

Equating Δ'_{AX} and Δ'_{BX} gives a pure quadratic equation in x as

$$x^2 = \frac{b(2a+b)}{3} \qquad \text{for } b > a$$

With numerical values of a and b, x may be determined. Then

$$\Delta_{\max} = \Delta'_{AX} = \frac{Wax^3}{3EIL} \qquad \text{for } b > a$$

Statically Determinate Beams—Slopes and Deflections Chap. 5

(d) *Alternate solution for deflection* Δ_C *at concentrated load.* Referring to Fig. 5.5.6, an alternative solution to obtain Δ_C is to subtract Δ'_{CB} from the proportion of Δ'_{AB} reduced to the location C. Thus,

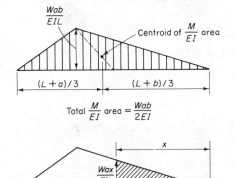

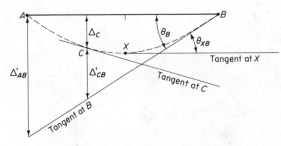

Figure 5.5.6 Alternate solution for Example 5.5.3.

$$\Delta_C = \Delta'_{AB}\left(\frac{b}{L}\right) - \Delta'_{CB}$$

$$= \left(\text{moment of }\frac{M}{EI}\text{ area on }AB\text{ about }A\right)\frac{b}{L} - \left(\text{moment of }\frac{M}{EI}\text{ area on }CB\text{ about }C\right)$$

$$= \left(\frac{Wab}{2EI}\right)\left(\frac{L+a}{3}\right)\left(\frac{b}{L}\right) - \frac{1}{2}\left(\frac{Wab}{EIL}\right)(b)\left(\frac{b}{3}\right)$$

$$= \frac{Wa^2b^2}{3EIL} \quad \text{(same as before)}$$

(e) *Alternate solution for maximum deflection,* $\Delta_{\max}$. Referring to Fig. 5.5.6, the distance x between the point of maximum deflection and the right end can be obtained from the condition

$$\theta_B = \theta_{XB}$$

$$\frac{\Delta'_{AB}}{L} = \theta_{XB}$$

$$\frac{\dfrac{Wab}{2EI}\left(\dfrac{L+a}{3}\right)}{L} = \frac{1}{2}\left(\frac{Wax}{EIL}\right)(x)$$

$$x^2 = \frac{b(2a+b)}{3} \quad \text{(same as before)}$$

The alternate solutions shown in parts (d) and (e) are particularly noteworthy because they essentially form the basis of the conjugate beam method to be presented in Section 5.6.

Example 5.5.4

By the moment area method, compute the slope and deflection at point A of the beam of Fig. 5.5.7. Assume constant flexural rigidity EI along the entire length of the beam.

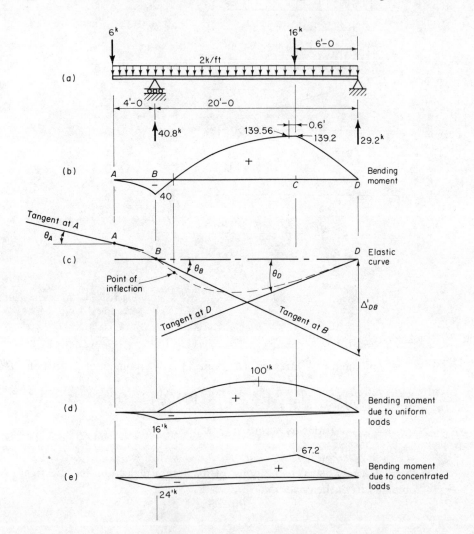

Figure 5.5.7 Beam of Example 5.5.4.

Solution

(a) *Reactions and M/EI diagram.* The reactions and bending moment diagram are shown in Fig. 5.5.7a and b. Based on the moment diagram, a freehand sketch of the elastic curve is made and shown in Fig. 5.5.7c. Although the elastic curve between the point A and the point of inflection has to be convex at the top, whether the slope at A is clockwise as shown, and whether the deflection at A is upward as shown, are yet to be confirmed.

For convenience in using the moment area method, the bending moment diagram is separated into component parts. The choice of component breakdown is arbitrary; in this case the separation is by considering AB as a cantilever and BCD as a simple span subject to the loads on the span and additional bending moment caused by the loads on the cantilever applied at B. This method of separation is preferable to that used in Fig. 5.3.3 whenever the uniform load covers the entire span.

(b) *Compute* θ_B. Since initially there are no known slopes, the slope θ_B is first determined by computing Δ'_{DB} and dividing by the span (20 ft in this example).

$$\theta_B = \frac{\Delta'_{DB}}{20}$$

$$\Delta'_{DB} = \text{moment of } \frac{M}{EI} \text{ area on } BD \text{ about } D$$

$$= \left[\frac{2}{3}(100)(20)(10) - \frac{1}{2}(16)(20)\left(\frac{40}{3}\right) \right. $$
$$+ \frac{1}{2}(67.2)(14)\left(\frac{14}{3} + 6\right) + \frac{1}{2}(67.2)(6)\left(\frac{12}{3}\right)$$
$$\left. - \frac{1}{2}(24)(20)\left(\frac{40}{3}\right) \right] \frac{1}{EI}$$

$$= \frac{13,824}{EI}$$

$$\theta_B = \frac{13,824}{20(EI)} = \frac{691.2}{EI} \text{ kip-ft}^2 \quad \text{(clockwise)}$$

Thus, θ_B is as assumed in Fig. 5.5.7c. Note that a positive moment diagram tends to cause a "sagging" angle at B; and since positive bending predominates on BD, the angle at B is obviously sagging with reference to BD.

(c) *Compute* θ_A. With the slope θ_B known, the first moment area theorem, Eq. (5.5.2), may be applied to obtain θ_A. Because the bending moment is negative between A and B, the tangent will rotate counterclockwise toward A from the angle of the tangent at B. Thus, θ_A is expected to be less than θ_B, as assumed in Fig. 5.5.7c. Thus,

$$\theta_A(\text{clockwise}) = \theta_B(\text{clockwise}) - \left(\text{absolute value of } \frac{M}{EI} \text{ area on } AB\right)$$
$$= \frac{691.2}{EI} - \frac{1}{EI}\left[\frac{1}{3}(16)(4) + \frac{1}{2}(24)(4)\right] = \frac{621.9}{EI} \text{ kip-ft}^2$$

The positive value of $621.9/EI$ confirms that θ_A is clockwise as predicted.

(d) *Compute* Δ_A. Referring to Fig. 5.5.8, Δ_A is obtained by multiplying θ_B by the distance (4 ft) from A to B and then subtracting the deflection Δ'_{AB} between the tangents at A and B. Thus,

$$\Delta_A = 4\theta_B - \Delta'_{AB}$$

$$\Delta_A = \theta_B(4) - \left(\text{moment of } \frac{M}{EI} \text{ area on } BA \text{ about } A\right)$$

$$= \frac{691.2}{EI}(4) - \left[\frac{64}{3EI}(3) + \frac{48}{EI}\left(\frac{8}{3}\right)\right] = \frac{2572.8}{EI} \text{ kip-ft}^3 \quad \text{(up)}$$

Note that only the absolute magnitudes of the M/EI areas are used, the quantities to be evaluated are definitely defined in Fig. 5.5.8.

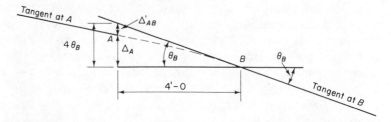

Figure 5.5.8 Computation of Δ_A for Example 5.5.4 by the moment area method.

Example 5.5.5

By the moment area method compute the slope of the left elastic curve at the internal hinge F, the slope of the right elastic curve at the internal hinge F, and the vertical deflection at the internal hinge F in the beam of Fig. 5.5.9. The moment of inertia of segment $ABCDE$ is $5I_c$ and that of segment $EFGH$ is $2I_c$, where I_c is a reference value.

Solution

(a) *Reactions, bending moment diagram, and sketch of elastic curve.* The reactions and bending moment diagram as computed in Example 5.3.3 (see Fig. 5.3.6b) are shown in Fig. 5.5.9. Based on the moment diagram a sketch of the elastic curve is done freehand, shown in Fig. 5.5.9c. It is noted that at this stage it is unknown whether point F is above or below the original beam axis. All that is known is the degree of concavity or convexity of the curvature of the beam, corresponding to the magnitude and sign of the bending moment. In other words, corrections of the first assumed sketch of the elastic curve should be continually made as the computations progress. Note, however, the discontinuities in the slope at points C and F.

The M/EI diagram in Fig. 5.5.9d is obtained by dividing the M-diagram of Fig. 5.5.9b by EI in terms of EI_c for the respective segments, keeping EI_c as a common divisor.

(b) *Compute Δ_C* (see Fig. 5.5.10a). It is appropriate to work from the known slope (i.e., zero) at A. Then

$$\Delta_C = \Delta'_{CA} = \text{moment of } \frac{M}{EI} \text{ area on } AC \text{ about } C$$

$$= \frac{1}{EI_c}\left[\frac{1}{2}(24)(5)\left(5 + \frac{10}{3}\right) + \frac{1}{2}(6)(10)(5)\right]$$

$$= \frac{650}{EI_c} \text{ kip- ft}^3 \text{ (downward)}$$

(c) *Compute θ_E* (see Fig. 5.5.10b). First, the span CE may be treated as a simply supported beam disregarding the deflection at C, as shown in the upper part of Fig. 5.5.10b, where the contributions to the end slopes for this simply supported beam are given the symbols ϕ_C and ϕ_E. Thus,

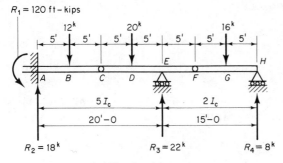

(a) The given beam

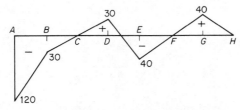

(b) Bending moment diagram, ft – kips

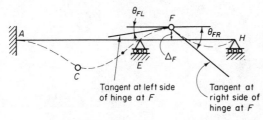

(c) Sketch of elastic curve

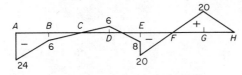

(d) M/EI diagram (in terms of $1/EI_c$)

Figure 5.5.9 Beam with internal hinges for Example 5.5.5.

$$\phi_C = \frac{\Delta'_{EC(1)} - \Delta'_{EC(2)}}{\text{span } CE}$$

$$\Delta'_{EC(1)} - \Delta'_{EC(2)} = \text{moment of } \frac{M}{EI} \text{ area on } CE \text{ about } E$$

$$= \frac{1}{EI_c}\left[\frac{1}{2}(6)(10)(5) - \frac{1}{2}(8)(5)\left(\frac{5}{3}\right)\right] = \frac{350}{3I_c} \text{ kip-ft}^3$$

$$\phi_C = \frac{35}{3EI_c} \text{ kip-ft}^2$$

This is clockwise measured from the straight line CE as shown in Fig. 5.5.10b.

$$\phi_E = \frac{\Delta'_{CE(1)} - \Delta'_{CE(2)}}{\text{span } CE}$$

$$= \frac{1}{10EI_c}\left[\frac{1}{2}(6)(10)(5) - \frac{1}{2}(8)(5)\left(10 - \frac{5}{3}\right)\right] = \frac{-5}{3EI_c} \text{ kip-ft}^2$$

The negative sign indicates ϕ_E is opposite to the direction shown in the upper part of Fig. 5.5.10b; that is, it is clockwise from the straight line CE as shown in the lower part of Fig. 5.5.10b.

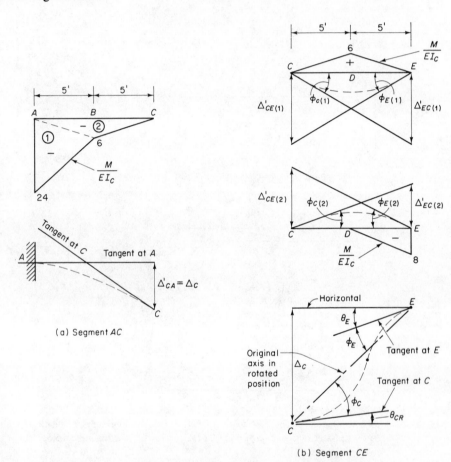

(a) Segment AC

(b) Segment CE

Figure 5.5.10 Segment $ABCDE$ in Example 5.5.5.

Next, the entire segment CE must be rotated counterclockwise about E to account for the downward deflection Δ_C. Thus, the true slope θ_E is

$$\theta_E = \frac{\Delta_C}{10}(\text{counterclockwise}) - \phi_E(\text{clockwise})$$

$$= \frac{650}{10EI_c} - \frac{5}{3EI_c} = \frac{190}{3EI_c} \text{ kip-ft}^2 \quad (\text{counterclockwise})$$

(d) *Compute* θ_{FL} (see Fig. 5.5.11a). The subscript L is used to identify the slope at the left side of the hinge at F. With the slope θ_E known, the first moment area theorem, Eq. (5.5.2), is used to obtain θ_{FL}. Thus,

$$\theta_{FL} = \theta_E - \int_E^F \frac{M\,dx}{EI}$$

$$= \frac{190}{3EI_c} - \frac{1}{EI_c}\left(\frac{1}{2}\right)(20)(5) = \frac{40}{3EI_c}\ \text{kip-ft}^2 \quad \text{(counterclockwise)}$$

Note again only the absolute value of M is used; all directions of slopes are clearly shown on the sketch of the elastic curve.

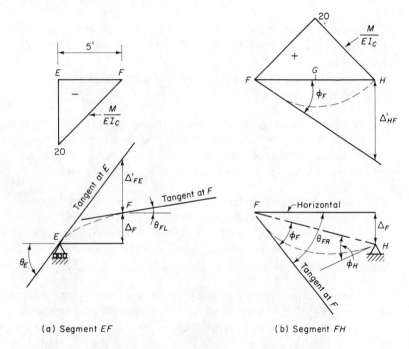

(a) Segment *EF* (b) Segment *FH*

Figure 5.5.11 Segment *EFGH* in Example 5.5.5.

(e) *Compute* Δ_F. From Fig. 5.5.11a,

$$\Delta_F = \theta_E(\text{span } EF) - \Delta'_{FE}$$

$$\Delta'_{FE} = \frac{1}{EI_c}\left(\frac{1}{2}\right)(20)(5)\left(\frac{10}{3}\right) = \frac{500}{3EI_c}$$

$$\Delta_F = \frac{190}{3EI_c}(5) - \frac{500}{3EI_c} = \frac{150}{EI_c}\ \text{kip-ft}^3 \quad \text{(up)}$$

(f) *Compute* θ_{FR} (see Fig. 5.5.11b). Using segment *FH*, first consider the segment a simple beam with supports F and H at the same level to obtain ϕ_F.

$$\phi_F = \frac{\Delta'_{HF}}{\text{span } FH}$$

$$= \frac{1}{10EI_c}\left(\frac{1}{2}\right)(20)(10)(5) = \frac{50}{EI_c} \quad \text{(clockwise)}$$

Next, consider the segment FH rotated clockwise through an angle created by the upward deflection Δ_F. Thus,

$$\theta_{FR} = \phi_F + \frac{\Delta_F}{\text{span } FH}$$

$$= \frac{50}{EI_c} + \frac{150}{EI_c(10)} = \frac{65}{EI_c} \text{ kip-ft}^2 \quad \text{(clockwise)}$$

5.6 CONJUGATE BEAM THEOREMS

The *conjugate beam method*, sometimes known as the *method of elastic weights*, was announced by Otto Mohr [4] in the 1860s. This method is closely related to the moment area method.

As applied between two points on a continuous elastic curve, the two conjugate beam theorems are as follows:

Conjugate Beam Theorem I.   *The clockwise angle between a straight line joining any two points on a continuous elastic curve and the tangent at a third intermediate point is equal to the shear at the third point in a simple beam equal in span to the distance between the two boundary points and loaded by the M/EI area.*

Conjugate Beam Theorem II.   *The distance of a third intermediate point from a straight line joining two points on a continuous elastic curve, measured in a direction perpendicular to the original axis of the beam, is equal to the bending moment at the third point in a simple beam equal in span to the distance between the two boundary points and loaded by the M/EI area.*

In many respects the conjugate beam concept is "obvious" once the relationships, as given in Fig. 5.1.2, between deflection, slope, bending moment, shear, and load intensity are firmly in mind. At this stage in the study of structural analysis, the procedure of beginning with loads and reactions, then drawing the shear diagram, and finally drawing the bending moment diagram should have been well established. One goes from load intensity q to shear V to moment M by the process of integration, as indicated in Fig. 5.1.2. In the same process one goes from M/EI to slope θ to deflection y. Comparing the equations

$$\frac{dV}{dx} = -q; \qquad \frac{dM}{dx} = +V; \qquad \frac{d^2M}{dx^2} = -q$$

derived in Chapter 4 with the equations

$$\frac{d\theta}{dx} = -\frac{M}{EI}; \qquad \frac{dy}{dx} = +\theta; \qquad \frac{d^2y}{dx^2} = -\frac{M}{EI}$$

in Fig. 5.1.2, one sees the analogy between (1) q (downward load is positive) and M/EI (positive M is for concavity at top of the elastic curve), (2) V (positive shear is upward force acting on left face of right free body) and θ (positive θ is clockwise), and (3) M (positive M for compression above the neutral axis) and y (positive y is downward deflection). Consequently, in more general terms, the two conjugate beam theorems can be stated as:

Alternate Conjugate Beam Theorem I. *Upon satisfying the boundary conditions, the clockwise slope at any point on an elastic curve is the positive shear at that point when the conjugate beam is loaded with a downward load equal to the M/EI area.*

Alternate Conjugate Beam Theorem II. *Upon satisfying the boundary conditions, the downward deflection at any point on an elastic curve is the positive bending moment at that point when the conjugate beam is loaded with a downward load equal to the M/EI area.*

In applying the alternate conjugate beam theorems, one may begin by considering M/EI as a "load," obtain the appropriate conjugate (or analogous) beam, compute the reactions for that "load", construct the shear diagram for that "load", and finally the bending moment diagram for that "load." The shear diagram shows the angle θ, and the bending moment diagram represents the deflection y. Thus, the term "conjugate" (meaning reciprocally related or interchangeable) beam has been used. Otto Mohr focused on the idea that M/EI was a load or "elastic weight" applied to a reciprocal or interchangeable beam. Careful examination of Fig. 5.6.1 shows the reciprocity as follows: (1) an exterior simple support on the real beam is still an exterior simple support on the conjugate beam, and vice versa; (2) a fixed support on the real beam becomes a free end in the conjugate beam, and vice versa; and (3) an internal hinge in the real beam becomes an intermediate support on the conjugate beam, and vice versa. When these rules are followed, the analyst can obtain the slope and deflection at any point without even making a sketch of the elastic curve. The method is elegant and laudatory; although the authors believe that beginners can benefit more by always having the sketch of the elastic curve, followed by computations referring to the geometry of the elastic curve.

Nevertheless, the chief advantage of the conjugate beam concept lies in its application between two points of known deflections (usually zero deflections), in which case the conjugate beam is always a simple beam. Thus the conjugate beam theorems are as stated at the beginning of this section. Rather than using the algebraic analogy, these two conjugate beam theorems can be demonstrated to be correct from the moment area theorems.

Referring to Fig. 5.6.2, the first conjugate beam theorem relating slope to the shear on the conjugate beam is

$$\theta_x = \text{shear at } x$$
$$= R_1' - \left(\frac{M}{EI} \text{ area between 1 and } x\right) \tag{5.6.1}$$

The second conjugate beam theorem relating deflection to the bending moment on the conjugate beam is

$$\Delta_x = \text{bending moment at } x$$
$$= R_1'x - \left(\text{moment of } \frac{M}{EI} \text{ area between 1 and } x \text{ about } x\right) \tag{5.6.2}$$

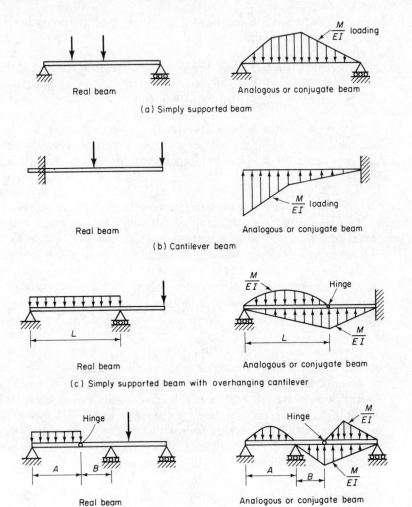

Figure 5.6.1 Typical real and conjugate beams.

Equations (5.6.1) and (5.6.2) may be proved by means of Eqs. (5.5.2) and (5.5.5). Using the second moment area theorem, Eq. (5.5.5),

$$\theta_1 = \frac{\Delta'_{21}}{\text{span 1-2}}$$

$$= \frac{\text{moment of } M/EI \text{ area between 1 and 2 about 2}}{\text{distance 1-2}} = R'_1$$

Using the first moment area theorem, Eq. (5.5.2),

$$\theta_x = \theta_1 - d\theta_{1x}$$

$$= R'_1 - \left(\frac{M}{EI} \text{ area between 1 and } X\right)$$

$$= \text{shear at } X$$

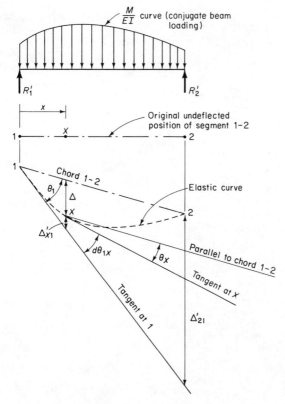

Figure 5.6.2 Conjugate beam method.

which is Eq. (5.6.1). From Fig. 5.6.2,

$$\Delta_x = \theta_1 x - \Delta'_{x1}$$

$$= R'_1 x - \left(\text{moment of } \frac{M}{EI} \text{ area between 1 and } X \text{ about } X\right)$$

$$= \text{bending moment at } X$$

which is Eq. (5.6.2).

Note in Fig. 5.6.2 that points 1 and 2 are *any* two points on a continuous elastic curve, and that θ and Δ are measured from the chord 1-2.

The conjugate beam method is the more convenient thought process when the chord between the two chosen points on the elastic curve does not rotate (or when occasionally the rotation has already been computed) as the structure is loaded with the real load. In this situation the conjugate beam is always a simple beam and the end reactions are the real angles between the chord and the tangents to the elastic curve at the ends. This is the most useful instance as the following examples will show.

Example 5.6.1

Use the conjugate beam method to compute the end slopes θ_A and θ_B and the midspan deflection for a simply supported uniformly loaded beam.

Solution

(a) *Establish the conjugate beam.* The conjugate beam with the parabolic loading (M/EI) is shown in Fig. 5.6.3.

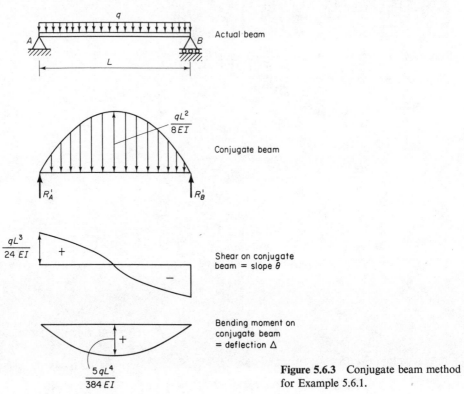

Figure 5.6.3 Conjugate beam method for Example 5.6.1.

(b) *Compute θ_A and θ_B.* These slopes are the reactions R'_A and R'_B.

$$R'_A = R'_B = \frac{1}{2}\left(\frac{2}{3}\right)\left(\frac{qL^2}{8EI}\right)L = \frac{qL^3}{24EI}$$

The complete shear diagram for the conjugate beam is shown in Fig. 5.6.3. Note that the shear is zero at midspan, indicating zero slope in the actual beam. The shear at the quarter point may be computed and found to be $\frac{11}{16}$ of the end shear.

(c) *Compute midspan deflection.* The bending moment on the conjugate beam at midspan is

$$M'_{\text{midspan}} = R'_A\left(\frac{L}{2}\right) - \frac{2}{3}\left(\frac{qL^2}{8EI}\right)\left(\frac{L}{2}\right)\left(\frac{3}{8}\right)\left(\frac{L}{2}\right)$$

$$= \frac{qL^3}{24EI}\left(\frac{L}{2}\right) - \frac{3qL^4}{24(16)} = \frac{5qL^4}{384EI}$$

$$\Delta_{\text{midspan}} = \frac{5qL^4}{384EI}$$

The complete bending moment diagram for the conjugate beam (shown purposely below the horizontal line to describe the actual elastic curve) is given in Fig. 5.6.3. The bending moment at the quarter point is 0.75 of the midspan value.

Example 5.6.2

By the conjugate beam method, compute the slopes at the ends, the deflection under the load W, and the maximum deflection for the beam of Fig. 5.6.4.

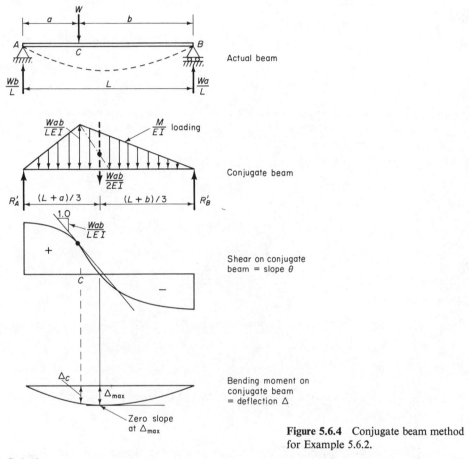

Figure 5.6.4 Conjugate beam method for Example 5.6.2.

Solution

(a) *Establish the conjugate beam.* The conjugate beam with the triangular loading (M/EI) is shown in Fig. 5.6.4.

(b) *Compute θ_A and θ_B.* The slopes are the reactions R'_A and R'_B. The centroid of a typical triangle as shown is located at $(L + a)/3$ from the left support and at $(L + b)/3$ from the right support. Thus,

$$\theta_A = R'_A = \frac{Wab}{2EI}\frac{L + b}{3}\left(\frac{1}{L}\right) = \frac{Wab(a + 2b)}{6LEI}$$

$$\theta_B = R'_B = \frac{Wab}{2EI}\frac{L + a}{3}\left(\frac{1}{L}\right) = \frac{Wab(2a + b)}{6LEI}$$

The complete shear diagram for the conjugate beam is shown in Fig. 5.6.4.

(c) *Compute deflection Δ_C under the load.* The deflection Δ_C is the bending moment at C of the conjugate beam. Using the left free body,

$$\Delta_C = M'_C = R'_A a - \frac{1}{2}\frac{Wab}{LEI}(a)\left(\frac{a}{3}\right) = \frac{Wa^2b(a+2b)}{6LEI} - \frac{Wa^3b}{6LEI}$$

$$= \frac{Wa^2b(a+2b-a)}{6LEI} = \frac{Wa^2b^2}{3LEI} \quad \text{(as in Example 5.5.3)}$$

Using the right free body,

$$\Delta_C = M'_C = R'_B b - \frac{1}{2}\frac{Wab}{LEI}(b)\left(\frac{b}{3}\right) = \frac{Wab^2(2a+b)}{6LEI} - \frac{Wab^3}{6LEI}$$

$$= \frac{Wab^2(2a+b-b)}{6LEI} = \frac{Wa^2b^2}{3LEI} \quad \text{(as before)}$$

(d) *Compute maximum deflection, $\Delta_{\max}$.* The maximum deflection occurs at a location where the shear in the conjugate beam is zero. Let $\Delta_{\max}$ occur at a distance x from the right support.

$$V'_x = R'_B - \frac{1}{2}\frac{Wax}{LEI}x = \frac{Wab(2a+b)}{6LEI} - \frac{Wax^2}{2LEI} = 0$$

$$ab(2a+b) = 3ax^2$$

$$x^2 = \frac{b(2a+b)}{3} \quad \text{for } b > a \quad \text{(as in Example 5.5.3)}$$

$$\Delta_{\max} = M'_x = R_B x - \frac{1}{2}\frac{Wax}{LEI}x\left(\frac{x}{3}\right)$$

$$= \frac{Wab(2a+b)}{6LEI}x - \frac{Wax^3}{6LEI}$$

$$= \frac{Wab(2a+b)}{6LEI}\sqrt{\frac{b(2a+b)}{3}} - \left(\frac{Wa}{6LEI}\right)\frac{b(2a+b)}{3}\sqrt{\frac{b(2a+b)}{3}}$$

$$= \sqrt{\frac{b(2a+b)}{3}}\left(\frac{Wab}{6LEI}\right)\left(\frac{2}{3}\right)(2a+b)$$

$$= \frac{Wab(2a+b)}{9LEI}\sqrt{\frac{b(2a+b)}{3}} \quad \text{for } b > a$$

Example 5.6.3

Using the conjugate beam method, determine the slopes at B and D along with the deflection at C for the beam of Fig. 5.6.5.

Solution

(a) *Establish the conjugate beam.* Since the bending moment diagram is known, the conjugate beam may be regarded as a simply supported one subject to positive bending moment due to the uniform loading on BD and the concentrated load of 16 kips. In addition, it is subject to negative bending due to the 40 ft-kip negative moment at B. For ease of solution the bending moment due to these actual loads is divided into three parts as shown in Fig. 5.6.5. The bending moment due to the 16-kip load, that due to the 2 kip/ft uniform load, and that due to the 40 ft-kip end moment when all divided by EI become the "loads" for the conjugate beam, which to emphasize again is a simple beam by itself supported at B and D only.

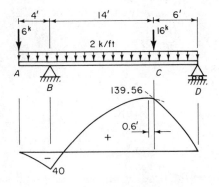

(a) Given beam and combined moment diagram

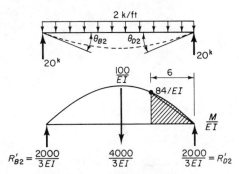

(c) Elastic curve and conjugate beam for uniform load – segment BD.

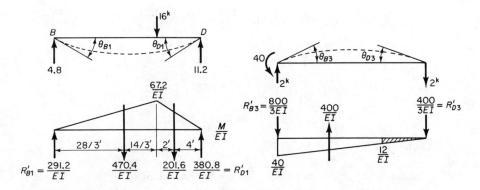

(b) Elastic curve and conjugate beam for 16 kip load – segment BD.

(d) Elastic curve and conjugate beam for 40 ft/kip end moment – segment BD.

Figure 5.6.5 Conjugate beam method applied to segment BD of the given beam, Example 5.6.3.

(b) *Obtain* θ_B. According to Conjugate Beam Theorem I, the slope θ_B equals the left reaction of the conjugate beam. In Fig. 5.6.5b, c, and d the M/EI loads and their centroids on the three conjugate beams are computed and shown with arrows, downward for M/EI from positive bending moment and upward for M/EI from negative bending moment. The reactions at each end of the three beams are also shown.

$$\theta_B = \theta_{B1} + \theta_{B2} - \theta_{B3} = R'_{B1} + R'_{B2} - R'_{B3}$$

$$= \frac{291.2}{EI} + \frac{2000}{3EI} - \frac{800}{3EI} = \frac{691.2}{EI} \text{ kip-ft}^2 \quad \text{(clockwise)}$$

(c) *Obtain* θ_D. The slope θ_D is the reaction at D. Thus,

$$\theta_D = R'_{D1} + R'_{D2} - R'_{D3}$$

$$= \frac{380.8}{EI} + \frac{2000}{3EI} - \frac{400}{3EI} = \frac{914.1}{EI} \text{ kip-ft}^2 \quad \text{(counterclockwise)}$$

(d) *Obtain* Δ_C. The deflection at C equals the bending moment at C on the conjugate beam. Taking moments using the forces on the right free body CD,

$$\Delta_C = M'_C = \frac{380.8}{EI}(6) - \frac{201.6}{EI}(2) + \frac{2000}{3EI}(6)$$

$$- \frac{1}{2}\left(\frac{84}{EI}\right)(6)(2) - \frac{2(6)^2}{8EI}\left(\frac{2}{3}\right)(6)(3)$$

$$- \frac{400}{3EI}(6) + \frac{1}{2}\left(\frac{12}{EI}\right)(6)(2)$$

$$= \frac{4541.6}{EI} \text{ kip-ft}^3 \quad \text{(down)}$$

Note that in the fourth and fifth terms of the bending moment computation, the 6-ft portion of the M/EI parabolic loading in Fig. 5.6.5c was broken into a triangle and a symmetrical parabolic segment that has its vertex at 3 ft from D (refer to Example 4.4.6). The parabolic segment has as its maximum ordinate $qL_1^2/8 = 2(6)^2/8$. Using the left free body,

$$\Delta_C = M'_C = \frac{291.2}{EI}(14) - \frac{470.4}{EI}\left(\frac{14}{3}\right) + \frac{2000}{3EI}(14) - \frac{1}{2}\left(\frac{84}{EI}\right)(14)\left(\frac{14}{3}\right)$$

$$- \frac{2(14)^2}{8EI}\left(\frac{2}{3}\right)(14)(7) - \frac{800}{3EI}(14) + \frac{1}{2}\frac{40}{EI}(14)\left(\frac{28}{3}\right)$$

$$+ \frac{1}{2}\frac{12}{EI}(14)\left(\frac{14}{3}\right) = \frac{4541.6}{EI} \text{ kip-ft}^3 \quad \text{(down)}$$

5.7 COMPARISON OF METHODS

In this chapter the basic tools have been presented to obtain deformations of beams using the classical differential equation method (i.e., double integration), the basic energy method using the principle of virtual work [i.e., the virtual work (unit load) method], the basic geometric method (i.e., the moment area or its modification the conjugate beam method).

It is true that traditionally there has been more interest in determining internal forces (such as shears and bending moments) in structures than in determining deformations. However, in current (1983) structural engineering practice many problems (and lawsuits) arise from inadequate attention to serviceability requirements, such as deflection control. Moreover, as will be treated in Chapters 7 and 8, computations for the deformations in statically determinate structures form the basis of obtaining solutions for statically indeterminate structures, when the force method is used. In that method the compatibility requirements of deformations furnish the additional "non-statics" conditions for the complete solution.

Regarding the choice of using the energy method (virtual work) or the geometric method (moment area, conjugate beam), the authors believe both are essential tools of structural analysis. In succeeding chapters, each method is applied in the situation where it seems most suitable. From the examples described in this chapter, the reader may note that certain types of problems lend themselves to certain solution methods.

The virtual work (unit load) method is useful when only one single deformation quantity at a single point is required. Other than the separate free-body diagrams of the beam under actual loading and under the unit load (or unit moment), no other diagrams (such as shear and moment diagrams, or sketch of the elastic curve) are necessary. Often this method can be used to verify correctness of the entire elastic curve computed by the geometric method by recomputing the deflection obtained in the final step of the geometric method.

The biggest reward in using the geometric method is that the analyst can get the "feel" of the behavior of the structure because a sketch of the elastic curve is essential before applying any of the moment area or conjugate beam theorems. This sketch may not be on the paper, but it has to be in the mind in any case. Some analysts prefer to apply "literally" the two alternate conjugate beam theorems once the "conjugate beam" equal in length to the total length of the actual beam is drawn on paper, thus bypassing the sketch of the elastic curve of the actual beam. The authors do not favor this approach; but certainly there is no objection if one wishes to obtain readily the slopes and deflections at many points along the beam.

In fact, the conjugate beam method is most convenient when it is applied between two points of known (or zero in the best situation) deflection on the actual beam. If the supports of the conjugate beam do translate, as for segment *CE* of Fig. 5.5.9b or segment *FH* of Fig. 5.5.10b, the analyst must recognize that the conjugate beam reactions are the end slopes measured with respect to the chord connecting the ends of the conjugate beam. To obtain the true slopes, the beam segment must undergo a rigid body rotation equal to the translation (up or down) of one end of the chord relative to the other divided by the chord length. This was illustrated in investigating the beam of Fig. 5.5.8 in Example 5.5.5 using the moment area method. Since the moment area and conjugate beam methods are inherently the same geometric method, many analysts use both in solving a given problem. To reiterate, the moment area method keeps the user always focusing on the basic concept of computing angles and deflections between tangents to the elastic curve and is useful as a basic tool. The conjugate beam method is useful to be applied between two points of zero deflection so that the conjugate beam is always a simple beam.

SELECTED REFERENCES

1. James Clerk Maxwell, "On the Calculation of the Equilibrium and Stiffness of Frames," *Philosophical Magazine* (4), *27* (1864), 294–299. See also *The Scientific Papers of James Clerk Maxwell*, Vol. 1, ed. W. D. Niven, Cambridge University Press, Cambridge, England, 1890, 598–604.

2. Heinrich F. B. Müller-Breslau, *Die Neuren Methoden der Festigkeitslehre und der Statik der Baukonstruktionen*, Berlin, 1886.

3. H. Müller-Breslau, "Beitrag zur Theorie des Fachwerks," *Zeitschrift des Architekten- und Ingenieur-Vereins zu Hannover*, *31* (1885), 418.

4. Otto Mohr, "Beitrag zur Theorie der Holz- und Eisenkonstruktionen," *Zeitschrift des Architekten- und Ingenieur-Vereins zu Hannover*, *6* (1860), 323–346, 407–442, plus Plate

175; 8 (1862), 245–280, plus Plate 232. See also Otto Mohr, *Abhandlungen aus dem Gebiete der Technischen Mechanik*, Wilhelm Ernst & Sohn, Berlin, 1906, 266–270.

5. Charles E. Greene, *Graphical Method for the Analysis of Bridge-Trusses*, Geo. H. Frost, Chicago, 1874. (See also Prefaces to Charles E. Greene, *Trusses and Arches Analyzed and Discussed by Graphical Methods*, Part II: Bridge-Trusses, 5th ed., John Wiley & Sons, Inc., New York, 1895.)

6. H. M. Westergaard, "One Hundred Fifty Years Advance in Structural Analysis" (presented at the meeting of the Structural Division, ASCE, Philadelphia, October 8, 1926), *Transactions*, ASCE, *94* (1930), 226–240.

PROBLEMS

For all problems, when shear and bending moment diagrams are used in solutions, draw them to reasonable scale and state critical ordinate values on the diagrams. Answers are to include rotational direction for slopes and up or down for deflections and are to be expressed in kip and ft units as well as *EI*, unless otherwise stated.

Problems by Differential Equation Method

5.1. For the beam of Prob. 4.1, obtain the equation (or equations) for deflection y and determine the maximum value.

5.2. For the beam of Prob. 4.2, obtain the equation for slope $\theta = dy/dx$ and deflection y (measure x from point A). Compute the maximum deflection.

Problems by Virtual Work (Unit Load) Method

5.3. For the beam of Prob. 4.1, compute the slope and deflection at point F using the virtual work (unit load) method. Assume constant *EI*. Show the bending moment diagrams used for the real and virtual systems.

5.4. For the beam of Prob. 4.2, compute the slope θ_A and deflection Δ_A at point A using the virtual work (unit load) method. Assume constant *EI*. Show the bending moment diagrams used for the real and virtual systems.

5.5. For the beam of Prob. 4.14, use the virtual work (unit load) method to compute the slopes θ_B and θ_C (left) and the deflection Δ_C at C. Show the bending moment diagrams for the real and virtual systems.

5.6. For the beam of Prob. 4.14, use the virtual work (unit load) method to compute the slopes θ_C (right), θ_E, θ_G, and the deflection Δ_F. Show the bending moment diagrams for the real and virtual systems.

5.7. For the beam of Prob. 4.3, compute the slopes and the deflection at the internal hinge B using the virtual work (unit load) method. Assume constant *EI*.

5.8. Compute the slope of the elastic curve at D (midspan between A and B), and the deflection at C for the beam in the accompanying figure, using the method of virtual work (unit load). Show the bending moment diagrams for the real and virtual systems.

5.9. Using the virtual work (unit load) method, write numerically the integrals that represent the slopes to the left side of B and at C of the elastic curve and for the deflection under the 18-kip load for the beam of the accompanying figure. Draw the bending

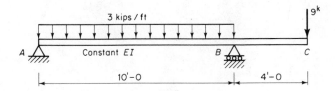

Probs. 5.8, 5.26, and 5.39

moment diagrams for both the real and virtual systems. (*Note:* Actual evaluation of the integrals is not required.)

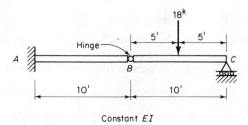

Constant *EI*

Probs. 5.9 and 5.27

5.10. Using the virtual work (unit load) method, determine the slopes at *A* and under the 9-kip load, and the deflection under the 9-kip load, for the beam of the accompanying figure.

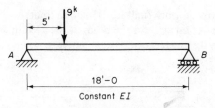

Constant *EI*

Probs. 5.10 and 5.40

5.11. Using the virtual work (unit load) method, determine the slopes at *A* and *B* (both left and right side of hinge) on the beam of the accompanying figure.

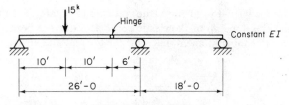

Probs. 5.11 and 5.28

5.12. For the overhanging beam shown in the accompanying figure, determine the deflection at midspan of *BC* and the slope at *C* using the virtual work (unit load) method. Show the bending moment diagrams for the real and virtual systems.

5.13. Use the virtual work (unit load) method to compute the deflections at *B* and *C* on the beam of the accompanying figure. Show the bending moment diagrams for the real and virtual systems.

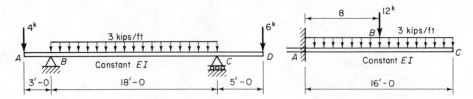

Probs. 5.12 and 5.42   **Probs. 5.13, 5.29, and 5.41**

5.14. Use the virtual work (unit load) method to compute the deflections at midspan and 6 ft from A and the slope at 6 ft from A for the beam of the accompanying figure. Show the bending moment diagrams for the real and virtual systems.

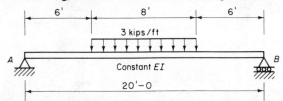

Probs. 5.14, 5.30, and 5.45

5.15. Use the virtual work (unit load) method to determine the slopes at A, D, and C, as well as the deflection at D for the beam of the accompanying figure. Show bending moment diagrams for the real and virtual systems.

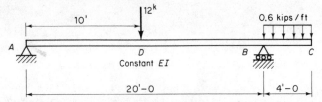

Probs. 5.15 and 5.31

5.16. Use the virtual work (unit load) method to obtain the deflection Δ_x at midspan of AB and the slope θ_C of the elastic curve at C for the beam of the accompanying figure. Show bending moment diagrams for the real and virtual systems.

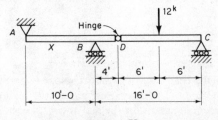

Constant EI **Probs. 5.16 and 5.32**

5.17. Use the virtual work (unit load) method to obtain the slope and deflection at C for the beam of the accompanying figure. Show all bending moment diagrams.

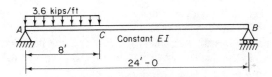

Probs. 5.17, 5.33, and 5.46

5.18. Use the virtual work (unit load) method to obtain the slope and deflection at A on the beam of the accompanying figure. Show all bending moment diagrams.

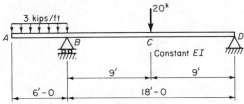

Probs. 5.18 and 5.34

5.19. Use the virtual work (unit load) method to obtain the slope and deflection at C on the beam of the accompanying figure. Show all the bending moment diagrams. By what percent is the deflection reduced by having $2EI_c$ over CB compared with having constant EI_c over the entire span?

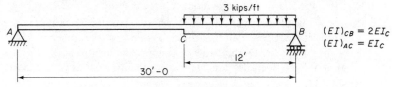

Probs. 5.19 and 5.47

5.20. Using the virtual work (unit load) method, determine the slope θ of the elastic curve at points B and C of the accompanying figure. Also find the deflection at D. Sketch the elastic curve.

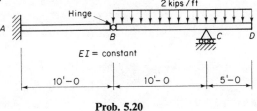

Prob. 5.20

Problems by Moment Area Method

Focus on the basic concept of working with tangents to the elastic curve, show all geometry being used, and avoid using the conjugate beam thought process.

5.21. For the beam of Prob. 4.1, use the moment area method to sketch the slope θ diagram and the deflection Δ diagram for the entire beam. Draw the bending moment diagram with respect to support D; that is, consider the total bending moment at D to be made up of the contributions (diagrammed separately) of R_1, the 1.2 kips/ft uniform load, and the 10-kip concentrated load. Similarly, on the right side of D the bending moment at D equals the contributions (diagrammed separately) of the 1.2 kips/ft uniform load and the 8-kip concentrated load. At each 5-ft interval compute the slope diagram ordinate (and scale and label it). Check that the slope and deflection diagrams are correctly related, such as when θ is zero the slope of the deflection curve actually is zero.

5.22. Solve Prob. 5.4 using the moment area method.

5.23. Solve Prob. 5.5 using the moment area method.

5.24. Solve Prob. 5.6 using the moment area method.

5.25. Solve Prob. 5.7 using the moment area method.

5.26. Solve Prob. 5.8 using the moment area method.

5.27. For the beam of Prob. 5.9, use the moment area method to determine θ_B (left), θ_C, and the deflection Δ under the 18-kip load. In addition, from the M/EI diagram compute the slopes at 5-ft intervals across the beam, plot them to scale, and draw the curve. Then determine and plot the deflection at 5-ft intervals and draw the curve.

5.28. Solve Prob. 5.11 using the moment area method.

5.29. Solve Prob. 5.13 using the moment area method.

5.30. Solve Prob. 5.14 using the moment area method.

5.31. Solve Prob. 5.15 using the moment area method.

5.32. Solve Prob. 5.16 using the moment area method.

5.33. Solve Prob. 5.17 using the moment area method.

5.34. Solve Prob. 5.18 using the moment area method.

5.35. For the beam of the accompanying figure, use the moment area method to sketch the elastic curve. Begin with the M/EI diagram and from that compute the slopes at points A through E. Draw the curve of θ approximately to scale. Next compute deflections at B and D; then use the θ curve to qualitatively draw the elastic curve (deflection curve).

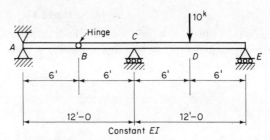

Prob. 5.35

5.36. For the beam of the accompanying figure, use the moment area method to compute the slopes of the elastic curve at points B and D, and the deflections at points C and under the 10-kip load. Sketch the curve of θ approximately to scale.

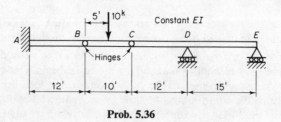

Prob. 5.36

5.37. For the beam of the accompanying figure, use the moment area method to obtain the midspan deflection Δ_C and the slopes θ_A and θ_B in terms of q, M_A, and M_B.

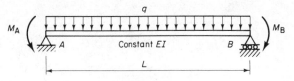

Prob. 5.37 and 5.44

Problems by Conjugate Beam Method

5.38. For the beam of Prob. 4.1, compute the slopes at A and D and the deflection at C using the conjugate beam method.

5.39. For the beam of Prob. 5.8, compute the slopes at A and B and midspan (of AB) deflection using the conjugate beam method.

5.40. For the beam of Prob. 5.10, compute the deflection under the 9-kip load by the conjugate beam method. By what percentage would the deflection reduce if the EI is doubled over the center 8 ft of the span?

5.41. For the beam of Prob. 5.13, compute the slope at C and the deflection at B by the conjugate beam method.

5.42. For the beam of Prob. 5.12, determine the slope at C and the deflection at midspan of BC by the conjugate beam method.

5.43. For the beam of the accompanying figure, show using the conjugate beam method that the midspan deflection is equal to $5qL^4/768EI$.

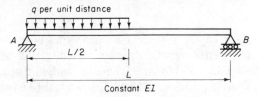

Prob. 5.43

5.44. For the beam of Prob. 5.37, develop the expression for midspan deflection using the conjugate beam method.

5.45. For the beam of Prob. 5.14, compute the midspan deflection using the conjugate beam method.

5.46. Solve Prob. 5.17 using the conjugate beam method.

5.47. Solve Prob. 5.19 using the conjugate beam method.

5.48. For the beam of the accompanying figure, use the conjugate beam method to obtain M_A as a function of θ_A, and then determine θ_B as a function of M_A. What would M_A have to be if θ_B were to be zero?

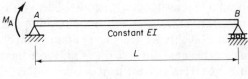

Prob. 5.48

[6]

Influence Lines and Movable Loads

6.1 INTRODUCTION: FIXED-POSITION AND MOVABLE LIVE LOADS

As described in Chapter 1, there are many types of loads that must be provided for when designing a structure. Of the various loads, such as gravity dead load, gravity live load, snow load, wind load, earthquake load, and vehicular load, only the gravity dead load is fixed in position. All other loads must be treated as variable in magnitude and location. The structural engineer has as a major part of the structural analysis preliminary to design the task of properly positioning the loads on the structure. For example, uniform gravity live load such as that due to occupancy (i.e., people and goods) must be treated as acting over the entire structure or only over certain portions of it, depending on the possibility of occurrence and the situation that will cause the most severe effect. Vehicular loads particularly require positioning, perhaps in many positions, to give the most severe effect at each different location.

How does the structural analyst know what arrangement of live load, uniform or concentrated, will cause the most severe effect? This high-priority topic is the subject of this chapter. The influence line (i.e., diagram) is the means of determining the effect of a load acting in various positions on the magnitude and sign of an internal force (i.e., bending moment, shear, and axial tension or compression) in a member. The structural engineer must establish the position (or positions) to be used for the live loads, either explicitly by using computations involving influence lines or implicitly by visualizing the qualitative shape of influence lines. The importance of influence lines cannot be overemphasized. In this chapter the subject is restricted to statically determinate beams and trusses so that the concept may be thoroughly developed. In

Chapter 8 the concept is extended to continuous beams, with additional treatment in Chapters 10 and 11. Without the knowledge of where to position loads for the most severe effect (as well as possible reversal of direction or sign), a knowledge of the various structural analysis methods and design criteria is useless.

6.2 DEFINITION OF AN INFLUENCE LINE

An influence line for a function (such as a support reaction, a bending moment at a particular location, or a shear at a particular location) is the graph showing the value of that function caused by a single concentrated unit load moving along the span of a structure, wherein the value of the function is plotted directly at the location of the unit load.

Without showing how to obtain the influence line shape shown in Fig. 6.2.1, note that the influence line ordinate at the location (B) of the unit load is the value of the bending moment at the location (A) for which the influence line has been drawn. Although the shape of the influence line for moment at A is the same as the bending moment diagram when a single concentrated load is at A, the meanings are different. The bending moment diagram gives the variation in bending moment along the span due to fixed loading, whereas the influence line gives the variation in bending moment at one point due to a unit load moving along the span.

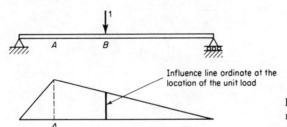

Influence line ordinate at the location of the unit load

Figure 6.2.1 Influence line for bending moment at A.

6.3 HISTORICAL BACKGROUND OF INFLUENCE LINES

The concept of using the influence of a moving unit load to determine placement of actual loads to give a maximum or minimum effect was published by Emil Winkler of Dresden, Germany, in 1868 [1, 2 (p. 28)]. Winkler's textbook [2] of 1886 provides an extensive treatment of influence lines. In this book Winkler claims [2 (p. 28)] to be the originator of the concept in 1868 but credits Jacob Weyrauch [3, 2 (p. 28)] with introducing the name *line of influence* in 1873, Actually, Winkler in an earlier work [4] in 1862 developed coefficients for maximum and minimum moments, shears, and support reactions for up to five continuous spans using a unit intensity of distributed load. Otto Mohr, then of Stuttgart, Germany, independently used the concept during the same period [5, 6]. The useful practical concept that the influence line may be treated as a deflection curve, known as the *Müller-Breslau principle*, was presented by Heinrich Müller-Breslau of the Technische Hochschule at Berlin in his classic textbook [7].

6.4 INFLUENCE LINES FOR SUPPORT REACTIONS ON STATICALLY DETERMINATE BEAMS

The influence lines for the two reactions R_A and R_B on the overhanging beam of Fig. 6.4.1a are shown in Fig. 6.4.1b and c; and those for the two reactions R_A and M_A on the cantilever beam of Fig. 6.4.1d, in Fig. 6.4.1e and f. In every case the influence line is the graph of the function (reaction in this case) as the unit load moves along the beam. For example, to verify Fig. 6.4.1b, it is only necessary to note that R_A is found by taking the moment of the unit load about point B and then dividing this moment by the distance L. The reader should think through the four influence lines shown in Fig. 6.4.1 slowly and carefully.

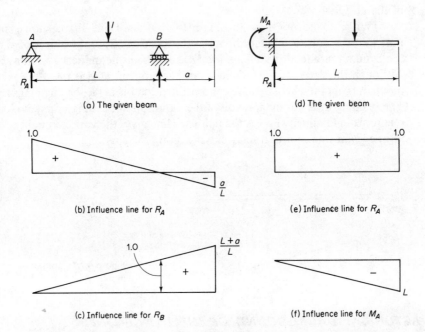

(a) The given beam

(b) Influence line for R_A

(c) Influence line for R_B

(d) The given beam

(e) Influence line for R_A

(f) Influence line for M_A

Figure 6.4.1 Influence lines for reactions on overhanging and cantilever beams.

Because the influence ordinate at any location is the value of the function due to a unit load at that location, the value of the function due to the influence of several concentrated loads such as W_1, W_2, and W_3 in Fig. 6.4.2a and b is

$$R_A = W_1 y_1 + W_2 y_2 + W_3 y_3 \tag{6.4.1}$$

Since the influence line ordinate is the reaction R_A (say kips) due to a unit load (say 1 kip) at a location, the influence line ordinate is dimensionless because it is force per unit force (such as kips per kip unit load or newtons per newton unit load).

For distributed loading such as q in Fig. 6.4.2c, the value of the function (say reaction R_A) due to distributed load from a to b is

$$R_A = \int_a^b (q \, dx)(y) = q \int_a^b y \, dx \tag{6.4.2}$$

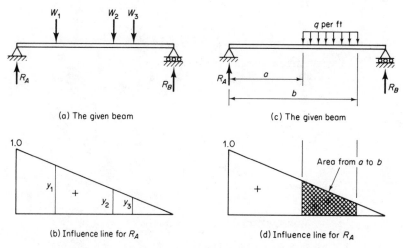

(a) The given beam

(c) The given beam

(b) Influence line for R_A

(d) Influence line for R_A

Figure 6.4.2 Computation for functions by influence lines.

Again, with regard to units, the influence ordinate has units such as kips per kip. Multiplying q kips per ft by an area that is kips per kip times feet gives the result in kips. Examination of Eq. (6.4.2) shows that R_A has the units of q multiplied by a length.

The influence line is primarily useful in determining, by visual inspection in most cases, the position causing maximum or minimum effects of a series of moving concentrated loads of known magnitudes at known spacings, or of a uniform live load of definite or indefinite length. Once the critical position has been obtained, the critical value of the function itself may be computed either by using Eqs. (6.4.1) and (6.4.2) or by simply applying the loads in the critical position and making the analysis without using the influence line values. The suggestion is to do this computation both ways in order to see that the same results are obtained.

Example 6.4.1

Construct the influence line for R_B on the overhanging beam of Fig. 6.4.3a; and determine the maximum positive and maximum negative values of R_B due to (a) the three concentrated loads of Fig. 6.4.3c moving along the beam in either direction, and (b) the uniform live load of indefinite length of Fig. 6.4.3c.

Solution

(a) *Influence line for R_B* (Fig. 6.4.3b). The influence line ordinates are determined by placing the unit load at various positions and computing R_B. That value of R_B is plotted directly under that position of the unit load. Once it is apparent that the reaction at B is a linear function of the position of the unit load, placing the unit load at B gives R_B equal to 1.0; thus 1.0 is the ordinate at R_B. Placing the unit load at R_A gives zero for R_B, making zero the influence line ordinate at R_A. The (-0.25) value is obtained by placing the unit load at the free end of the cantilever and computing R_B.

(b) *Maximum effect due to the moving concentrated loads.* The maximum effect is obtained when the sum of the products of the concentrated load times the influence line ordinate at that load gives the greatest result. The critical position for the maximum positive value of R_B could be either Fig. 6.4.3d or e; its values may be computed

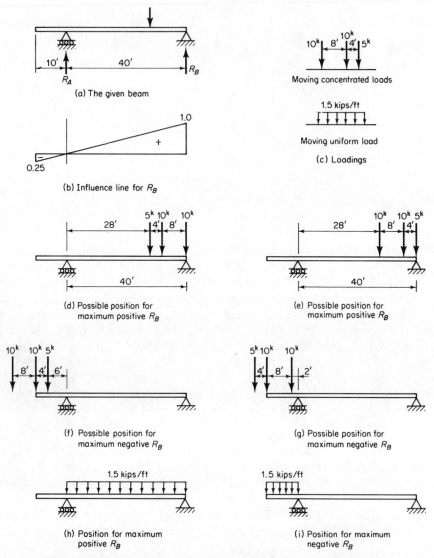

Figure 6.4.3 Maximum reaction on an overhanging beam, Example 6.4.1.

by Eq. (6.4.1) to be

$$\text{Max. pos. } R_B \text{ (Fig. 6.4.3d)} = \sum Wy = 10(1.0) + 10(32/40) + 5(28/40)$$
$$= 10 + 8 + 3.5 = +21.5 \text{ kips}$$
$$\text{Max. pos. } R_B \text{ (Fig. 6.4.3e)} = 5(1.0) + 10(36/40) + 10(28/40)$$
$$= 5 + 9 + 7 = +21 \text{ kips}$$

or by considering the free body of Fig. 6.4.3d or e,

Max. pos. R_B (Fig. 6.4.3d) $= \dfrac{\text{moments about } A}{40} = \dfrac{5(28) + 10(32) + 10(40)}{40}$

$$= \frac{860}{40} = +21.5 \text{ kips} \quad \text{(check)}$$

Max. pos. R_B (Fig. 6.4.3e) $= \dfrac{10(28) + 10(36) + 5(40)}{40} = \dfrac{840}{40} = +21 \text{ kips} \quad \text{(check)}$

Thus,

$$\text{Max. pos. } R_B = +21.5 \text{ kips}$$

The critical position for the maximum negative value of R_B could be either Fig. 6.4.3f or g; its values are -3.25 and -3.00 kips, respectively. Thus,

$$\text{Max. neg. } R_B = -3.25 \text{ kips}$$

(c) *Maximum effect from uniform live load.* For maximum positive (upward) reaction at B, as much uniform load must be put on the portion of the beam indicated by the positive portion of the influence line. In this case, the entire 40-ft span must be loaded as indicated in Fig. 6.4.3h. Thus, multiplying the intensity of uniform load by the positive area of the influence line, or by directly computing the reaction for the loading in the position shown, gives

$$\text{Max. pos. } R_B = \frac{1.5(40)}{2} = +30 \text{ kips}$$

Then loading the portion of the beam indicated by the negative portion of the influence line gives (Fig. 6.4.3i)

$$\text{Max. neg. } R_B = -\frac{1.5(10)^2/2}{40} = -\frac{75}{40} = -1.875 \text{ kips}$$

This latter may be verified by using Eq. (6.4.2).

6.5 INFLUENCE LINES FOR SHEAR AND BENDING MOMENT IN STATICALLY DETERMINATE BEAMS

Shear

To explain the concept, the influence line for shear at point C of the beam of Fig. 6.5.1a is developed. The influence line is the plot of the value of shear at C as a function of the position x of the unit load. For any position of the unit load, a three-step thought process must be used to compute V_C: (1) the beam is divided into two free bodies by cutting at point C; (2) the reactions must be computed using the whole beam as the free body; and (3) the equilibrium requirement that $\sum F_v = 0$ (summation of transverse forces on either free body equals zero) can be applied to obtain V_C. No matter how rapid it may seem, these three steps are used for each position of the unit load.

Example 6.5.1

Determine the influence line for shear at C on the beam of Fig. 6.5.1a.

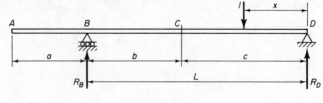

(a) The given beam

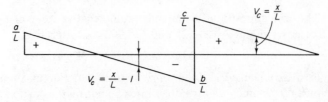

(b) Influence line for shear at C

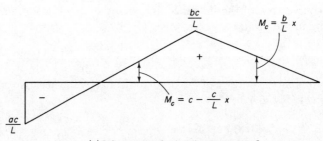

(c) Influence line for bending moment at C

Figure 6.5.1 Influence lines for shear and bending moment.

Solution

(a) *Unit load on segment CD.* For the unit load on CD, select segment AC as the free body. Thus, for $0 \le x < c$,

$$V_C = +R_B \tag{a}$$

Taking the whole beam as the free body,

$$R_B = \frac{x}{L} \tag{b}$$

Then

$$V_C = \frac{x}{L} \tag{c}$$

which is the equation of the influence line. When x is an infinitesimal amount less than c, the unit load is still on free body CD. When x is an infinitesimal amount greater than c, the unit load has shifted to free body AC. The unit load has to be *on* one free body or the other.

(b) *Unit load on segment AC.* For $x > c$,

$$V_C = +R_B - 1 \tag{d}$$

Since the equation for R_B is the same as given in Eq. (b), the influence line equation is

$$V_C = \frac{x}{L} - 1 \qquad (e)$$

as shown in Fig. 6.5.1b. Once again the reader is reminded that the shear *at* a concentrated load is undefined; one must be considering a section an infinitesimal distance to either side of the concentrated load.

From Eqs. (c) and (e) it is shown that the influence lines are straight, indicative of statically determinate situations. When it is recognized that straight lines result, one may simply position the unit load at critical locations, compute V_C, and then connect the points by straight lines.

The discontinuity at point C always occurs at the point for which the shear influence line is drawn, because as the unit load shifts from one free body to the other, the value of the shear changes by the unit amount. Equation (e) is a unit amount different from Eq. (c). The slope of the shear influence line on either side of the discontinuity is always identical (in this case, $1/L$).

Bending Moment

To illustrate the procedure and concept, the influence line for bending moment at point C of the beam of Fig. 6.5.1a is developed. The influence line is the value of bending moment at C as a function of the position x of the unit load. For any position, a three-step thought process is used to compute M_C: (1) the beam is divided into two free bodies by a cut at point C; (2) the reactions are computed using the whole beam as the free body, and (3) the equilibrium requirement that $\sum M = 0$ (summation of moments on either free body equals zero) can be applied to obtain M_C.

Example 6.5.2
Determine the influence line for bending moment at C on the beam of Fig. 6.5.1a.

Solution
(a) *Unit load on segment CD.* For the unit load on CD, use segment AC as the free body. Thus, for $0 \leq x \leq c$,

$$M_C = R_B\, b \qquad (a)$$

$$R_B = \frac{x}{L} \qquad (b)$$

$$M_C = \frac{b}{L} x \qquad (c)$$

(b) *Unit load on segment AC.* For this case, it is convenient to use the right free body (CD). For $x \geq c$,

$$M_C = R_D c \qquad (d)$$

$$R_D = \frac{L - x}{L} \qquad (e)$$

$$M_C = \left(\frac{L - x}{L}\right)c = c - \frac{c}{L}x \qquad (f)$$

Both Eqs. (c) and (f) are straight lines, as expected for influence lines for statically determinate beams. The results are in Fig. 6.5.1c. Knowing that straight lines are expected, one may apply the unit load at critical locations, directly obtain the ordinates, and then connect with straight lines.

Example 6.5.3

Construct the influence lines for shear and bending moment at section B of the overhanging beam $ABCD$ shown in Fig. 6.5.2a; compute the maximum and minimum shears and bending moments at B due to a dead load of 0.6 kip/ft and a live load of 1.5 kips/ft, assuming that the live load may be broken into segments of any length.

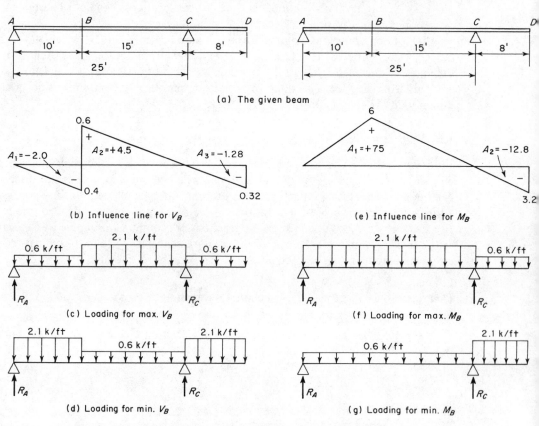

Figure 6.5.2 Maximum and minimum shears and bending moments.

Solution

(a) *Maximum and minimum shears at B.* The influence line for V_B is constructed as shown in Fig. 6.5.2b; its component areas A_1, A_2, and A_3 are

$$A_1 = \frac{1}{2}(-0.4)(10) = -2.0$$

$$A_2 = \frac{1}{2}(+0.6)(15) = +4.5$$

$$A_3 = \frac{1}{2}(-0.32)(8) = -1.28$$

The influence line indicates that live load should be placed over the length from B to C for maximum positive effect, and over the lengths AB and CD for maximum negative effect. Since dead load acts always as a fixed position loading, it applies in this case over the entire length AD. Thus, loading the beam as shown in Fig. 6.5.2c and d by the positive and negative portions of the influence line gives the range of values (maximum and minimum effects) for which the member must be designed. The minimum value may actually be a minimum positive value or it may be a negative value, depending on the relative proportions of dead and live loads.

The maximum shear may be obtained by multiplying the intensity of load by the influence diagram area, Eq. (6.4.2), or it may be obtained by a direct computation of the desired quantity after using the influence line to indicate where to place the live load. Using Eq. (6.4.2), with the loading of Fig. 6.5.2c,

$$\text{Max. } V_B = 0.6A_1 + 2.1A_2 + 0.6A_3$$
$$= 0.6(-2.0) + 2.1(+4.5) + 0.6(-1.28) = +7.482 \text{ kips*}$$

This value should be checked directly by computing the shear at B; thus,

$$\text{Max. } V_B = R_A - 0.6(10) = \frac{0.6(33)(8.5) + 1.5(15)(7.5)}{25} - 6$$

$$= +7.482 \text{ kips} \quad \text{(check)}$$

In similar manner, applying Eq. (6.4.2) to the loading pattern of Fig. 6.5.2d,

$$\text{Min. } V_B = 2.1A_1 + 0.6A_2 + 2.1A_3$$
$$= 2.1(-2.0) + 0.6(+4.5) + 2.1(-1.28) = -4.188 \text{ kips*}$$

Directly computing the shear at B without using the influence line areas,

$$\text{Min. } V_B = R_A - 2.1(10) = \frac{0.6(33)(8.5) + 1.5(10)(20) - 1.5(8)(4)}{25} - 21.0$$

$$= -4.188 \text{ kips} \quad \text{(check)}$$

(b) *Maximum and minimum bending moments at B.* The influence line for M_B is constructed as shown in Fig. 6.5.2e; its component areas A_1 and A_2 are

$$A_1 = \frac{1}{2}(+6)(25) = +75$$

$$A_2 = \frac{1}{2}(-3.2)(8) = -12.8$$

Applying Eq. (6.4.2) to the loading pattern of Fig. 6.5.2f,

$$\text{Max. } M_B = 2.1A_1 + 0.6A_2$$
$$= 2.1(+75) + 0.6(-12.8) = +149.82 \text{ ft-kips*}$$

Check by directly computing the bending moment at B for the beam loaded as in Fig. 6.5.2f,

---

*Note that the number of significant figures used here is only for the purpose of showing that the two methods of computation indeed give exactly the same results. In no way does it mean that such accuracy is needed in structural analysis for the purpose of providing values for design.

$$\text{Max. } M_B = R_A(10) - \frac{1}{2}(2.1)(10)^2$$

$$= \frac{0.6(33)(8.5) + 1.5(25)(12.5)}{25}(10) - 105$$

$$= +149.82 \text{ ft-kips} \quad \text{(check)}$$

Applying Eq. (6.4.2) to the loading pattern of Fig. 6.5.2g,

$$\text{Min. } M_B = 0.6A_1 + 2.1A_2$$

$$= 0.6(+75) + 2.1(-12.8) = +18.12 \text{ ft-kips*}$$

Check by directly computing the bending moment at B for the beam loaded as in Fig. 6.5.2g,

$$\text{Min. } M_B = R_A(10) - \frac{1}{2}(0.6)(10)^2$$

$$= \frac{0.6(33)(8.5) - 1.5(8)(4)}{25}(10) - 30$$

$$= +18.12 \text{ ft-kips} \quad \text{(check)}$$

The beam must be designed to resist a maximum positive bending moment of 149.8 ft-kips. The minimum value in this case is also positive. The designer has, however, learned that no negative moment can occur at point B. Had there been a reversal in sign between the maximum and minimum values of M_B, the design must provide for tension to occur at both top and bottom of the cross section at B, particularly of importance when the member is of reinforced concrete, a material having relatively low tensile strength.

6.6 INFLUENCE LINE AS A DEFLECTION DIAGRAM —MÜLLER-BRESLAU PRINCIPLE

The Müller-Breslau principle of considering the influence line as a deflection diagram is a valuable technique for visualizing the shape of the influence diagram. First, the technique is to be described, after which the proof is presented.

Support Reaction

The principle for the linear reaction at a support may be stated as follows:

1. Remove support and give a unit displacement at the reaction for which the influence line is desired.
2. The area between the original and final positions is the influence diagram.

The influence lines for several statically determinate reactions are shown in Fig. 6.6.1.

*Note that the number of significant figures used here is only for the purpose of showing that the two methods of computation indeed give exactly the same results. In no way does it mean that such accuracy is needed in structural analysis for the purpose of providing values for design.

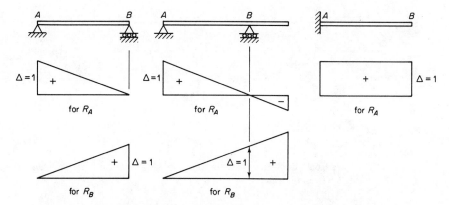

Figure 6.6.1 Influence lines for support reactions for various statically determinate beams—Müller-Breslau principle.

When the unit displacement is made, there are no loads on the beam; therefore, the only restraint to displacement arises from the remaining supports. Thus, when $\Delta = 1$ displacement is made, the beam as a rigid body rotates about the other knife-edge or hinged support, as for the simple and overhanging beams in Fig. 6.6.1. In the case of the cantilever of Fig. 6.6.1, the unit vertical displacement ($\Delta = 1$) is made; however, the moment restraint remains so that no rotation occurs and the deflection diagram is the horizontal beam remaining horizontal but displaced an amount $\Delta = 1$.

Shear

The Müller-Breslau principle applied to an influence line for shear at a section may be stated as follows:

1. Cut the beam at the section.
2. Open a unit displacement between the two sides of the cut (with the right cut end higher than the left cut end), without introducing relative rotation of the two sides of the cut.
3. The deflected shape that results is the influence diagram for shear at the cut location.

The influence lines for shear in several statically determinate situations are presented in Fig. 6.6.2.

For the simple and overhanging beams the second step described above requires rotations about the knife-edge, hinge, or roller supports. For the cantilever beam the segment with the fixed end has to stay in the original position and the segment (if it is the righthand segment) with the free end goes up one unit and remains in the horizontal position. For statically determinate situations these deflection diagrams consist of straight lines.

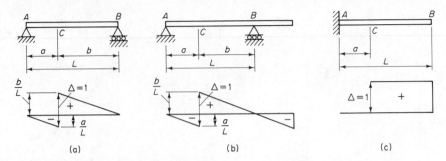

Figure 6.6.2 Influence lines for shear at C for statically determinate beams—Müller-Breslau principle.

Bending Moment

The Müller-Breslau principle applied to an influence line for bending moment at a section may be stated as follows:

1. Insert a hinge at the section.
2. Introduce a unit relative rotation of the beam at each side of the hinge.
3. The deflection diagram that results is the influence diagram.

The influence lines for bending moment in several statically determinate situations are presented in Fig. 6.6.3.

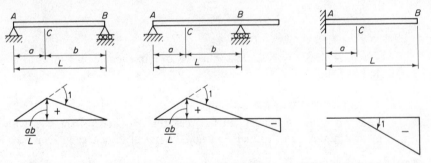

Figure 6.6.3 Influence lines for the bending moment at C for statically determinate beams—Müller-Breslau principle.

Proof for Müller-Breslau Principle

The proof is made using the influence line for bending moment.

The simple beam in Fig. 6.6.4 is cut into two parts AC and CB. The free-body diagrams for AC and CB are shown below the whole beam. When the segment AC moves as a rigid body into the position AC', the total virtual work (sum of products of force and displacement) done by all the forces acting on it must be zero, because the resultant of these forces is zero. Thus, for segment AC,

$$R_A(0) - V_C(CC') + M_C(\theta) = 0 \tag{6.6.1}$$

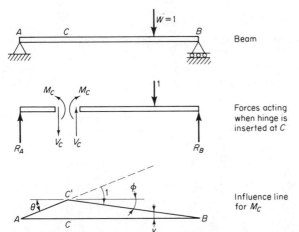

Figure 6.6.4 Müller-Breslau principle.

In the equation above, the virtual work of V_C is negative because V_C is downward while CC' is upward.

Similarly, the virtual work done by the forces on segment CB, when it moves from the original position to that of $C'B$, is

$$R_B(0) - 1.0(y) + V_C(CC') + M_C(\phi) = 0 \tag{6.6.2}$$

Adding Eqs. (6.6.1) and (6.6.2) and canceling out the virtual work of V_C,

$$M_C(\theta + \phi) - 1.0(y) = 0 \tag{6.6.3}$$

But

$$\theta + \phi = \text{unit angle} \tag{6.6.4}$$

Substituting Eq. (6.6.4) in Eq. (6.6.3),

$$M_C(\text{unit angle}) = (\text{unit load})y$$

$$M_C = y \tag{6.6.5}$$

Thus, the influence line ordinate is a deflection diagram. Similar proofs may be made for shear and reaction influence diagrams.

6.7 MAXIMUM BENDING MOMENT IN A SIMPLE BEAM DUE TO MOVING CONCENTRATED LOADS

When a series of concentrated loads passes over a simple beam in either direction, its critical position effecting the maximum bending moment at a chosen cross section may be obtained by trial and error with the aid of Eq. (6.4.1), but it may be more desirable to derive an algebraic criterion.

First, it can be shown that if the influence line is a straight-line segment, the value of a function due to a group of wheel loads is equal to the product of the influence ordinate at the location of the resultant force G, and the value of G itself. Referring to Fig. 6.7.1, and by definition of the resultant force,

$$G\bar{x} = W_1x_1 + W_2x_2 + W_3x_3 + W_4x_4 \tag{6.7.1}$$

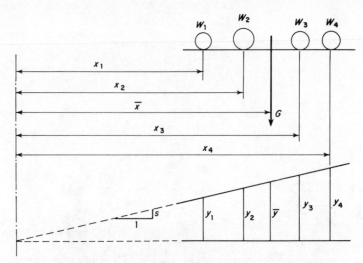

Figure 6.7.1 Influence line ordinate under resultant force.

Multiplying every term in Eq. (6.7.1) by the slope s of the influence line segment,

$$G s \bar{x} = W_1 s x_1 + W_2 s x_2 + W_3 s x_3 + W_4 s x_4 \tag{6.7.2}$$

which, using $y = sx$, gives

$$G \bar{y} = W_1 y_1 + W_2 y_2 + W_3 y_3 + W_4 y_4 = \sum W y \tag{6.7.3}$$

Now consider the primary wheel position of Fig. 6.7.2 to be advanced a distance Δx toward the left. The change in the bending moment at C is

$$\Delta M_C = G_1(-s_1 \, \Delta x) + G_2(+s_2 \, \Delta x) = -\frac{G_1 b}{L} \, \Delta x + \frac{G_2 a}{L} \, \Delta x$$

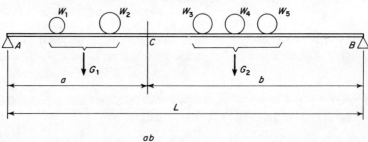

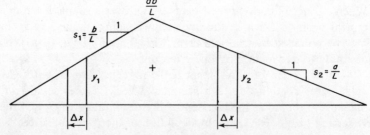

Figure 6.7.2 Influence line for bending moment at C.

                 Influence Lines and Movable Loads   Chap. 6

and substituting $G_2 = G - G_1$ gives

$$\Delta M_c = \left(\frac{Ga}{L} - G_1\right)\Delta x \qquad (6.7.4)$$

from which

$$\frac{dM_C}{dx} = \frac{Ga}{L} - G_1 \qquad (6.7.5)$$

Equation (6.7.5) is the loading position criterion for maximum bending moment at a section of a simple beam; it requires that the value of the expression $(Ga/L - G_1)$ change from positive to negative when a load placed at C is first considered to be at right of C and then at left of C.

Once the critical loading position has been obtained, the value of the maximum bending moment itself may be computed either by using Eq. (6.4.1), or by applying the conventional method for fixed position loads.

The position criterion, Eq. (6.7.5), has been derived for the load series as advancing onto the span from the right toward the left. In order to obtain the maximum bending moment at a section C due to passage of the load series from left to right, it is only necessary to determine the maximum bending moment at a section C' (symmetrical to section C with respect to the center line of the beam) due to passage of load from right to left. Employing this method one need not draw another sketch showing the load series in reverse order.

Example 6.7.1

Determine the maximum bending moment at a section 15 ft from the left support in a 40-ft span simple beam shown in Fig. 6.7.3a, due to the passage in either direction of the six concentrated loads shown in Fig. 6.7.3b.

Solution

(a) *Maximum M_C from wheels moving from right to left.* The position criterion, Eq. (6.7.5), requires that the value of the expression $(Ga/L - G_1)$ go from positive to negative when a wheel load placed at the section in question is first excluded from G_1 and then included in G_1. This dictates that the value of Ga/L must lie between the two extreme values of G_1. The procedure is to move one wheel at a time up to section C.

Wheel 1 at C:

$$G\frac{a}{L} = 50\frac{15}{40} = 18.75 \quad \text{(all six loads are on the beam)}$$

$$\frac{Ga}{L} - G_1 = 18.75 - 0 = +18.75 \quad \text{(wheel 1 right of C)}$$

$$\frac{Ga}{L} - G_1 = 18.75 - 5 = +13.75 \quad \text{(wheel 1 left of C)}$$

There is no change of sign as wheel 1 shifts from not being included in G_1 to being included; thus, the criterion is not satisfied.

Wheel 2 at C:

$$\frac{Ga}{L} - G_1 = 18.75 - 5 = +13.75 \quad \text{(wheel 2 right of C)}$$

$$\frac{Ga}{L} - G_1 = 18.75 - 13 = +5.75 \quad \text{(wheel 2 left of C)}$$

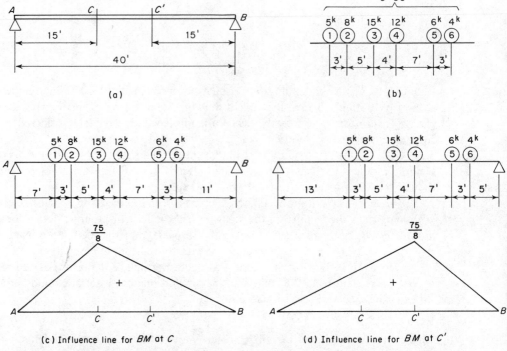

Figure 6.7.3 Maximum bending moment at a section of a simple beam.

There is no change in sign as wheel 2 shifts from not being included in G_1 to being included; thus, the criterion is not satisfied.

Wheel 3 at C:

$$\frac{Ga}{L} - G_1 = 18.75 - 13 = +5.75 \quad \text{(wheel 3 right of } C)$$

$$\frac{Ga}{L} - G_1 = 18.75 - 28 = -9.25 \quad \text{(wheel 3 left of } C)$$

Criterion is satisfied since $Ga/L - G_1$ changes sign as wheel 3 shifts from right of C to left of C.

Should a wheel move off the span, that wheel must be excluded from G. Sometimes as a wheel moves the infinitesimal amount from just right of C to just left of C, a wheel may just move off the left end (or onto the beam at the right end) of the span. In such cases, consistent values of G and G_1 must be used in the evaluation of the expression, $Ga/L - G_1$.

Sometimes the criterion may be satisfied by more than one wheel position, especially when loads get on and off the span as they advance from right to left. Thus, it is desirable to make sure that further advancement will cause M_C to decrease, that is, the value of $Ga/L - G_1$ is negative whether the load at C is excluded from or included in G_1. For the present case, with wheel 4 at C,

$$\frac{Ga}{L} - G_1 = 18.75 - 28 = -9.25 \quad \text{(wheel 4 right of } C\text{)}$$

$$\frac{Ga}{L} - G_1 = 18.75 - 40 = -21.25 \quad \text{(wheel 4 left of } C\text{)}$$

Thus, dM_C/dx is between -9.25 and -21.25; there is no chance for it to become zero by considering only a part of wheel 4 inside the segment AC.

In this example the maximum M_C occurs with wheel 3 at C, and the value of M_C may be computed using influence line ordinates, as follows,

$$\text{Max. } M_C = 5\left(\frac{7}{15}\right)\left(\frac{75}{8}\right) + 8\left(\frac{10}{15}\right)\left(\frac{75}{8}\right) + 15\left(\frac{75}{8}\right) + 12\left(\frac{21}{25}\right)\left(\frac{75}{8}\right) + 6\left(\frac{14}{25}\right)\left(\frac{75}{8}\right)$$
$$+ 4\left(\frac{11}{25}\right)\left(\frac{75}{8}\right)$$
$$= 21.875 + 50.000 + 140.625 + 94.500 + 31.500 + 16.500 = 355 \text{ ft-kips}$$

By correctly positioning the loads as shown in Fig. 6.7.3c and directly computing the value of M_C without using influence line ordinates,

$$\text{Max. } M_C = R_A(15) - 5(8) - 8(5)$$
$$= \frac{4(11) + 6(14) + 12(21) + 15(25) + 8(30) + 5(33)}{40}(15) - 80$$
$$= 29(15) - 80 = 355 \text{ ft-kips} \quad \text{(check)}$$

(b) *Maximum M_C from wheels moving left to right.* For maximum bending moment at a section C' 15 ft from the right support, the position criterion, Eq. (6.7.5), is satisfied by placing wheel 4 at C', shown by the following tabulation (see Fig. 6.7.3d):

Wheel at C'	G_1 on AC' (excluding wheel at C')	$\frac{Ga}{L} = \frac{G(25)}{40} = \frac{5}{8}G$	G_1 on AC' (including wheel at C')	Criterion satisfied?
No. 3	13	31.25	28	No
No. 4	28	31.25	40	Yes
No. 5	40	31.25	46	No

Applying Eq. (6.4.1) to the influence line of Fig. 6.7.3d,

$$\text{Max. } M_{C'} = 5\left(\frac{13}{25}\right)\left(\frac{75}{8}\right) + 8\left(\frac{16}{25}\right)\left(\frac{75}{8}\right) + 15\left(\frac{21}{25}\right)\left(\frac{75}{8}\right) + 12\left(\frac{75}{8}\right) + 6\left(\frac{8}{15}\right)\left(\frac{75}{8}\right)$$
$$+ 4\left(\frac{5}{15}\right)\left(\frac{75}{8}\right)$$
$$= 24.375 + 48.000 + 118.125 + 112.500 + 30.000 + 12.500$$
$$= 345.5 \text{ ft-kips}$$

By the conventional method for fixed position loads,

$$\text{Max. } M_{C'} = R_A(25) - 5(12) - 8(9) - 15(4)$$

$$= \frac{4(5) + 6(8) + 12(15) + 15(19) + 8(24) + 5(27)}{40}(25) - 192$$

$$= 21.5(25) - 192 = 345.5 \text{ ft-kips} \quad \text{(check)}$$

Thus, the maximum bending moment at C is 355 ft-kips due to the passage of the load series from right to left; the maximum bending moment at C' is also 355 ft-kips, but it is due to the passage of the load series from left to right.

6.8 ABSOLUTE MAXIMUM BENDING MOMENT IN A SIMPLE BEAM

In the preceding section, a method has been presented for obtaining the maximum bending moment at a chosen location along the span of a simple beam due to a series of fixed distance apart moving concentrated loads. Suppose now that the beam is to be of uniform cross section and the largest bending moment which may happen anywhere in the beam is to be determined. One might assume offhand that the appropriate location to make this computation would be midspan. This *may* give a result that is close to the absolute maximum bending moment occurring due to the passage of the series of loads. It can be shown, however, that the exact absolute maximum bending moment occurs *at a wheel load* when that wheel load and the center of gravity (i.e., location of resultant) of the series of loads are equidistant from the ends of the span.

Consider the problem of finding the *position* of wheel 3 in Fig. 6.8.1 so that the bending moment under it is maximum. Using the loading of Fig. 6.8.1, the bending moment at wheel 3 is

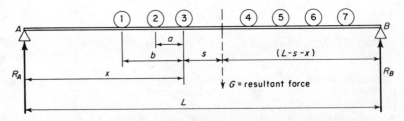

Figure 6.8.1 Absolute maximum bending moment in a simple beam.

$$M = R_A x - (\text{moment of wheels 1 and 2 about wheel 3})$$

$$= \frac{G(L - s - x)}{L}x - W_1 b - W_2 a \qquad (6.8.1)$$

Differentiating Eq. (6.8.1) with respect to x,

$$\frac{dM}{dx} = \frac{G}{L}(L - s - 2x) \qquad (6.8.2)$$

Setting Eq. (6.8.2) equal to zero gives

$$x = \frac{L - s}{2} \tag{6.8.3}$$

which makes

$$L - s - x = \frac{L - s}{2} \tag{6.8.4}$$

Equations (6.8.3) and (6.8.4) show that wheel 3 and the resultant force G should be equidistant from the ends of the span.

To determine the absolute maximum bending moment in a simple beam due to a series of moving concentrated loads, the maximum bending moment at the midspan of that beam should be determined first. Then the absolute maximum bending moment should occur under the wheel which had been placed at the midspan to yield maximum bending moment there. Of course, this wheel and the resultant force must take the position defined by Eq. (6.8.3) or (6.8.4).

Example 6.8.1

Determine the maximum bending moment at the center of the 40-ft simple beam shown in Fig. 6.8.2a, as well as the absolute maximum bending moment in the beam due to the passage of the six concentrated loads shown in Fig. 6.8.2b.

Solution

(a) *Critical wheel at midspan.* For maximum bending moment at midspan, the position criterion, Eq. (6.7.5), is satisfied by placing wheel 3 at midspan as shown by the following tabulation:

Wheel at C	G_1 on AC (excluding wheel at C)	$\frac{Ga}{L} = \frac{1}{2}G$	G_1 on AC (including wheel at C)	Criterion satisfied?
No. 2	5	25	13	No
No. 3	13	25	28	Yes
No. 4	28	25	40	No

(b) *Maximum moment at midspan.* Using the influence line (Fig. 6.8.2c) ordinates to compute M_C,

$$\text{Max. } M_C = 5\left(\frac{12}{20}\right)(10) + 8\left(\frac{15}{20}\right)(10) + 15(10) + 12\left(\frac{16}{20}\right)(10)$$

$$+ 6\left(\frac{9}{20}\right)(10) + 4\left(\frac{6}{20}\right)(10)$$

$$= 30 + 60 + 150 + 96 + 27 + 12 = 375 \text{ ft-kips}$$

Directly computing M_C for the loads positioned as in Fig. 6.8.2c without using values from the influence line,

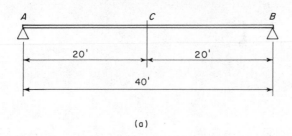

(a)

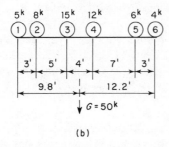

(b)

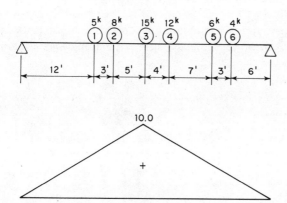

(c) Influence line for moment at midspan

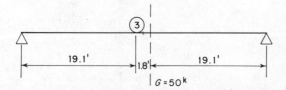

(d) Load position for absolute maximum bending moment

Figure 6.8.2 Absolute maximum bending moment in the simply supported beam of Example 6.8.1.

$$\text{Max. } M_C = R_A(20) - 5(8) - 8(5)$$

$$= \frac{4(6) + 6(9) + 12(16) + 15(20) + 8(25) + 5(28)}{40}(20) - 80$$

$$= 455 - 80 = 375 \text{ ft-kips} \quad (\text{check})$$

(c) *Location of wheel 3 to give absolute maximum bending moment.* From Fig. 6.8.2b, the resultant force G is at a distance of

$$\frac{6(3) + 12(10) + 15(14) + 8(19) + 5(22)}{50} = 12.2 \text{ ft}$$

from wheel 6. The distance s between wheel 3 and $G = 50$ kips is 1.8 ft; thus,

$$\frac{L - s}{2} = \frac{40 - 1.8}{2} = 19.1 \text{ ft}$$

(d) *Magnitude of absolute maximum bending moment.* From Fig. 6.8.2d, the absolute maximum bending moment at wheel 3 is

$$\text{Absolute max. moment} = \frac{50(19.1)^2}{40} - 8(5) - 5(8)$$

$$= 456.0125 - 80 = 376.0125 \text{ ft-kips}$$

It should be noted that the absolute maximum bending moment of 376.0125 ft-kips is only slightly larger than the maximum bending moment of 375 ft-kips at midspan. This closeness happens frequently, although not always. It depends to a large extent on the spacing of the wheels in the series. There is more divergence between the absolute maximum bending moment and the maximum moment at midspan when there are only two or three wheels in the series at relatively large spacings apart. This is the case of vehicle loadings on highway bridges.

6.9 BENDING MOMENT ENVELOPE AND SHEAR ENVELOPE

So far, emphasis has been placed on constructing the influence lines for shear and bending moment at preselected locations, and then on computing the maximum and minimum values of the function due to given concentrated loads and variable-length uniform loading.

The designing engineer must have the maximum and minimum values of the function not only at one point but all along the entire span. An *envelope* is defined as the curve which gives the extreme (upper and lower bounds) values for the function at each point along the span.

Bending Moment Envelope

Referring to the beam of Fig. 6.9.1, the influence line for bending moment at a location x from the left end is shown qualitatively. Whatever the value of x between A and B the positive portion of the influence diagram extends from A to B. This shows that to obtain maximum positive moment at *every point between A and B*, uniform live load must be placed over the entire span AB (all of the positive region of the influence line) and none of it on the cantilever (the negative portion of the influence line). Dead load of course is wherever it occurs—it is fixed-position load. For the minimum positive effect (conversely, maximum negative effect) only the cantilever is loaded with live load. When both extremes are plotted to show the range of bending moment values, the *moment envelope* as shown in Fig. 6.9.1 is obtained.

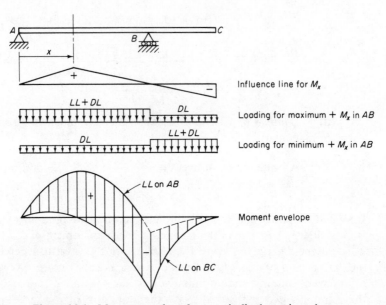

Figure 6.9.1 Moment envelope for a statically determinate beam.

If the influence line for bending moment at any point in segment BC is drawn, it will show that loading to the right of the chosen point will cause negative bending moment at the point, while loading to the left of the point has no effect. Of course, the same conclusion can be reached without the use of influence line by thinking through the idea that the bending moment at any point on BC can only be negative (for downward loading) and can only be caused by load between that point and the free end of the cantilever.

One of the most common situations where the moment envelope is needed is in reinforced concrete design. Since concrete without steel reinforcement has little tensile capacity, the envelope tells the designer that tension can occur in the bottom of the beam over the distance where the envelope has positive values, that is, nearly all of span AB in Fig. 6.9.1. There is tension in the top of the beam over the length where the envelope has negative values, that is, all of BC and a good part inside BA in Fig. 6.9.1.

Since, in general, live load is always considered to have variable length and must be positioned for design where it will cause the most severe effect, the structural engineer must determine where to put such load by making use of the knowledge of influence lines.

Shear Envelope

In contrast to the moment envelope where usually the entire span is loaded to give maximum effect, the maximum shear at a point x along a simply supported span (Fig. 6.9.2a) is obtained by using partial span loading. The influence line of Fig.

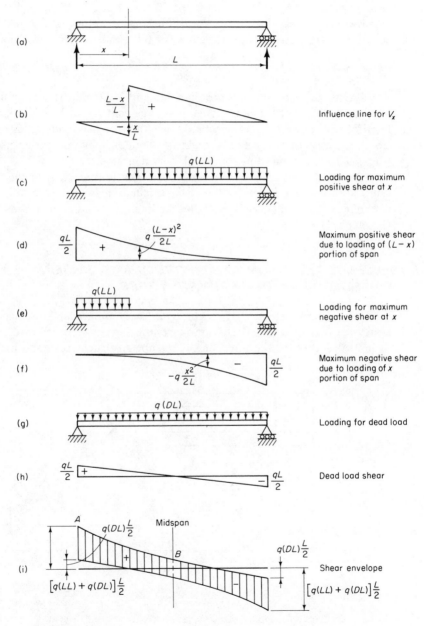

Figure 6.9.2 Shear envelope for uniform live load on a simply supported beam.

6.9.2b shows that the live load must be placed over the distance $(L - x)$ to the right of the section for which the influence line is drawn in order to obtain maximum positive shear, and over the segment of length x to the left of the section to obtain maximum negative shear. In Fig. 6.9.2c to f are shown the loading diagrams and maximum value equations for both positive and negative live load shear. The dead load shear shown in Fig. 6.9.2g and h is due to full span loading, as it has to be always present. When the maximum positive and negative live load shears are combined with the shear due to dead load, the shear envelope of Fig. 6.9.2i results. The equation of the positive shear envelope for this simply supported beam is

$$V_{max} = q(LL)\frac{(L - x)^2}{2L} + q(DL)\frac{L}{2} - q(DL)x$$

For design purposes, it is the absolute value of shear that is usually used; thus, the maximum shear to be used in design is the segment AB of Fig. 6.9.2i, and it is symmetrical with respect to the midspan. Once the true maximum shear curve is understood, the designer may approximate it, commonly by computing points A and B (Fig. 6.9.2i) and connecting them with a straight line.

Moment and shear envelopes are particularly important in the design of continuous beams and frames, where their calculation involves multiple loading cases determined from influence lines. Influence lines for continuous beams is a subject of Chapter 8.

6.10 INFLUENCE LINES FOR AXIAL FORCES IN MEMBERS OF A STATICALLY DETERMINATE TRUSS

Statically determinate trusses, such as those in industrial buildings and offshore structures, may need to be designed to carry moving concentrated loads. The treatment of influence lines for the forces in the members of such trusses is presented here to

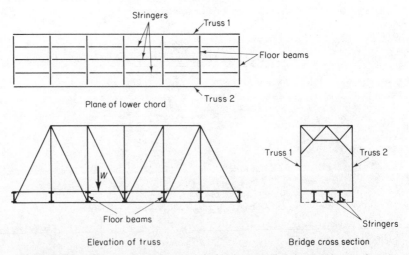

Figure 6.10.1 Floor system of truss such that intermediate load W is transferred to truss panel points.

enable the designer to understand the effect on a given member from a load placed at an arbitrary panel point. Further, the correct visualization of a qualitative influence line helps to indicate the proper loading positions for maximum and minimum effect.

Wherever trusses are used, continuous, three-dimensional, or simple statically determinate, the structural arrangement is such that loads are transmitted into the truss at the panel points. Commonly, for bridge trusses longitudinal stringers parallel to the chord members (see Fig. 6.10.1) carry the loads to floor beams. These floor beams span between trusses at the sides of the roadway, and they are connected to the trusses at the panel points. Thus, when a concentrated load is shown in a side view of the structure as acting between panel points, actually the simple beam reactions of that concentrated load act at the two adjacent panel points of the truss.

6.11 INFLUENCE LINE FOR SHEAR IN A PANEL OF A PARALLEL CHORD TRUSS

Since web members (diagonals and verticals) of a truss primarily carry shear, the influence line for shear in a truss panel is a prerequisite for obtaining the influence line for axial force in a web member. Referring to Fig. 6.11.1, the influence line for shear in a panel of a parallel chord truss is identical to the influence line for shear at a section on a beam, except for the portion within the panel. Loads within a panel are carried at the adjacent panel points.

Examine the shape of the influence line between panel points. Take any influence line for a statically determinate system (say, for example, shear in a panel) as in Fig. 6.11.2. A unit load at panel point 2 causes a shear of y_2 and a unit load at panel point 3 causes a shear of y_3. If the load W acts on the stringer at an intermediate location within panel 2–3, that load W has reactions Wb/p and Wa/p at panel points 2 and 3, respectively. The shear in the panel then becomes

$$Wy = \frac{Wb}{p}y_2 + \frac{Wa}{p}y_3 \tag{6.11.1}$$

$$y = y_3 + (y_2 - y_3)\frac{b}{p} \tag{6.11.2}$$

which is the equation of a straight line. Thus, for trusses where the reactions from loads in locations between panel points are carried to the panel points by simply supported stringers, the influence lines are straight lines between panel points.

Returning to Fig. 6.11.1, the influence line for shear in panel 2–3 is established by determining the ordinates at panel points 2 and 3 and connecting by a straight line. From the geometry one may note that point i divides the entire span L in the same ratio as it divides panel 2–3. For instance, in Fig. 6.11.1, the initial influence ordinates are in the ratio of 2 to 3; so are the two short parts of the panel length between L_2 and L_3; so are the distances L_0L_2 and L_3L_6; and finally, so are the distances L_0i and iL_6.

To convert from shear in a panel to force in a web member, cut a section through panel 2–3 noting that positive shear puts a pull on member U_2L_3 and thus puts tension

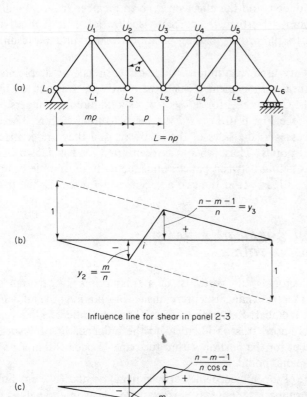

(a)

mp p

$L = np$

(b)

$\dfrac{n-m-1}{n} = y_3$

$+$

$y_2 = \dfrac{m}{n}$

Influence line for shear in panel 2-3

(c)

$\dfrac{n-m-1}{n\cos\alpha}$

$+$

$\dfrac{m}{n\cos\alpha}$

Influence line for force in member U_2L_3

Figure 6.11.1 Influence lines for shear and force in web member of a statically determinate truss.

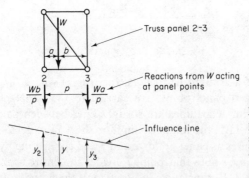

Truss panel 2-3

Reactions from W acting at panel points

$\dfrac{Wb}{p}$ p $\dfrac{Wa}{p}$

Influence line

y_2 y y_3

Figure 6.11.2 Influence line between panel points 2 and 3 of a truss is a straight line when stringers transmit loads as simple beam reactions to panel points.

in the member. The vertical component, $U_2L_3 \cos\alpha$, equals the panel shear. Thus, the influence line for the force in member U_2L_3 equals the influence line for shear divided by $\cos\alpha$. Positive sign indicates tension in the member.

6.12 INFLUENCE LINE FOR BENDING MOMENT AT A PANEL POINT IN THE LOADED CHORD OF A TRUSS

The influence line for bending moment at a panel point is identical to the influence line for bending moment at a location in a simple beam. This information is of interest because the internal force in a chord member is a function of the bending moment at a panel point (see Method of Moments and Shears, Section 2.4). The influence line for bending moment at panel point 2 is shown in Fig. 6.12.1. The bending moment M_2 may be expressed as follows:

$$M_2 = R_R b = \frac{x}{L} b \qquad \text{for } 0 \le x \le a$$

$$M_2 = R_L a = \frac{L - x}{L} a \qquad \text{for } a \le x \le L$$

This is identical to the influence line for bending moment in a simple beam (Section 6.5). From such an influence line for bending moment at a panel point, the internal force in a chord member is obtained by direct proportion. For instance, the internal force in $U_2 U_3$ is obtained by cutting a section through panel 2–3; then

$$-U_2 U_3 (h) = M_2$$

$$U_2 U_3 = -\frac{M_2}{h}$$

Thus, the influence line for the internal force in $U_2 U_3$ is the influence line for bending moment M_2 (Fig. 6.12.1) multiplied by $(-1/h)$. Positive bending moment causes compression in upper chord members.

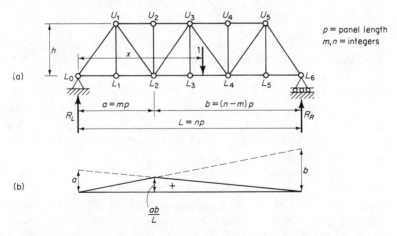

Figure 6.12.1 Influence line for bending moment at panel point 2 in the loaded chord of a truss.

Example 6.12.1

For the truss of Fig. 6.12.2, construct the influence lines for internal forces in members U_2L_3, U_2L_2, U_2U_3, and U_1L_1.

Solution

(a) *Influence line for force in U_2L_3.* The force in this web member is proportional to the shear in panel 2–3 of Fig. 6.12.2a. Influence lines for reactions R_L and R_R are shown in Fig. 6.12.2b. For the influence line for shear in panel 2–3, note that when the unit load is at panel points 1 and 2, $V_{2-3} = -R_R$, and when the unit load is at panel points 3 through 7, $V_{2-3} = +R_L$. Thus, the influence line for shear in panel 2–3 shown in Fig. 6.12.2c is obtained by simply connecting the influence ordinate at panel point 2 on the R_R influence line (positive below the horizontal base line) with the influence ordinate at panel point 3 on the R_L influence line.

Cutting section ①–① shows that

$$V_{2-3} = \text{vertical component of } U_2L_3 = U_2L_3\left(\frac{h}{d}\right)$$

Then

$$U_2L_3 = V_{2-3}\left(\frac{d}{h}\right)$$

The influence line for internal force in U_2L_3 is given in Fig. 6.12.2d.

(b) *Influence line for force in U_2L_2.* Cutting section ②–②, one may see that force U_2L_2 equals the shear V_{2-3} in magnitude but acts in compression when the shear is positive. Thus,

$$U_2L_2 = -V_{2-3}$$

and the influence line for force in U_2L_2 is given as Fig. 6.12.2e.

(c) *Influence line for force in U_2U_3.* Cutting a section ①–① and taking moments about L_3 indicates

$$-U_2U_3(h) = \text{bending moment at } L_3$$

$$U_2U_3 = \frac{-\text{bending moment at } L_3}{h}$$

The influence line for bending moment at L_3 is given in Fig. 6.12.2f, and the influence line for force in U_2U_3 is in Fig. 6.12.2g. Positive bending moment causes compression (−) in the top chord.

(d) *Influence line for force in U_1L_1.* The method of joints shows that U_1L_1 has an internal force in it only when the unit load is located within panels 0–1 and 1–2. When the unit load is at L_1, the internal force U_1L_1 equals unity. The influence line is shown as Fig. 6.12.2h.

For unusual trusses having subdivided panels (see, for instance, Prob. 6.37) and for curved chord trusses, a direct approach of computing the influence line ordinates for the unit load at the panel points adjacent to the panel in question, and then connecting the critical ordinates by straight lines, is the most expeditious method. If only qualitative influence lines are needed, then the qualitative influence lines for shear or bending moment in a panel can be used to advantage for those members whose internal forces are proportional to shear or bending moment.

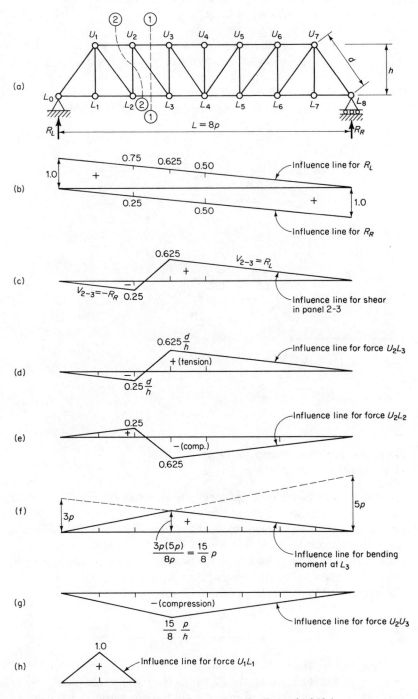

Figure 6.12.2 Influence lines for Example 6.12.1.

6.13 MAXIMUM AXIAL FORCE IN A WEB MEMBER OF A PARALLEL CHORD TRUSS

Criteria for determining the position of a series of loads to give maximum axial force in a member may be developed for any shaped influence diagram. Truss web members, and some verticals, have internal forces proportional to the shear in a panel, where the influence line is as shown in Fig. 6.13.1 for parallel chord trusses. This is one of the more common shapes for influence lines relating to trusses.

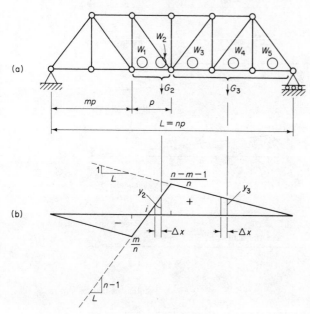

Figure 6.13.1 Development of criterion for maximum shear in a panel of a parallel chord truss.

Criterion for Maximum Shear in a Panel of a Parallel Chord Truss

For maximum positive shear in panel p of the truss of Fig. 6.13.1, as many loads as possible must be placed between point i and the right end of the span. No load should pass point i unless the increase in shear due to increase in left reaction by loads moving forward is greater than the decrease in shear due to increase in panel point load at panel point to left of point i by loads moving to the left within the panel.

The slope s_3 of the segment of the influence line to the right of the panel is $1/L$. The slope s_2 of the segment within the panel is

$$s_2 = \frac{\dfrac{n-m-1}{n} + \dfrac{m}{n}}{p} = \frac{\dfrac{n-1}{n}}{L/n} = \frac{n-1}{L} \tag{6.13.1}$$

Letting G_2 equal the resultant of the loads within the panel and G_3 equal the resultant of the loads on all the panels to the right of the panel in question. At any one trial position of the moving loads, the positive shear in the panel is

$$V = G_2 y_2 + G_3 y_3 \tag{6.13.2}$$

As the wheels advance Δx, there will be a change in shear ΔV,

$$\Delta V = G_3 s_3 \, \Delta x - G_2 s_2 \, \Delta x \tag{6.13.3}$$

Substituting $s_3 = 1/L$ and the expression for s_2 of Eq. (6.13.1) in the equation above,

$$\frac{\Delta V}{\Delta x} = G_3\left(\frac{1}{L}\right) - G_2\left(\frac{n-1}{L}\right) \tag{6.13.4}$$

For the usual case with no loads over the length mp, let $G = G_2 + G_3$. Noting that $L = np$, Eq. (6.13.4) becomes

$$\frac{\Delta V}{\Delta x} = \frac{G - G_2}{np} - \frac{G_2(n-1)}{np} = \frac{1}{p}\left(\frac{G}{n} - G_2\right) \tag{6.13.5}$$

Since p is a constant, the criterion may be stated

$$\frac{G}{n} - G_2 = \gtrless 0 \tag{6.13.6}$$

For the criterion of Eq. (6.13.6) to be satisfied, it must change sign as a given wheel is shifted an infinitesimal amount from the segment having slope $1/L$ to the segment having the slope $(n-1)/L$. This requires a wheel load to be placed at the panel point with the peak influence ordinate so that when this load is excluded from G_2, the criterion expression is positive and when this load is included in G_2, the criterion expression becomes negative.

Obviously, when only uniform load (or a few concentrated loads) is involved, it is unnecessary to use the criterion, Eq. (6.13.6), to determine the correct position of loads for maximum tension or maximum compression in a given member. Inspection of the influence line gives the answer directly. When a series of concentrated loads must be used, the criterion may be found useful.

Example 6.13.1

For the loading in Fig. 6.13.2b acting on the truss of Fig. 6.13.2a, determine the maximum compressive force and the maximum tensile force in member U_2L_3.

Solution

(a) *Construct influence line for shear in panel 2–3.* This influence line is shown in Fig. 6.13.2c.

(b) *Determine location of moving loads for maximum positive shear in panel 2–3.* Using the criterion given by Eq. (6.13.6), try load 2 at L_3:

For load 2 right of L_3,

$$G = 45 + 1.2(54 - 20) = 85.8$$

$$\frac{G}{n} - G_2 = \frac{85.8}{6} - 5 = +9.3$$

For load 2 left of L_3,

$$\frac{G}{n} - G_2 = \frac{85.8}{6} - 15 = -0.7$$

Since the sign of the criterion changes from $+$ to $-$, load 2 at L_3 gives the maximum positive effect.

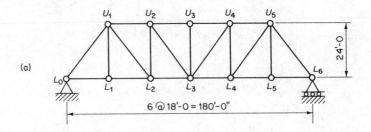

(a)

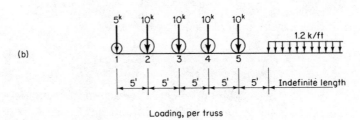

(b)

Loading, per truss

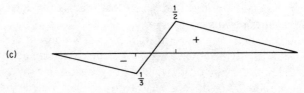

(c)

Influence line for shear in panel 2-3

Figure 6.13.2 Truss and loading for Example 6.13.1.

(c) *Determine maximum tensile force in member* U_2L_3. Cutting a section through panel 2–3 gives

$$U_2L_3\left(\frac{24}{30}\right) = V_{2-3}$$

$$U_2L_3 = V_{2-3}\left(\frac{30}{24}\right)$$

Placing the loads with load 2 at L_3 gives

$$V_{2-3} = 5\left[\frac{13}{18}(0.833) - 0.333\right] + 10(0.5)\left(1 + \frac{49}{54} + \frac{44}{54} + \frac{39}{54}\right)$$

$$+ 1.2\left[\frac{34}{54}(0.5)(0.5)(34)\right]$$

$$= 24.99 \text{ kips}$$

$$U_2L_3 = 24.99\left(\frac{30}{24}\right) = 31.23 \text{ kips} \quad \text{(tension)}$$

Without using the influence line ordinates, the shear in the panel can be computed by

$$V_{2-3} = R_L - (\text{panel point load at } L_2)$$

Taking moments about the right end,

$$R_L = \frac{0.5(1.2)(34)^2 + 10(39 + 44 + 49 + 54) + 5(59)}{108} = 26.38 \text{ kips}$$

$$\text{Panel point load at } L_2 = \frac{5(5)}{18} = 1.39 \text{ kips}$$

Thus,

$$V_{2-3} = 26.38 - 1.39 = 24.99 \text{ kips} \quad \text{(check)}$$

(d) *Determine location of moving loads for maximum negative shear in panel 2–3.* Reverse the loads to move along the negative portion of the influence line. Since the slopes involved are the same as for the positive portion of the influence line, the criterion for maximum is the same. Thus, try load 2 at L_2:

For load 2 left of L_2,

$$G = 45 + 1.2(36 - 20) = 64.2$$

$$\frac{G}{n} - G_2 = \frac{64.2}{6} - 5 = +5.7$$

For load 2 right of L_2,

$$\frac{G}{n} - G_2 = \frac{64.2}{6} - 15 = -4.3$$

Thus, maximum negative shear occurs with load 2 at L_2 with loads moving from left to right.

(e) *Determine maximum compressive force in member U_2L_3.* As in part (c),

$$U_2L_3 = V_{2-3}\left(\frac{30}{24}\right)$$

Placing the loads with load 2 at L_2 and using the influence line ordinates to make the computation gives

$$V_{2-3} = 5\left[\frac{5}{18}(0.833) - 0.333\right] + 10(-0.333)\left(1 + \frac{31}{36} + \frac{26}{36} + \frac{21}{36}\right)$$

$$+ 1.2\left[\frac{16}{36}(-0.333)(0.5)(16)\right]$$

$$= -12.49 \text{ kips}$$

$$U_2L_3 = (-12.49)\frac{30}{24} = -15.61 \text{ kips} \quad \text{(negative means compression)}$$

Without using the influence line ordinates, the shear in the panel can be computed by

$$V_{2-3} = -(R_R - \text{panel point load at } L_3)$$

Taking moments about the left end,

$$R_R = \frac{0.5(1.2)(16)^2 + 10(21 + 26 + 31 + 36) + 5(41)}{108} = 13.88 \text{ kips}$$

$$\text{Panel point load at } L_3 = \frac{5(5)}{18} = 1.39 \text{ kips}$$

$$V_{2-3} = -(13.88 - 1.39) = -12.49 \text{ kips} \quad \text{(check)}$$

6.14 MAXIMUM AXIAL FORCE IN A CHORD MEMBER OF A PARALLEL CHORD TRUSS

The influence line for the axial force in a chord member has the same shape as that for the bending moment in a simple beam, as has been discussed in Section 6.12. Thus the criterion for maximum axial force in a chord member is identical with that used for maximum bending moment in a beam, Eq. (6.7.5).

Example 6.14.1

Determine the maximum compressive force in member U_4U_5 of the truss of Fig. 6.13.2a due to the moving loads of Fig. 6.13.2b.

Solution

(a) *Determine the influence line for force in member U_4U_5.* Cutting a section through panel 4–5 shows that the force in U_4U_5 is proportional to the bending moment at L_4.

$$-U_4U_5(24) = \text{bending moment at } L_4$$

$$U_4U_5 = -\frac{\text{bending moment at } L_4}{24}$$

The influence line ordinate for bending moment at L_4 equals

$$M = R_L(72) = \frac{1}{3}(72) = 24 \text{ ft-kips/kip}$$

which when converted into an influence line ordinate for the force in U_4U_5 gives

$$U_4U_5 = -\frac{M}{24} = -\frac{24}{24} = -1.00 \text{ kips/kip}$$

The influence line for the force in U_4U_5 is shown in Fig. 6.14.1c.

(b) *Determine the position of the loads to cause the maximum force in member U_4U_5.* The average load per unit distance under the concentrated loads is 2 kips/ft. Therefore, these loads should be placed near the peak influence ordinate, and then as much as possible of the uniform load of indefinite length should be placed on the span. For the reasons stated above, the maximum is expected to occur due to passage of the loads from left to right. Check the criterion, Eq. (6.7.5), to determine at which load position $Ga/L - G_1$ changes sign. Note that it is more convenient to consider G_1 in the distance L_4L_6 because the loads are going from left to right.

Try load 2 at L_4:

Load 2 just left of L_4 gives

$$G = 45 + 1.2(72 - 20) = 107.4$$

$$\frac{Ga}{L} - G_1 = \frac{107.4(36)}{108} - 5 = +30.8$$

Load 2 just right of L_4 gives

$$\frac{Ga}{L} - G_1 = \frac{107.4(36)}{108} - 15 = +20.8$$

The criterion is not satisfied.

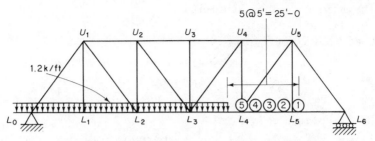

(a) Loading position for maximum effect, loads moving left to right

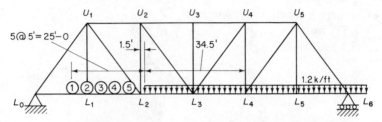

(b) Loading position for maximum effect, loads moving right to left

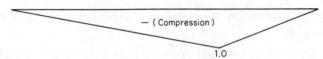

(c) Influence line for force in member $U_4 U_5$

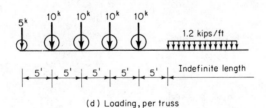

(d) Loading, per truss

Figure 6.14.1 Maximum compressive force in chord member $U_4 U_5$, Example 6.14.1.

Try load 3 at L_4:
Load 3 just left of L_4 gives

$$G = 45 + 1.2(72 - 15) = 113.4$$

$$\frac{Ga}{L} - G_1 = \frac{113.4(36)}{108} - 15 = +22.8$$

Load 3 just right of L_4 gives

$$\frac{Ga}{L} - G_1 = \frac{113.4(36)}{108} - 25 = +12.8$$

The criterion is still not satisfied.

Try load 4 at L_4:

Load 4 just left of L_4 gives

$$G = 45 + 1.2(72 - 10) = 119.4$$

$$\frac{Ga}{L} - G_1 = \frac{119.4(36)}{108} - 25 = +14.8$$

Load 4 just right of L_4 gives

$$\frac{Ga}{L} - G_1 = \frac{119.4(36)}{108} - 35 = +4.8$$

The criterion is still not satisfied.

Try load 5 at L_4:

Load 5 just left of L_4 gives

$$G = 45 + 1.2(72 - 5) = 125.4$$

$$\frac{Ga}{L} - G_1 = \frac{125.4(36)}{108} - 35 = +6.8$$

Load 5 just right of L_4 gives

$$\frac{Ga}{L} - G_1 = \frac{125.4(36)}{108} - 45 = -3.2$$

The criterion is satisfied with load 5 at L_4 (see Fig. 6.14.1a) for passage of loads from left to right.

Check wheel 3 at L_4 for passage from right to left: Load 3 just right of L_4 gives

$$G = 45 + 1.2(36 - 15) = 70.2$$

$$\frac{Ga}{L} - G_1 = \frac{70.2(72)}{108} - 15 = +31.8$$

Load 3 just right of L_4 gives

$$\frac{Ga}{L} - G_1 = \frac{70.2(72)}{108} - 25 = +21.8$$

The criterion is not satisfied.

Try load 4 at L_4:

Load 4 just right of L_4 gives

$$G = 45 + 1.2(36 - 10) = 76.2$$

$$\frac{Ga}{L} - G_1 = \frac{76.2(72)}{108} - 25 = +25.8$$

Load 4 just left of L_4 gives

$$\frac{Ga}{L} - G_1 = \frac{76.2(72)}{108} - 35 = +15.8$$

The criterion is not satisfied.

Try load 5 at L_4:
Load 5 just right of L_4 gives

$$G = 45 + 1.2(36 - 5) = 82.2$$

$$\frac{Ga}{L} - G_1 = \frac{82.2(72)}{108} - 35 = +19.8$$

Load 5 just left of L_4 gives

$$\frac{Ga}{L} - G_1 = \frac{82.2(72)}{108} - 45 = +9.8$$

The criterion is not satisfied.

Let the uniform load pass L_4 by a distance x. Then

$$G = 45 + 1.2(x + 36) = 88.2 + 1.2x$$

$$G_1 = 45 + 1.2x$$

$$\frac{Ga}{L} - G_1 = \frac{(88.2 + 1.2x)(72)}{108} - (45 + 1.2x) = 0$$

Solving the equation above,

$$x = 34.5 \text{ ft}$$

The position of loads giving maximum effect is shown in Fig. 6.14.1b.

(c) *Maximum force in member U_4U_5.* Using the influence line ordinates, and taking passage of loads from left to right (load 5 at L_4),

$$U_4U_5 = 10(-1.0) + 10\left(-\frac{31}{36} - \frac{26}{36} - \frac{21}{36}\right) + 5\left(-\frac{16}{36}\right) + 1.2(0.5)(-1.0)\left(\frac{67}{72}\right)(67)$$

$$= -71.30 \text{ kips (negative means compression)}$$

Without the use of influence line ordinates,

Bending moment at $L_4 = R_R(36) - $ (moments of loads 1 to 5 about load 5)

$$R_R = \frac{0.5(1.2)(67)^2 + 10(72 + 77 + 82 + 87) + 5(92)}{108} = 58.64 \text{ kips}$$

Bending moment at $L_4 = 58.64(36) - 10(5 + 10 + 15) - 5(20)$

$$= 1711.1 \text{ ft-kips}$$

$$U_4U_5 = -\frac{1711.1}{24} = -71.30 \text{ kips} \quad \text{(check)}$$

For passage of loads from right to left, using the influence line ordinates gives

$$U_4U_5 = 5\left(-\frac{12.5}{72}\right) + 10\left(-\frac{17.5}{72} - \frac{22.5}{72} - \frac{27.5}{72} - \frac{32.5}{72}\right) + 1.2(0.5)(-1.0)\left(\frac{37.5}{72}\right)(34.5)$$

$$+ 1.2(0.5)(-1.0)(34.5) + 1.2(0.5)(-1.0)(36)$$

$$= -67.84 \text{ kips} \quad \text{(negative means compression)}$$

Without the use of influence line ordinates,

Bending moment at $L_4 = R_R(36) - $ (moment of 36 ft of uniform load about L_4)

$$R_R = \frac{5(12.5) + 10(17.5 + 22.5 + 27.5 + 32.5) + 1.2(70.5)(37.5 + 35.25)}{108}$$

$$= 66.82 \text{ kips}$$

Bending moment at $L_4 = 66.82(36) - 1.2(36)(18) = 1628.1$ ft-kips

$$U_4U_5 = -\frac{1628.1167}{24} = -67.84 \text{ kips}$$

Thus the largest compression force in U_4U_5 is -71.30 kips when load 5 is at L_4 as the load series passes from left to right.

SELECTED REFERENCES

1. E. Winkler, "Spannungskurven," *Zeitschrift des böhmischen Ingenieuren- und Architekten Vereins*, 1868.

2. Emil Winkler, *Vorträge über Brückenbau*, Vol. I: *Theorie der Brücken*, 3rd ed., Carl Gerold's Sohn, Vienna, 1886, pp. 28, 134.

3. Jacob I. Weyrauch, *Allgemeine Theorie und Berechnung der continuirlichen und einfachen Träger*, B. G. Teubner, Leipzig, 1873, 175 pp.

4. E. Winkler, "Beiträge zur Theorie der continuirlichen Brüchentrager," *Der Civilingenieur*, *8* (1862), 135–182, plus Appendix Table 10.

5. Hans Straub, *A History of Civil Engineering* (English translation by Erwin Rockwell), The MIT Press, Cambridge, Mass., 1964, p. 201.

6. Otto Mohr, "Beitrag zur Theorie der Holz- und Eisen-Constructionen," *Zeitschrift des Architekten- und Ingenieur-Vereins zu Hannover*, *14* (1868), 19–50, plus Plates 397–400.

7. Heinrich Müller-Breslau, *Die neuren Methoden der Festigkeitslehre*, Baumgartner's Buchhandlung, Leipzig, 1886.

PROBLEMS

When any influence line is asked for, critical ordinate values and their sign ($+$ or $-$) of each portion must always be shown. Reference to maximum value means the highest absolute value (either sign); minimum value means the highest value of opposite sign or if the value cannot be of opposite sign the minimum is the least value of the same sign. Whenever uniform live load is referred to, it must always be assumed to be of indefinite length and that loaded segments of the span need not be contiguous.

6.1. Referring to the beam of the accompanying figure, draw to scale the influence lines for (a) the reaction R_B at B; (b) the bending moment M_B at B. Compute their maximum and minimum values due to the passage in either direction of the given concentrated loads. Then compute the maximum and minimum values due to the given uniform *dead* load of 1.5 kips/ft.

6.2. For the beam of the accompanying figure, draw to scale the influence lines for: (a) the reaction at B; (b) the bending moment at section E-E. Compute the maximum reaction at B and the maximum and minimum values of the bending moment at section E-E due to the passage of the given loads in either direction.

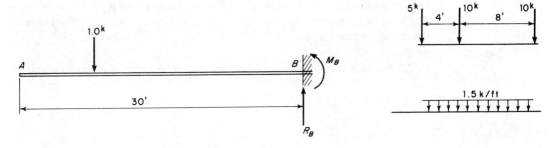

Prob. 6.1

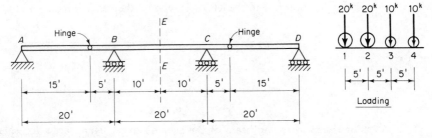

Prob. 6.2

6.3. For the overhanging beam of the accompanying figure, draw to scale the influence lines for (a) the bending moment M_B at B; (b) the shear V_C at C; (c) the bending moment M_C at C. Compute the maximum and minimum values of M_B, V_C, and M_C for the loading of Prob. 6.1. The concentrated loads may move in either direction across the spans, and the uniform load is *live* load. The concentrated and uniform loads may act simultaneously.

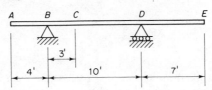

Prob. 6.3

6.4. For the overhanging beam of the accompanying figure, draw to scale the influence lines for (a) the shear and (b) the bending moment at section E. Next, (c) compute the shear and bending moment at E if the entire beam is loaded with a dead load of 6.0 kips/ft, first by using the influence line area method and then by the free body method. Finally, (d) for a live load of 7.2 kips/ft, compute the maximum and minimum values of the shear and bending moment at E.

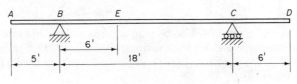

Prob. 6.4

6.5. Draw the influence lines to scale for: (a) the bending moment at B; (b) the shear at E; (c) the bending moment at E; and (d) the bending moment at C, for the beam of the accompanying figure.

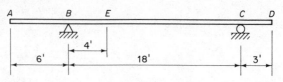

Prob. 6.5

6.6. For the overhanging beam of the accompanying figure, draw to scale the influence lines for: (a) the shear at C; (b) the bending moment at C; (c) the shear just to the left of B; (d) the shear just to the left of D; and (e) the bending moment at D.

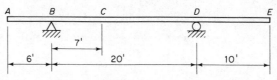

Prob. 6.6

6.7. For the overhanging beam of the accompanying figure, draw to scale the influence lines for: (a) the shear at E and (b) the bending moment at C. Then compute the maximum and minimum values for the shear at E and the bending moment at C caused by a uniform live load of 1.5 kips/ft. Compute values by both the influence line method and by the free body method.

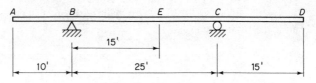

Prob. 6.7

6.8. Draw to scale the influence lines for: (a) the shear at C; (b) the bending moment at B; and (c) the shear at the right of D, for the beam of the accompanying figure. Apply the loading of Prob. 6.1 as live loading to obtain the maximum and minimum values for V_C, M_B, and V_D (right).

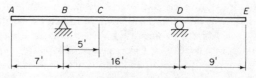

Prob. 6.8

6.9. For the beam of the accompanying figure, draw influence lines giving values for all critical ordinates for the following: (a) shear at the right side of support A; (b) shear 5 ft to the right of A; (c) shear at the hinge; (d) bending moment at B; (e) bending moment at C.

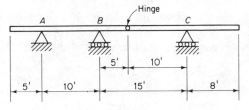

Prob. 6.9

6.10. For the beam of the accompanying figure, assuming that the load travels on the portion A through G: (a) draw the influence line for bending moment at E; (b) indicate what portions of the span AG that must be loaded with uniform live load to give the maximum bending moment at E (give sign); (c) determine the maximum moment at E caused by a moving concentrated load of 25 kips; (d) draw the influence line for shear in panel DE; (e) compute the maximum shear if there is a 25-kip concentrated load together with uniform live load of 4.5 kips/ft.

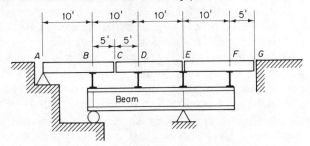

Prob. 6.10

6.11. For the beam of the accompanying figure, draw the influence line for shear at X-X and then compute the maximum positive and negative shear due to passage of the given wheel loads.

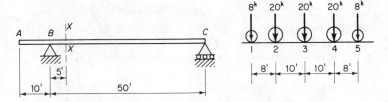

Prob. 6.11

6.12. For the beam of the accompanying figure, draw to scale the influence lines for shear and bending moment at section C; then compute the maximum and minimum shears and bending moments at C due to a dead load of 0.8 kip/ft and a live load of 2 kips/ft.

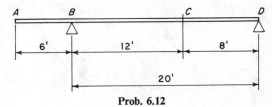

Prob. 6.12

6.13. For the loading shown in the accompanying figure, determine the maximum bending moment at a section 20 ft from the left support in a 50-ft-span simply supported beam due to the passage in either direction of the seven concentrated loads.

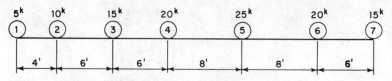

Probs. 6.13 and 6.14

6.14. Determine the maximum bending moment at midspan of a 50-ft simple beam, as well as the absolute maximum bending moment in the beam due to the passage of the seven concentrated loads of Prob. 6.13.

6.15. For two moving concentrated loads of equal magnitude W separated a fixed distance a apart, moving along a simply supported span, determine (a) the location of the critical wheel when the span is loaded to give absolute maximum bending moment; (b) the equation for the absolute maximum bending moment; (c) the maximum ratio of a to span length L for which the absolute maximum bending moment occurs with both concentrated loads on the span.

6.16. Repeat the requirements of (a) and (b) for Prob. 6.15 except use three equal concentrated loads equidistant a apart. For part (c), find the maximum ratio a/L for which the absolute maximum bending moment occurs with three loads on the span.

6.17. Determine the absolute maximum bending moment that may be induced in a 25-ft span simply supported beam due to the passage of the series of four concentrated loads as shown in the accompanying figure.

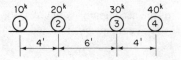

Probs. 6.17 and 6.27

6.18. Determine the maximum bending moment at midspan, and the absolute maximum bending moment in a simply supported beam of 50-ft span due to the passage of the four wheels as shown in the accompanying figure.

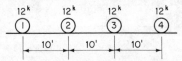

Probs. 6.18 and 6.28

6.19. Compute the maximum bending moment at the midspan of the simple beam shown in the accompanying figure due to the passage of the five concentrated loads in either direction. Also compute the absolute maximum bending moment.

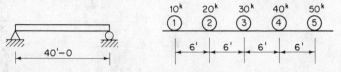

Probs. 6.19 and 6.29

6.20. Determine the maximum bending moment at the quarter point and at midspan of the simple beam shown in the accompanying figure due to the four concentrated loads moving across the span in either direction. Also determine the absolute maximum bending moment.

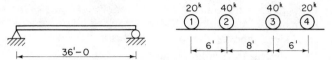

Prob. 6.20

6.21. Compute the absolute maximum bending moment in the simple beam shown in the accompanying figure due to the passage of the two concentrated loads shown.

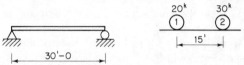

Probs. 6.21 and 6.30

6.22. For the beam of the accompanying figure, compute the maximum bending moment at C and at C' due to the passage of the series of concentrated loads from right to left.

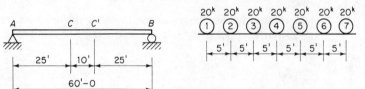

Prob. 6.22

6.23. Due to the passage of the four concentrated loads as shown in the accompanying figure over the simple beam AB in either direction, determine (a) the maximum bending moment at C; (b) the maximum bending moment at midspan; (c) the absolute maximum bending moment.

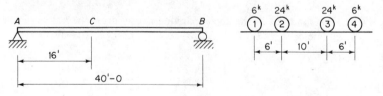

Prob. 6.23

6.24. For the beam and loading of the accompanying figure, determine the maximum bending moment at a section that is at 18 ft from the left support, and at the midspan, due to the passage of the four wheels in either direction. Also determine the absolute maximum bending moment.

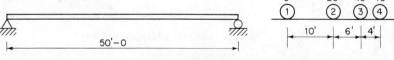

Prob. 6.24

6.25. Draw to scale the shear and bending moment envelopes for a simply supported beam of 30-ft span to carry uniform dead load of 1.7 kips/ft and live load of 2.2 kips/ft. For both envelopes, give computed values at $\frac{1}{4}$ points along the span in addition to any critical values.

6.26. Draw to scale the bending moment and shear envelopes for a 25-ft simply supported beam having an additional 10-ft overhanging cantilever at its right end. The dead load is 2.5 kips/ft and the live load is 3.0 kips/ft.

6.27. Draw to scale the bending moment and shear envelopes for the 25-ft simply supported beam carrying the live loading of Prob. 6.17. In addition to the given concentrated live loads, there is uniform dead load of 1.5 kips/ft.

6.28. Draw to scale the live load bending moment and shear envelopes for the beam and loading of Prob. 6.18. Compute values at the quarter points and at midspan.

6.29. Draw to scale the live load bending moment and shear envelopes for the beam and loading of Prob. 6.19.

6.30. Draw to scale the live load bending moment and shear envelopes for the beam and loading of Prob. 6.21.

Influence Lines for Axial Force in Truss Members

For all problems assume that the load moves along a support system that frames into the bottom chord.

6.31. For the truss of the accompanying figure, draw the influence lines for the axial force in members U_1L_2, U_2L_2, U_2U_3, L_3L_4, and U_1L_2 (see also Probs. 6.38 and 6.39).

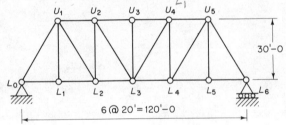

Probs. 6.31, 6.38, and 6.39

6.32. For the truss of the accompanying figure, draw the influence lines for the axial force in members L_0U_1, L_2L_3, U_3U_4, U_4L_5, and L_5L_6 (see also Probs. 6.40 and 6.41).

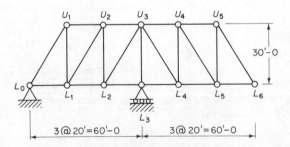

Probs. 6.32, 6.40, and 6.41

6.33. For the truss of the accompanying figure, draw the influence lines for the axial force in members U_1U_2, M_1U_2, M_1L_2, L_2L_3, and U_3L_3 (see also Prob. 6.42).

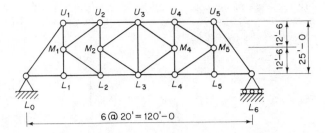

Probs. 6.33 and 6.42

6.34. For the truss of the accompanying figure, draw the influence lines for the axial force in members U_1L_2, U_2L_2, L_2M_2, the vertical from M_2, U_3M_3, and M_3U_4 (see also Prob. 6.43).

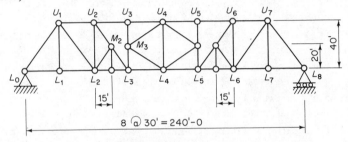

Probs. 6.34 and 6.43

6.35. For the truss of the accompanying figure, draw the influence lines for the axial force in members U_0U_1, M_0U_1, U_1M_1, M_1L_1, U_2M_1, U_2L_2, and U_2U_3.

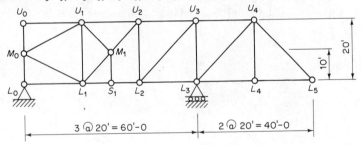

Prob. 6.35

6.36. For the truss of the accompanying figure, draw the influence lines for the axial force in members U_1L_2, U_3U_4, L_3U_4, and U_4L_4.

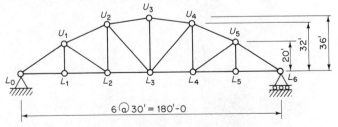

Prob. 6.36

6.37. For the truss of the accompanying figure, draw the influence lines for the axial force in members U_2L_2, U_1U_2, M_1U_2, M_1S_1, U_1M_1, U_1L_1, and L_1S_1.

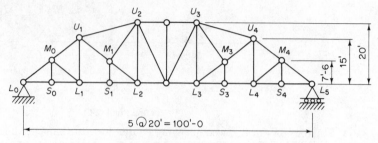

Prob. 6.37

6.38. For the truss and members indicated in Prob. 6.31, determine the maximum and minimum axial forces due to the loads of Prob. 6.21.

6.39. For the truss and members indicated in Prob. 6.31, determine the maximum and minimum axial forces due to the series of loads given in Prob. 6.23.

6.40. For the truss and members indicated in Prob. 6.32, determine the maximum and minimum axial forces due to the loads in Prob. 6.21.

6.41. For the truss and members indicated in Prob. 6.32, determine the maximum and minimum axial forces due to the series of loads in Prob. 6.23.

6.42. For the truss and members indicated in Prob. 6.33, determine the maximum and minimum axial forces due to the concentrated loads in Prob. 6.21, in addition to a uniform live load of 1.5 kips/ft and uniform dead load of 1.0 kip/ft.

6.43. For the truss and members indicated in Prob. 6.34, determine the maximum and minimum axial forces due to the concentrated loads of Prob. 6.21, in addition to a uniform live load of 1.5 kips/ft and uniform dead load of 1.0 kip/ft.

[7]

Statically Indeterminate Trusses

7.1 DEGREE OF STATICAL INDETERMINACY

When the laws of statics are insufficient to allow complete solution for all the unknown external reactions and internal forces, a truss is said to be *statically indeterminate*. For such cases equations relating to compatible deformations, known as *compatibility conditions* or equations, are needed in addition to the laws of statics. A truss may be statically indeterminate with respect to only the internal forces in the members, to only the external reactions, or to both internal forces and external reactions.

The reader should be alerted to the improper casual use of the term "indeterminate structures." The situations are only indeterminate without the use of compatibility conditions in addition to the conditions of statics. Thus, to be correct, structures can be *statically* indeterminate, but certainly they are all determinable when both equilibrium and compatibility conditions are fully invoked.

For systematic treatment in structural analysis it is convenient to establish certain nomenclature and symbols. As previously stated, a truss is a framed structure in which loads are assumed to be applied only at frictionless pins that join straight bars of uniform cross section. For a planar truss (the only kind dealt with in this elementary text) two components define any *random* force acting at a joint; correspondingly, two components of the movement of a joint define its displacement. Let P be the symbol used to denote a *possible* joint force and F to denote an internal force in a member; then NP may be used to indicate the number of possible directions in which forces may be applied at the joints. In the aforementioned, it must be emphasized that it is the *random* force or displacement that requires two components for definition; that is, the force or displacement that does not have its direction restricted. For instance, there would be only one force or displacement component at a roller support; the external force and corresponding displacement would have to be parallel to the roller support. In general, then, there are always the same number (NP) of possible external force components as there are joint displacement components. Since the joint displacements represent the ways in which a framed structure may *freely* move as a result of any external disturbance, the number of unknown joint displacements NP (or possible directions of joint forces) may be referred to as the *degree of freedom*.

It is recalled from Section 2.1 that the statically determinate truss has no more unknown external reactions and internal forces than the available equations of statics provided by the two equilibrium equations per joint. Using the symbol NJ for the number of joints and NR for the number of reactions in a truss, the requirement for statical determinacy becomes

$$NR + NF = 2\,(NJ) \tag{7.1.1}$$

Further, using the symbol NI for the degree of statical indeterminacy,

$$NI = (NR + NF) - 2(NJ) \tag{7.1.2}$$

The fact that

$$NP + NR = 2(NJ) \tag{7.1.3}$$

must remain true for a planar truss. Substituting Eq. (7.1.3) into Eq. (7.1.2),

$$NI = NF - NP \tag{7.1.4}$$

Thus, the degree of statical indeterminacy NI is equal to the number of bars in the truss minus the number of possible displacement components (i.e., degree of freedom). Note that NR does not appear in Eq. (7.1.4); it is usually equal to three but may be larger than three in special situations.

One must note that by making NF equal to NP, the truss can only meet the *necessary* condition for statical determinacy, but it may still not have the *sufficient* condition for statical stability. For statical stability, the truss must be able to resist a *random* force applied at any one of its joints. For instance, the truss of Fig. 7.1.1a can be made statically determinate and stable by removing any one of the six bars around the center panel, but it would be statically unstable if, say, the bar $L_0 U_1$ is removed. The truss of Fig. 7.1.1b can be made statically determinate and stable by

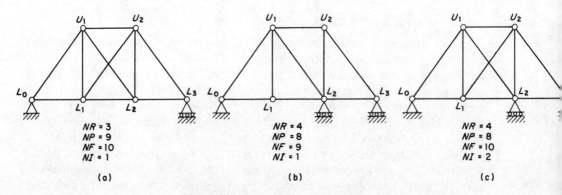

Figure 7.1.1 Statically indeterminate trusses.

removing one of the two supports at L_2 or L_3, or by removing, say, the bar U_1U_2, but not by removing, say, the bar U_1L_1. In Fig. 7.1.1c the removal of two *appropriate* members, or one *appropriate* member and one *appropriate* reaction, would render the structure statically determinate. The *appropriate* removals must only be those that will not make the remaining structure unstable.

7.2 CONSISTENT DEFORMATION (MAXWELL–MOHR) METHOD—GENERAL CONCEPT

The classical approach to solving statically indeterminate framework problems has been to remove members and/or support reactions such as to create a residual (here-after called *basic*) statically determinate structure. This "basic" structure is then analyzed (computing forces in all members and displacements in the direction of the removed supports or members). Using the laws of statics, the members or reactions removed from the statically indeterminate structure may be thought of as loads acting on the "basic" statically determinate structure. In place of each of the removed members or reactions, a unit force is applied and the displacements of the "basic" structure in the directions of the removed members and/or reactions are determined. The unknown forces in the removed members and/or reactions are determined such that the displacements in the directions of the removed members and/or reactions due to the known loads and the unknown forces are *compatible* with the unknown forces themselves. For example, if removal of a roller reaction would render a structure statically determinate, its replacement by a unit force in the known direction will cause a deflection in that direction. Multiplying the unknown force by the deflection due to a unit force would give the actual deflection due to the unknown force, which can then be used to cancel the deflection opposite to the unknown reaction due to the known applied loads. Thus, the force in question and the displacement (zero in case of a reaction) due to that force and the applied loads would then be compatible, or one could say there is *consistent deformation*. This general approach may be called the *method of consistent deformation*.

The term *redundant* or *redundant force* is used to refer to either an internal member force or a support reaction that is removed to change a structure from statically indeterminate to statically determinate. Such a removed force is redundant (i.e., not necessary) in the sense that the structure without the force is still stable and in equilibrium. Hence, the method of consistent deformation may also be called the *redundant force method*. In Fig. 7.2.1a the force F_x in bar L_1U_2 is chosen as the redundant; in Fig. 7.2.1b, the reaction R_3; and in Fig. 7.2.1c, the force F_x in bar L_1U_2 and the reaction R_3.

The use of redundant forces in the consistent deformation method derives from the original work on reciprocal deflection by James Clerk Maxwell [1] in 1864. However, Maxwell's work was largely unnoticed by engineers. Otto Mohr [2], without knowledge of Maxwell's work, used the principle of virtual work to arrive at the law of reciprocal deflections (see Section 5.4) and applied the redundant force method to statically indeterminate trusses, utilizing the virtual work principle to compute the required deflections. This redundant force method using virtual work as applied to

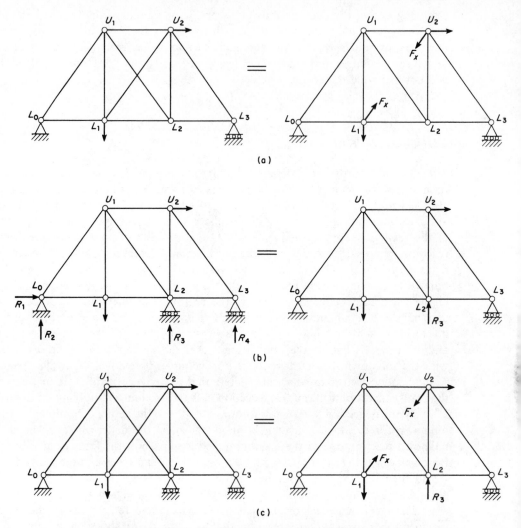

Figure 7.2.1 Redundant forces in statically indeterminate trusses.

trusses has frequently been called the Maxwell–Mohr method. Since the publication of Mohr's work in 1874–1875, and particularly since publication of the classic textbook by Heinrich Müller-Breslau [3] in 1886, this method was one of the two major methods for treating statically indeterminate trusses until the use of matrix methods with the digital computer. The other method is that of Castigliano treated in Sections 7.7 and 7.8. For details on the historical background of the Maxwell–Mohr redundant force method (or consistent deformation method), see Westergaard [4] and Timoshenko [5].

In applying the compatibility conditions, the deflections in the directions of the redundants must be the sum of (1) those due to the known loads, and (2) those due to each redundant force, as applied to the "basic" (residual or degenerate) determinate truss. These conditions yield a system of linear simultaneous equations from which the redundant forces may be solved. Such an approach of solving for forces as the basic unknowns is known as the *force method*. The alternative of considering the joint displacements as the basic unknowns to be solved for is called the *displacement method*. Because of the large number of simultaneous equations involved, even in simple structures, matrix formulation of the displacement method (treated in Chapters 13 to 20) is essential so that the digital computer can be utilized. For statically determinate structures and statically indeterminate structures with a small degree of statical indeterminacy (say, less than 4 or 5), the force method has traditionally been the preferable method of solution.

7.3 APPLICATION OF THE CONSISTENT DEFORMATION (MAXWELL–MOHR) METHOD

In the consistent deformation (Maxwell–Mohr) method, there are usually alternative choices for redundants, although their number must always be equal to the degree of statical indeterminacy. The only important consideration is that after the redundants are removed, the remaining "basic" structure is not only statically determinate, but also statically stable.

Once the redundants are decided upon, the primary computational requirement is to determine deflections due to the known applied loads and due to each of the redundants. Thus, any computational procedure for determining deflections may be used, such as the energy approach using virtual work (unit load) method or the geometric approach using the joint displacement equation (or graphical representation).

Because of the fact that only the deflections in the directions of the redundants are desired in forming the compatibility conditions, the preferred choice is the virtual work (unit load) method. After all the reactions and member forces are solved, if one still desires the final configuration of the statically indeterminate structure, the joint displacement equation or the Williot–Mohr graphical method can still be applied.

Support Reaction as a Redundant

In this situation, the displacements due to the known applied loads and the redundant reaction forces are set equal to the actual displacements of the supports; to zero in the usual case of unyielding supports.

Example 7.3.1

Analyze the statically indeterminate truss of Fig. 7.3.1a by the redundant force method, using the reaction R_3 at L_2 as the redundant. Compute all axial forces and joint displacements, and show that all statics and compatibility conditions are satisfied.

Solution. The compatibility condition requires that the vertical displacement of joint L_2 due to the combined action of the applied loads and the redundant reaction R_3 at joint L_2 be equal to zero.

(a) *Compute vertical deflection X_1 at L_2 due to loads on basic truss (i.e., with the redundant removed).* Using Eq. (3.2.3) in the virtual work (unit load) method with the real (Fig. 7.3.1b) and virtual (Fig. 7.3.1d) systems,

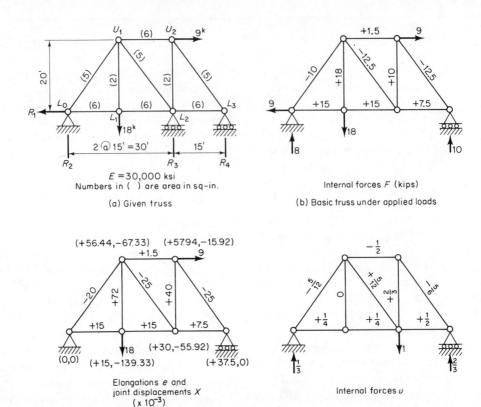

Figure 7.3.1 Deflection due to applied loads in Example 7.3.1.

$$X_1 = \sum u_i e_i$$

$$= \sum (u \text{ in Fig. 7.3.1d})(e \text{ in Fig. 7.3.1c})$$

$$= \left(-\frac{1}{2}\right)(+1.5) + \left(+\frac{1}{4}\right)(15 + 15) + \left(+\frac{1}{2}\right)(+7.5) + \left(-\frac{5}{20}\right)(-20)$$

$$+ \left(+\frac{5}{12}\right)(-25) + \left(-\frac{5}{6}\right)(-25) + \left(+\frac{2}{3}\right)(+40)$$

$$= +55.92 \text{ units} \quad \text{(positive means downward)}$$

(b) *Compute vertical deflection X_2 at L_2 due to R_3 acting upward.* In this computation one may think of computing the deflection due to a unit load and then multiplying by R_3, or as illustrated here, the deflection due to R_3 upward is thought of as being directly computed. Referring to the real and virtual systems in Fig. 7.3.2,

$$X_2 = \sum u_i e_i$$

$$= \sum (u \text{ in Fig. 7.3.2c})(e \text{ in Fig. 7.3.2b})$$

$$= R_3\left[2\left(\frac{1}{2}\right)^2 + 2\left(\frac{1}{4}\right)^2 + 2\left(\frac{5}{12}\right)\left(\frac{5}{6}\right) + \left(\frac{5}{6}\right)\left(\frac{5}{3}\right) + \left(\frac{2}{3}\right)\left(\frac{8}{3}\right)\right]$$

$$= +4.486R_3 \text{ units} \quad \text{(positive means upward)}$$

Statically Indeterminate Trusses   Chap. 7

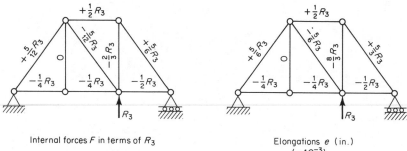

Internal forces F in terms of R_3 Elongations e (in.)
($\times 10^{-3}$)

(a) Basic truss under action of redundant   (b) Basic truss under action of redundant

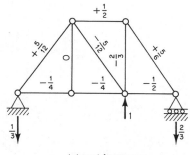

Internal forces u

(c) Virtual system with unit load at R_3

Figure 7.3.2   Deflection due to redundant reaction R_3 in Example 7.3.1.

(c) *Solve for R_3 using the compatibility condition.* The superposition of the basic statically determinate truss (Fig. 7.3.1b) with the truss under the action of the redundant R_3 (Fig. 7.3.2a) must actually give zero deflection at L_2. Thus, setting the total deflection, X_1(down) $- X_2$(up), equal to zero,

$$X = X_1 - X_2 = +55.92 - 4.486 R_3 = 0$$

$$R_3 = +12.46 \text{ kips} \quad (\text{positive means upward})$$

The forces on the truss due to the redundant acting alone (Fig. 7.3.2a) are given in Fig. 7.3.3a after multiplying by the numerical value of R_3. The addition of (1) the internal forces in the basic statically determinate truss under the applied loads (Fig. 7.3.1b), and (2) the internal forces due to the redundant R_3 (Fig. 7.3.3a) is given in Fig. 7.3.3b.

(d) *Check the result.* Eight statics checks are made along the eight degrees of freedom using the forces F in Fig. 7.3.3b. A check of statics alone is not sufficient because the original problem could not be solved by means of statics alone.

Joint displacements of the given truss are computed in Fig. 7.3.3d by means of the joint displacement equation. The compatibility check is obtained from the fact that the vertical displacements of joints L_0, L_2, and L_3 are all equal to zero.

One may note that the diagrams in Figs. 7.3.1 to 7.3.3 are used to show the

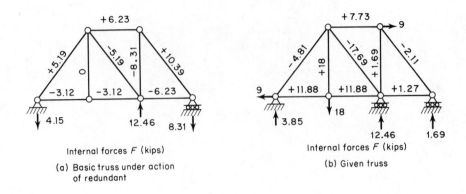

Internal forces F (kips)

(a) Basic truss under action of redundant

Internal forces F (kips)

(b) Given truss

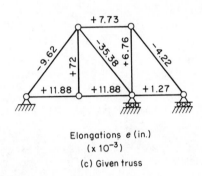

Elongations e (in.)
($\times 10^{-3}$)

(c) Given truss

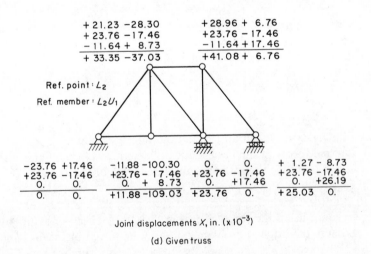

Joint displacements X, in. ($\times 10^{-3}$)

(d) Given truss

Figure 7.3.3 Solution for the statically indeterminate truss in Example 7.3.1.

214

solution procedure in detail. Alternatively, all the required computations could have been presented in tabular form, or automated by a computer program [6].

Truss Member as a Redundant

In this situation, the compatibility condition requires that the *relative* displacement between the ends of the redundant bar must be equal to the elongation of the redundant bar due to the combined action of the applied loads and the action of a pair of redundant forces acting on the joints at the ends of the removed redundant bar.

The relative displacement X_i (or increase in the distance) between two joints of a truss may be computed by the virtual work (unit load) method. This can be accomplished by using a *pair* of unit Q_i-forces as the virtual loading system. Then, applying Eq. (3.2.3) to the real and virtual systems of Fig. 7.3.4,

$$\text{Relative displacement } X_i \text{ (apart)} = \sum u_i e_i \qquad (7.3.1)$$

In the event that all the joint displacements in the real system have been computed as shown in Fig. 7.3.5, the relative displacement between two typical joints such as L_1 and U_2 becomes

$$\text{Relative displacement } X_i \text{ (apart)} = (X_3 - X_1) \cos \alpha + (X_4 - X_2) \sin \alpha \qquad (7.3.2)$$

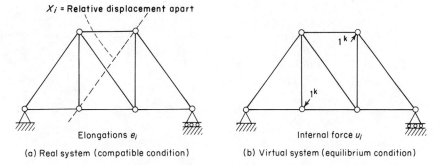

(a) Real system (compatible condition)   (b) Virtual system (equilibrium condition)

Figure 7.3.4 Relative displacement by the virtual work (unit load) method.

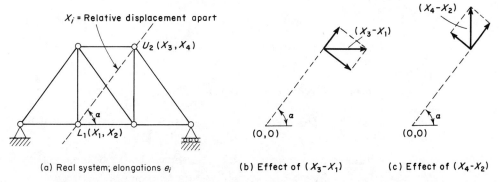

(a) Real system; elongations e_i (b) Effect of (X_3-X_1) (c) Effect of (X_4-X_2)

Figure 7.3.5 Relative displacement obtained from horizontal and vertical displacements.

The equation above can be easily substantiated by taking the components of the relative horizontal and vertical displacements, $(X_3 - X_1)$ and $(X_4 - X_2)$, in the direction of an imaginary line joining L_1 and U_2. In fact, the values of X's used in Eq. (7.3.2) may be associated with any reference point and reference member through it, because the relative displacement between any two joints is not affected by rigid-body translation or rotation of the entire truss.

Example 7.3.2

Analyze the statically indeterminate truss of Fig. 7.3.6a by the redundant force method, using the force in bar L_1U_2 as the redundant. Compute all axial forces and joint displacements, and show that all statics and compatibility conditions are satisfied.

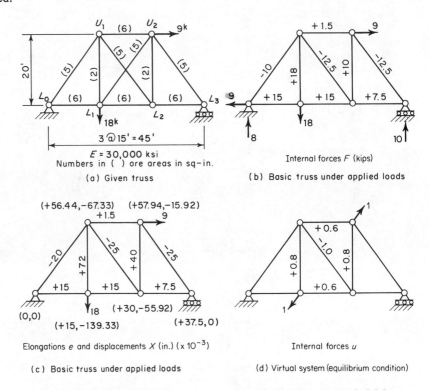

Figure 7.3.6 Relative displacement due to applied loads in Example 7.3.2.

Solution

(a) *Compute the relative displacement X_1 of joints L_1 and U_2 in the direction along the line connecting them for the basic statically determinate truss with the selected redundant removed.* Using the pair of unit forces in the virtual system of Fig. 7.3.6d with the real system of Fig. 7.3.6b and applying Eq. (3.2.3),

$$X_1 \text{ (due to applied loads)} = \sum (u \text{ in Fig. 7.3.6d})(e \text{ in Fig. 7.3.6c})$$

$$= (+0.6)(+1.5 + 15) + (+0.8)(+72 + 40) + (-1)(-25)$$

$$= +124.5 \text{ units} \quad \text{(positive means increase in distance)}$$

Since the joint displacements are available in Fig. 7.3.6c, the relative displacement between joints L_1 and U_2 may be also computed by the geometric method of joint displacement equation as

X_1 (due to applied loads)

$$= (+57.94 - 15)(+0.6) + (-15.92 + 139.33)(+0.8)$$

$$= +124.5 \text{ units} \quad \text{(check)}$$

Note that if joint U_1 and member U_1L_2 are used as the reference point and reference member, the joint displacements at L_1 and U_2 can be obtained by inspection and used to compute the relative displacement.

(b) *Compute the relative displacement X_2 of joints L_1 and U_2 in the direction along the line connecting them due to the pair of redundant forces F_x acting on the basic statically determinate truss.* This time the real system is that given in Fig. 7.3.7a while the virtual system is that in Fig. 7.3.7c. Thus,

X_2 (due to the pair of F_x) $= \sum (u$ in Fig. 7.3.7c)(e in Fig. 7.3.7b)

$$= 2(-0.6)(-0.6F_x) + 2(-0.8)(-3.2F_x) + (+1)(+2F_x)$$

$$= +7.84F_x \text{ units} \quad \text{(positive means decrease in distance)}$$

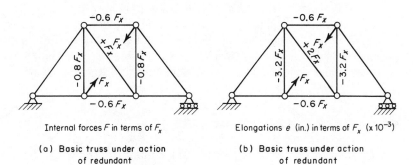

Internal forces F in terms of F_x

(a) Basic truss under action of redundant

Elongations e (in.) in terms of F_x ($\times 10^{-3}$)

(b) Basic truss under action of redundant

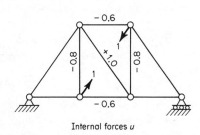

Internal forces u

(c) Virtual system with pair of unit loads

Figure 7.3.7 Relative displacement due to pair of redundant forces in Example 7.3.2.

(c) *Solve for F_x using the compatibility equation.* Equating the elongation of the redundant bar with the total relative displacement,

$$\frac{F_x L}{EA} = X_1 - X_2$$

$$\frac{F_x(300)}{(30)(5)} = +124.5 - 7.84F_x$$

$$F_x = \frac{124.5}{9.84} = +12.65 \text{ kips}$$

The actual forces in the truss members due to F_x equal to 12.65 kips tension are shown in Fig. 7.3.8a. The final result is the superposition of the effect of the given loads acting on the basic statically determinate truss shown in Fig. 7.3.6b and the effect on the same basic statically determinate truss acted upon by the redundant force shown in Fig. 7.3.8a. The final answer appears as Fig. 7.3.8c.

(d) *Check the result.* To verify that the answer is correct, a check must be made. The check consists of an investigation of equilibrium in statics and compatibility of deformation. Nine statics checks must be made for the internal forces F in Fig. 7.3.8b, one along each degree of freedom.

Joint displacements of the given truss are computed in Fig. 7.3.8d by means of the joint displacement equation omitting a different bar such as $U_2 L_2$. The compatibility check is obtained from the fact that the vertical displacements of joints U_2 and L_2 in Fig. 7.3.8d are consistent with the negative elongation in bar $U_2 L_2$.

Thus, the shortening ($-41.86 + 41.36$) of the distance $U_2 L_2$, which equals -0.50, should be equal to the elongation of member $U_2 L_2$, given as -0.48. This is considered adequately close, as round-off error of 0.02 does occur in the last significant figure used in all the displacement values. Note that the number of compatibility checks is one, as is the degree of statical indeterminacy of the given truss.

7.4 DISPLACEMENT METHOD—CONCEPT

Instead of using forces as the primary unknowns, known as the *force method*, the alternative is to use displacements as the primary unknowns, known as the *displacement method*. From the joint displacements, the elongation of each member can be determined and then from the stiffness (EA/L for axially loaded members) the force in the member can be computed. Since the stiffness (at least the relative values) of members is involved in the computational procedure, this approach is frequently also called the *stiffness method*.

Although the displacement method is not used in this text until the slope deflection method is presented in Chapter 9 and extended using matrix applications in the latter part of the book, the concept is deemed important at this stage of study.

Examine the statically indeterminate three-bar truss (called a truss even though there is no closed triangle in it because all bars are only axially stressed) with three hinged supports, as shown in Fig. 7.4.1. A solution by the force method would require one of the bars to be removed as the redundant. In the displacement method, the structure never needs to be modified but is examined for degree of freedom. The joint at A can have both horizontal and vertical movement; that is, this is a two-degree-of-freedom situation. Next, subject the structure to separate displacements

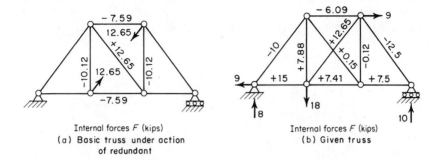

Internal forces F (kips)
(a) Basic truss under action
of redundant

Internal forces F (kips)
(b) Given truss

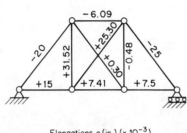

Elongations e (in.) ($\times 10^{-3}$)
(c) Given truss

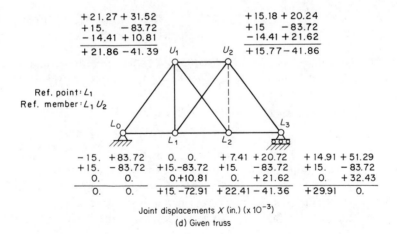

Joint displacements X (in.) ($\times 10^{-3}$)
(d) Given truss

Figure 7.3.8 Solution for the statically indeterminate truss in Example 7.3.2.

along each degree of freedom as in Fig. 7.4.1c and d. In Fig. 7.4.1c with the joint at A displaced X_1, the elongation of each member can be expressed in terms of X_1. Then, using the member axial stiffness (EA/L), the internal forces may also be so expressed. In order that equilibrium can exist, there must be external forces P_1 and P_2 acting at joint A. These external forces must again be linear functions of X_1. Using the proportionality coefficients K_{ij} to denote the force in the direction of the ith degree of freedom due to a displacement in the direction of the jth degree of freedom,

$$P_1 = K_{11}X_1; \qquad P_2 = K_{21}X_1 \qquad (7.4.1)$$

Similarly, in Fig. 7.4.1d, if the joint at A is displaced X_2, the forces P_1 and P_2 for equilibrium to exist at joint A are

$$P_1 = K_{12}X_2; \qquad P_2 = K_{22}X_2 \qquad (7.4.2)$$

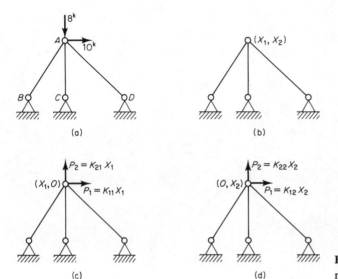

Figure 7.4.1 Concept of displacement method.

To solve the problem shown in Fig. 7.4.1a, the total horizontal force at A due to both X_1 and X_2 occurring is equated to $+10$; and the total vertical force, to -8. Thus,

$$K_{11}X_1 + K_{12}X_2 = +10 \qquad (7.4.3a)$$

$$K_{21}X_1 + K_{22}X_2 = -8 \qquad (7.4.3b)$$

The solution of the two equations, Eqs. (7.4.3a) and (7.4.3b), for X_1 and X_2 gives the actual joint displacements caused by the known applied loads. With X_1 and X_2 the internal forces can be determined, each as a function of X_1 and X_2. Dimensionally, the coefficients K_{ij} are forces per unit displacement and thus are termed *stiffness coefficients*. In Section 7.6 it will be shown by the Reciprocal Energy Theorem that K_{ij} is always equal to K_{ji}.

The concept of the displacement method (stiffness method) is relatively simple, but its application requires the solution of a set of simultaneous equations equal in number to the degree of freedom regardless of whether or not the structure is statically determinate. Using the force method, together with numerous statical analyses, the number of simultaneous equations is equal to the number of redundants in a statically indeterminate structure. These simultaneous equations involve the compatibility of deformations, computations for which require the use of the member flexibility $L/(EA)$. Thus, the redundant force method has also been called the *flexibility method*.

The reader may note that most of the classical structural analysis methods are force methods, indicative of necessity in the precomputer era. Conversely, most, if not all of the computer methods make use of the matrix formulation of the displacement method.

7.5 RECIPROCAL ENERGY THEOREM

Maxwell's presentation [1] in 1864 of the law of reciprocal deflections (treated in Section 5.4) may be viewed as a narrow application of a general reciprocal energy theorem. E. Betti [7] provided the general proof for this general theorem in 1872; however, the extensive application of the method for both static and dynamic systems was due to Lord Rayleigh [8–10].

The *reciprocal energy theorem* may be stated: *The external or internal work, done by the external forces or internal stress resultants in a P-system, in going through the corresponding displacements or deformations in a Q-system, must be equal to the external or internal work done by the external forces or internal stress resultants in the Q-system, in going through the corresponding displacements or deformations in the P-system.* In order that this theorem may be held to be true, the first-order method of analysis must be assumed and the material must be linearly elastic (see Sections 1.2 and 1.3).

To prove the reciprocal energy theorem, one imagines that the P-system is first applied to the linear elastic body, causing deformations (ΔP), and the resulting external or internal work may be symbolically represented by $\frac{1}{2}P(\Delta P)$. Then the Q-system is added, causing additional deformations (ΔQ); and the resulting external or internal work becomes

$$W = \frac{1}{2}P(\Delta P) + \frac{1}{2}Q(\Delta Q) + P(\Delta Q) \tag{7.5.1}$$

because the P-forces would have a "free ride" through the deformations (ΔQ). If, however, the Q-system had been applied first and then the P-system is added, the resulting external or internal work is

$$W = \frac{1}{2}Q(\Delta Q) + \frac{1}{2}P(\Delta P) + Q(\Delta P) \tag{7.5.2}$$

Since the total internal work within the linear elastic body under both the P- and Q-systems must be the same regardless of the order of application, equating Eq. (7.5.1) to Eq. (7.5.2) gives

$$P(\Delta Q) = Q(\Delta P) \tag{7.5.3}$$

which is the *reciprocal energy theorem*.

The reciprocal energy theorem, Eq. (7.5.3), can be used to prove that the proportionality coefficient K_{ij} in the displacement method is always equal to K_{ji}.

Using Fig. 7.4.1c and d as the P- and Q-systems, respectively,

$$P(\Delta Q) \overset{\perp}{=} Q(\Delta P)$$

$$K_{11}X_1(0) + K_{21}X_1(X_2) = K_{12}X_2(X_1) + K_{22}X_2(0)$$

from which

$$K_{21} = K_{12}$$

In general, then,

$$K_{ij} = K_{ji} \tag{7.5.4}$$

If there is only one unit load in each of the P- and Q-systems and the notation δ_{PQ} is used to denote the deflection at P due to a unit load at Q, then by applying the reciprocal energy theorem to each case of Fig. 5.4.1, one would obtain

$$(1.0)(\delta_{QP}) = (1.0)(\delta_{PQ})$$

or

$$\delta_{QP} = \delta_{PQ} \tag{7.5.5}$$

Equation (7.5.5) is the *law of reciprocal deflections*, which states that *the deflection at Q due to a unit load at P is equal to the deflection at P due to a unit load at Q.*

Thus the law of reciprocal deflections, Eq. (7.5.5), is derived as a simple extension of the general reciprocal energy theorem, Eq. (7.5.3). Further, Eq. (7.5.4) can be stated as the *Law of Reciprocal Forces*: *The force at P to cause unit deflection at Q is equal to the force at Q to cause a unit deflection at P.*

Example 7.5.1

Show that the reciprocal energy theorem holds between the force systems of Fig. 7.5.1a and b.

Solution. Figure 7.5.1a is the basic statically determinate truss of Example 7.3.1 with the redundant reaction removed and is here called the P-system. Figure 7.5.1b is the given statically indeterminate truss of Example 7.3.1 after the solution for forces and displacements has been completed, here called the Q-system. The elastic structure in each case is the same; the reaction force of 12.46 kips may be viewed as an applied load. Thus, applying the reciprocal energy theorem, first with P-system loads going through the Q-system displacements,

$$P(\Delta Q) = [(9)(41.08) + (18)(109.03)](10^{-3}) = 2.332 \text{ in.-kips}$$

and then the Q-system forces going through the P-system displacements,

$$Q(\Delta P) = [(9)(57.94) + (18)(139.33) - (12.46)(55.92)](10^{-3}) = 2.332 \text{ in.-kips}$$

In terms of internal work,

$$\begin{aligned}
P(\Delta Q) &= (-10)(-9.62) + (+1.5)(+7.73) + (-12.5)(-4.22) \\
&\quad + (+15)(+11.88) + (+15)(+11.88) + (+7.5)(+1.27) \\
&\quad + (+18)(+72) + (-12.5)(-35.38) + (+10)(+6.76) \\
&= 2332 \text{ units or } 2.332 \text{ in.-kips}
\end{aligned}$$

$$\begin{aligned}
Q(\Delta P) &= (-4.81)(-20) + (+7.73)(+1.5) + (-2.11)(-25) \\
&\quad + (+11.88)(+15) + (+11.88)(+15) + (+1.27)(+7.5) \\
&\quad + (+18)(+72) + (-17.69)(-25) + (+1.69)(+40) \\
&= 2332 \text{ units or } 2.332 \text{ in.-kips}
\end{aligned}$$

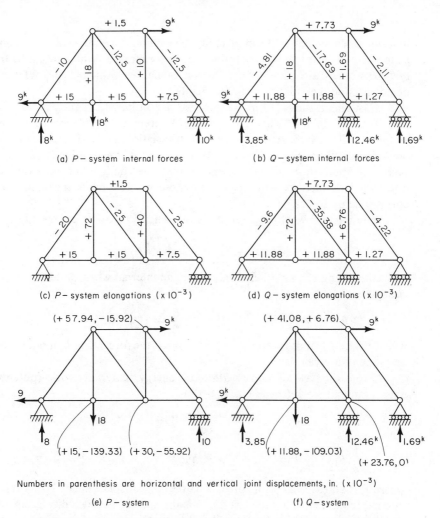

(a) P – system internal forces

(b) Q – system internal forces

(c) P – system elongations ($\times 10^{-3}$)

(d) Q – system elongations ($\times 10^{-3}$)

Numbers in parenthesis are horizontal and vertical joint displacements, in. ($\times 10^{-3}$)

(e) P – system

(f) Q – system

Figure 7.5.1 The reciprocal energy theorem applied to a truss. (See Example 7.3.1, Figs. 7.3.1 and 7.3.3, for the solution used to obtain these values.)

Thus, the principle of virtual work itself is, in fact, saying that the external work $Q(\Delta P)$ must be equal to the internal work $Q(\Delta P)$, where Q represents the virtual system and P represents the real system. In symbols,

$$\text{External } (Q\,\Delta P) = \text{internal } (Q\,\Delta P)$$

$$(1.0)(X_i) = \sum u_i e_i$$

7.6 CASTIGLIANO'S THEOREM—PART I (CASTIGLIANO'S FIRST THEOREM)

As an alternative to the redundant force method using consistent deformation, the Italian engineer Carlo Alberto Castigliano presented [11, 12] in 1875 an energy method for solving statically indeterminate elastic frameworks. In fact, L. F. Ménabréa [13] first suggested in 1858 that forces acting in redundant members of a truss must be such as to make the internal work (strain energy) a minimum, which he called the *principle of least work*. Castligiano examined the Ménabréa principle and was successful in demonstrating the validity of the principle.

Castigliano states [12, p. 32 (see also p. 15)] Part I (frequently referred to as the first theorem) of his "Theorems of the Differential Coefficients of Internal Work" as follows (from E. S. Andrews' English translation) [12]: "If we express the internal work of a body or elastic structure as a function of the relative displacements of the points of application of the external forces, we shall obtain an expression whose differential coefficients with regard to these displacements give the values of the corresponding forces."

The internal work W_{int} (elastic strain energy) may be expressed

$$W_{\text{int}} = \sum \left(\frac{1}{2} F_j e_j \right) \tag{7.6.1}$$

and substituting $e_j = F_j L_j / E_j A_j$ gives for the internal work, or as it is usually called, *strain energy*,

$$W_{\text{int}} = \frac{1}{2} \sum \frac{F_j^2 L_j}{E_j A_j} \tag{7.6.2}$$

where F_j, L_j, E_j, and A_j represent, respectively, the force, length, modulus of elasticity, and area of a given bar.

From the law of conservation of energy, external work (potential energy) equals internal work (strain energy); thus,

$$W_{\text{ext}} = W_{\text{int}} = W \tag{7.6.3}$$

and

$$\frac{1}{2}[P_1 X_1 + P_2 X_2 + \cdots + P_m X_m + \ldots + P_{\text{NP}} X_{\text{NP}}] = \frac{1}{2} \sum_{j=1}^{\text{NF}} F_j e_j \tag{7.6.4}$$

where P is an external joint force, X is the joint displacement, NF is the number of internal forces (say, bars in a truss), m is the specific force or displacement for which the differential coefficient is desired, and NP is the degree of freedom (i.e., the number of possible joint forces or displacements).

As a specific displacement X_m is changed slightly, there would be a small change in the energy W. This small change in energy W due to a small change in X_m can be found by taking $\partial W / \partial X_m$, for each side of Eq. (7.6.4), recognizing that F_j can be expressed as a *linear* function of the displacements X_m. Taking the partial derivative of each side of Eq. (7.6.4) with respect to X_m,

$$\frac{1}{2} P_m = \frac{1}{2} \sum \frac{\partial F_j}{\partial X_m} e_j$$

$$P_m = \sum \frac{\partial F_j}{\partial X_m} e_j \tag{7.6.5}$$

and then noting that $e_j = F_j L_j / E_j A_j$, Eq. (7.6.5) becomes

$$P_m = \sum_{j=1}^{\text{NF}} \frac{\partial F_j}{\partial X_m} \frac{F_j L_j}{E_j A_j} \tag{7.6.6}$$

from which

$$P_m = \frac{\partial \sum \frac{F_j^2 L_j}{2E_j A_j}}{\partial X_m} \qquad (7.6.7)$$

Noting that the numerator of Eq. (7.6.7) contains the strain energy expression for internal work,

$$P_m = \frac{\partial W_{\text{int}}}{\partial X_m} \qquad (7.6.8)$$

which is Castigliano's theorem, part I, or first theorem.

Since the application of Castigliano's first theorem requires that the strain energy expression be expressed in terms of displacements, Eq. (7.6.8) corresponds to the displacement method as described in Section 7.4. Since except for elementary problems or problems solved by the digital computer, the displacement method is generally impractical because of the large number of simultaneous equations involved, Eq. (7.6.8) was not much used until recent years. Later in Section 7.7 it will be shown that Castigliano's second theorem corresponds to the force method.

Example 7.6.1
Solve for the forces in the truss of Fig. 7.6.1a using Castigliano's first theorem.

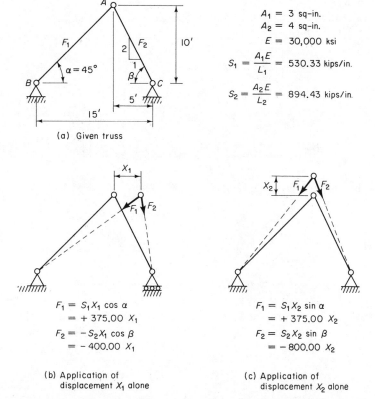

(a) Given truss

$A_1 = 3$ sq-in.
$A_2 = 4$ sq-in.
$E = 30,000$ ksi

$S_1 = \frac{A_1 E}{L_1} = 530.33$ kips/in.

$S_2 = \frac{A_2 E}{L_2} = 894.43$ kips/in.

$F_1 = S_1 X_1 \cos \alpha$
$\quad = +375.00\ X_1$
$F_2 = -S_2 X_1 \cos \beta$
$\quad = -400.00\ X_1$

(b) Application of displacement X_1 alone

$F_1 = S_1 X_2 \sin \alpha$
$\quad = +375.00\ X_2$
$F_2 = S_2 X_2 \sin \beta$
$\quad = -800.00\ X_2$

(c) Application of displacement X_2 alone

Figure 7.6.1 Application of Castigliano's Theorem, Part I (Castigliano's first theorem).

Solution. One can count that for this truss the degree of freedom NP is 2 and the number of bars NF is also 2. So this structure is statically determinate. However, application of Castigliano's first theorem requires no determination of statical determinacy or indeterminacy.

(a) *Determine internal strain energy as a function of displacements.* Referring to Fig. 7.6.1b, by application of displacement X_1 alone, F_1 and F_2 are determined as a function of X_1. Similarly referring to Fig. 7.6.1c, by application of displacement X_2 alone, F_1 and F_2 are determined as a function of X_2. Thus, from superposition,

$$F_1 = +375X_1 + 375X_2$$
$$F_2 = -400X_1 + 800X_2$$

Using the strain energy equation, Eq. (7.6.2),

$$W = \frac{1}{2} \sum \frac{F_j^2 L_j}{E_j A_j} = \frac{1}{2} \sum \frac{F_j^2}{S_j}$$

if $S_j = E_j A_j / L_j$. Then substituting the F's in terms of the X's, the strain energy equation becomes

$$W = \frac{1}{2} \left(\frac{F_1^2}{S_1} + \frac{F_2^2}{S_2} \right) = \frac{1}{2} \left[\frac{(+375X_1 + 375X_2)^2}{S_1} + \frac{(-400X_1 + 800X_2)^2}{S_2} \right]$$

(b) *Apply Castigliano's first theorem. Using the equations*

$$P_1 = \frac{\partial W}{\partial X_1}$$

$$P_2 = \frac{\partial W}{\partial X_2}$$

$$\frac{\partial W}{\partial X_1} = \frac{(+375X_1 + 375X_2)(375)}{S_1} + \frac{(-400X_1 + 800X_2)(-400)}{S_2}$$

$$\frac{\partial W}{\partial X_2} = \frac{(+375X_1 + 375X_2)(375)}{S_1} + \frac{(-400X_2 + 800X_2)(800)}{S_2}$$

Combining terms and taking $S_1 = 530.33$ and $S_2 = 894.43$,

$$P_1 = \frac{\partial W}{\partial X_1} = 444.05X_1 - 92.60X_2$$

$$P_2 = \frac{\partial W}{\partial X_2} = -92.60X_1 + 980.70X_2$$

(c) *Alternative method to express P's in terms of X's.* When P-forces can be explicitly determined from the F-forces, use of Castigliano's first theorem may be bypassed. In the present case, applying the equations of equilibrium to joint A, from $\sum F_x = 0$,

$$P_1 = +F_1 \cos \alpha - F_2 \cos \beta$$
$$= +(375X_1 + 375X_2)(0.70711) - (-400X_1 + 800X_2)(0.44721)$$
$$= +444.05X_1 - 92.60X_2$$

and from $\sum F_y = 0$,

$$P_2 = +F_1 \sin \alpha + F_2 \cos \beta$$
$$= +(375X_1 + 375X_2)(0.70711) + (-400X_1 + 800X_2)(0.89443)$$
$$= -92.60X_1 + 980.70X_2$$

However, when the internal forces, such as bending moments, are distributed throughout the elastic body the joint forces are not easily ascertained from direct application of statics. Thus, Castigliano's first theorem involves a more subtle way of attaining equilibrium through partial differentiation of the internal strain energy. This approach is a necessity in the finite element method of stress analysis.

(d) *Solve the displacements X_1 and X_2*. Since the applied loads are known, the simultaneous equations are:

$$444.05X_1 - 92.60X_2 = 10 \text{ kips}$$
$$-92.60X_1 + 980.70X_2 = 0$$

Solving for X_1 and X_2 gives

$$X_1 = +0.022972 \text{ in.}$$
$$X_2 = +0.002169 \text{ in.}$$

(e) *Obtain F-values from the X-values*. The values of X are then substituted into the equations for internal forces F; thus,

$$F_1 = +9.427 \text{ kips}$$
$$F_2 = -7.453 \text{ kips}$$

7.7 CASTIGLIANO'S THEOREM—PART II (CASTIGLIANO'S SECOND THEOREM)

Castigliano's second theorem will be shown to correspond to the force method, the usual procedure of solving structural analysis problems before the advent of using matrix methods on the digital computer. This may be the reason why the first theorem was used relatively little and the second theorem was considered the important one.

Castigliano stated [12, p. 32 (see also p. 15)] Part II of his "Theorems of the Differential Coefficients of Internal Work" as follows:* "If we express the internal work of a body or elastic structure as a function of the external forces, the differential coefficient of this expression, with regard to one of the forces, gives the relative displacement of its point of application."

The proof is similar to that used for the first theorem (see Section 7.6). From the law of conservation of energy, external work (potential energy) equals internal work (strain energy); thus, repeating Eq. (7.6.3),

$$W_{\text{ext}} = W_{\text{int}} = W \tag{7.7.1}$$

and repeating Eq. (7.6.4),

$$\frac{1}{2}[P_1X_1 + P_2X_2 + \cdots + P_mX_m + \ldots + P_{\text{NP}}X_{\text{NP}}] = \frac{1}{2}\sum_{j=1}^{\text{NF}} F_j e_j \tag{7.7.2}$$

*From E. S. Andrews' English translation [12].

where P is an external joint force, X is the joint displacement, NF is the number of internal forces (say, bars in a truss), m is the specific force or displacement for which the differential coefficient is desired, and NP is the degree of freedom (i.e., the number of possible joint forces or displacements).

As a specific force P_m is changed slightly, there would be a small change in the energy W. This small change in W due to a small change in P_m is found by taking the partial derivative, $\partial W/\partial P_m$, for each side of Eq. (7.7.2), assuming that F_j is to be expressed as a *linear* function of the forces P_m. Taking the partial derivative of each side of Eq. (7.7.2) with respect to P_m,

$$\frac{1}{2} X_m = \frac{1}{2} \sum \frac{\partial F_j}{\partial P_m} e_j$$

$$X_m = \sum \frac{\partial F_j}{\partial P_m} e_j \tag{7.7.3}$$

and noting that $e_j = F_j L_j / E_j A_j$, Eq. (7.7.3) becomes

$$X_m = \sum_{j=1}^{NF} \frac{\partial F_j}{\partial P_m} \frac{F_j L_j}{E_j A_j} \tag{7.7.4}$$

from which

$$X_m = \frac{\partial \sum \frac{F_j^2 L_j}{2 E_j A_j}}{\partial P_m} \tag{7.7.5}$$

Noting that the numerator of Eq. (7.7.5) contains the strain energy expression for internal work,

$$X_m = \frac{\partial W_{int}}{\partial P_m} \tag{7.7.6}$$

which is Castigliano's second theorem.

The application of Eq. (7.7.6) requires that the strain energy expression be expressed in terms of external forces; that is, the F_j's will be expressed in terms of the applied loads. These applied loads include the actual loads plus the redundants treated as applied loads; thus, since external forces (the redundants) acting on the basic statically determinate truss are the quantities solved for, Eq. (7.7.6) corresponds to the force method.

Redundant Force Method Using Castigliano's Second Theorem

Examine the truss shown in Fig. 7.7.1a, which is statically indeterminate to the second degree. Select the two redundants to be removed (in this case diagonal member number 6 and the vertical reaction below member 8) and obtain a basic statically determinate truss. The internal forces F_j' are determined for the given loads on the basic truss, as in Fig. 7.7.1b. The redundants R_1 and R_2 are each applied to the statically determinate truss, as in Fig. 7.7.1c and d.

The total internal forces F_j may be expressed as

$$F_j = F_j' + R_1 u_{j1} + R_2 u_{j2} \tag{7.7.7}$$

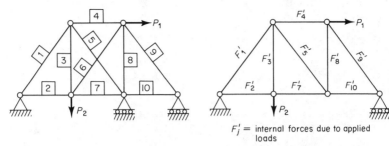

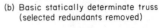

F'_j = internal forces due to applied loads

(a) Given truss

(b) Basic statically determinate truss (selected redundants removed)

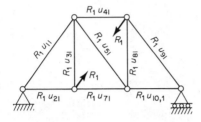

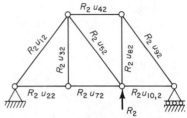

u_{j1} = internal forces due to unit load applied at the location of R_1

u'_{j2} = internal forces due to unit load applied at the location of R_2

(c) Basic statically determinate truss acted upon by redundant R_1

(d) Basic statically determinate truss acted upon by redundant R_2

Figure 7.7.1 Redundant force method using Castigliano's second theorem.

where F'_j = internal forces due to applied loads on the truss with redundants removed, Fig. 7.7.1b

u_{j1} = internal forces due to a unit load applied at location of redundant R_1, Fig. 7.7.1c

u_{j2} = internal forces due to unit load (a pair of unit loads in the present case) applied at location of redundant R_2, Fig. 7.7.1d

R_1, R_2 = redundants (Fig. 7.7.1c and d)

The total strain energy in the truss is therefore equal to

$$W_{\text{int}} = \sum_{j=1}^{NF} \frac{F_j^2 L_j}{2E_j A_j} = \sum \frac{(F'_j + R_1 u_{j1} + R_2 u_{j2})^2 L_j}{2E_j A_j} \tag{7.7.8}$$

Applying Castigliano's second theorem to Eq. (7.7.8) gives

$$\frac{\partial W_{\text{int}}}{\partial R_1} = \sum \frac{(F'_j + R_1 u_{j1} + R_2 u_{j2}) u_{j1} L_j}{E_j A_j} = -e_6 \tag{7.7.9}$$

where $(-e_6)$ is the relative displacement (together) between the ends of member number 6 when it is removed;

$$\frac{\partial W_{\text{int}}}{\partial R_2} = \sum \frac{(F'_j + R_1 u_{j1} + R_2 u_{j2}) u_{j2} L_j}{E_j A_j} = \Delta \tag{7.7.10}$$

where Δ is the displacement (upward deflection) of the joint where R_2 is applied.

For the example used for this discussion, the elongation e_6 is equal to $+R_1 L_6/E_6 A_6$, so that the relative displacement (together) between the ends of member 6 should be $-e_6$. The upward deflection Δ is zero since R_2 is the support reaction. Thus, Eqs. (7.7.9) and (7.7.10) become

$$\left. \begin{aligned} \sum \frac{(F'_j + R_1 u_{j1} + R_2 u_{j2})u_{j1} L_j}{E_j A_j} &= -\frac{R_1 L_6}{E_y A_y} \\ \sum \frac{(F'_j + R_1 u_{j1} + R_2 u_{j2})u_{j2} L_j}{E_j A_j} &= 0 \end{aligned} \right\} \qquad (7.7.11)$$

Equations (7.7.11) give two equations with two unknowns (R_1 and R_2) which can be solved simultaneously. These equations are in fact identical to the compatibility conditions described in Section 7.3. There will be as many such equations as there are redundants.

Example 7.7.1

Analyze the statically indeterminate truss of Example 7.3.1 (Fig. 7.3.1a) using Castigliano's second theorem.

Solution

(a) *Select redundants to establish basic statically determinate truss and state the internal forces in terms of the external forces.* As in the Maxwell–Mohr approach of Example 7.3.1, a redundant is chosen (in this case, reaction R_3 given by the problem statement) to make the truss statically determinate. The basic statically determinate truss and the internal forces F'_j due to the known applied loads are shown in Fig. 7.3.1b. The redundant R_3 is also applied to the basic truss and the internal forces u_j are given in Fig. 7.3.2c. The total internal forces are

$$F_j = F'_j + R_3 u_j$$

(b) *Apply Castigliano's second theorem.* The strain energy is

$$W_{\text{int}} = \sum \frac{(F'_j + R_3 u_j)^2 L_j}{2 E_j A_j}$$

Then

$$\frac{\partial W_{\text{int}}}{\partial R_3} = \sum \frac{(F'_j + R_3 u_j)u_j L_j}{E_j A_j} = 0$$

This equation above equals zero because the deflection at the joint where R_3 acts, in the direction of R_3, equals zero. The equation above then becomes

$$\sum \frac{F'_j u_j L_j}{E_j A_j} + \sum \frac{R_3 u_j^2 L_j}{E_j A_j} = 0$$

from which R_3 may be solved.

To correlate with the Maxwell–Mohr method of Example 7.3.1, one may interpret the first term of the preceding equation as the upward deflection at joint L_2 occurring on the basic truss without R_3 acting; therefore, it is the opposite of X_1 in Example 7.3.1. The second term, according to the virtual work (unit load) method, is the upward deflection due to R_3 acting on the basic truss; therefore, it is the same as X_2 in Example 7.3.1. This equation is then identical to the consistent deformation equation from which R_3 was solved in Example 7.3.1, part (c).

The practical way of solving R_3 using the equation in part (b) is to arrange computations in tabular form as follows:

Member (Fig. 7.3.1a)	S_j (EA/L)	F'_j (Fig. 7.3.1b)	u_j (Fig. 7.3.2c)	$\dfrac{F'_j u_j}{S_j}$ ($\times 10^{-6}$)	$\dfrac{u_j^2}{S_j}$ ($\times 10^{-6}$)
$L_0 U_1$	500	-10	$+\frac{5}{12}$	$-$ 8,333.33	347.22
$L_0 L_1$	1000	$+15$	$-\frac{1}{4}$	$-$ 3,750.00	62.50
$U_1 L_1$	250	$+18$	0	0	0
$U_1 U_2$	1000	$+ 1.5$	$+\frac{1}{2}$	$+$ 750.00	250.00
$U_1 L_2$	500	-12.5	$-\frac{5}{12}$	$+10,416.67$	347.22
$L_1 L_2$	1000	$+15$	$-\frac{1}{4}$	$-$ 3,750.00	62.50
$U_2 L_2$	250	$+10$	$-\frac{2}{3}$	$-26,666.67$	1,777.78
$U_2 L_3$	500	-12.5	$+\frac{5}{6}$	$-20,833.33$	1,388.89
$L_2 L_3$	1000	$+ 7.5$	$-\frac{1}{2}$	$-$ 3,750.00	250.00
				$-55,916.66$	4,486.11

Thus,

$$\sum \frac{F'_j u_j L}{E_j A_j} + \sum \frac{R_3 u_j^2 L_j}{E_j A_j} = 0$$

$$-55,916.66 + R_3(4486.11) = 0$$

$$R_3 = +12.46 \text{ kips} \quad \text{(acting upward)}$$

This solution is identical to that of Example 7.3.1.

Example 7.7.2

Analyze the statically indeterminate truss of Fig. 7.7.2a using Castigliano's second theorem.

Solution

(a) *Establish the basic statically determinate truss.* Taking member 6 (Fig. 7.7.2) and the interior roller support as the redundants leaves the basic statically determinate truss as shown in Fig. 7.7.2c.

(b) *Apply a unit load to the basic truss for each of the redundants.* For member 6 as the redundant R_1, a pair of unit loads is applied to obtain u_{j1}, as given in Fig. 7.7.2d. For the interior roller reaction as the redundant R_2, a unit load is applied to obtain u_{j2} as in Fig. 7.7.2e.

(c) *Apply Castigliano's second theorem.* The total internal forces are

$$F_j = F'_j + R_1 u_{j1} + R_2 u_{j2}$$

The strain energy is

$$W_{\text{int}} = \sum \frac{(F'_j + R_1 u_{j1} + R_2 u_{j2})^2 L_j}{2 E_j A_j}$$

Then, applying Eq. (7.7.6) for each selected redundant,

$$\frac{\partial W_{\text{int}}}{\partial R_1} = \sum \frac{(F'_j + R_1 u_{j1} + R_2 u_{j2}) u_{j1} L_j}{E_j A_j} = -e_6$$

$$\frac{\partial W_{\text{int}}}{\partial R_2} = \sum \frac{(F'_j + R_1 u_{j1} + R_2 u_{j2}) u_{j2} L_j}{E_j A_j} = 0$$

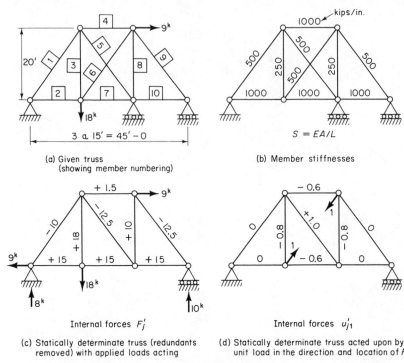

(a) Given truss
(showing member numbering)

(b) Member stiffnesses

$S = EA/L$

Internal forces F_j'

(c) Statically determinate truss (redundants removed) with applied loads acting

Internal forces u_{j1}'

(d) Statically determinate truss acted upon by unit load in the direction and location of R_1

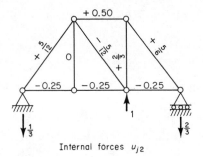

Internal forces u_{j2}

(e) Statically determinate truss acted upon by unit load in the direction and location of R_2

Figure 7.7.2 Statically indeterminate truss analysis of Example 7.7.2.

The first equation is set equal to the decrease in the distance between the ends of member 6 when it is removed. Thus, the term on the right-hand side of the equation is negative. In the second equation the deflection at R_2 is zero. Expanding the two equations above gives

$$\sum \frac{F_j' u_{j1} L_j}{E_j A_j} + R_1 \sum \frac{u_{j1}^2 L_j}{E_j A_j} + R_2 \sum \frac{u_{j1} u_{j2} L_j}{E_j A_j} = \frac{-R_1 L_6}{E_6 A_6}$$

$$\sum \frac{F_j' u_{j2} L_j}{E_j A_j} + R_1 \sum \frac{u_{j1} u_{j2} L_j}{E_j A_j} + R_2 \sum \frac{u_{j2}^2 L_j}{E_j A_j} = 0$$

The terms of the equations are evaluated in tabular form as follows:

Member	S_j (EA/L)	F'_j Fig. 7.7.2c	u_{j1} Fig. 7.7.2d	u_{j2} Fig. 7.7.2e	$\dfrac{F'_j u_{j1}}{S_j}$ ($\times 10^{-6}$)	$\dfrac{F'_j u_{j2}}{S_j}$ ($\times 10^{-6}$)	$\dfrac{u_{j1}^2}{S_j}$ ($\times 10^{-6}$)	$\dfrac{u_{j2}^2}{S_j}$ ($\times 10^{-6}$)	$\dfrac{u_{j1}u_{j2}}{S_j}$ ($\times 10^{-6}$)
1 L_0U_1	500	-10	0	$+\frac{5}{12}$	0	$-8,333$	0	347.2	0
2 L_0L_1	1000	$+15$	0	$-\frac{1}{4}$	0	$-3,750$	0	62.5	0
3 U_1L_1	250	$+18$	-0.8	0	$-57,600$	0	2,560	0	0
4 U_1U_2	1000	$+1.5$	-0.6	$+\frac{1}{2}$	-900	$+750$	360	250.0	-300.0
5 U_1L_2	500	-12.5	$+1.0$	$-\frac{5}{12}$	$-25,000$	$+10,417$	2,000	347.2	-833.3
6	500								
7 L_1L_2	1000	$+15$	-0.6	$-\frac{1}{4}$	$-9,000$	$-3,750$	360	62.5	$+150.0$
8 U_2L_2	250	$+10$	-0.8	$-\frac{2}{3}$	$-32,000$	$-26,667$	2,560	1,777.8	$+2,133.3$
9 U_2L_3	500	-12.5	0	$+\frac{5}{6}$	0	$-20,833$	0	1,388.9	0
10 L_2L_3	1000	$+7.5$	0	$-\frac{1}{2}$	0	$-3,750$	0	250.0	0
					$-124,500$	$-55,916$	7,840	4,486.1	$+1,150.0$

The two equations then become

$$-124,500 + R_1(7840) + R_2(1150.0) = -R_1(2000)$$
$$-55,916 + R_1(1150.0) + R_2(4486.1) = 0$$

or

$$9840R_1 + 1150.0R_2 = 124,500$$
$$1150.0R_1 + 4486.1R_2 = 55,916$$

(d) *Solve for the redundants.* Solving the two equations obtained in part (c) for R_1 and R_2 gives

$$R_1 = F_6 = 11.54 \text{ kips} \quad \text{(tension)}$$
$$R_2 = 9.51 \text{ kips} \quad \text{(up)}$$

(e) *Make statics checks.* Although not shown, the final internal forces may be computed by using $F_j = F'_j + u_{j1}R_1 + u_{j2}R_2$. Then the eight statics checks along the eight degrees of freedom may be made.

(f) *Make compatibility checks.* The compatibility checks may be made in one of two ways. In the first approach, one may solve the problem over again by choosing, say, member 5 and the reaction at L_3 as the redundants, then establish the two compatibility conditions by means of Castigliano's second theorem, and see that the two equations are satisfied by the now already known answers for F_5 and R at L_3. In the second approach, the same new redundants, member 5 and reaction at L_3, may be removed; but by means of the joint displacement equation, all joint displacements can be determined using the final elongations of the members without member 5. The two compatibility checks are then: (1) the change in distance between the ends of member 5 must be equal to the change in length of that member, and (2) the deflection at L_3 must come out zero. These checks are shown in Fig. 7.7.3.

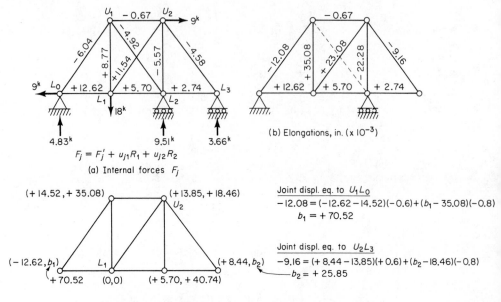

(b) Elongations, in. (× 10⁻³)

$$F_j = F_j' + u_{j1}R_1 + u_{j2}R_2$$

(a) Internal forces F_j

Joint displ. eq. to $U_1 L_0$

$-12.08 = (-12.62 - 14.52)(-0.6) + (b_1 - 35.08)(-0.8)$

$b_1 = +70.52$

Joint displ. eq. to $U_2 L_3$

$-9.16 = (+8.44 - 13.85)(+0.6) + (b_2 - 18.46)(-0.8)$

$b_2 = +25.85$

(c) Joint displacements using reference point at L_1 and reference member $L_1 U_2$

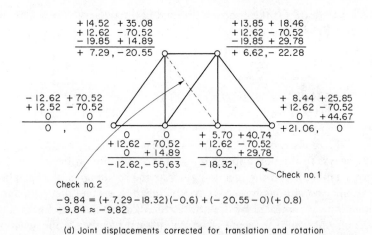

Check no. 2

$-9.84 = (+7.29 - 18.32)(-0.6) + (-20.55 - 0)(+0.8)$

$-9.84 \approx -9.82$

(d) Joint displacements corrected for translation and rotation

Figure 7.7.3 Statics and compatibility checks for Example 7.7.2.

234

7.8 THEOREM OF LEAST WORK

An extension of Castigliano's energy theorems is the "Theorem of Least Work," a term given by Ménabréa [13] and proved by Castigliano [12, p. 35]. The theorem as stated by Castigliano is as follows:* "The stresses which occur between the molecular couples of a body or structure after strain are such as to render the internal work a minimum, regard being had to the equations which express equilibrium between these stresses around each molecule." This statement may seem obscure by the use of the terms "molecular couples" and "molecules." From the point of view of trusses, the "molecules" or "molecular couples" refer to truss members, since they link the nodes (joints) together. Castigliano therefore stated [12, p. 35] the theorem in another way as follows:* "Whatever may be the unknown quantities in terms of which the internal work of a structure is expressed, the values which they must have after strain of the structure are those which render this internal work a minimum, having regard to the fundamental equations which obtain between them."

Referring to Eq. (7.7.6), Castigliano's second theorem,

$$X_m = \frac{\partial W_{\text{int}}}{\partial P_m} \qquad [7.7.6]$$

one may note that when applied to a redundant truss member, such as $P_m = F_i$ in Fig. 7.8.1, Eq. (7.7.6) becomes

$$\frac{\partial W_{\text{int}}}{\partial F_i} = -\frac{F_i L_i}{E_i A_i} \qquad (7.8.1)$$

This condition has been used as a part of Example 7.7.2. Next note that the right-hand side of Eq. (7.8.1) represents the derivative with respect to F_i of the strain energy, $F_i^2 L_i / 2 E_i A_i$, of the redundant bar. The original strain energy expression W_{int} included all bars except the redundant; thus, if the expression for strain energy included *all* the bars in the truss, Eq. (7.8.1) would become

$$\frac{\partial W}{\partial F_i} = 0 \qquad (7.8.2)$$

which is mathematically the *Theorem of Least Work*.

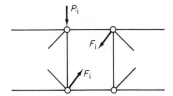

Figure 7.8.1 Portion of statically indeterminate truss with redundant internal force F_i.

For a linear elastic structure, the theorem states that the redundant forces within a statically indeterminate structure are such as to make the total strain energy a minimum. Mathematically, the application of Eq. (7.8.2) is identical to Eq. (7.7.6) (Castigliano's second theorem) for statically indeterminate trusses (see Example 7.7.2); thus, no examples are added here.

*From E. S. Andrews' English translation [12].

SELECTED REFERENCES

1. James Clerk Maxwell, "On the Calculation of the Equilibrium and Stiffness of Frames," *Philosophical Magazine* (4), *27* (1864), 294–299. (See also *The Scientific Papers of James Clerk Maxwell*, Vol. 1, ed. W. D. Niven, Cambridge University Press, Cambridge, 1890, pp. 598–604).

2. Otto Mohr, "Beitrag zur Theorie des Fachwerks," *Zeitschrift des Architekten- und Ingenieur-Vereins zu Hannover*, *20* (1874), 509–526, plus Plates 609–610; *21* (1875), 17–35, plus Plates 611–613.

3. Heinrich F. B. Müller-Breslau, *Die neueren Methoden der Festigkeitslehre und der Statik der Baukonstruktionen*, Baumgartner's Buchhandlung, Leipzig, 1886, 191 pp.

4. H. M. Westergaard, "One Hundred Fifty Years Advance in Structural Analysis," *Transactions*, ASCE, *94* (1930), 226–240.

5. Stephen P. Timoshenko, *History of Strength of Materials*, McGraw-Hill Book Company, New York, 1953, pp. 204–208, 304–322.

6. Wang, C. K., *Matrix Methods of Structural Analysis*, 2nd ed., American Publishing Company, Madison, Wisconsin, 1970, Chap. 8 and Appendix F.

7. E. Betti, *Il Nuovo Cimento* (2), *7–8* (1872).

8. John William Strutt, Lord Rayleigh, "On Some General Theorems Relating to Vibrations," *Proceedings of the London Mathematical Society*, 1873.

9. John William Strutt, Lord Rayleigh, "A Statical Theorem," *The London, Edinburgh, and Dublin Philosophical Magazine and Journal of Science*, 4th Series, *48*, December 1874, 452–456; *49*, March 1875, 183–185.

10. John William Strutt, Lord Rayleigh, *The Theory of Sound*, Vol. 1, 1st ed., London, 1877, 2nd ed., 1894. (Reprinted by Dover Publications, Inc., 1945.) (See Secs. 72, 77, 78, and 107–111a of the 1894 edition.)

11. C. A. Castigliano, "Intorno all'equilibrio dei sistemi elastici" and "Nuova teoria intorno all'equilibrio dei sistemi elastici," *Atti della Reale Accademia delle Scienze di Torino*, Vols. X, XI, Turin, 1875.

12. Carlo Alberto Pio Castigliano, *Théorie de l'équilibre des systèmes élastiques et ses applications*, A. F. Negro, Turin, 1879. (English translation by E. S. Andrews, *Elastic Stresses in Structures*, Scott, Greenwood & Son, London, 1919; reprinted under the title, *The Theory of Equilibrium of Elastic Systems and Its Application*, Dover Publications, Inc., New York, 1966.)

13. L. F. Ménabréa, "Nouveau principe sur la distribution des tensions dans les systèmes élastiques," *Comptes Rendus des Séances de l'Académie des Sciences*, Paris, *46* (1858), 1056–1060.

PROBLEMS

When axial forces in the members are required by the problem statement, a line drawing of the truss is to be made stating all internal forces directly on the members together with horizontal and vertical components, and the external reactions are to be shown.

7.1. For the truss of the accompanying figure, use the consistent-deformation method (Maxwell–Mohr method) to do the following: (a) determine the horizontal displacement of joint C if bar CD is removed; (b) determine the horizontal displacement of joint C, if bar CD is removed and a 1-kip force is applied horizontally to the right at C; (c) determine the forces in all members for the original truss and loading; (d) check that the compatibility condition is satisfied.

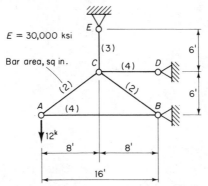

Probs. 7.1 and 7.20

7.2. For the truss of the accompanying figure, determine the reactions R_1 through R_4 and the axial forces in all bars. Also determine the deflections of all joints. Use R_3 as the redundant force. Check the results.

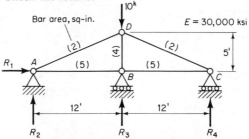

Prob. 7.2

7.3. For the truss of the accompanying figure, use the axial force in AB as the redundant to find the axial forces in all members.

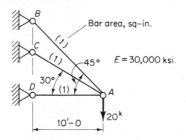

Probs. 7.3 and 7.23

7.4. Analyze the statically indeterminate truss as shown in the accompanying figure by the redundant force method using reaction R_3 at L_1 as the redundant. Compute all axial forces and joint displacements, and show that all statics and compatibility conditions are satisfied.

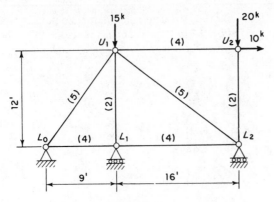

E = 30,000 ksi
Numbers in ( ) are areas in sq-in.   **Prob. 7.4**

7.5. For the truss shown in part (a) of the accompanying figure the relative displacement
between joints A and D is 40×10^{-3} in. toward each other. For the truss shown in
part (b), the relative displacement between joints A and D is 9.2×10^{-3} in. toward
each other. Determine the axial forces in all bars of the truss shown in part (c).

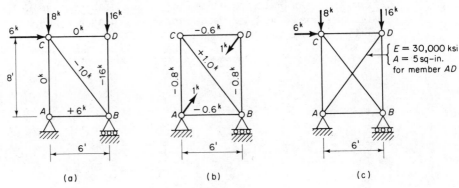

(a)   (b)   (c)

Prob. 7.5

7.6. The solution for the axial forces in the members of the statically indeterminate truss
shown in the accompanying figure has been obtained and these forces are shown on
the members. Check if the compatibility condition is satisfied.

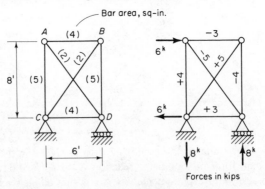

Prob. 7.6

7.7. Analyze the truss of the accompanying figure by using the reaction at D as the redundant. Compute all axial forces and show that statics and compatibility conditions are satisfied.

7.8. Repeat Prob. 7.7 except use the reaction at B as the redundant.

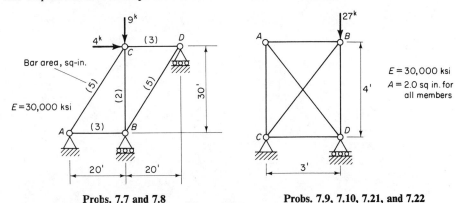

Probs. 7.7 and 7.8   Probs. 7.9, 7.10, 7.21, and 7.22

7.9. For the truss of the accompanying figure, use the consistent deformation (Maxwell–Mohr) concept of the redundant force method with the axial force in bar AB as the redundant force to determine the axial forces in all bars of the truss.

7.10. Repeat Prob. 7.9 using the axial force in member BD as the redundant force.

7.11. Analyze the statically indeterminate truss as shown in the accompanying figure by the redundant force method using the force in bar L_1U_2 as the redundant. Compute all axial forces and joint displacements, and show that all statics and compatibility conditions are satisfied.

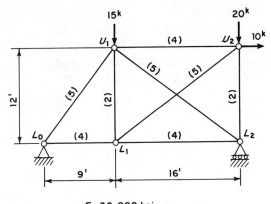

$E = 30,000$ ksi
Numbers in ( ) are areas in sq-in.

Prob. 7.11

7.12. Analyze the truss of the accompanying figure using the axial force in member AB as the redundant force. Compute all axial forces and show that statics and compatibility conditions are satisfied.

7.13. Analyze the truss of the accompanying figure using the axial force in member AC as the redundant force. Compute all axial forces and show that statics and compatibility conditions are satisfied.

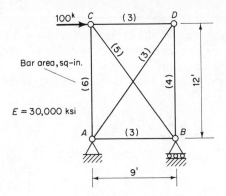

Prob. 7.12

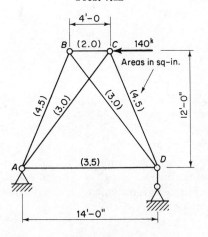

Prob. 7.13

7.14. Analyze the truss of the accompanying figure using the axial force in member $L_1 U_2$ as the redundant force. Solve for a 1-kip vertically downward load at L_2. Check statics and compatibility conditions.

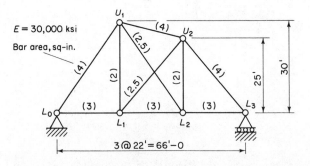

Probs. 7.14 to 7.17 and 7.24 to 7.26

7.15. Repeat Prob. 7.14 for a fabrication error of $\frac{1}{2}$ in. too long in member $L_0 U_1$, instead of the 1-kip load.

7.16. Repeat Prob. 7.14 for a fabrication error of $\frac{1}{2}$ in. too long in member $U_1 L_2$ instead of the 1-kip load.

7.17. Repeat Prob. 7.14 for a support settlement of $\frac{1}{2}$ in. at L_3, instead of the 1-kip load.

7.18. Analyze the truss of the accompanying figure using the axial force in one of the following members (as assigned by the instructor) as the redundant force: (a) L_1U_2, (b) U_1L_2, (c) U_1U_2, and (d) L_1L_2. Make all the statics and compatibility checks.

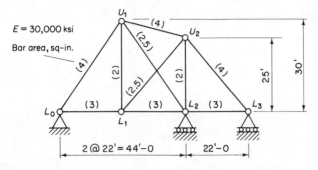

Prob. 7.18

7.19. Solve for the forces in all members of the truss in the accompanying figure using Castigliano's theorem, part I.

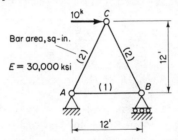

Prob. 7.19

7.20. Solve Prob. 7.1 using Castigliano's theorem, part II instead of the consistent deformation (Maxwell–Mohr) method.

7.21. Solve Prob. 7.9 using Castigliano's theorem, part II, instead of the consistent deformation (Maxwell–Mohr) method.

7.22. Solve Prob. 7.10 using Castigliano's theorem, part II, instead of the consistent deformation (Maxwell–Mohr) method.

7.23. Solve Prob. 7.3 using Castigliano's theorem, part II, instead of the consistent deformation (Maxwell–Mohr) method.

7.24. Solve Prob. 7.14 using Castigliano's theorem, part II, instead of the consistent deformation (Maxwell–Mohr) method.

7.25. Solve Prob. 7.15 using Castigliano's theorem, part II, instead of the consistent deformation (Maxwell–Mohr) method.

7.26. Solve Prob. 7.17 using Castigliano's theorem, part II, instead of the consistent deformation (Maxwell–Mohr) method.

[8]

Statically Indeterminate Beams

8.1 DEFINITION OF STATICAL INDETERMINACY

A statically indeterminate beam is one for which the bending moment and shear diagrams cannot be obtained by using statics alone. *Continuous beam* is the common term for a statically indeterminate beam which has more than two supports. The degree of statical indeterminacy of a beam is equal to the number of external reactions minus two when no internal hinges are present. For a continuous beam each additional span adds one additional reaction (roller type, giving rise to an additional unknown coplanar parallel force). Thus a two-span continuous beam with two exterior simple supports requires one compatibility condition in addition to the two independent equations of statics.

When multiple-span beams have internal hinges, the hinged locations provide points of known (zero) moment which are effectively additional conditions. Actual hinges are, in fact, frequently used in multispan bridge construction for various reasons. A common reason is to create a statically determinate system without having to provide the space at an interior support for two separate adjacent bearings (reactions) as would be necessary if two simply supported spans were used. In structural steel design the behavior of beams having hinges is utilized in the "plastic" design of continuous beams. Such *plastic hinges* are sections where yielding has occurred on every fiber of the cross section, in which case rotation may occur with constant bending moment at the "hinge." As long as the moment is *known* at the location, the extra statics condition is available, whether the known value is zero as for a real hinge or constant as for the so-called "plastic hinge."

242

Since each internal hinge in a multiple-span beam provides an additional equation of statics, a general expression for the degree NI of statical indeterminacy for a continuous beam is

$$NI = NR - NIH - 2 \qquad (8.1.1)$$

in which NR and NIH are the numbers of reactions and internal hinges, respectively. For example, the degree of statical indeterminacy of the beam in Fig. 8.1.1a is three, but that in Fig. 8.1.1b is two.

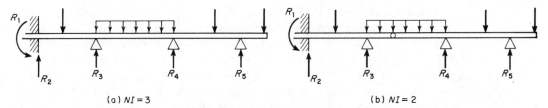

(a) $NI = 3$ (b) $NI = 2$

Figure 8.1.1 Statically indeterminate beams.

8.2 FORCE METHOD—CONSISTENT DEFORMATION USING REACTIONS AS REDUNDANTS

The division of structural analysis procedures into the two general categories of force methods and displacement methods, as discussed previously in relation to trusses, applies equally well to beams. In the force method either external reactions or internal bending moments in statically indeterminate beams can be used as redundants which are to be solved for as the basic unknowns. Having evaluated the "redundants" such that deformation compatibility is satisfied, the shear and bending moment variation for the entire beam can be determined.

The earliest reported solution for continuous beams was the force method applied by the German engineer Johann Albert Eytelwein [1] in 1808. He treated the beam with three and four supports by using the differential equation (see the double-integration method discussed in Section 5.2). The equation of the elastic curve was established in terms of the reactions taken as redundants. This deflection equation was evaluated at the locations of the redundant reactions, these deflections were set equal to zero, and the equations were then solved for the unknown reactions. Louis M. H. Navier [2] in 1826 extended the same procedure of using the elastic curve to a beam fixed at both ends and also to a beam fixed at one end and simply supported at the other. In general, these differential equation applications of the force method were extremely cumbersome. A breakthrough in concepts came in 1857 with the French engineer Émile P. B. Clapeyron's paper [3] giving the "Theorem of Three Moments," wherein the support bending moments were taken as the redundants. The theorem of three moments is treated in the next section. Although historically the theorem of three moments was the predominate method for analyzing statically indeterminate beams for the next 60 years after its introduction, the consistent deformation method is treated first here because of its prior introduction for use with statically indeterminate trusses.

The consistent deformation procedure for analysis of statically indeterminate beams and frames involves the imagined removal of restraints in the structure such that the remaining structure is statically determinate. The remaining statically determinate structure may be referred to as the "basic" statically determinate structure. Since there are usually choices of restraints to be removed, the "basic" structure is really *any* statically determinate structure remaining after the chosen redundant restraints have been removed.

Examine the fixed-ended beam of Fig. 8.2.1a. Choosing the reacting moments R_1 and R_2 as redundants means that the removal of the rotational restraints at the fixed ends leaves the "basic" statically determinate simply supported beam of Fig. 8.2.1b acted upon by the applied forces consisting of the uniform loading and the moments R_1 and R_2 at the ends. These forces acting on the "basic" beam cause end slopes which may be computed by the methods of Chapter 5. Since the end slopes of the actual structure are zero for fixed ends, the values of R_1 and R_2 can be determined such that zero end slopes result.

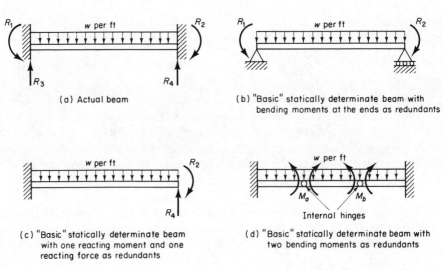

(a) Actual beam

(b) "Basic" statically determinate beam with bending moments at the ends as redundants

(c) "Basic" statically determinate beam with one reacting moment and one reacting force as redundants

(d) "Basic" statically determinate beam with two bending moments as redundants

Figure 8.2.1 Redundant restraints for a statically indeterminate fixed-ended beam.

Alternatively, one might choose the reacting moment R_2 and the reacting force R_4 as redundants, as shown in Fig. 8.2.1c. If this choice is made, the "basic" statically determinate beam is a cantilever beam acted on by the applied forces consisting of the uniform loading, the moment R_2, and the concentrated load R_4. Computing the deflection and the slope at the end of the cantilever using the methods of Chapter 5, the magnitudes of R_2 and R_4 can be determined such that at the free end both deflection and slope are zero.

Actually, the forces or moments to be taken as redundants and their corresponding restraints removed to give the basic statically determinate beam can be *anywhere* in the structure. For instance, the fixed-ended beam of Fig. 8.2.1a could have the rotational restraint at *any* two locations along the span removed by the insertion of internal hinges to give a "basic" statically determinate structure. In such a case as shown in Fig. 8.2.1d, the redundants are the bending moments M_a and M_b. In fact, the use of bending moments as redundants, with the continuity of slopes as conditions to solve for them, is the approach illustrated in the next section on the theorem of three moments.

In this section, treatment is limited to using *external* reactions as redundants. Once the redundant reacting forces have been found, the remaining reactions are determined from the conditions of statics after which the shear and bending moment diagrams for the actual beam may be obtained.

Generally, a solution can be first obtained by using one convenient set of redundants, after which a check is made by confirming that the compatibility conditions are satisfied using a different "basic" statically determinate beam. In the case of the fixed-ended beam of Fig. 8.2.1a, R_1 and R_2 may be obtained using the simply supported beam of Fig. 8.2.1b as the "basic" statically determinate beam, followed by a check verifying that R_2 and R_4 will cause zero slope and zero deflection at the free end of the "basic" cantilever beam of Fig. 8.2.1c.

Example 8.2.1

Solve for the reactions to the fixed-ended beam subject to uniform load shown in Fig. 8.2.2a using the method of consistent deformation taking the two reacting moments as redundants.

Solution

(a) *Use reacting moments at the ends as redundants.* This choice makes the "basic" statically determinate beam the one shown in Fig. 8.2.2b. Superposition may be used to permit computation of the end slopes separately due to the uniform load and due to the end moments R_1 and R_2, and then add the results.

Referring to Fig. 8.2.2b, the simply supported "basic" beam is treated in two parts. From symmetry, R_1 and R_2 are equal. The computation of the slope may be done using either the virtual work (unit load) method or the moment area method (or its modification, conjugate beam method). Using the moment area method, and referring to Fig. 8.2.2b,

$$\theta_A = \theta_{A1} - \theta_{A2} = \frac{\Delta'_{BA(1)} - \Delta'_{BA(2)}}{L} \tag{a}$$

where $\Delta'_{BA(1)}$ and $\Delta'_{BA(2)}$ are the moments of the M/EI diagrams, for the uniform loading and the end moment loading, respectively, taken about B. These Δ'_{BA} values are the deflections between tangents at A and B measured at B. Thus,

$$\Delta'_{BA(1)} = \frac{2}{3}\left(\frac{wL^2}{8EI}\right)(L)\left(\frac{L}{2}\right) = \frac{wL^4}{24EI}$$

and

$$\Delta'_{BA(2)} = \frac{R_1}{EI}(L)\left(\frac{L}{2}\right) = \frac{R_1L^2}{2EI}$$

which gives, according to Eq. (a),

$$\theta_A = \frac{wL^3}{24EI} - \frac{R_1L}{2EI}$$

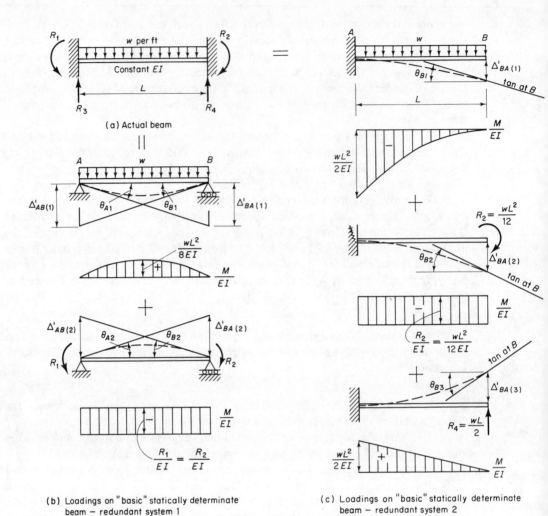

Figure 8.2.2 Solution of fixed-ended uniformly loaded beam of Example 8.2.1.

Applying the consistent deformation requirement that the actual $\theta_A = 0$,

$$R_1 = \frac{wL^2}{12} = R_2$$

By symmetry,

$$R_3 = R_4 = \frac{wL}{2}$$

In later chapters the concept of *fixed-end moment* will be used. Here the fixed-end moment for uniform load is $wL^2/12$. Note that the answers for R_1 and R_2 are positive. This is due to the fact that R_1 and R_2 were each shown as unknowns in the correct direction. The bending moments at the ends of the fixed-ended beams are, however, negative because the moments R_1 and R_2 each cause curvature that is concave downward (i.e., tension at the top of the beam).

(b) *Check correctness of solution by using different redundants.* Using the cantilever beam of Fig. 8.2.2c as the "basic" statically determinate beam, the moment area method will be used to compute θ_B and Δ_B. Since the tangent to the elastic curve at A is horizontal, the deflection Δ'_{BA} between tangents equals the actual deflection Δ_B, and the angle between tangents at A and B is the actual slope θ_B. Thus, using directions as shown in Fig. 8.2.2,

$$\theta_B = \theta_{B1} + \theta_{B2} - \theta_{B3} = \frac{M}{EI} \text{ area between } A \text{ and } B$$

$$= +\frac{1}{3}\left(\frac{wL^2}{2EI}\right)(L) + \frac{wL^2}{12EI}(L) - \frac{1}{2}\left(\frac{wL^2}{2EI}\right)(L)$$

$$= \frac{wL^3}{EI}\left(+\frac{1}{6} + \frac{1}{12} - \frac{1}{4}\right) = 0 \quad \text{(check)}$$

$$\Delta_B = \Delta'_{BA(1)} + \Delta'_{BA(2)} - \Delta'_{BA(3)}$$

$$= \text{moment of } \frac{M}{EI} \text{ area between } A \text{ and } B \text{ about } B$$

$$= +\frac{wL^3}{6EI}\left(\frac{3}{4}L\right) + \frac{wL^3}{12EI}\left(\frac{L}{2}\right) - \frac{wL^3}{4EI}\left(\frac{2}{3}L\right)$$

$$= \frac{wL^4}{EI}\left(+\frac{1}{8} + \frac{1}{24} - \frac{1}{6}\right) = 0 \quad \text{(check)}$$

Thus, the answer is verified to be correct.

Example 8.2.2

Solve for the reactions to a fixed-ended beam subject to a concentrated load using the method of consistent deformation taking the reacting moments at the ends as redundants.

Solution

(a) *Obtain solution using a simply supported beam as the "basic" statically determinate beam.* In this case, the "basic" simply supported beam is loaded with a concentrated load W and two end moments R_1 and R_2, as shown in Fig. 8.2.3b.

Using the moment area method to compute θ_A and θ_B in Fig. 8.2.3b,

$$\theta_A = \theta_{A1} - \theta_{A2} - \theta_{A3} = \frac{\Delta'_{BA(1)} - \Delta'_{BA(2)} - \Delta'_{BA(3)}}{L}$$

$$= \frac{1}{L}\left(\text{moment of } \frac{M}{EI} \text{ areas between } A \text{ and } B \text{ about } B\right)$$

$$= \frac{1}{L}\left[\frac{1}{2}\left(\frac{Wab}{LEI}\right)(L)\left(\frac{L+b}{3}\right) - \frac{1}{2}\left(\frac{R_1}{EI}\right)(L)\left(\frac{2L}{3}\right) - \frac{1}{2}\left(\frac{R_2}{EI}\right)(L)\left(\frac{L}{3}\right)\right]$$

$$= \frac{Wab}{6EI}\left(\frac{L+b}{L}\right) - \frac{R_1 L}{3EI} - \frac{R_2 L}{6EI} \tag{a}$$

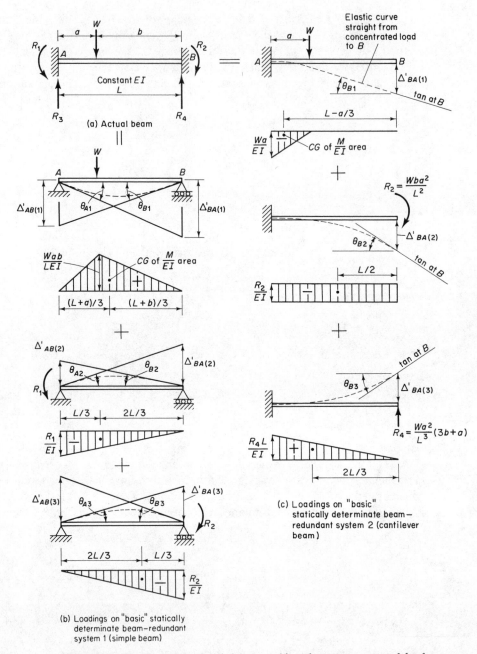

Figure 8.2.3 Solution of fixed-ended beam subjected to a concentrated load, Example 8.2.2.

$$\theta_B = \theta_{B1} - \theta_{B2} - \theta_{B3} = \frac{\Delta'_{AB(1)} - \Delta'_{AB(2)} - \Delta'_{AB(3)}}{L}$$

$$= \frac{1}{L}\left(\text{moment of } \frac{M}{EI} \text{ areas between } A \text{ and } B \text{ about } A\right)$$

$$= \frac{1}{L}\left[\frac{1}{2}\left(\frac{Wab}{LEI}\right)(L)\left(\frac{L+a}{3}\right) - \frac{1}{2}\left(\frac{R_1}{EI}\right)(L)\left(\frac{L}{3}\right) - \frac{1}{2}\left(\frac{R_2}{EI}\right)(L)\left(\frac{2L}{3}\right)\right]$$

$$= \frac{Wab}{6EI}\left(\frac{L+a}{L}\right) - \frac{R_1 L}{6EI} - \frac{R_2 L}{3EI} \tag{b}$$

Using the requirements of consistent deformation, θ_A and θ_B are actually zero. Setting Eqs. (a) and (b) equal to zero and solving simultaneously for R_1 and R_2,

$$R_1 = \frac{Wab^2}{L^2}; \qquad R_2 = \frac{Wba^2}{L^2}$$

These answers for R_1 and R_2 are positive; thus, the directions taken for them as unknowns are found to be correct. Referring to the M/EI diagrams of Fig. 8.2.3b, it is clear that there is negative bending moment at each end of the beam.

Using the values of R_1 and R_2, the remaining reactions, R_3 and R_4, can be determined using the two available equations of statics. Thus, referring to Fig. 8.2.3a, using $\sum M$ about end B gives

$$R_3 = \frac{Wb}{L} + \frac{R_1 - R_2}{L} = \frac{Wb^2}{L^3}(3a + b)$$

and $\sum M$ about end A gives

$$R_4 = \frac{Wa}{L} - \frac{R_1 - R_2}{L} = \frac{Wa^2}{L^3}(3b + a)$$

(b) *Check solution using a different system of redundants.* Using the cantilever beam of Fig. 8.2.3c as the "basic" statically determinate beam, the moment area method will be illustrated to compute θ_B and Δ_B. Any method for computing deformations is acceptable in the consistent deformation method.

Referring to Fig. 8.2.3c, and noting that the tangent to the elastic curve at A is the original line AB,

$$\theta_B = \theta_{B1} + \theta_{B2} - \theta_{B3}$$

$$= \frac{M}{EI} \text{ area between } A \text{ and } B$$

$$= \frac{1}{2}\left(\frac{Wa}{EI}\right)a + \frac{R_2}{EI}(L) - \frac{1}{2}\left(\frac{R_4 L}{EI}\right)(L)$$

Substituting the values of R_2 and R_4 found in part (a),

$$\theta_B = \frac{Wa^2}{2EI} + \frac{Wba^2}{LEI} - \frac{Wa^2(3b + a)}{2LEI}$$

$$= \frac{Wa^2}{2LEI}(L + 2b - 3b - a) = 0 \quad \text{(check)}$$

The deflection Δ_B equals for this case Δ'_{BA}, the moment of the M/EI area between A and B about B. Thus,

$$\Delta_B = \Delta'_{BA(1)} + \Delta'_{BA(2)} - \Delta'_{BA(3)}$$

$$= \frac{Wa^2}{2EI}\left(L - \frac{a}{3}\right) + \frac{Wba^2}{LEI}\left(\frac{L}{2}\right) - \frac{Wa^2(3b + a)}{2LEI}\left(\frac{2L}{3}\right)$$

$$= \frac{Wa^2}{6EI}(3L - a + 3b - 6b - 2a) = 0 \quad \text{(check)}$$

8.3 FORCE METHOD—THEOREM OF THREE MOMENTS

For the beam continuous over more than three supports, the use of reactions as redundants with the method of consistent deformation becomes a cumbersome process. In 1857, Émile Clapeyron presented [3] his *Theorem of Three Moments*, which can effectively relate the bending moments over three consecutive supports. Although actually first published [4] by the French engineer Bertot in 1855, the method had been used by Clapeyron in 1849, in connection with the reconstruction of a bridge near Paris and then used for several years before Clapeyron presented it to the Academy of Sciences [5].

The use of the bending moments at the supports as redundants in the application of the force method requires the insertion of hinges in the beam over the supports to obtain the basic statically determinate structure. The theorem of three moments is a way of grouping the unknown internal bending moments rather than treating one redundant at a time. Historically, Clapeyron presented his equation at a time when the only available method was the differential equation (see double integration method of Section 5.2). The theorem of three moments was derived by using the differential equation procedure to obtain the equations of deflection and slope in each of two adjacent spans and then satisfying the compatibility requirement that the slope at the right end of the left span must be equal to the slope at the left end of the right span. The virtual work (unit load) and moment area methods for computing deflections and slopes were not known in 1857. Thus, the consistent deformation (compatibility) requirement used to derive the theorem of three moments is the same, regardless of whatever procedure is used to compute the slope of the elastic curve.

Consider the two adjacent spans AB and BC in Fig. 8.3.1. Each span, when taken as a separate free body, is subjected to the load applied on the span and the bending moments at its ends. The bending moment diagram for a span in a continuous beam may always be considered as the sum of the bending moment diagrams of each of the loads considered separately. In this case, to remove the restraint offered by the internal bending moments, which are now the redundants chosen for the solution, it is necessary to imagine that hinges are inserted at the location of the redundants. The bending moments M_A, M_B, and M_C are then applied *at* the hinges. Recall from Chapter 4 (Fig. 4.2.1, for example) that the bending moment at a location in a beam appears on the adjacent free bodies as a pair of rotational arrows with opposite directions. Thus, the applied bending moment M_B, for example, consists of the counterclockwise arrow acting on the left side of the hinge inserted at B along with the clockwise arrow acting on the right side of the hinge.

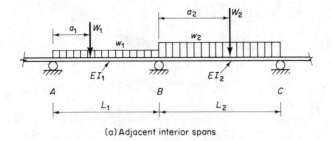

(a) Adjacent interior spans

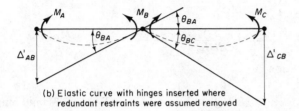

(b) Elastic curve with hinges inserted where redundant restraints were assumed removed

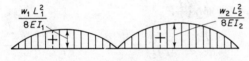

(c) M/EI diagrams due to uniform loading

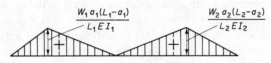

(d) M/EI diagrams due to concentrated loads

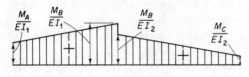

(e) M/EI diagrams due to moments M_A, M_B, and M_C inserted to replace redundant restraints

Figure 8.3.1 Development of the theorem of three moments for uniform full span loading and for concentrated loads.

Using the moment area method, the angles θ_{BA} and θ_{BC} are to be computed as follows:

$$\theta_{BA} = \frac{\Delta'_{AB}}{L_1} = \frac{1}{L_1}\left(\text{moment of } \frac{M}{EI} \text{ area between } A \text{ and } B \text{ about } A\right)$$

$$\Delta'_{AB} = \frac{2}{3}\left(\frac{w_1 L_1^2}{8EI_1}\right)(L_1)\left(\frac{L_1}{2}\right) + \frac{1}{2}\left(\frac{W_1 a_1(L_1-a_1)}{L_1 EI_1}\right)(L_1)\left(\frac{L_1+a_1}{3}\right)$$

$$+ \frac{1}{2}\left(\frac{M_A}{EI_1}\right)(L_1)\left(\frac{L_1}{3}\right) + \frac{1}{2}\left(\frac{M_B}{EI_1}\right)(L_1)\left(\frac{2L_1}{3}\right)$$

$$\theta_{BA} = \frac{w_1 L_1^3}{24EI_1} + \frac{W_1 a_1(L_1^2 - a_1^2)}{6EI_1 L_1} + \frac{M_A L_1}{6EI_1} + \frac{M_B L_1}{3EI_1} \qquad (8.3.1)$$

$$\theta_{BC} = \frac{\Delta'_{CB}}{L_2} = \frac{1}{L_2}\left(\text{moment of }\frac{M}{EI}\text{ area between }B\text{ and }C\text{ about }C\right)$$

$$\Delta'_{CB} = \frac{2}{3}\left(\frac{w_2 L_2^2}{8EI_2}\right)(L_2)\left(\frac{L_2}{2}\right) + \frac{1}{2}\left(\frac{W_2 a_2(L_2-a_2)}{L_2 EI_2}\right)(L_2)\left(\frac{L_2+L_2-a_2}{3}\right)$$

$$+ \frac{1}{2}\left(\frac{M_B}{EI_2}\right)(L_2)\left(\frac{2}{3}L_2\right) + \frac{1}{2}\left(\frac{M_C}{EI_2}\right)(L_2)\left(\frac{L_2}{3}\right)$$

$$\theta_{BC} = \frac{w_2 L_2^3}{24EI_2} + \frac{W_2 a_2(L_2 - a_2)(2L_2 - a_2)}{6EI_2 L_2} + \frac{M_B L_2}{3EI_2} + \frac{M_C L_2}{6EI_2} \qquad (8.3.2)$$

For consistent deformation,

$$\theta_{BA} + \theta_{BC} = 0 \qquad (8.3.3)$$

Thus, substituting Eqs. (8.3.1) and (8.3.2) into Eq. (8.3.3) and multiplying through by 6, the three-moment equation becomes

$$M_A\left(\frac{L_1}{EI_1}\right) + 2M_B\left(\frac{L_1}{EI_1} + \frac{L_2}{EI_2}\right) + M_C\left(\frac{L_2}{EI_2}\right)$$

$$= -\frac{w_1 L_1^3}{4EI_1} - \frac{w_2 L_2^3}{4EI_2} - \frac{W_1 a_1(L_1 - a_1)(L_1 + a_1)}{EI_1 L_1}$$

$$- \frac{W_2 a_2(L_2 - a_2)(2L_2 - a_2)}{EI_2 L_2} \qquad (8.3.4)$$

In Eq. (8.3.4), if more than one concentrated load acts in the span, there will be additional terms of the same form as those containing W_1 and W_2. Also, a_1 and a_2 are each measured from the left end of the span; a definition that aids in the use of the theorem of three moments for developing the equations for influence lines.

The unknown bending moments M_A, M_B, and M_C have been taken as applied in the positive direction so that the correct bending moment sign will automatically result.

Whereas Eq. (8.3.4) applies for uniform full-span loading and for concentrated loads, the formula may be generalized to become applicable for any loading. Noting from the derivation of Eq. (8.3.4) that the right-hand side of that equation is actually the contribution to the discontinuity angle $(\theta_{BA} + \theta_{BC})$ at B, as shown in Fig. 8.3.1b, multiplied by (-6). Thus, referring to Fig. 8.3.2, the right-hand side of the equation is $-6(\theta_{BA(1)} + \theta_{BC(1)})$. The *theorem of three moments* may then be generalized to be

$$M_A\left(\frac{L_1}{EI_1}\right) + 2M_B\left(\frac{L_1}{EI_1} + \frac{L_2}{EI_2}\right) + M_C\left(\frac{L_2}{EI_2}\right) = -6(\theta_{BA(1)} + \theta_{BC(1)}) \qquad (8.3.5)$$

Let A_{s1} and A_{s2} shown in Fig. 8.3.2c be the bending moment diagram areas on the left and right spans, respectively, due to any pattern of applied loads. The distances from the centroids of areas A_{s1} and A_{s2} to ends A and C, respectively, are designated $\bar{x}_1$ and $\bar{x}_2$, as shown in Fig. 8.3.2c. Then, using the moment area method,

$$\theta_{BA(1)} = \frac{\Delta'_{AB}}{L_1} = \frac{A_{s1}\bar{x}_1}{EI_1 L_1} \qquad (8.3.6)$$

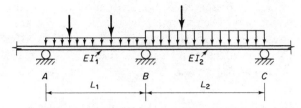

(a) Adjacent interior spans of a continuous beam

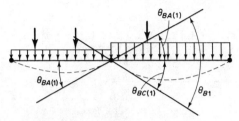

(b) Redundants removed; hinges inserted with
   only applied loads acting

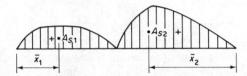

(c) Bending moment diagram due to applied loads
   on simply supported beams

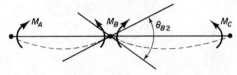

(d) Moments applied where redundant restraints
   were removed

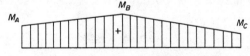

(e) Bending moment diagram due to moments
   inserted to replace redundant restraints

Figure 8.3.2 General theorem of three-
moments equation for any loading.

$$\theta_{BC(1)} = \frac{\Delta'_{CB}}{L_2} = \frac{A_{s2}\bar{x}_2}{EI_2 L_2} \tag{8.3.7}$$

Substitution of Eqs. (8.3.6) and (8.3.7) into Eq. (8.3.5) gives the generalized *theorem of three moments* as

$$M_A\left(\frac{L_1}{EI_1}\right) + 2M_B\left(\frac{L_1}{EI_1} + \frac{L_2}{EI_2}\right) + M_C\left(\frac{L_2}{EI_2}\right) = -\frac{6A_{s1}\bar{x}_1}{EI_1 L_1} - \frac{6A_{s2}\bar{x}_2}{EI_2 L_2} \tag{8.3.8}$$

where A_{s1}, A_{s2}, $\bar{x}_1$, and $\bar{x}_2$ are bending moment diagram areas and centroid locations as shown in Fig. 8.3.2c.

When Clapeyron originally presented [3] the theorem of three moments it was for full-span uniform loading only. The theorem may also be extended to cases where the beam supports are at different levels, as first presented [6] by Otto Mohr in 1860. Although not shown in this introductory text, the derivation can be made by using the same compatibility equation, Eq. (8.3.3).

Example 8.3.1

Analyze the continuous beam shown in Fig. 8.3.3 using the theorem of three moments.

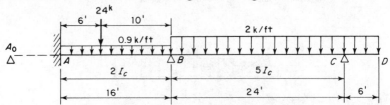

Figure 8.3.3 Continuous beam for Example 8.3.1.

Solution

(a) *Establish the degree of statical indeterminacy.* Using Eq. (8.1.1),

$$NI = NR - NIH - 2 = 4 - 0 - 2 = 2$$

The beam is statically indeterminate to the second degree.

(b) *Redundants for use in theorem of three moments.* The bending moments at A and B are used as redundants. The application of bending moments M_A and M_B at supports A and B is to satisfy the consistent deformation requirements for the continuity of slopes to the elastic curve at each side of those supports.

(c) *Establish the three-moment equations and solve for the bending moments required to satisfy conditions of consistent deformation.* Application of the theorem of three moments utilizes sequences of two adjacent spans. For this problem one imagines an infinitely stiff fictitious span A_0A to the left of support A, as in Fig. 8.3.3. Then two "three-moment" equations may be written, first for spans A_0A and AB, and then for spans AB and BC. The subscripts A, B, and C in the three-moment equation always refer to the moments at the left, center, and right supports, respectively, for the two adjacent spans. The cantilever portion CD is statically determinate; thus, the bending moment at C is

$$M_C = -\frac{2(6)^2}{2} = -36 \text{ ft-kips}$$

Applying the theorem of three moments for uniform loading and for concentrated loads, Eq. (8.3.4), to spans A_0A and AB,

$$M_{A0}\left(\frac{L_0}{\infty}\right) + 2M_A\left(\frac{L_0}{\infty} + \frac{16}{E(2I_c)}\right) + M_B\left(\frac{16}{E(2I_c)}\right)$$

$$= -\frac{w_0 L_0^3}{4(\infty)} - \frac{0.9(16)^3}{4E(2I_c)} - \frac{24(6)(10)(26)}{E(2I_c)(16)}$$

and then to spans AB and BC,

$$M_A\left(\frac{16}{E(2I_c)}\right) + 2M_B\left(\frac{16}{E(2I_c)} + \frac{24}{E(5I_c)}\right) + M_C\left(\frac{24}{E(5I_c)}\right)$$

$$= -\frac{0.9(16)^3}{4E(2I_c)} - \frac{2(24)^3}{4E(5I_c)} - \frac{24(6)(10)(22)}{E(2I_c)(16)}$$

Noting that anything divided by infinity is zero and canceling EI_c from the denominator of each term, the two three-moment equations become

$$16M_A + \quad 8M_B \qquad\qquad = -1630.80$$
$$8M_A + 25.6M_B + 4.8M_C = -2833.20$$

The value of $M_C = -36.0$ as determined previously is substituted and the two equations with the two unknowns M_A and M_B become

$$16M_A + \quad 8M_B = -1630.80$$
$$8M_A + 25.6M_B = -2660.40$$

Solving the two equations above,

$$M_A = -59.22 \text{ ft-kips}$$
$$M_B = -85.42 \text{ ft-kips}$$

(d) *Obtain answer in the form of shear and moment diagrams.* While the solution for M_A and M_B eliminates the statical indeterminacy, the real objective in structural analysis is to provide the internal forces (shear and bending moment diagrams) so that the members can be designed. Thus, the answers for this example are the shear and moment diagrams for the entire beam, $ABCD$, as shown in Fig. 8.3.4a.

For span AB, the free-body diagram is drawn as in Fig. 8.3.4, containing the end moments 59.22 and 85.42 ft-kips in addition to the gravity loads. The reactions at each end are computed as the sum of the contributions of uniform load, concentrated load, and end moments (taking actual values without signs), as follows:

$$R_{AB} = \frac{wL}{2} + \frac{Wb}{L} + \frac{M_A - M_B}{L}$$

$$= \frac{0.9(16)}{2} + \frac{24(10)}{16} + \frac{59.22 - 85.42}{16}$$

$$= +7.20 + 15.00 - 1.638 = +20.562 \text{ kips}$$

$$R_{BA} = \frac{wL}{2} + \frac{Wa}{L} - \frac{M_A - M_B}{L}$$

$$= +7.20 + \frac{24(6)}{16} + 1.638 = +17.838 \text{ kips}$$

In similar fashion the end reactions R_{BC} and R_{CB} are computed to be $+26.059$ kips and $+21.941$ kips, respectively. The end reaction R_{CD} is obtained from summation of vertical forces to be $+12.00$ kips.

The external reaction at a support equals the sum of the end reactions to the spans adjacent to that support; thus,

$$R_A = R_{AB} = +20.562 \text{ kips} \quad \text{(up)}$$
$$R_B = R_{BA} + R_{BC} = +17.838 + 26.059 = +43.897 \text{ kips} \quad \text{(up)}$$
$$R_C = R_{CB} + R_{CD} = +21.941 + 12.000 = +33.941 \text{ kips} \quad \text{(up)}$$

The shear diagrams may now be drawn as shown in Fig. 8.3.4, from which the bending moment diagrams may be drawn using Eq. (4.4.1), that is, $\int dM = \int V\,dx$, which means that the area under the shear diagram between two points along the beam

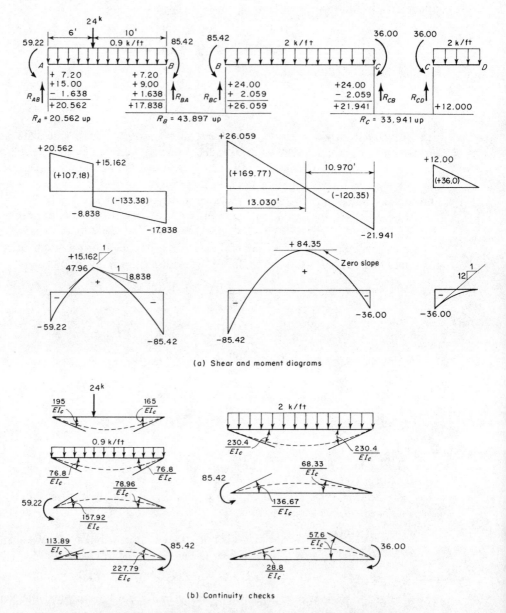

(a) Shear and moment diagrams

(b) Continuity checks

Figure 8.3.4 Solution by the three-moment equation, Example 8.3.1.

equals the change in bending moment between those points. The areas of the parts of the shear diagrams are given in parentheses in Fig. 8.3.4 and the complete bending moment diagrams are drawn, showing sign and giving numerical values for critical ordinates.

Instead of using the areas of the shear diagram, the bending moment diagrams may be drawn by superposition of the bending moment diagrams of the component loadings, as shown in Fig. 8.3.5. This procedure is frequently the most expeditious,

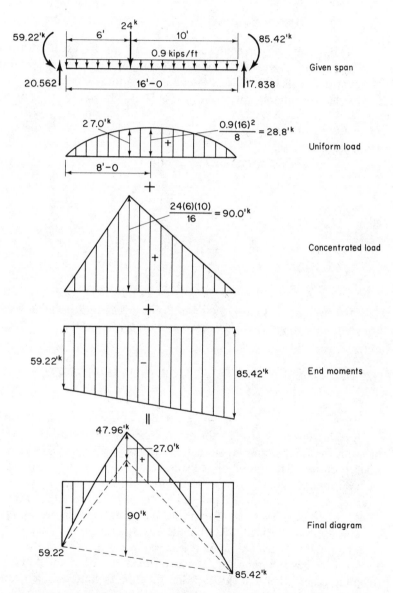

Figure 8.3.5 Bending moment diagram for span of a continuous beam, established by superposition.

both for plotting the diagrams from which the needed information may be scaled to design the members, and for ease in checking the statically indeterminate solution.

(e) *Check the results obtained for M_A and M_B.* Actually, this should be done before completely drawing the final shear and moment diagrams. Making use of the component bending moment diagrams provides the easiest means of checking consistent deformation. The two checks are as follows:

1. $\theta_A(\text{span } AB) = 0$
2. $\theta_B(\text{span } AB) = \theta_B(\text{span } BC)$

The moment area or conjugate beam methods are used on each of the simple beams loaded with one component of the loading at a time to determine the end slopes, as shown in Fig. 8.3.4b. These component loadings include not only the applied loads on the span but also the bending moments acting at the ends of each span, such as $M_A = 59.22$ ft-kips and $M_B = 85.42$ ft-kips acting on AB, and $M_B = 85.42$ ft-kips acting on BC. In other words, the answers are used to verify the consistent deformation requirements.

Thus, using clockwise slopes as positive,

$$\theta_A = \frac{1}{EI_c}(+195 + 76.8 - 157.92 - 113.89) = \frac{1}{EI_c}(271.80 - 271.81) \approx 0 \quad \text{(check)}$$

and

$$\theta_B(\text{span } AB) = \frac{1}{EI_c}(-165 - 76.8 + 78.96 + 227.79) = \frac{1}{EI_c}(+64.95)$$

$$\theta_B(\text{span } BC) = \frac{1}{EI_c}(+230.40 - 136.67 - 28.8) = \frac{1}{EI_c}(+64.93)$$

Thus, the two checks that $\theta_A = 0$ and $\theta_B(\text{span } AB) = \theta_B(\text{span } BC)$ are obtained and the solution for M_A and M_B is verified as correct.

8.4 INFLUENCE LINES FOR STATICALLY INDETERMINATE BEAMS

The analysis of beams, either statically determinate or statically indeterminate, is useful primarily to satisfy the objective of providing the shear and bending moment diagrams for use in design when the appropriate loads have been placed in the appropriate position. If the gravity live loads, which are always considered movable, have not been applied in the correct position (or positions) to cause maximum (or minimum) shear or bending moment the analysis effort has been wasted. The reader at this stage should review the discussion of movable loads, the definition of an influence line, and its application to statically determinate beams as given in Chapter 6.

To restate the definition, an influence line is the graph of the value of a function (say shear or bending moment) at *one location* on a beam due to a unit load moving along the beam, when the value of that function is plotted directly at the location of the unit load.

The influence lines for functions in a statically indeterminate structure, unlike those for functions in statically determinate structures, do not consist of straight segments, but instead are curves. These influence lines are curves because the value of the function is not obtainable from the laws of statics, from which for the statically determinate beam the function is a linear variation with respect to the position of the unit load along the span. For continuous beams, an equation of the function must be derived for each portion of the influence line between sections of discontinuity using any convenient method of statically indeterminate analysis. Alternatively, of course, the influence line could be obtained by computing values at sufficient individual points along the beam to permit drawing a smooth curve.

Influence lines for continuous beams were first treated in detail by W. Fränkel [7] in 1876 and then later extensively treated by Winkler, the orginator of the concept, in his 1886 textbook [8].

Influence Line for Reaction

Consider first as a qualitative example the influence line for the reaction R_B on the two-span continuous beam of Fig. 8.4.1a. The loading on the beam is a unit load at location x, a variable distance between A and C measured from A. Selecting the reaction at B as the redundant, the "basic" statically determinate beam is a simply supported one from A to C, as in Fig. 8.4.1b with the unit load acting and Fig. 8.4.1c with R_B acting where the redundant support has been removed. The symbol δ is used to represent the deflection due to a 1-kip load, with the first subscript to identify the location of the deflection and the second subscript to identify the location of the 1-kip load. Thus, δ_{BB} is the deflection at B due to 1 kip applied at B, in which case when R_B is applied the deflection is $R_B\delta_{BB}$.

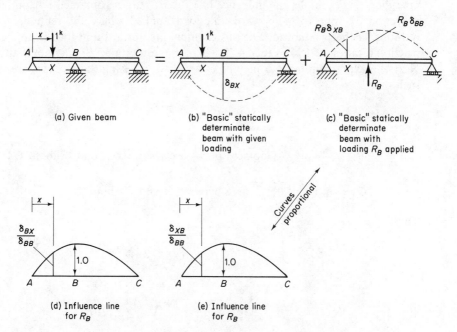

(a) Given beam

(b) "Basic" statically determinate beam with given loading

(c) "Basic" statically determinate beam with loading R_B applied

(d) Influence line for R_B

(e) Influence line for R_B

Figure 8.4.1 Influence line for reaction on a two-span continuous beam.

Consistent deformation requires the total deflection at B to equal zero. Thus,

$$\delta_{BX} - R_B \delta_{BB} = 0 \qquad (8.4.1)$$

from which

$$R_B = \frac{\delta_{BX}}{\delta_{BB}} \qquad (8.4.2)$$

The influence line ordinate given by Eq. (8.4.2) is plotted directly under the unit load location to obtain the influence line for R_B in Fig. 8.4.1d.

Alternatively, using Maxwell's theorem of reciprocal deflections, as first presented in Section 5.4 and later in Section 7.5, it is known that

$$\delta_{BX} = \delta_{XB} \qquad (8.4.3)$$

Substituting Eq. (8.4.3) into Eq. (8.4.2) gives the influence line ordinate for R_B as

$$R_B = \frac{\delta_{XB}}{\delta_{BB}} \qquad (8.4.4)$$

which is shown in Fig. 8.4.1e. Comparison of Fig. 8.4.1e and c shows the curves therein to be geometrically proportional. The influence line curve of Fig. 8.4.1e is obtained by dividing the ordinates of the deflection diagram of Fig. 8.4.1c by $R_B \delta_{BB}$.

Example 8.4.1

Determine the equation for the influence line for reaction R_B on the two-span continuous beam of Fig. 8.4.2.

Solution

(a) *Use deflection diagram method, with Eq. (8.4.4).* As shown by the development preceding this example, the influence line for R_B may be obtained by computing the deflection δ_{XB} due to an upward unit load applied where the redundant reaction restraint has been removed. For this example the upward unit load is applied at the reaction location of R_B in Fig. 8.4.2b where the resulting M/EI diagram is also shown. Next the expression for δ_{XB} may be established by using the moment area method, such that

$$\delta_{XB} = \theta_A x - \Delta'_{XA} \qquad (a)$$

where

$$\theta_A = \frac{\Delta'_{CA}}{L} = \frac{1}{L}\left(\text{moment of } \frac{M}{EI} \text{ area between } A \text{ and } C \text{ about } C\right)$$

$$\Delta'_{XA} = \text{moment of } \frac{M}{EI} \text{ area between } A \text{ and } X \text{ about } X$$

Thus,

$$\theta_A = \frac{1}{50EI}\left[\frac{1}{2}(12)(20)\left(30 + \frac{20}{3}\right) + \frac{1}{2}(12)(30)\left(\frac{60}{3}\right)\right] = \frac{160}{EI}$$

and for $x \le 20$ ft,

$$\Delta'_{XA} = \frac{1}{EI}\left[\frac{1}{2}\left(\frac{x}{20}\right)(12)(x)\left(\frac{x}{3}\right)\right] = \frac{x^3}{10EI}$$

and for $x > 20$ ft,

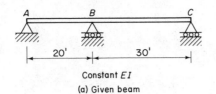

20' 30'

Constant EI

(a) Given beam

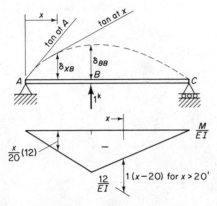

$\dfrac{x}{20}(12)$

$\dfrac{12}{EI}$ $1(x-20)$ for $x > 20'$

(b) "Basic" statically determinate
beam with unit load applied
where redundant restraint was removed

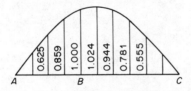

0.625 0.859 1.000 1.024 0.944 0.781 0.555

(c) Influence line for R_B

Figure 8.4.2 Influence line for reaction on the two-span continuous beam of Example 8.4.1.

$$\Delta'_{XA} = \frac{x^3}{10EI} - \frac{1}{EI}\left[\frac{1}{2}(x-20)(x-20)\left(\frac{x-20}{3}\right)\right]$$

$$= \frac{x^3}{10EI} - \frac{(x-20)^3}{6EI}$$

Thus, for $x \le 20$,

$$\delta_{XB} = \frac{160x}{EI} - \frac{x^3}{10EI} = \frac{x}{EI}\left(160 - \frac{x^2}{10}\right) \tag{b}$$

and, for $x > 20$,

$$\delta_{XB} = \frac{160x}{EI} - \frac{x^3}{10EI} + \frac{(x-20)^3}{6EI} \tag{c}$$

Next, evaluate δ_{BB}, the value of δ_{XB} when $x = 20$,

$$\delta_{BB} = \frac{160(20)}{EI} - \frac{(20)^3}{10EI} = \frac{2400}{EI} \tag{d}$$

Thus, according to Eq. (8.4.4), the influence line ordinates for R_B are

$$R_B = \frac{\delta_{XB}}{\delta_{BB}} \tag{e}$$

For $x \leq 20$,

$$R_B = \frac{x}{2400}\left(160 - \frac{x^2}{10}\right) \tag{f}$$

For $x > 20$,

$$R_B = \frac{1}{2400}\left[160x - \frac{x^3}{10} + \frac{(x-20)^3}{6}\right] \tag{g}$$

The values of the influence line ordinates are computed at 5-ft intervals and plotted in Fig. 8.4.2c.

(b) *Use the theorem of three moments to establish the influence line for R_B.* Using Eq. (8.3.4) without the uniform loading terms,

$$M_A\left(\frac{20}{EI}\right) + 2M_B\left(\frac{20}{EI} + \frac{30}{EI}\right) + M_C\left(\frac{30}{EI}\right)$$
$$= -\frac{1(a_1)(20 - a_1)(20 + a_1)}{EI(20)} - \frac{(a_2)(30 - a_2)(60 - a_2)}{EI(30)} \tag{h}$$

In the equation above, the right-hand term involving a_1 is used when the unit load is on span AB and the term involving a_2 is used when the unit load is on span BC. For this two-span beam M_A and M_B are zero. Thus, Eq. (h) becomes

$$M_B = -\frac{a_1(20 - a_1)(20 + a_1)}{2000} - \frac{a_2(30 - a_2)(60 - a_2)}{3000} \tag{i}$$

The equation for reaction R_B may be obtained from the free-body diagrams of Fig. 8.4.3, as follows:

$$R_B = R_{BA} + R_{BC} \tag{j}$$

For the unit load on span AB (Fig. 8.4.3a),

$$R_B = \frac{a_1}{20} - \frac{M_B}{20} - \frac{M_B}{30} = \frac{a_1}{20} - \frac{M_B}{12} \tag{k}$$

and for the unit load on span BC (Fig. 8.4.3b),

$$R_B = -\frac{M_B}{20} + \frac{30 - a_2}{30} - \frac{M_B}{30} = \frac{30 - a_2}{30} - \frac{M_B}{12} \tag{l}$$

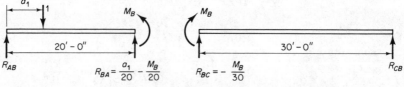

(a) Unit load on span AB

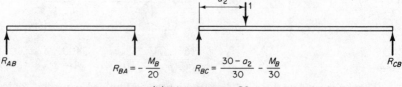

(b) Unit load on span BC

Figure 8.4.3 Three moment equation solution for two-span continuous beam acted upon by unit loads.

Substitution of Eq. (i) into Eqs. (k) and (l) gives:

for load on span AB ($a_1 \leq 20$),

$$R_B = \frac{a_1}{20} + \frac{a_1(20 - a_1)(20 + a_1)}{24{,}000} \qquad \text{(m)}$$

and for load on span $BC(a_2 \leq 30)$,

$$R_B = \frac{30 - a_2}{30} + \frac{a_2(30 - a_2)(60 - a_2)}{36{,}000} \qquad \text{(n)}$$

Equations (m) and (n) are identical to Eqs. (f) and (g). In Eqs. (f) and (g) x is measured from A for the unit load in either span, whereas for Eqs. (m) and (n) a_1 and a_2 are defined as in Fig. 8.4.3.

Conceptually, treating the influence line as a deflection diagram, as for the solution in part (a), is an important concept that allows the influence line to be sketched qualitatively. To achieve this goal, it is necessary to use the function for which the influence line is to be drawn as the redundant. This subject is treated further in Section 8.5.

The reader may note from observing Fig. 8.4.1e that the influence line for reaction on a beam may be obtained: (1) by removing the support for that reaction, and (2) by *applying an arbitrary force in the direction of that reaction sufficient to cause a unit deflection* at the reaction location. In Fig. 8.4.4 are shown several influence lines for reaction drawn as deflection diagrams with a unit deflection at the support for which the influence line is drawn.

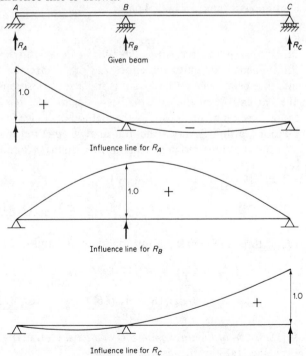

Figure 8.4.4 Influence lines for reactions as deflection diagrams.

In the analysis of statically indeterminate beams, once the reactions are obtained, the remainder of the analysis (i.e., the computation and drawing of the shear and bending moment diagrams) is readily done using the laws of statics. Thus, one might believe that one needs *only* the influence lines for reactions, and such reactions caused by movable loads are readily available from which to obtain shears and bending moments. However, the loading that may cause maximum positive or maximum negative reaction may not be the same loading to cause maximum positive or maximum negative shear or bending moment at a particular location. Thus, influence lines for shear and bending moment are equally, if not more, important and are not as easily visualized as influence lines for reactions.

Influence Line for Bending Moment

The development of several typical influence lines for bending moment in a continuous beam will provide the technique for computation of influence line ordinates as well as conclusions about how to load continuous beams for maximum positive and negative bending moment. Notwithstanding the following illustrations of computing influence line ordinates, the chief practical use of the influence line is to use the qualitative result to decide where to put the live load for its most severe effect.

Example 8.4.2

Determine the equations for and draw to scale the influence lines for the support bending moment M_B and the bending moment at midspan of beam BC, for the three-span continuous beam of Fig. 8.4.5a.

Solution

(a) *Write the two theorem-of-three-moments equations needed to solve for M_B and M_C.* The influence ordinates for either M_B or M_C are the values for those bending moments that occur when a unit load is in various positions, designated a_1, a_2, or a_3 from the left end when the unit load is on span AB, BC, or CD, respectively. The theorem of three moments is usually the method of choice for establishing the influence line equations, although any structural analysis method may be used.

Writing the three-moment equation for spans AB and BC,

$$M_A\left(\frac{22}{EI}\right) + 2M_B\left(\frac{22}{EI} + \frac{16}{EI}\right) + M_C\left(\frac{16}{EI}\right)$$
$$= -\frac{1(a_1)(22 - a_1)(22 + a_1)}{22EI} - \frac{1(a_2)(16 - a_2)(32 - a_2)}{16EI} \quad \text{(a)}$$

The second three-moment equation for spans BC and CD is

$$M_B\left(\frac{16}{EI}\right) + 2M_C\left(\frac{16}{EI} + \frac{22}{EI}\right) + M_D\left(\frac{22}{EI}\right)$$
$$= -\frac{1(a_2)(16 - a_2)(16 + a_2)}{16EI} - \frac{1(a_3)(22 - a_3)(44 - a_3)}{22EI} \quad \text{(b)}$$

(b) *Solution of the three-moment equations.* Noting that M_A and M_D are zero for this beam, Eqs. (a) and (b) may be written,

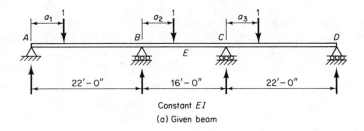

(a) Given beam

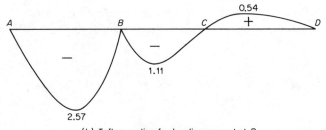

(b) Influence line for bending moment at B

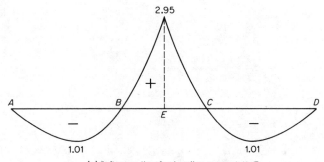

(c) Influence line for bending moment at E

Figure 8.4.5 Influence lines for Example 8.4.2.

$$76M_B + 16M_C = -\frac{a_1(22-a_1)(22+a_1)}{22} - \frac{a_2(16-a_2)(32-a_2)}{16}$$

$$16M_B + 76M_C = -\frac{a_2(16-a_2)(16+a_2)}{16} - \frac{a_3(22-a_3)(44-a_3)}{22}$$

Solving for M_B and M_C,

$$M_B = -\frac{4.75a_1(22-a_1)(22+a_1)}{7590} - \frac{a_2(16-a_2)(136-5.75a_2)}{5520}$$

$$+ \frac{a_3(22-a_3)(44-a_3)}{7590} \tag{c}$$

$$M_C = +\frac{a_1(22-a_1)(22+a_1)}{7590} - \frac{a_2(16-a_2)(44+5.75a_2)}{5520}$$

$$- \frac{4.75a_3(22-a_3)(44-a_3)}{7590} \tag{d}$$

(c) *Evaluation of the ordinates and use of the diagram.* The values of the influence line ordinates are given in Table 8.4.1 for the unit load at 0.2 points along the span. Since the influence line represents the action of a *single* unit load on the entire structure, the three terms of Eqs. (c) and (d) really are the three separate equations for spans AB, BC, and CD; thus, only one of the variables a_1, a_2, and a_3 can be involved at one time. The influence line for M_B, Eq. (c), is drawn as Fig. 8.4.5b. This influence line has the shape typical of bending moment at an interior support of a continuous beam. The diagram tells the designer that live load should be placed on the two adjacent spans AB and BC but not on span CD to cause maximum bending moment at B, and that the maximum will be of negative sign. For maximum positive bending moment due to live load, only span CD should be loaded. Of course, dead load is always everywhere so that the net result of placing live load on span CD is usually still a negative moment.

TABLE 8.4.1 INFLUENCE LINE ORDINATES FOR BEAM OF FIG. 8.4.5a

$a_1/22$	M_B*	M_C*	M_E*	$M_F†$	$M_G‡$	$V_{BA}§$	$V_{BC}§$	$V_{0.4}§$
0	0	0	0	0	0	0	0	0
0.2	−1.2794	+0.2693	−0.5051	−1.0101	−1.1245	−0.258154	+0.096794	+0.096794
0.4	−2.2390	+0.4714	−0.8838	−1.7680	−1.9680	−0.501773	+0.169400	+0.169400
0.6	−2.5589	+0.5387	−1.0101	−2.0202	−2.2491	−0.716314	+0.193600	+0.193600
0.8	−1.9192	+0.4040	−0.7576	−1.5152	−1.6869	−0.887236	+0.145200	+0.145200
1.0	0	0	0	0	0	−1.000000	0	0

$a_2/16$								
0	0	0	0	0	0	0	+1.000000	0
0.1	−0.5293	−0.2221	——	——	+0.9415	——	——	——
0.17	−0.7995	−0.3998	——	+1.5686	——	——	——	——
0.2	−0.8726	−0.4630	+0.9322	+1.4247	+0.4484	−0.039664	+0.825600	−0.174400
0.4	−1.1041	−0.8993	+2.1983	+0.6011	−0.1236	−0.050186	+0.612800	−0.387200
0.5	−1.0435	−1.0435	+2.9565	+0.3478	−0.2435	−0.047432	+0.500000	+0.500000
0.6	−0.8993	−1.1041	+2.1983	+0.1781	−0.2798	−0.040877	+0.387200	+0.387200
0.8	−0.4630	−0.8726	+0.9322	+0.0223	−0.1840	−0.021045	+0.174400	+0.174400
1.0	0	0	0	0	0	0	0	0

$a_3/22$								
0	0	0	0	0	0	0	0	0
0.2	+0.4040	−1.9192	−0.7576	0	+0.1717	+0.018364	−0.145200	−0.145200
0.4	+0.5387	−2.5589	−1.0101	0	+0.2289	+0.024486	−0.193600	−0.193600
0.6	+0.4714	−2.2390	−0.8838	0	+0.2004	+0.021427	−0.169400	−0.169400
0.8	+0.2693	−1.2794	−0.5051	0	+0.1144	+0.012241	−0.096794	−0.096794
1.0	0	0	0	0	0	0	0	0

*See Example 8.4.2 for equations (Fig. 8.4.5 for graph).
†See Example 8.4.3 for equations (Fig. 8.4.8 for graph).
‡See Example 8.4.4 for equations (Fig. 8.4.8 for graph).
§See Example 8.4.5 for equations (Fig. 8.4.10 for graph).

(d) *Influence line for the midspan bending moment* M_E. From the free-body diagram of span BC shown in Fig. 8.4.6, the bending moment M_E is

$$M_E = R_{BC}(8) + M_B - 1(8 - a_2)$$

or, substituting R_{BC} from Fig. 8.4.6,

$$M_E = \underbrace{0.5M_B + 0.5M_C}_{\substack{\text{applies for unit} \\ \text{load anywhere}}} + \underbrace{(8 - 0.5a_2)}_{\substack{\text{applies for} \\ \text{unit load on} \\ \text{span } BC}} - \underbrace{(8 - a_2)}_{\substack{\text{applies for} \\ \text{unit load} \\ \text{between } B \\ \text{and } E}}$$

The evaluation of the equation above is given in Table 8.4.1 and the influence line for M_E is drawn as Fig. 8.4.5c. This influence line is typical for the middle portion of any continuous span; thus, to position live load for maximum positive moment, which occurs somewhere in the middle portion of a span, the influence line indicates that the span in question should be loaded with adjacent spans unloaded. For more than three spans, alternate spans should be loaded.

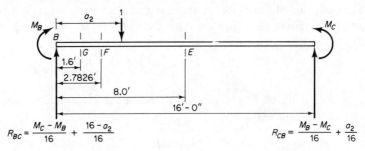

Figure 8.4.6 Free-body diagram of span BC used in Examples 8.4.2d and 8.4.3.

Concept of Fixed Point

From a study of Fig. 8.4.5b and c, one may note that influence lines for points between points B and E would have a gradual transition from that for point B to that for point E. The slope of the influence line at C must change from a slope upward to the right for the influence line for M_B to a slope downward to the right for the influence line for M_E. There will be some location F between B and E where the slope at C of the influence line for M_F will be zero. Such a location F is called a *fixed point.*[*] When the slope at C is zero for the influence line for M_F the entire influence line to the right of C will have zero ordinate, indicating that wherever a load is placed to the right of point C the bending moment at F will be zero. Thinking in another way, F is a *unique location* at which *any loading* on *any span* to the right of point C will cause zero moment. Hence, it is called a fixed point. Similarly, between points E and C there will be a unique location where the bending moment will be zero for any loading on any span to the *left* of point B. Before noting the usefulness of fixed points, the fol-

*The name "fixed point" was given by Otto Mohr, "Beitrag zur Theorie der Holz- und Eisen-Constructionen," *Zeitschrift des Architekten- und Ingenieur-Vereins zu Hannover*, 1868, pp. 19–50.

lowing example of locating a fixed point and illustration of the influence line shape is presented.

Example 8.4.3

For the beam of Fig. 8.4.5, locate F, the fixed point near the left support of span BC. Then compute the influence ordinates and draw to scale the influence line for the bending moment M_F.

Solution

(a) *Locate the fixed point F near the left end of span BC.* By definition the desired fixed point is a point of zero moment near the left end of an unloaded span when loads are located in spans to the right of the span in question. Thus, the fixed point may be located by writing the three-moment equation, Eq. (8.3.4), for spans AB and BC in Fig. 8.4.5 as follows:

$$M_A\left(\frac{22}{EI}\right) + 2M_B\left(\frac{22}{EI} + \frac{16}{EI}\right) + M_C\left(\frac{16}{EI}\right) = 0 \tag{a}$$

The right side of the equation above is zero because there is no load on spans AB and BC, although there *is* load on span CD. Setting M_A in Eq. (a) to zero, the ratio of M_C to M_B is obtained as

$$\frac{M_C}{M_B} = -\frac{2(22 + 16)}{16} = -\frac{76}{16} = -4.75 \tag{b}$$

Referring to Fig. 8.4.7, the equation for bending moment at any location a distance x from point B is

$$M_x = M_B + \frac{(M_C - M_B)x}{16} \tag{c}$$

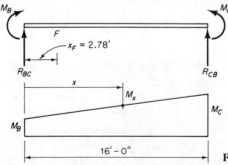

Figure 8.4.7 Locating a fixed point.

For Eq. (c) to be at the fixed point F the bending moment must be zero; thus,

$$M_x = M_F = 0 = M_B + \frac{(M_C - M_B)x_F}{16}$$

which gives for the fixed-point location

$$x_F = \frac{-16M_B}{M_C - M_B} = \frac{16}{1 - M_C/M_B} \tag{d}$$

Statically Indeterminate Beams   Chap. 8

Substituting Eq. (b) in Eq. (d) gives the location of the fixed point F as

$$x_F = \frac{16}{1 - (-4.75)} = \frac{16}{5.75} = 2.78 \text{ ft}$$

(b) *Evaluate the influence line ordinates for bending moment at fixed point F.* Since the values of M_B and M_C are already available in Table 8.4.1 for various positions of the unit load on all three spans of the continuous beam, the values of M_F may be computed by considering the free-body diagram (see Fig. 8.4.6) of span BC by itself using the values of M_B and M_C. When the unit load is on spans AB or CD, M_F is equal to, from Eq. (c),

$$M_F = M_B + \frac{(M_C - M_B)(2.78)}{16} \tag{e}$$

When the unit load is on span BC, M_F is equal to the sum of Eq. (e) and the simple-span moment at F due to the unit load. Tabulation of values of M_F is also given in Table 8.4.1, and the influence line is drawn in Fig. 8.4.8b.

Example 8.4.4

Calculate the ordinates and draw to scale the influence line for the bending moment at point G (Fig. 8.4.8c), which lies between support B and fixed point F, at the one-tenth point of span BC.

Solution. Referring to Fig. 8.4.7 when the unit load is not on span BC, the bending moment at point G (one-tenth of distance from B to C) is equal to

$$M_G = M_B + \frac{1}{10}(M_C - M_B)$$

or

$$M_G = 0.90M_B + 0.10M_C \tag{a}$$

When the unit load is on span BC, the bending moment at G is equal to the simple-span moment at G due to the unit load and that of Eq. (a). The influence line ordinates are given in Table 8.4.1 and the graph is shown in Fig. 8.4.8c.

Partial Span Loading for Maximum Bending Moment

Several important conclusions can be drawn from a study of the influence lines in Fig. 8.4.8. The fixed point is the closest toward the support where *full span* loadings give the maximum and minimum bending moments for the moment envelope. Maximum and minimum bending moments for points between the fixed point and the support are obtained only by loading a portion of the span; for maximum positive moment loading the portion of the span indicated by the positive portion of the influence line; and for maximum negative moment loading the portion indicated by the negative influence line.

When the loading is uniform and the spans are relatively short, as for the design of typical beams in buildings, the extra labor of using partial span loadings for the bending moment envelope is not usually justified; thus, rarely are partial span loadings used in obtaining the bending moment envelope in building design. For the design of bridges involving longer spans and heavy concentrated loads, the correct maximum and minimum bending moments are computed making use of the information

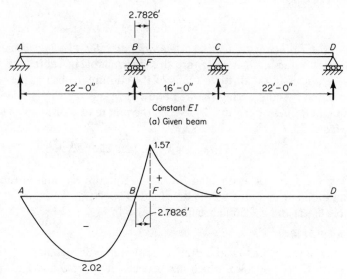

(a) Given beam

(b) Influence line for bending moment M_F at fixed point F

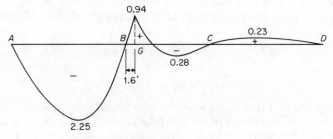

(c) Influence line for bending moment M_G at location
between fixed point F and the support at B.

Figure 8.4.8 Influence lines for a "fixed point" and for a location between a
"fixed point" and a support, Examples 8.4.3 and 8.4.4.

provided by the influence line, including the correct positioning of loads according to
the positive and negative portions of the influence line.

Regarding the computation of ordinates of the influence line for bending moment
at a point within a span such as BC, Examples 8.4.2(c), 8.4.3, and 8.4.4 show that the
influence ordinates in the spans outside the span containing the point in question are
equal to a proportion of M_B and M_C. The influence ordinates within the span (such as
BC in the examples) containing the point to which the influence line refers are equal
to a proportion of M_B and M_C superimposed with the simple-span moment due to the
unit load. Actually, once the influence lines for negative moments M_B and M_C are
plotted along with the curve representing the locus of maximum positive ordinates
for the influence line (such as values 1.57 and 0.94 in Fig. 8.4.8) at points along the
span being considered, a whole series of influence lines for bending moment (say, for
the 0.1 points along the span) may be drawn by graphical construction. Originally

presented by Winkler in German, the construction was detailed and proved by Salmon and Buettner [9].

Influence Line for Shear

As for the influence line for reaction and for bending moment, the influence line for shear is of interest primarily to tell the designer where to place the live load to cause the most severe effect. The following examples will show the computation technique and the influence line shape.

Example 8.4.5

For the beam of Examples 8.4.2 to 8.4.4 (see Figs. 8.4.5 and 8.4.8), compute the ordinates and draw to scale the influence lines for (a) the shear V_{BA} to the left of support B; (b) the shear V_{BC} to the right of support B; and (c) at the 0.4 point of span BC.

Solution

(a) *Influence line for shear V_{BA}.* Taking a free-body diagram of span AB as in Fig. 8.4.9, the shear V_{BA} is

$$V_{BA} = V_{AB} - 1 = \frac{M_B}{22} - \frac{a_1}{22} \qquad \text{(a)}$$

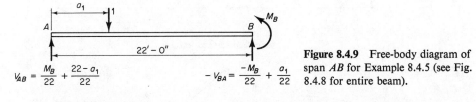

$$V_{AB} = \frac{M_B}{22} + \frac{22 - a_1}{22} \qquad -V_{BA} = \frac{-M_B}{22} + \frac{a_1}{22}$$

Figure 8.4.9 Free-body diagram of span AB for Example 8.4.5 (see Fig. 8.4.8 for entire beam).

where the second term applies only when the unit load is on span AB. The evaluation of the influence line ordinates for V_{BA} is given in Table 8.4.1 and the graph is plotted in Fig. 8.4.10.

(b) *Influence line for shear V_{BC}.* From the free-body diagram of span BC given in Fig. 8.4.6,

$$V_{BC} = R_{BC} = \frac{M_C - M_B}{16} + \frac{16 - a_2}{16} \qquad \text{(b)}$$

where the second term is applicable only when the unit load is on span BC. The evaluation of the influence line ordinates for V_{BC} is given in Table 8.4.1 and the graph is plotted in Fig. 8.4.10.

(c) *Influence line for shear $V_{0.4}$ at the 0.4 point of span BC.* The free-body diagram of span BC is Fig. 8.4.6, showing

$$V_{BC} = R_{BC} = \frac{M_C - M_B}{16} + \frac{16 - a_2}{16}$$

when the unit load is on span BC.

The shear $V_{0.4}$ at the 0.4 point is as follows:

1. Unit load not on span BC,

$$V_{0.4} = V_{BC} = \frac{M_C - M_B}{16}$$

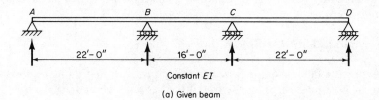

(a) Given beam

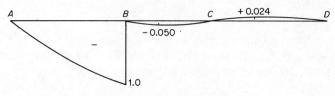

(b) Influence line for shear V_{BA}

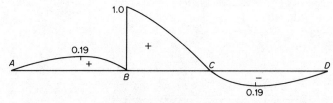

(c) Influence line for shear V_{BC}

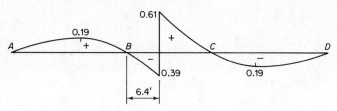

(d) Influence line for shear at 0.4 point of span BC

Figure 8.4.10 Influence lines for shear at several locations, Example 8.4.5.

2. Unit load on span BC between B and the 0.4 point,

$$V_{0.4} = V_{BC} - 1 = \frac{M_C - M_B}{16} + \frac{16 - a_2}{16} - 1$$

$$= \frac{M_C - M_B}{16} - \frac{a_2}{16}$$

3. Unit load on span BC to the right of the 0.4 point,

$$V_{0.4} = V_{BC} = \frac{M_C - M_B}{16} + \frac{16 - a_2}{16}$$

$$= \frac{M_C - M_B}{16} - \frac{a_2}{16} + 1$$

The results are given in Table 8.4.1 and the graph is shown in Fig. 8.4.10.

Statically Indeterminate Beams   Chap. 8

The influence lines for shear shown in Fig. 8.4.10 are of typical shape. Again, as stated previously, the most important use of the influence line concept is to tell the designer where to position loads for maximum effect, either positive or negative. The influence lines for shear in Fig. 8.4.10 show that for maximum shear (negative) an infinitesimal distance to the left of support B (Fig. 8.4.10b) adjacent spans AB and BC should have live load on them but span CD should have no live load. Fig. 8.4.10c shows that the same conclusion is valid for a location an infinitesimal distance to the right of support B; this time the maximum effect will be a positive shear. It will be true that maximum shear at an infinitesimal distance from any support will occur from loading of entire span (i.e., no partial span loading).

The maximum positive or negative shear for any point except at the ends of a span will occur under partial loading. The influence line for shear at the 0.4 point of span BC (Fig. 8.4.10d) is typical of all interior locations.

8.5 MÜLLER-BRESLAU PRINCIPLE OF THE INFLUENCE LINE AS A DEFLECTION DIAGRAM

As discussed for the statically determinate beams in Section 6.6, the Müller-Breslau principle [10] of considering the influence line as a deflection diagram equally applies for statically indeterminate beams. The principle may be restated as follows: The value of a function due to a unit downward concentrated load moving across the span is equal to the upward deflection at the position of the unit load, the cause of the deflection being an arbitrary force or a pair of forces applied to the structure at the location of the function, when the resistance offered by the function is removed and replaced by a unit deflection in the direction of the function.

The result of the long definition may be accomplished by the following steps:

1. *Remove the restraint representing the function for which the influence line is desired.* Referring to Fig. 8.5.1a, remove the displacement restraint at X (i.e., remove the support at X) to obtain the influence line for the reaction X; referring to Fig. 8.5.1b, remove the restraint to relative vertical displacement at X (i.e., cut vertically through the section at X) to obtain the influence line for the shear X; referring to Fig. 8.5.1c, remove the resistance to relative angular displacement at X (i.e., imagine a hinge inserted at X) to obtain the influence line for the bending moment X. The same notation X is used for a support reaction, a shear force, and a bending moment to facilitate making the general proof later.

2. *Give the structure with the restraint removed a unit displacement in the direction of the function (i.e., the reaction, the pair of shear forces, or the pair of bending moments) by treating the removed restraint as an applied load.* Thus, in Fig. 8.5.1a the reaction force F displaces its point of application upward a unit amount; in Fig. 8.5.1b the pair of shear forces F separates the two sides of the section at X a unit amount with the pair of moments M and the angles ϕ remaining constant; in Fig. 8.5.1c the pair of bending moments F displaces the two sides of the beam at section X angularly a unit amount ($\phi_1 + \phi_2$ equals unity) with the pair of shear forces V remaining constant.

3. *The resulting deflection diagram is the influence line.* In Fig. 8.5.1 are shown the influence lines (deflection diagrams) for a reaction, shear, and bending moment at a location on a continuous beam.

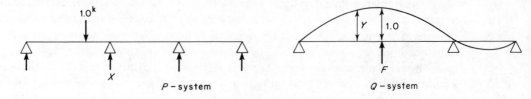

(a) Influence line for reaction X

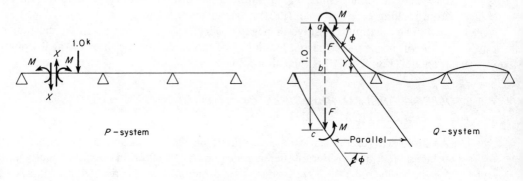

(b) Influence line for shear force X

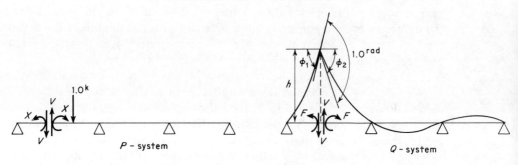

(c) Influence line for bending moment X

Figure 8.5.1 Müller-Breslau principle applied to continuous beams.

The Müller-Breslau principle may be expressed mathematically by the simple equation

$$Y = X$$

where Y on the deflection diagram can be proved to be equal to the influence value (the reaction, shear, or moment) at X due to the unit load applied where Y is measured.

Proof

In each of Figs. 8.5.1a, b, and c there are two sets of forces in equilibrium, the P-system (corresponding to the actual structure) and the Q-system (corresponding to the structure deflected to give the influence diagram). According to the reciprocal energy theorem developed in Section 7.6, the work done by a P-system, in going through the displacements of a Q-system, is equal to the work done by the Q-system in going through the displacements of the P-system.

In Fig. 8.5.1a,

$$\text{Work of } P\text{-forces} = \text{Work of } Q\text{-forces}$$

$$-1.0(Y) + X(1.0) = F(0)$$

$$Y = X$$

There is no displacement in the P-system for the force F in the Q-system.

In Fig. 8.5.1b,

$$\text{Work of } P\text{-forces} = \text{Work of } Q\text{-forces}$$

$$-M(\phi) + M(\phi) + X(bc) + X(ab) - 1.0(Y) = M(0) + F(0)$$

$$Y = X(ab + bc) = X$$

There is *no* discontinuity of slope and no relative deflection in the P-system for the pair of moments M and the pair of shear forces F in the Q-system.

In Fig. 8.5.1c,

$$\text{Work of } P\text{-forces} = \text{Work of } Q\text{-forces}$$

$$-V(h) + V(h) + X(\phi_1) + X(\phi_2) - 1.0(Y) = V(0) + F(0)$$

$$Y = X(\phi_1 + \phi_2) = X$$

Again, there is no slope discontinuity and no relative deflection in the P-system for the pair of moments F and the pair of shear forces V in the Q-system.

For a qualitative determination of the influence line for a given function, the Müller-Breslau principle is extremely valuable. From the qualitative influence line for reaction, shear, or bending moment, the positioning of loads for maximum positive or negative value of the function may be established. The values on the shear and bending moment envelopes may then be computed directly.

8.6 CASTIGLIANO'S THEOREM—PART I

As discussed in Section 7.6, the theorem of Carlo Alberto Castigliano [Ref. 11 of Chapter 7] published in 1875 (his book in 1879) provided an energy method alternative to the method of consistent deformation as applied to trusses. These theorems are equally applicable to beams. Mathematically, the theorem, Part I, also known as the first theorem, has been given previously as Eq. (7.6.8); or

$$\frac{\partial W_i}{\partial X_i} = P_i \tag{8.6.1}$$

where W_i is the internal work, expressible in terms of internal forces; X_i is the displacement of the structure at the location of and in the direction of the external force P_i.

For trusses the internal work is the sum of the work done on each of the finite number of bars (members), expressible in terms of the internal forces F_i and elongations e_i,

$$W_i = \sum \frac{1}{2} F_i e_i \tag{8.6.2}$$

In terms of unit stress σ_x and unit strain ϵ_x on an element of cross-sectional area $b\, dy$ for a slice of beam of length dx,

$$F_i = \sigma_x b\, dy \left.\vphantom{\begin{array}{c}a\\b\end{array}}\right\}$$
$$e_i = \epsilon_x\, dx \qquad (8.6.3)$$

For the beam, the stress at any location y from the neutral axis is $\sigma_x = My/I$ and the strain is $\epsilon_x = \sigma_x/E = My/EI$. Thus Eq. (8.6.2) becomes for the beam element $b\, dy\, dx$,

$$dW_i = \frac{1}{2} F_i e_i$$

$$= \frac{1}{2}\left(\frac{My}{I} b\, dy\right)\left(\frac{My}{EI}\, dx\right) \qquad (8.6.4)$$

To obtain W_i for the entire beam, integration must be performed with respect to y over the entire cross-sectional area, and with respect to x over the entire length of the beam. Equation (8.6.4) may be written

$$W_i = \int_0^L \frac{M^2}{2EI} \frac{\int_A y^2\, dA}{I}\, dx \qquad (8.6.5)$$

and noting that $\int_A y^2\, dA = I$,

$$W_i = \int_0^L \frac{M^2\, dx}{2EI} \qquad (8.6.6)$$

Equation (8.6.6) is usually referred to as the *strain energy* due to bending.

The application of Eq. (8.6.1) with W_i from Eq. (8.6.6) requires that M, the bending moment equations, be expressed in terms of the displacements (angular, θ, or linear, Δ). This would be the displacement method of analysis (as opposed to the force method). For purposes of the displacement method of analysis, the strain energy [Eq. (8.6.6)] may more conveniently be expressed in terms of displacement by replacing M with $EI\, d^2y/dx^2$; thus, the strain energy becomes

$$W_i = \int_0^L \frac{[EI(d^2y/dx^2)]^2\, dx}{2EI}$$

or

$$W_i = \int_0^L \frac{EI}{2}\left(\frac{d^2y}{dx^2}\right)^2\, dx \qquad (8.6.7)$$

Since the second derivative is squared it is not necessary to use the negative sign ($M = -EI\, d^2y/dx^2$).

As will be seen in later chapters, the displacement method will be the method of choice for computer methods; otherwise, Eq. (8.6.1) is only of academic interest. Thus, no examples are presented until later in Chapter 14.

8.7 CASTIGLIANO'S THEOREM—PART II

The Castigliano theorem, part II, as first discussed in Section 7.7, represents an energy method expression of the force method. Also known as Castigliano's second theorem, the theorem may be written

$$\frac{\partial W_i}{\partial P_i} = X_i \qquad (8.7.1)$$

where W_i is the internal work, expressible in terms of internal forces; X_i is the displacement of the structure at the location of and in the direction of the external force P_i.

For beams the internal work, called strain energy, is given by Eq. (8.6.6),

$$W_i = \int_0^L \frac{M^2\,dx}{2EI} \qquad [8.6.6]$$

where M is the bending moment along the beam. For Eq. (8.7.1), M must be expressed in terms of the applied loads.

Example 8.7.1

Determine the fixed-end moments on a uniformly loaded beam, as shown in Fig. 8.7.1, using Castigliano's theorem, part II.

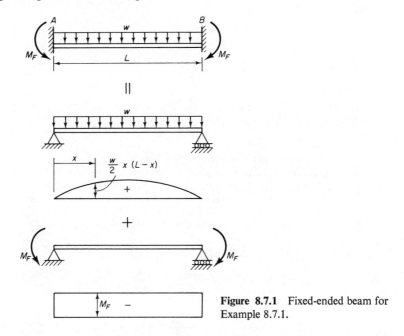

Figure 8.7.1 Fixed-ended beam for Example 8.7.1.

Solution

(a) *Establish the "basic" statically determinate structure.* Choose the end moments as redundant forces, equal from symmetry and called M_F in Fig. 8.7.1.

(b) *Write the bending moment equation for the actual structure.* The bending moment is the sum of that due to uniform loading and that due to the end moments M_F. Thus,

$$M_x = \frac{w}{2}(x)(L - x) - M_F$$

(c) *Apply Castigliano's theorem, part II;*

$$\frac{\partial W_i}{\partial P_i} = X_i$$

$$\frac{\partial}{\partial M_F}\left(\int_0^L \frac{M^2\,dx}{2EI}\right) = \theta_A$$

$$\frac{\partial}{\partial M_F}\left[\int_0^L \frac{\left[\frac{w}{2}x(L-x)-M_F\right]^2 dx}{2EI}\right] = \theta_A$$

$$\frac{1}{EI}\int_0^L \left[\frac{w}{2}x(L-x)-M_F\right](-1)\,dx = \theta_A$$

$$\frac{1}{EI}\left[-\frac{wLx^2}{4}+\frac{wx^3}{6}+M_F x\right]_0^L = \theta_A$$

$$\frac{L}{EI}\left(-\frac{wL^2}{12}+M_F\right) = \theta_A$$

For fixed ends, $\theta_A = 0$; thus,

$$M_F = \frac{wL^2}{12}$$

The answer has a positive $(+)$ sign, indicating that the assumed direction for M_F is correct.

Example 8.7.2

Determine the reaction at B for the uniformly loaded two-span continuous beam of Fig. 8.7.2 using Castigliano's theorem, part II.

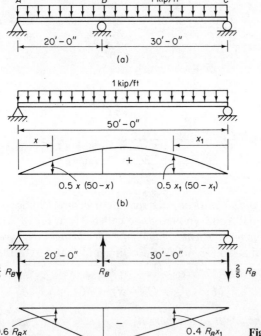

Figure 8.7.2 Two-span continuous beam of Example 8.7.2 using Castigliano's theorem, part II.

Statically Indeterminate Beams   Chap. 8

Solution

(a) *Establish the "basic" statically determinate structure.* Choose the reaction R_B as the redundant force. The simply supported uniformly loaded beam of Fig. 8.7.2b becomes the "basic" statically determinate structure.

(b) *Write the bending moment equations for the actual structure.* The bending moment is the sum of the positive value due to uniform loading as in Fig. 8.7.2b plus the negative moment due to R_B as in Fig. 8.7.2c. Thus, for $x \le 20$,

$$M_x = \frac{x}{2}(50 - x) - \frac{3}{5}R_B x$$

and for $20 \le x \le 50$,

$$M_x = \frac{x}{2}(50 - x) - \frac{3}{5}R_B x + R_B(x - 20)$$

(c) *Apply Castigliano's theorem, part II.*

$$\frac{\partial W_i}{\partial R_B} = X_B = 0$$

$$W_i = \int_0^L \frac{M^2 \, dx}{2EI}$$

In this case it will be convenient to integrate on span AB from the left end for $x = 0$ to 20, and integrate on span BC from the right end for x_1 from 0 to 30.

$$W_i = \int_0^{20} \frac{[0.5x(50 - x) - 0.6R_B x]^2 \, dx}{2EI}$$

$$+ \int_0^{30} \frac{[0.5x_1(50 - x_1) - 0.4R_B x_1]^2 \, dx_1}{2EI}$$

$$\frac{\partial W_i}{\partial R_B} = \int_0^{20} \frac{[0.5x(50 - x) - 0.6R_B x](-0.6x) \, dx}{EI}$$

$$+ \int_0^{30} \frac{[0.5x_1(50 - x_1) - 0.4R_B x_1](-0.4x_1) \, dx_1}{EI}$$

$$EI\frac{\partial W_i}{\partial R_B} = \int_0^{20} (-15x^2 + 0.3x^3 + 0.36R_B x^2) \, dx$$

$$+ \int_0^{30} (-10x_1^2 + 0.2x_1^3 + 0.16R_B x_1^2) \, dx_1$$

$$EI\frac{\partial W_i}{\partial R_B} = \left[-15\frac{x^3}{3} + 0.3\frac{x^4}{4} + 0.36R_B\frac{x^3}{3} \right]_0^{20}$$

$$+ \left[-10\frac{x_1^3}{3} + 0.2\frac{x_1^4}{4} + 0.16R_B\frac{x_1^3}{3} \right]_0^{30}$$

$$EI\frac{\partial W_i}{\partial R_B} = (-28,000 + 960R_B) + (-49,500 + 1440R_B)$$

$$= -77,500 + 2400R_B = 0$$

$$R_B = +32.29 \text{ kips}$$

SELECTED REFERENCES

1. Johann Albert Eytelwein, *Handbuch der Statik fester Körper*, 3 vols., Berlin, 1808. (This credit is given by Emil Winkler, Ref. 8 below, p. 97.)

2. Louis M. H. Navier, *Résumé des Leçons données a l'École des ponts et chaussées sur l'application de la mécanique à l'établissement des constructions et des machines*, 3rd ed., Dunod, Éditeur, Paris, 1864.

3. E. P. B. Clapeyron, "Calcul d'une poutre élastique reposant librement sur des appuis inégalement espacés," *Comptes Rendus des Séances de l'Académie des Sciences*, Paris, *45*, December 28, 1857, 1076–1080.

4. Bertot, [untitled item], *Mémoires et Compte-rendu des Travaux de la Société des Ingénieurs Civils*, Paris, 1855, 278–280.

5. Stephen P. Timoshenko, *History of Strength of Materials*, McGraw-Hill Book Company, New York, 1953, pp. 144–146.

6. Otto Mohr, "Beiträge zur Theorie der Holz- und Eisen-Constructionen," *Zeitschrift des Architekten- und Ingenieur-Vereins zu Hannover*, *6* (1860), 323–346, plus Plate 175.

7. W. Fränkel, "Theorie des enfachen Sprengwerkes," *Der Civilingenieur*, *22* (1876), 21–32.

8. Emil Winkler, *Vorträge über Brückenbau*, Vol. 1: *Theorie der Brucken*, 3rd ed., Carl Gerold's Sohn, Vienna, 1886.

9. Charles G. Salmon and Donald R. Buettner, "Influence Line Spectrum by Graphical Construction," *Civil Engineering*, July 1964, pp. 54–55.

10. Heinrich F. B. Müller-Breslau, *Die neuren Methoden der Festigkeitslehre und der Statik der Baukonstruktionen*, Baumgartner's Buchhandlung, Leipzig, 1886, pp. 33–40.

PROBLEMS

For all problems (except 8.31 through 8.36), after solving for the reactions on the statically indeterminate structure, draw the resulting shear and bending moment diagrams, either qualitatively when numerical values are not involved or to scale for the numerical problems.

8.1. Using the method of consistent deformation, determine the reactions for the beam of the accompanying figure. Solve using a simply supported beam as the "basic" statically determinate structure and check by using a cantilever as the "basic" statically determinate structure.

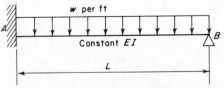

Probs. 8.1, 8.22, and 8.37

8.2. For the beam of the accompanying figure, follow the requirements of Prob. 8.1.

8.3. Using the method of consistent deformation, determine the reactions for the beam of the accompanying figure. Solve using a simply supported beam *AC* as the "basic"

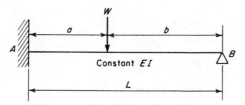

Probs. 8.2, 8.22, and 8.38

statically determinate structure, and check using an overhanging cantilever supported at A and B as the "basic" statically determinate structure.

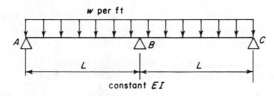

Prob. 8.3

8.4. For the beam of the accompanying figure, follow the requirements of Prob. 8.3.

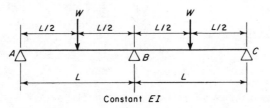

Prob. 8.4

8.5. In the accompanying figure, apply the method of consistent deformation using a simply supported beam AB as the "basic" statically determinate structure to solve for the reactions at A and B.

8.6. Solve Prob. 8.5 except use a cantilever beam supported at A as the "basic" statically determinate structure.

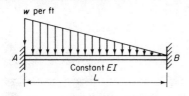
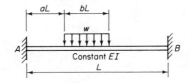

Probs. 8.5 and 8.6   **Probs. 8.7 and 8.8**

8.7. In the accompanying figure, apply the method of consistent deformation using a simply supported beam AB as the "basic" statically determinate structure to solve for the reactions at A and B.

8.8. Solve Prob. 8.7 except use a cantilever beam supported at A as the "basic" statically determinate structure.

8.9. For the beam of the accompanying figure, apply the method of consistent deformation to determine the reactions at A and B. Use one "basic" statically determinate beam to solve the problem and then use a different "basic" beam to make the check.

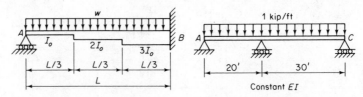

Probs. 8.9 and 8.39   **Probs. 8.10, 8.21, and 8.40**

8.10. For the beam of the accompanying figure, apply the method of consistent deformation to determine the reactions. Use the span *AC* as the "basic" statically determinate structure.

8.11. For the beam of the accompanying figure, apply the consistent deformation method to evaluate the reactions. Use a cantilever beam supported at *A* as the "basic" statically determinate beam.

8.12. Solve Prob. 8.11, except instead of the cantilever use a beam simply supported at *A* and *B* as the "basic" statically determinate beam.

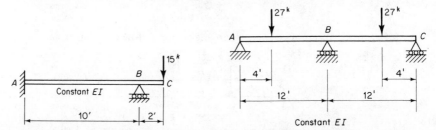

Probs. 8.11, 8.12, and 8.41   **Probs. 8.13 and 8.21**

8.13. For the beam of the accompanying figure, apply the method of consistent deformation to evaluate the reactions. Use a simply supported beam *AC* as the "basic" statically determinate structure. Check the compatibility at *C* by using as the "basic" statically determinate structure a beam simply supported at *A* and *B* with overhang.

8.14. For the beam of the accompanying figure, apply the method of consistent deformation to evaluate the reactions. Use a simply supported beam *AC* as the "basic" statically determinate structure.

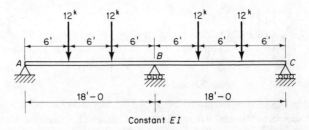

Prob. 8.14

8.15. For the beam of the accompanying figure, apply the method of consistent deformation to evaluate the reactions due to the simultaneous action of the applied loads and a support settlement of 1 in. at *B*. Use a cantilever beam supported at *A* as the "basic" statically determinate structure. Determine all deformations by the virtual work (unit load) method.

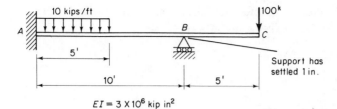

$EI = 3 \times 10^6$ kip in^2

Probs. 8.15 and 8.31

8.16. For the beam of the accompanying figure, apply the method of consistent deformation to evaluate the reactions for the applied loads when the support at B settles 1 in.
   (a) Take the reaction R_B as the redundant and solve for all deformations using the virtual work (unit load) method.
   (b) Take the bending moment M_B as the redundant and solve for all deformations using the moment area method.

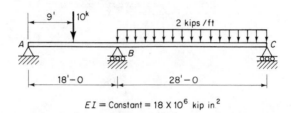

$EI = $ Constant $= 18 \times 10^6$ kip in^2

Prob. 8.16

8.17. For the beam of Fig. 8.3.3 (Example 8.3.1), apply the method of consistent deformation to evaluate the reactions. Take a cantilever beam supported at A as the "basic" statically determinate structure. Solve for all deformations using moment area method.

8.18. Re-solve Prob. 8.17 except solve for all deformations using the virtual work (unit load) method.

8.19. Re-solve Prob. 8.17 except assume the supports at B and C settle 0.5 in. and 1 in., respectively, together with the applied loads.

8.20. For the case assigned by the instructor, for the three-span continuous beam of the accompanying figure, apply the method of consistent deformation using the simply supported beam over the span AD as the "basic" statically determinate structure. The cases are as follows: **(a)** $\alpha = 0.75$; **(b)** $\alpha = 1.0$; **(c)** $\alpha = 1.25$; **(d)** $\alpha = 1.50$; and **(e)** $\alpha = 1.75$.

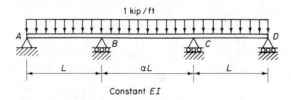

Constant EI

Probs. 8.20, 8.34, 8.35 and 8.36

8.21. Apply the theorem of three moments to determine the reactions for (a) the beam of Prob. 8.10; (b) the beam of Prob. 8.13.

8.22. Apply the theorem of three moments to determine the reactions for the following beams: (a) beam of Prob. 8.1; (b) the beam of Prob. 8.2.

8.23–8.30. Apply the theorem of three moments to analyze the beam of each accompanying figure. Make continuity checks to verify your result.

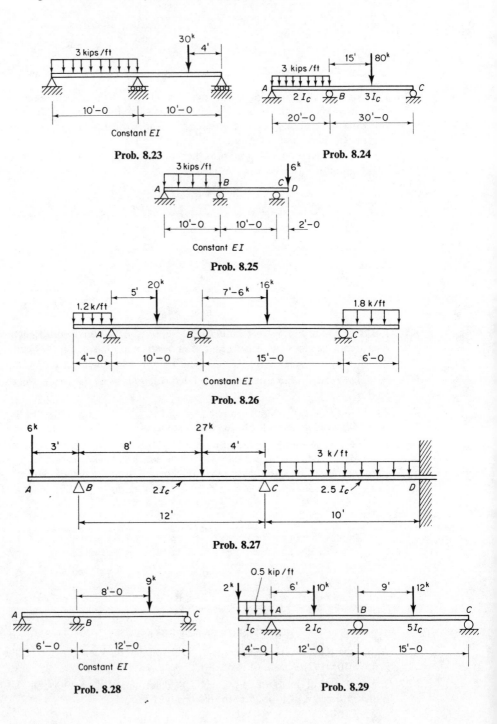

Prob. 8.23

Prob. 8.24

Prob. 8.25

Prob. 8.26

Prob. 8.27

Prob. 8.28

Prob. 8.29

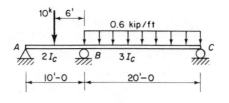

Prob. 8.30

8.31. Determine the equations for and plot to scale the influence line for the reaction at B on the beam of Prob. 8.15. Use the influence line to determine the reaction at B caused by the loading of Prob. 8.15.

8.32. For a two-span (40 ft–50 ft) continuous beam ABC of constant EI, determine the equations for and plot to scale the influence line for the reaction at the interior support B. Using the influence line determine the maximum and minimum values of R_B for the following loads: dead load, 1.0 kip/ft; live load, 2.5 kips/ft; one moving concentrated load of 24 kips.

8.33. For the beam of Prob. 8.32, determine the equations for and plot to scale the influence line for the bending moment at the interior support B, and then determine the maximum and minimum values for M_B for the loads given in Prob. 8.32.

     In addition, determine the equations for and plot to scale the influence lines for whichever of the following functions may be assigned by the instructor. Use the equations already determined for M_B to determine the equations for the other influence line(s). L refers to the 40 ft span AB.

(a) reaction at A;

(b) shear BA at the left side of support B;

(c) shear BC at the right side of support B;

(d) shear at $0.25L$ from support A;

(e) shear at $0.5L$ from support A;

(f) bending moment at $0.5L$ from support A;

(g) bending moment at $0.7L$ from support A;.

(h) bending moment at $0.9L$ from support A.

8.34. For the three-span continuous beam of Prob. 8.20 with $\alpha = 1.25$, determine the equations for and plot to scale the following influence lines, then determine the maximum and minimum values for the functions due to the passage of two 24-kip concentrated loads spaced 4 ft apart, when the span length L is 40 ft:

(a) bending moment M_B at support B;

(b) bending moment $M_{0.5}$ at $0.5L$ from support A in span AB;

(c) bending moment at the fixed point near end B in span BC;

(d) shear $V_{0.5}$ at $0.5L$ from end A in span AB.

8.35. Solve for the requirements of Prob. 8.34 except use $\alpha = 1.5$.

8.36. For the beam of Prob. 8.20 with $\alpha = 1.0$ determine the equation for and plot to scale the influence lines for the bending moments at B and C, and for the bending moment $M_{0.1}$ at $0.1L$ from B on span BC. If the live load consists of one moving concentrated load of 25 kips and uniform load of 2.1 kips/ft of indefinite length, and the dead load is 0.8 kip/ft everywhere, determine the maximum and minimum bending moment $M_{0.1}$.

8.37. Solve Prob. 8.1 by Castigliano's theorem: (a) using the reaction at B as the redundant force, and (b) using the moment reaction at A as the redundant force.

8.38. Use the requirements of Prob. 8.37 to solve Prob. 8.2 using Castigliano's Theorem.

8.39. Solve Prob. 8.9 using Castigliano's theorem taking the reaction at A as the redundant.

8.40. Solve Prob. 8.10 using Castigliano's theorem, taking the reaction at A as the redundant.

8.41. Solve Prob. 8.11 using Castigliano's theorem, taking the reaction at B as the redundant.

[9]

Slope Deflection Method
—Application to Beams

9.1 DISPLACEMENT METHOD—CONCEPT

First briefly mentioned in Section 1.16 and later discussed for a simple truss analysis in Section 7.4, the displacement method is the direct link with present-day computer methods of structural analysis. In Section 7.4 it was noted that in the displacement method of analysis, the joint displacements X_i are taken as the primary unknowns. Every analysis will involve the determination of all possible joint displacements. Since joint displacement indicates a "freedom" to move, the total number of possible joint displacements is referred to as the "degree of freedom."

In the case of a beam, or segment of a beam, the knowledge of the displacements (deflection Δ and slope θ) at each end of the beam segment permits determination of the internal forces (shears and bending moments). When the deflections and slopes are taken as the primary unknowns, the procedure is known as the *displacement method*.

9.2 SLOPE DEFLECTION EQUATIONS

The method known as the *slope deflection method* originally was developed because of a need to consider the secondary effects in truss members; that is, the internal bending moments and shears in truss members resulting from rotational restraint inherent in bolted or welded truss joints. Truss analysis as discussed in Chapters 2, 3, and 7 was based on frictionless pins connecting truss members at joints.

Heinrich Manderla [1] presented in 1879–1880 the first satisfactory solution for the secondary bending moments in trusses. In 1892–1893, Otto Mohr [2] presented an improved version of a slope and deflection procedure for the secondary analysis of a truss, a method essentially identical to what is now called the slope deflection method. Axel Bendixen of Berlin published [3] in 1914, and George A. Maney of the University of Minnesota published [4] in 1915, their developments of the method. Extensive treatment of the method followed in the works of Wilson and Maney [5], Wilson, Richart, and Weiss [6], Parcel and Maney [7], and Ostenfeld [8]. During the period from the development of the slope deflection method until the development of the moment distribution method, the slope deflection method became the method of choice for continuous beams and was the only practical procedure for analyzing rigid frames.

To gain a conceptual idea of what the method is about, refer to Fig. 9.2.1a. First consider the unloaded segment, AB, which in its deformed position caused by loading on the structure will have slopes θ_A and θ_B of the elastic curve and deflections Δ_A and Δ_B at its two ends. These four displacements θ_A, θ_B, Δ_A, and Δ_B completely determine the deformed shape of segment AB, as shown in Fig. 9.2.1c. The forces M'_A, M'_B, V'_A, and V'_B are the forces necessary to cause the four displacements.

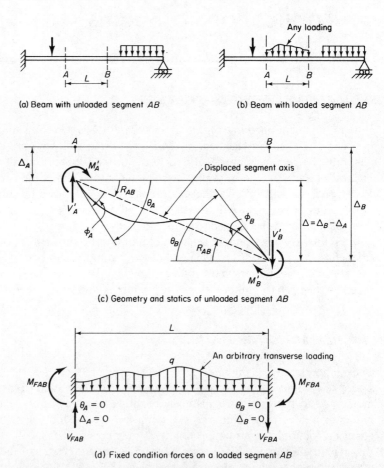

(a) Beam with unloaded segment AB

(b) Beam with loaded segment AB

(c) Geometry and statics of unloaded segment AB

(d) Fixed condition forces on a loaded segment AB

Figure 9.2.1 Forces and displacements for the slope deflection method.

When the segment also has transverse loading, as for AB in Fig. 9.2.1b, super-position of effects may be applied. If the loaded segment, as given separately in Fig. 9.2.1d with all reactions shown, is considered to have *zero* displacements in θ_A, θ_B, Δ_A, and Δ_B, restraining forces M_{FAB}, M_{FBA}, V_{FAB}, and V_{FBA} are necessary to keep the deformations zero. This is the so-called *fixed* condition, and the restraining moments M_{FAB} and M_{FBA} are called *fixed-end moments*.

By the principle of superposition, the end moments M_A and M_B on any loaded segment AB may be expressed as the sum of those in the fixed (zero displacement) condition of Fig. 9.2.1d and those in the displaced condition of Fig. 9.2.1c; thus,

$$M_A = M_{FAB} + M'_A \tag{9.2.1a}$$

$$M_B = M_{FBA} + M'_B \tag{9.2.1b}$$

Referring to Fig. 9.2.1c, let ϕ_A and ϕ_B be the *clockwise* end rotations measured from the axis connecting the ends of the displaced segment AB to the tangents to the elastic curve at A and B, respectively. Further, let R_{AB} be the *clockwise* rotation of the displaced segment axis measured from its original position in the unloaded structure to its displaced position. Then,

$$R_{AB} = \frac{\Delta_B - \Delta_A}{L} \tag{9.2.2}$$

$$\theta_A = \phi_A + R_{AB} \tag{9.2.3a}$$

$$\theta_B = \phi_B + R_{AB} \tag{9.2.3b}$$

Determination of the angles ϕ_A and ϕ_B may be made by the use of the moment area method (see Section 5.5). Referring to Fig. 9.2.2,

$$\phi_A = \frac{\Delta'_{BA}}{L} = \frac{\text{moment of } M/EI \text{ area between } A \text{ and } B \text{ about } B}{L}$$

$$= \frac{1}{L}\left[\frac{M'_A}{EI}\left(\frac{1}{2}\right)(L)\left(\frac{2L}{3}\right) - \frac{M'_B}{EI}\left(\frac{1}{2}\right)(L)\left(\frac{L}{3}\right)\right]$$

$$= +\frac{L}{3EI}M'_A - \frac{L}{6EI}M'_B \tag{9.2.4a}$$

Similarly,

$$\phi_B = \frac{\Delta'_{AB}}{L} = -\frac{L}{6EI}M'_A + \frac{L}{3EI}M'_B \tag{9.2.4b}$$

Solving Eqs. (9.2.4) for M'_A and M'_B in terms of ϕ_A and ϕ_B gives

$$\left.\begin{array}{l} M'_A = \dfrac{4EI}{L}\phi_A + \dfrac{2EI}{L}\phi_B \\[3mm] M'_B = \dfrac{2EI}{L}\phi_A + \dfrac{4EI}{L}\phi_B \end{array}\right\} \tag{9.2.5}$$

Substituting Eqs. (9.2.3) into Eqs. (9.2.5) gives

$$\left.\begin{array}{l} M'_A = \dfrac{4EI}{L}(\theta_A - R_{AB}) + \dfrac{2EI}{L}(\theta_B - R_{AB}) \\[3mm] M'_B = \dfrac{2EI}{L}(\theta_A - R_{AB}) + \dfrac{4EI}{L}(\theta_B - R_{AB}) \end{array}\right\} \tag{9.2.6}$$

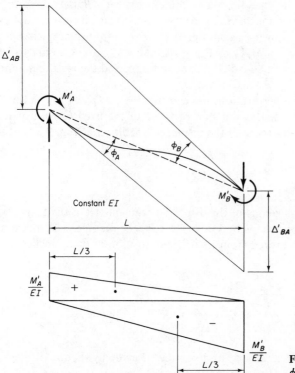

Figure 9.2.2 Determination of ϕ_A and ϕ_B by the moment area method.

Finally, substituting Eqs. (9.2.6) into Eqs. (9.2.1) gives

$$\left. \begin{array}{l} M_A = M_{FAB} + \dfrac{4EI}{L}(\theta_A - R_{AB}) + \dfrac{2EI}{L}(\theta_B - R_{AB}) \\[3mm] M_B = M_{FBA} + \dfrac{2EI}{L}(\theta_A - R_{AB}) + \dfrac{4EI}{L}(\theta_B - R_{AB}) \end{array} \right\} \qquad (9.2.7)$$

which are the *slope deflection equations*.

The slope deflection equations may be rewritten combining terms involving the member rotation R_{AB}, which actually is Δ/L. (Note that Δ is equal to $\Delta_B - \Delta_A$.) Thus,

$$\left. \begin{array}{l} M_A = M_{FAB} + \dfrac{4EI}{L}\theta_A + \dfrac{2EI}{L}\theta_B - \dfrac{6EI}{L^2}\Delta \\[3mm] M_B = M_{FBA} + \dfrac{2EI}{L}\theta_A + \dfrac{4EI}{L}\theta_B - \dfrac{6EI}{L^2}\Delta \end{array} \right\} \qquad (9.2.8)$$

Philosophically, one may visualize the equation of M_A, for example, as the superposition of four separate loading cases.: (a) $M_{A1} = M_{FAB}$, the so-called fixed-end moment due to transverse loads, as in Fig. 9.2.3b; (b) $M_{A2} = 4EI\theta_A/L$ acting to cause θ_A on member AB with end B fixed, as in Fig. 9.2.3c; (c) $M_{A3} = 2EI\theta_B/L$ at fixed end A when end B has the slope θ_B, as in Fig. 9.2.3d; and (d) $M_{A4} = -6EI\Delta/L^2$ acting at fixed end A when both ends are fixed but are displaced by the amount Δ, as in Fig. 9.2.3e. In other words, M_A is the sum of the moments caused by the effect of the

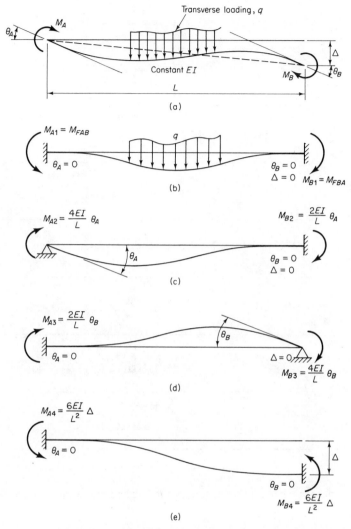

Figure 9.2.3 Slope deflection equations as superposition of four loading cases.

applied transverse load with all three displacements equal to zero, plus the separate application of each of the three displacements θ_A, θ_B, and Δ with the other two equal to zero.

In order to use the slope deflection equations to make the analysis of a continuous beam or a rigid frame, the fixed-end moments M_{FAB} and M_{FBA} are required. These may be determined from any method of statically indeterminate analysis, such as by consistent deformation or the theorem of three moments (by adding one fictitious span on each side) as treated in Chapter 8. A summary of various fixed-end moments is given in Fig. 9.2.4. Since in the derivation of the slope deflection equations clockwise end moments are positive, gravity loading on a beam segment will give rise

M_{FAB}	Load condition	M_{FBA}
$-WLk(1-k)^2$	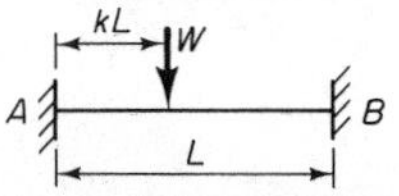	$+WLk^2(1-k)$
$-\dfrac{wL^2}{12}$	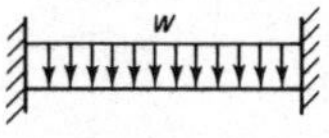	$+\dfrac{wL^2}{12}$
$-\dfrac{wL^2}{12}\left[k^2(3k^2-8k+6)\right]_{k_1}^{k_2}$	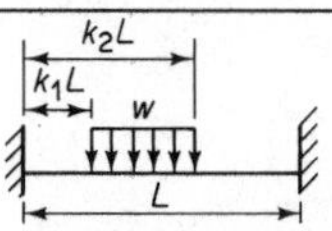	$+\dfrac{wL^2}{12}\left[k^3(4-3k)\right]_{k_1}^{k_2}$
$-\dfrac{L^2}{60}(3w_1+2w_2)$	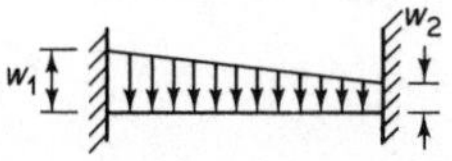	$+\dfrac{L^2}{60}(2w_1+3w_2)$
$-\dfrac{wL^2}{20}$	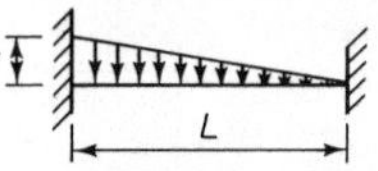	$+\dfrac{wL^2}{30}$
$-M(1-k)(1-3k)$	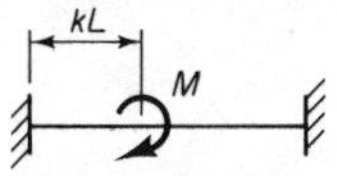	$+Mk(2-3k)$
$-\dfrac{wL^2}{60}k^2(10-10k+3k^2)$	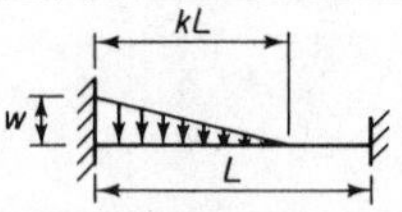	$+\dfrac{wL^2}{60}k^3(5-3k)$
$-\dfrac{5wL^2}{96}$	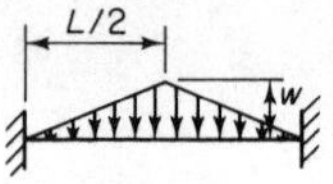	$+\dfrac{5wL^2}{96}$

Figure 9.2.4 Fixed-end moments (clockwise moments positive).

to a negative fixed-end moment at the left end and a positive fixed-end moment at the right end.

9.3 APPLICATION TO STATICALLY INDETERMINATE BEAMS

In the displacement method of beam analysis, the beam segments used are usually entire spans between external supports. The beam span member end moments are expressed in terms of the unknown slopes and deflections at the joints (using Eq. 9.2.7 or 9.2.8). Then the equilibrium conditions at the joints, corresponding directly to the degrees of freedom, are used to establish a system of simultaneous equations in which the joint displacements are the unknowns. Thus the degree of statical indeterminacy has no relationship to the slope deflection method, which can be used to analyze statically determinate as well as statically indeterminate beams. The complexity depends on the number of joints used by the analyst. In calculations where the simultaneous solution of the set of linear equations must be performed without the aid of the digital computer, there is no reason to use the slope deflection method for any statically determinate analysis, and in statically indeterminate analysis there is little reason to use a degree of freedom larger than the minimum necessary. When the digital calculator is used, the analyst may treat as a joint any location where information on slope, deflection, shear, or bending moment is desired as a part of the output. The use of additional joints to get extra output is at the expense of additional input data and computer time.

In the event that only entire spans in a continuous beam are used as the beam segments for which slope deflection equations are written, all the *unknown* joint displacements are slopes to the elastic curve (or rotations of the cross section) at the external supports. When there are no support settlements, the quantity R in Eqs. (9.2.7) or Δ in Eqs. (9.2.8) is zero, which shortens the slope deflection equations. However, when analysis is to be made for *known* (or anticipated) support settlements, the net effect of substituting the known value of R in Eqs. (9.2.7), or of Δ in Eqs. (9.2.8), is to give an additional fixed-end moment value to be added to M_{FAB} or M_{FBA}, which are the fixed-end moments due to the applied transverse load.

The following three examples illustrate the application of the slope deflection method to statically indeterminate beams.

Example 9.3.1

Analyze the two span continuous beam shown in Fig. 9.3.1 using the slope deflection method.

Solution

(a) *Fixed-end moments.* Referring to Fig. 9.2.4, the fixed-end moments are computed and shown in Fig. 9.3.1.

$$M_{F1} = -\frac{wL_1^2}{12} = -\frac{2(20)^2}{12} = -66.67 \text{ ft-kips}$$

$$M_{F2} = +\frac{wL_1^2}{12} = +\frac{2(20)^2}{12} = +66.67 \text{ ft-kips}$$

$$M_{F3} = -\frac{wL_2^2}{12} = -\frac{2(30)^2}{12} = -150.00 \text{ ft-kips}$$

$$M_{F4} = +\frac{wL_2^2}{12} = +\frac{2(30)^2}{12} = +150.00 \text{ ft-kips}$$

(b) *Slope deflection equations for end moments.* Using Eqs. (9.2.7) the end moments are (see Fig. 9.3.1 for numbering)

$$M_1 = -66.67 + \frac{4EI}{20}\theta_A + \frac{2EI}{20}\theta_B$$

$$M_2 = +66.67 + \frac{2EI}{20}\theta_A + \frac{4EI}{20}\theta_B$$

$$M_3 = -150.00 + \frac{4EI}{30}\theta_B + \frac{2EI}{30}\theta_C$$

$$M_4 = +150.00 + \frac{2EI}{30}\theta_B + \frac{4EI}{30}\theta_C$$

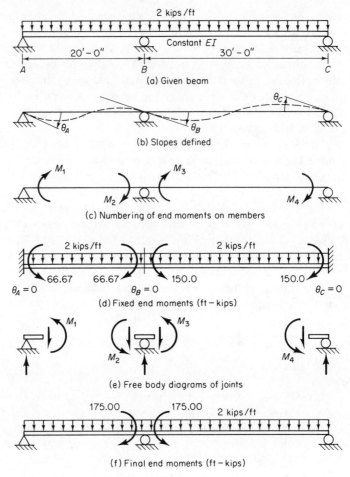

2 kips/ft

Constant EI

20'–0"

30'–0"

A

B

C

(a) Given beam

θ_C

θ_A

θ_B

(b) Slopes defined

M_1

M_3

M_2

M_4

(c) Numbering of end moments on members

2 kips/ft

2 kips/ft

66.67   66.67   150.0   150.0

$\theta_A = 0$ $\theta_B = 0$ $\theta_C = 0$

(d) Fixed end moments (ft – kips)

M_1

M_3

M_2

M_4

(e) Free body diagrams of joints

175.00   175.00   2 kips/ft

(f) Final end moments (ft – kips)

Figure 9.3.1 Slope deflection solution for Example 9.3.1.

Note that the R values are zero because there is no relative deflection at the ends of the members.

(c) *Equilibrium equations at joints.* Since this two-span beam has rotation possible at all three joints but has zero deflection at those joints, the structure is said to have three degrees of freedom, one rotation at each joint. The number of joint equilibrium equations must equal the degree of freedom. Each joint is shown in Fig. 9.3.1e, where the forces involved in rotational equilibrium are shown. Writing the equilibrium equations for moments at each joint gives

$$M_1 = 0 \tag{a}$$

$$M_2 + M_3 = 0 \tag{b}$$

$$M_4 = 0 \tag{c}$$

(d) *Solution for θ values.* Substituting the equations for M_1 to M_4 into Eqs. (a) to (c) gives

$$\frac{4EI}{20}\,\theta_A + \frac{2EI}{20}\,\theta_B \qquad\qquad = +66.67 \tag{d}$$

$$\frac{2EI}{20}\,\theta_A + \frac{EI}{3}\,\theta_B + \frac{2EI}{30}\,\theta_C = +83.33 \tag{e}$$

$$\frac{2EI}{30}\,\theta_B + \frac{4EI}{30}\,\theta_C = -150.00 \tag{f}$$

Solving Eqs. (d), (e), and (f) for θ_A, θ_B, and θ_C gives

$$EI\theta_A = +83.33$$
$$EI\theta_B = +500.00$$
$$EI\theta_C = -1375.00$$

(e) *Back substitution to obtain final end moments.* With all joint displacements already determined, they can be substituted into the equations for M_1 to M_4 of part (b). Thus,

$$M_1 = -66.67 + \frac{83.33}{5} + \frac{500}{10} = 0$$

$$M_2 = +66.67 + \frac{83.33}{10} + \frac{500}{5} = +175.00$$

$$M_3 = -150.00 + \frac{500}{7.5} + \frac{-1375}{15} = -175.00$$

$$M_4 = +150.00 + \frac{500}{15} + \frac{-1375}{7.5} = 0$$

The positive sign for M_2 above shows that it acts on BA at end B in the clockwise direction, and the negative sign for M_3 above shows that it acts on BC at end B in the counterclockwise direction. The actual directions of these member end moments are shown in Fig. 9.3.1f. One may note that the moments acting on the joint at B are such as to both cause tension at the top and compression at the bottom of the beam; thus, the *bending moment* at B is -175.00 ft-kips. For design use, the pair of end moments acting in opposite rotational directions on each side of a joint must be converted to the appropriate bending moment sign. With the bending moment at support B known, the reactions, the shear diagram, and the moment diagram may be obtained as shown in Chapter 8.

Example 9.3.2

Analyze the continuous beam of Fig. 9.3.2 (previously solved in Example 8.3.1) using the slope deflection method.

Solution

(a) *Establish number of segments to be used.* In this problem, the load on the overhang CD may be transferred to act on joint C, as the combination of a downward load of 12 kips and a clockwise moment of 36 ft-kips. If the beam ABC without the overhang is thus analyzed and the rotation θ_C is found, the elastic curve for CD, as well as the shear and moment diagrams for it, are all obtainable using statics. Thus the minimum number of segments that can be used is two, spans AB and BC, and the minimum number of joints is three, at A, B, and C. The member ends are designated

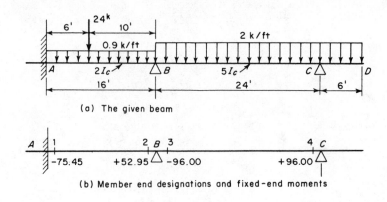

(a) The given beam

A ⫽|1   2 B 3   4 C
⫽ −75.45   +52.95 △ −96.00   +96.00 △

(b) Member end designations and fixed-end moments

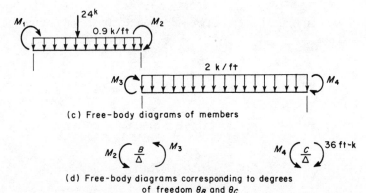

(c) Free-body diagrams of members

$M_2 \left(\dfrac{B}{\Delta} \right) M_3$ $M_4 \left(\dfrac{C}{\Delta} \right)$ 36 ft-k

(d) Free-body diagrams corresponding to degrees
of freedom θ_B and θ_C

Figure 9.3.2 Solution by the slope deflection method for Example 9.3.2.

1 through 4 as in Fig. 9.3.2b. Since joint A is fixed and has no freedom to rotate, there are only two freedoms possessed by the three joints, those of rotation at joints B and C.

(b) *Fixed-end moments.* Referring to Fig. 9.2.4, the fixed-end moments are computed and shown in Fig. 9.3.2.

$$M_{F1} = -\frac{24(6)(10)^2}{(16)^2} - \frac{0.9(16)^2}{12} = -56.25 - 19.20 = -75.45 \text{ ft-kips}$$

$$M_{F2} = +\frac{24(10)(6)^2}{(16)^2} + \frac{0.9(16)^2}{12} = +33.75 + 19.20 = +52.95 \text{ ft-kips}$$

$$M_{F3} = -\frac{2(24)^2}{12} = -96.0 \text{ ft-kips}$$

$$M_{F4} = +\frac{2(24)^2}{12} = +96.0 \text{ ft-kips}$$

(c) *Slope deflection equations.* Use Eqs. (9.2.7) with R equal to zero because the member axes do not rotate. Noting that θ_A is zero,

$$M_1 = -75.45 + \frac{4E(2I_c)}{16}(0) + \frac{2E(2I_c)}{16} \theta_B = -75.45 + \frac{1}{4} EI_c\theta_B$$

$$M_2 = +52.95 + \frac{2E(2I_c)}{16}(0) + \frac{4E(2I_c)}{16}\theta_B = +52.95 + \frac{1}{2}EI_c\theta_B$$

$$M_3 = -96.00 + \frac{4E(5I_c)}{24}\theta_B + \frac{2E(5I_c)}{24}\theta_C = -96.00 + \frac{5}{6}EI_c\theta_B + \frac{5}{12}EI_c\theta_C$$

$$M_4 = +96.00 + \frac{2E(5I_c)}{24}\theta_B + \frac{4E(5I_c)}{24}\theta_C = +96.00 + \frac{5}{12}EI_c\theta_B + \frac{5}{6}EI_c\theta_C$$

(d) *Joint equilibrium equations.* The two equilibrium equations corresponding to the degrees of freedom in rotation θ_B and θ_C are, from Fig. 9.3.2d,

$$+M_2 + M_3 \quad = 0$$
$$+M_4 - 36.00 = 0$$

Note that since M_1, M_2, M_3, and M_4 are all unknowns, they must be considered to act clockwise on member ends and counterclockwise on the joints.

Substituting the expressions for M_1, M_2, M_3, and M_4 into the two equilibrium equations, the following two equations are obtained:

$$\frac{4}{3}EI_c\theta_B + \frac{5}{12}EI_c\theta_C = +43.05$$

$$\frac{5}{12}EI_c\theta_B + \frac{5}{6}EI_c\theta_C = -60.00$$

Solving,

$$EI_c\theta_B = +64.93 \text{ kip-ft}^2$$
$$EI_c\theta_C = -104.47 \text{ kip-ft}^2$$

(e) *Back substitution to obtain member end moments.* Substituting the known values of $EI_c\theta_B$ and $EI_c\theta_C$ into the slope deflection equations,

$$M_1 = -75.45 + \frac{1}{4}(+64.93) = -59.22 \text{ ft-kips}$$

$$M_2 = +52.95 + \frac{1}{2}(+64.93) = +85.42 \text{ ft-kips}$$

$$M_3 = -96.00 + \frac{5}{6}(+64.93) + \frac{5}{12}(-104.47) = -85.42 \text{ ft-kips}$$

$$M_4 = +96.00 + \frac{5}{12}(+64.93) + \frac{5}{6}(-104.47) = +36.00 \text{ ft-kips}$$

The shear and moment diagrams for AB and BC may now be obtained (see Fig. 8.3.4 for computation details), and the results are shown in Fig. 9.3.3 for use in the next example.

(f) *Equilibrium and compatibility checks.* To ensure the correctness of the solution, the obvious checks are for satisfaction of the equilibrium conditions equal in number to the degree of freedom. In this example they are $M_2 + M_3 = 0$ and $M_4 - 36.00 = 0$. In addition, compatibility checks equal in number to the degree of statical indeterminacy should be made. Using the bending moment diagrams (actually M/EI) with moment area method the slope θ_A may be shown equal to zero and the slope θ_B in span BA may be shown equal to the slope θ_B in span BC. This was illustrated in Example 8.3.1(e). One may note, however, that the compatibility checks make use of the simple beam moment diagrams which were used in establishing the theorem of three moments used in Example 8.3.1. Because these simple beam moment diagrams (especially the part due to transverse load applied on the span) were not used in developing the slope deflection equations, the compatibility checks made after the slope deflection solution are truly independent.

Example 9.3.3

Using the results from Example 9.3.2, together with the bending moment diagram previously computed and drawn for Example 8.3.1 (Fig. 8.3.4), compute by the moment area method the slope and deflection at the free end of the beam shown in Fig. 9.3.3a. Then, considering the entire beam to have four joints (at A, B, C, and D) and three members (AB, BC, and CD), write the slope deflection equations for M_5 and M_6 in terms of θ_C, θ_D, and Δ_D. Finally, confirm that these two equations are numerically satisfied by the values of M_5, M_6, θ_C, θ_D, and Δ_D already obtained.

Solution. (a) *Determine θ_D, Δ_D, and R_{CD} using the moment area method.* The value of θ_C depends only on the flexural rigidity EI of members AB and BC and has been determined in Example 9.3.2 to be $-104.47/(EI_c)$ kip-ft². The slope and deflection at point D, however, depend not only on θ_C but also on the flexural rigidity of overhang CD, whose moment of inertia is now assumed to be $3I_c$ as shown in Fig. 9.3.3. By the moment area method,

$$\theta_D = \theta_C + \left[\frac{M}{EI}\text{ area on } CD\right] = \frac{1}{EI_c}\left[-104.47 + \frac{1}{3}\left(\frac{36}{3}\right)(6)\right]$$

$$= -\frac{80.47}{EI_c}\text{ kip-ft}^2$$

$$\Delta_D = 6\theta_C + \left(\text{moment of }\frac{M}{EI}\text{ area on } CD \text{ about } D\right)$$

$$= \frac{1}{EI_c}\left[6(-104.47) + \frac{1}{3}\left(\frac{36}{3}\right)(6)(4.5)\right] = -\frac{518.8}{EI_c}\text{ kip-ft}^3$$

$$R_{CD} = +\frac{\Delta_D}{6} = -\frac{86.47}{EI_c}\text{ kip-ft}^2$$

Note that θ_C, θ_D, and R_{CD} are positive if clockwise, and Δ_D is positive if downward; thus end D actually deflects upward and the slope θ_D is actually counterclockwise.

(b) *Write the slope deflection equations for M_5 and M_6.* The slope deflection equations for end moments M_5 and M_6 acting clockwise on member CD are

$$M_5 = M_{F5} + \frac{4EI}{L}(\theta_C - R_{CD}) + \frac{2EI}{L}(\theta_D - R_{CD})$$

$$= -\frac{2(6)^2}{12} + \frac{4(3)}{6}(-104.47 + 86.47) + \frac{2(3)}{6}(-80.47 + 86.47)$$

$$= -6 - 36 + 6 = -36.0\text{ ft-kips}\quad\text{(check)}$$

$$M_6 = M_{F6} + \frac{2EI}{L}(\theta_C - R_{CD}) + \frac{4EI}{L}(\theta_D - R_{CD})$$

$$= +\frac{2(6)^2}{12} + \frac{2(3)}{6}(-104.47 + 86.47) + \frac{4(3)}{6}(-80.47 + 86.47)$$

$$= +6 - 18 + 12 = 0\quad\text{(check)}$$

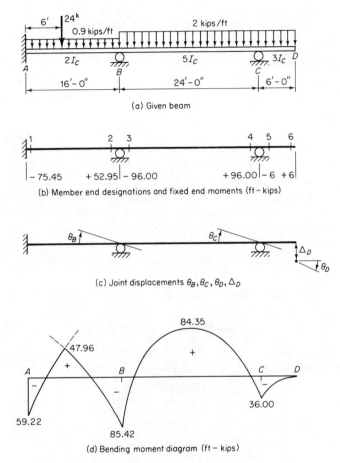

Figure 9.3.3 Slope deflection solution for Example 9.3.3.

9.4 USE OF AN ARBITRARY JOINT

In the use of the slope deflection method, the beam may be divided into as many beam segments as desired. For instance, in Example 9.3.2 the minimum of two segments (*AB* and *BC*) has been used to analyze the beam of Fig. 9.3.3. However, the solution could have been made using three segments (*AB*, *BC*, and *CD*), as partially demonstrated by the slope deflection equations for M_5 and M_6 in Example 9.3.4. The additional segment *CD* gives two additional degrees of freedom in displacements θ_D and Δ_D; thus, the slope deflection solution would have required solving four equations instead of two for the displacements. In fact, except for computation time and expense, a beam could be divided into many short segments and the solution would provide the slope and deflection of the elastic curve at each of the joints (i.e., the segment ends). Each additional segment used adds two degrees of freedom and therefore increases the number of equations by two.

To illustrate the treatment using an arbitrary joint, the following example is presented.

Example 9.4.1

Using the slope deflection method, compute the midspan deflection of a simply supported uniformly loaded beam. Utilize a joint at midspan in making the solution.

Solution

(a) *Select segments and establish degrees of freedom.* The slope deflection method may be used for either statically determinate or statically indeterminate structures. For the simply supported beam of Fig. 9.4.1 a joint will be taken at midspan point B in addition to the joints at A and C. This means that for the two segments AB and BC there are degrees of freedom in rotation at A, B, and C, involving θ_A, θ_B, and θ_C as unknowns. Additionally, there is a deflection degree of freedom at B involving the rotations R_{AB} and R_{BC} of the member axes AB and BC. In this example the rotation of member axis AB is $R_{AB} = +\Delta_B/(L/2)$ but the rotation of member axis BC is $R_{BC} = -\Delta_B/(L/2)$, noting that the sign convention takes clockwise rotation of member axis as positive.

(b) *Slope deflection equations.* Using the numbering in Fig. 9.4.1b and applying

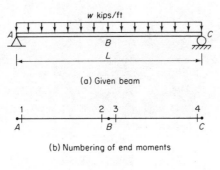

(a) Given beam

(b) Numbering of end moments

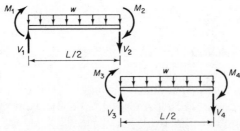

(c) Free body diagrams of elements

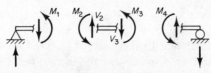

(d) Free body diagrams of joints

Figure 9.4.1 Free-body diagrams for the elements and joints used in the slope deflection method, Example 9.4.1.

Eqs. (9.2.7), by taking R as the absolute value of $\Delta_B/(L/2)$,

$$M_1 = -\frac{w(L/2)^2}{12} + \frac{4EI}{L/2}(\theta_A - 2\Delta_B/L) + \frac{2EI}{L/2}(\theta_B - 2\Delta_B/L)$$

$$M_2 = +\frac{w(L/2)^2}{12} + \frac{2EI}{L/2}(\theta_A - 2\Delta_B/L) + \frac{4EI}{L/2}(\theta_B - 2\Delta_B/L)$$

$$M_3 = -\frac{w(L/2)^2}{12} + \frac{4EI}{L/2}(\theta_B + 2\Delta_B/L) + \frac{2EI}{L/2}(\theta_C + 2\Delta_B/L)$$

$$M_4 = +\frac{w(L/2)^2}{12} + \frac{2EI}{L/2}(\theta_B + 2\Delta_B/L) + \frac{4EI}{L/2}(\theta_C + 2\Delta_B/L)$$

Note that since R is defined as the absolute value of $2\Delta_B/L$, it is therefore put into Eqs. (9.2.7) as positive for segment AB and negative for segment BC.

(c) *Joint equilibrium equations.* Referring to Fig. 9.4.1d, the following equilibrium equations must be satisfied:

$$M_1 = 0 \tag{a}$$

$$M_2 + M_3 = 0 \tag{b}$$

$$M_4 = 0 \tag{c}$$

$$+V_2 - V_3 = 0 \tag{d}$$

This last equation may be expressed in terms of the end moments by using the free-body diagrams of Fig. 9.4.1c. Thus,

$$V_2 = -\frac{M_1 + M_2}{L/2} - \frac{wL}{4}$$

$$V_3 = -\frac{M_3 + M_4}{L/2} + \frac{wL}{4}$$

Thus, Eq. (d) may be written

$$-\frac{2(M_1 + M_2)}{L} + \frac{2(M_3 + M_4)}{L} = +\frac{wL}{2} \tag{e}$$

(d) *Set of linear equations.* Substitution of the slope deflection equations of part (b) into the joint equilibrium equations of part (c) gives

$$\frac{8EI}{L}\theta_A + \frac{4EI}{L}\theta_B \qquad\qquad - \frac{24EI}{L^2}\Delta_B = +\frac{wL^2}{48}$$

$$\frac{4EI}{L}\theta_A + \frac{16EI}{L}\theta_B + \frac{4EI}{L}\theta_C \qquad\qquad = 0$$

$$\frac{4EI}{L}\theta_B + \frac{8EI}{L}\theta_C + \frac{24EI}{L^2}\Delta_B = -\frac{wL^2}{48}$$

$$\frac{-24EI}{L^2}\theta_A \qquad\qquad + \frac{24EI}{L^2}\theta_C + \frac{96EI}{L^2}\Delta_B = +\frac{wL^2}{2}$$

Note that the off-diagonal coefficients in the system of equations above are symmetrical with respect to the main diagonal (downward to the right), a fact that will be later proved in Chapter 16. In order to preserve this symmetry (although not necessary for obtaining the correct solution), however, the positive sense in the left side of Eqs. (a) to (d) must be opposite to the direction of the degree of freedom, that is, counterclockwise moments acting on joints A, B, and C in the free-body diagrams of Fig. 9.4.1d, and upward force acting on joint B in the free-body diagram of Fig. 9.4.1d.

Solving for θ_A, θ_B, θ_C, and Δ_B,

$$\theta_A = \frac{+wL^3}{24EI}$$

$$\theta_B = 0$$

$$\theta_C = \frac{-wL^3}{24EI}$$

and

$$\Delta_B = \frac{5wL^4}{384EI}$$

In this example the general procedure has been shown for the case when there is a degree of freedom in displacement, giving rise to the unknown deflection Δ_B together with the rotations θ of the joints.

SELECTED REFERENCES

1. Heinrich Manderla, "Die Berechnung der Sekundärspannungen, welche im einfachen Fachwerke in Folge starrer Knotenverbindungen auftreten," *Forster's Bauzeitung*, *45* (1880), p. 34. (See also *Annual Report of the Technische Hochschule, Munich*, 1879.)
2. Otto Mohr, "Die Berechnung der Fachwerke mit starren Knotenverbindungen," *Der Civilingenieur*, *38* (1892), 577–594; *39* (1893), 67–70.
3. Axel Bendixen, *Die Methode der Alpha-Gleichungen zur Berechnung von Rahmenkonstruktionen*, Berlin, 1914.
4. George A. Maney, *Studies in Engineering*, No. 1, University of Minnesota, March 1915.
5. W. M. Wilson and G. A. Maney, "Wind Stresses in the Steel Frames of Office Buildings," *University of Illinois Bulletin*, No. 80, Engineering Experiment Station, 1915.
6. W. M. Wilson, F. E. Richart, and Camillo Weiss, "Analysis of Statically Indeterminate Structures by the Slope Deflection Method," *University of Illinois Bulletin*, No. 108, Engineering Experiment Station, November 1918, 218 pp.
7. John I. Parcel and George A. Maney, *An Elementary Treatise on Statically Indeterminate Stresses*, John Wiley & Sons, Inc., New York, 1926, pp. 147–157.
8. A. Ostenfeld, *Die Deformationsmethode*, Springer-Verlag, Berlin, 1926.

PROBLEMS

For all problems, draw the resulting shear and bending moment diagrams after making the slope deflection method analysis. Unless otherwise specified by the statement, use the minimum number of degrees of freedom for making the solution.

9.1–9.4. Analyze the beams of Prob. 8.1 through 8.4, respectively, using the slope deflection method.

9.5. Analyze the beam of Prob. 8.13 using the slope deflection method.

9.6. Analyze the beam of Prob. 8.14 using the slope deflection method.

9.7. Analyze the beam of Prob. 8.16 using the slope deflection method.

9.8–9.12. Analyze the beam of each accompanying figure using the slope deflection method.

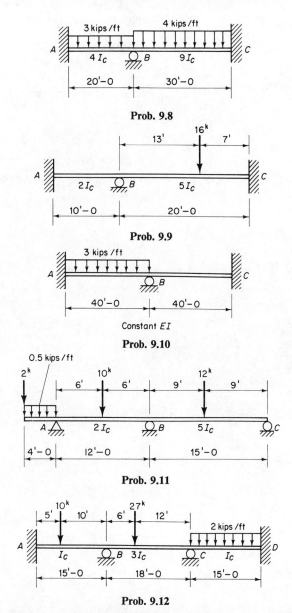

Prob. 9.8

Prob. 9.9

Prob. 9.10

Prob. 9.11

Prob. 9.12

9.13. Analyze the beam of the accompanying figure using the slope deflection method. (a) Solve using the minimum number of beam segments. (b) Solve using beam segments *AB* and *BC*. (c) Solve using beam segments *AD* and *DB*.

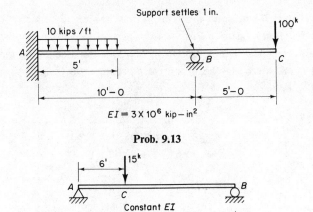

Prob. 9.13

Prob. 9.14

9.14. Analyze the simply supported beam of the accompanying figure using slope deflection method by taking segments *AC* and *CB*.

[10]

Moment Distribution Method
—Application to Beams

10.1 GENERAL

By the 1920s the use of reinforced concrete as a major structural material was well advanced, bringing with its increased use the greater need for analyzing monolithic statically indeterminate beams and rigid frames. All the analytical methods then available involved solving simultaneous equations. Hardy Cross of the University of Illinois introduced to his students in 1924 the procedure of analyzing continuous structures by a method of successive approximations leading to the "exact" answer when enough iterations have been used. This method now known generally as *moment distribution* was published first in 1929 in a sketchy form [1, 2] for use as a practical design tool in reinforced concrete design and then in 1930 [3] giving the complete procedure for the analysis of continuous frames.

The latter paper, which though complete is only 10 pages in length, provoked 145 pages of discussion. The method rapidly became the most widely used statically indeterminate analysis method until the use of the digital computer around 1970 caused a renewed emphasis on the slope deflection method in the matrix format.

A significant feature of moment distribution is that the process may be visualized as one that physically brings a structure into equilibrium through an iterative procedure. Many analysts still consider moment distribution as the method of choice, especially in the preliminary stages of the design. With the increasing capacity of hand-held electronic calculators, the moment distribution by itself can, in fact, be programmed and the designer can conveniently get the "feel" of the structural response through the stages of trial design. For the most extensive treatment of the subject, the reader is referred to the book by Gere [4].

10.2 CONCEPT OF MOMENT DISTRIBUTION

Before delving into the technical development of the method, the philosophical idea should be understood. Referring to the two-span beam of Fig. 10.2.1a, the process begins by locking all joints so that no rotation can occur. This means that fixed-end

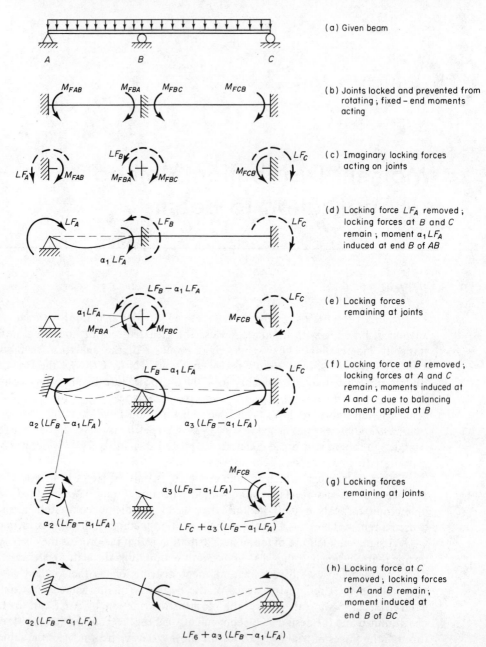

(a) Given beam

(b) Joints locked and prevented from rotating; fixed-end moments acting

(c) Imaginary locking forces acting on joints

(d) Locking force LF_A removed; locking forces at B and C remain; moment $\alpha_1 LF_A$ induced at end B of AB

(e) Locking forces remaining at joints

(f) Locking force at B removed; locking forces at A and C remain; moments induced at A and C due to balancing moment applied at B

(g) Locking forces remaining at joints

(h) Locking force at C removed; locking forces at A and B remain; moment induced at end B of BC

Figure 10.2.1 Concept of moment distribution.

moments are induced to act at the ends of the spans as in Fig. 10.2.1b. Free bodies of the joints as in Fig. 10.2.1c show that to lock the joints, that is, to prevent rotation, the "locking forces" (moments designated LF_A, LF_B, and LF_C) must be acting on the joints equal in magnitude to the algebraic sum of the fixed-end moments but in the opposite direction. In Fig. 10.2.1b, the arrows show the actual directions of the fixed-end moments. The locking moments LF_A and LF_C in Fig. 10.2.1c are obviously counterclockwise and clockwise, respectively, as shown. The locking moment LF_B in Fig. 10.2.1c is shown counterclockwise, consistent with the fact that span BC is shown longer than span AB, which means that M_{FBC} is numerically larger than M_{FBA}.

Next, *one* locking force is removed while the other joints remain locked. For example, Fig. 10.2.1d shows joint A unlocked; thus it rotates clockwise as if the counterclockwise moment required to lock it were applied in the clockwise direction. The application in the opposite direction of the moment required to lock a joint removes the rotational restraint and the joint may be said to be released, or "balanced," that is, in equilibrium. The application of LF_A to joint A in Fig. 10.2.1d with joint B still locked induces a new locking moment $\alpha_1 LF_A$ at end B of member AB. Consequently, the moment required to "lock" joint B changes; in this case it becomes $(LF_B - \alpha_1 LF_A)$, as shown in Fig. 10.2.1e.

The next step is to relock joint A in its new rotated position, and remove the locking force from joint B. Joint B then rotates clockwise as if the moment $(LF_B - \alpha_1 LF_A)$ were applied in the direction opposite to that for locking the joint, as in Fig. 10.2.1e. While joint B rotates to bring itself into equilibrium, the other joints remain locked. This means that the application of the balancing (or releasing) moment at B will induce new locking moments at end A of member AB and at end C of member BC. The inducement of the moment $\alpha_3(LF_B - \alpha_1 LF_A)$ at joint C will change the moment required to lock joint C, as shown in Fig. 10.2.1g.

Upon unlocking joint C, counterclockwise rotation occurs of an amount corresponding to the application of a counterlockwise moment equal in magnitude to the clockwise locking moment, as in Fig. 10.2.1h. Joint C is now balanced but a new locking moment has been induced at the clamped joint B.

At this stage of the process, there are still locking forces acting at joints A and B to keep those joints from rotating *from the positions they have assumed after the first release of locking forces to balance the joints*. But the remaining locking forces after the first pass from left to right are generally smaller than the original locking forces because the joints are now locked into a position closer to their true positions when no locking forces will be acting.

The concept of the moment distribution process may be summarized:

1. Lock all joints to prevent joint rotation. Compute fixed-end moments, and determine magnitudes of the locking moments.

2. Release one joint. The balancing moment at that joint will induce moments at adjacent joints (which are still locked).

3. Lock the balanced joint in its new rotated position.

4. Release another joint so that the balancing moment is applied as in step 2.

5. Lock the joint in its new rotated position.

6. Repeat steps 4 and 5 for each joint.

7. Repeat steps 1 to 6 as many times as may be necessary until the moments induced at adjacent locked joints are smaller than the preset tolerance such as two decimal places.

10.3 STIFFNESS, CARRY-OVER, AND DISTRIBUTION FACTORS

In carrying out the iterative procedure of moment distribution, several concepts are utilized, the definitions of which must be kept closely in mind.

Stiffness

The *stiffness* is defined as the moment acting on one end of a beam necessary to produce unit rotation at that end when the far end is fixed. The supports of both ends are assumed to have zero deflection. Referring to Fig. 10.3.1, the stiffness S_{AB} of the beam is the magnitude of moment M_A required to make $\theta_A = 1$. The expression for S_{AB} may be obtained by computing the slope θ_A in terms of M_A and M_B, using the moment area method. Alternatively, the slope deflection equations, Eqs. (9.2.7), may be used, or any method of computing the relationships between moments and displacements. Using the moment area method with the bending moment diagram of Fig. 10.3.1,

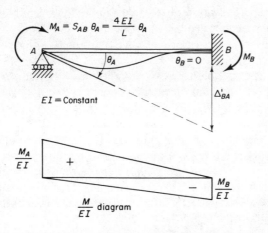

Figure 10.3.1 Stiffness and carry-over factors.

$$\theta_A = \frac{\Delta'_{BA}}{L} = \frac{\text{moment of } M/EI \text{ area between } A \text{ and } B \text{ about } B}{L}$$

$$= \frac{1}{L}\left[\frac{1}{2}\frac{M_A}{EI}(L)\left(\frac{2L}{3}\right) - \frac{1}{2}\frac{M_B}{EI}(L)\left(\frac{L}{3}\right)\right] \quad \text{(clockwise)}$$

$$\theta_B = \frac{\Delta'_{AB}}{L} = \frac{\text{moment of } M/EI \text{ area between } A \text{ and } B \text{ about } A}{L}$$

$$= \frac{1}{L}\left[\frac{1}{2}\frac{M_A}{EI}(L)\left(\frac{L}{3}\right) - \frac{1}{2}\frac{M_B}{EI}(L)\left(\frac{2L}{3}\right)\right] \quad \text{(counterclockwise)}$$

The displacement requirements to accord with the stiffness definition are $\theta_A = 1$ and

$\theta_B = 0$. Thus,

$$\theta_A \text{ (clockwise)} = +\frac{M_A L}{3EI} - \frac{M_B L}{6EI} = 1 \tag{10.3.1}$$

$$\theta_B \text{ (clockwise)} = -\frac{M_A L}{6EI} + \frac{M_B L}{3EI} = 0 \tag{10.3.2}$$

From Eq. (10.3.2),

$$M_B = \frac{1}{2} M_A \tag{10.3.3}$$

and substitution of Eq. (10.3.3) into Eq. (10.3.1) gives

$$M_A = \frac{4EI}{L} \theta_A = \frac{4EI}{L}(1) = S_{AB} \tag{10.3.4}$$

Thus, *stiffness* S_{AB} of a prismatic member *AB fixed* at end *B* is

$$S_{AB} = \frac{4EI}{L} \tag{10.3.5}$$

Frequently, the *modified stiffness* S_{AB} of a member *AB hinged* at end *B*, as shown by Fig. 10.3.2, is desired for use in the method. In that case, Eq. (10.3.1) may be simply used by setting M_B equal to zero. Thus, for the prismatic member *AB*,

$$(S_{AB})_{\text{modified}} = \frac{3EI}{L} \tag{10.3.6}$$

In other words, the basic stiffness for a member fixed at the far end may be modified by using three-fourths of it for a member hinged at the far end.

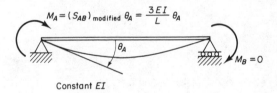

$$M_A = (S_{AB})_{\text{modified}} \,\theta_A = \frac{3EI}{L}\,\theta_A$$

$M_B = 0$

Constant *EI*

Figure 10.3.2 Modified stiffness at end *A*.

Carry-Over Factor

In the sequence of locking and releasing joints in the moment distribution method, the release of a joint allows rotation at that joint until equilibrium is achieved, an action identical to applying a moment opposite to that holding the joint in the locked position. The application of a moment M_A to end *A* of a member *AB* induces a moment M_B at end *B* when end *B* is fixed. The induced moment M_B may be thought of as being carried over from end *A*, and the ratio of M_B to M_A is then referred to as the *carry-over factor* C_{AB}. For the prismatic beam *AB*, Eq. (10.3.3) has already shown that the *carry-over factor* C_{AB} is

$$C_{AB} = \frac{M_B}{M_A} = +\frac{1}{2} \tag{10.3.7}$$

Note that the carry-over factor is positive, indicating that the induced moment M_B at the far end has the *same rotational direction* (clockwise positive) as the applied moment M_A. The clockwise (positive) moment M_A on the left end causes positive (tension at the bottom fiber) bending moment, whereas the clockwise (positive) moment M_B on the right end causes negative (tension at the top fiber) bending moment. In the moment distribution process, the rotational direction sign convention is used, after which the conversion to bending moment sign convention is made for use in design.

Distribution Factor

In the release of a locked joint during the moment distribution process, a joint rotates while the far ends of the members meeting at that joint are fixed. This rotation to reach equilibrium is equivalent to the application of a balancing moment at the joint. The applied balancing moment is *distributed* to each of the members meeting at the joint in proportion to the stiffness of each member. The proportion of the balancing moment taken by each member is called its *distribution factor*.

Referring to the rigid joint i of Fig. 10.3.3a, for each of the members the moment at end i with the far ends fixed, Eq. (10.3.4) gives

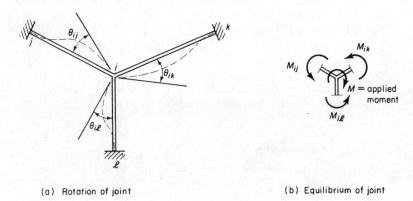

(a) Rotation of joint   (b) Equilibrium of joint

Figure 10.3.3   Distribution factors.

$$\left.\begin{aligned} M_{ij} &= S_{ij}\theta_{ij} \\ M_{ik} &= S_{ik}\theta_{ik} \\ M_{il} &= S_{il}\theta_{il} \end{aligned}\right\} \qquad (10.3.8)$$

Since joint i is "rigid," all members at the joint rotate through the same angle θ; thus,

$$\theta_{ij} = \theta_{ik} = \theta_{il} = \theta_i \qquad (10.3.9)$$

Then from Eqs. (10.3.8),

$$\theta_i = \frac{M_{ij}}{S_{ij}} = \frac{M_{ik}}{S_{ik}} = \frac{M_{il}}{S_{il}} \qquad (10.3.10)$$

Next, equilibrium of joint i as shown in Fig. 10.3.3b requires

$$M_{ij} + M_{ik} + M_{il} = M \qquad (10.3.11)$$

Substitution of Eq. (10.3.10) into Eq. (10.3.11) gives

$$\theta_i(S_{ij} + S_{ik} + S_{il}) = M \qquad (10.3.12)$$

or in general for any number N of members framing together at a joint,

$$\theta_i \sum_{n=1}^{N} S_{in} = M \tag{10.3.13}$$

The moment on an individual member may be obtained from Eqs. (10.3.8) after substituting θ_i from Eq. (10.3.13). Then

$$M_{ij} = \frac{S_{ij}}{\sum\limits_{n=1}^{N} S_{in}} M \tag{10.3.14}$$

$$= (DF)_{ij} M \tag{10.3.15}$$

Thus, $(DF)_{ij}$ is the *distribution factor.*

10.4 APPLICATION OF THE BASIC MOMENT DISTRIBUTION METHOD

The following example is presented to explain the basic method in detail.

Example 10.4.1

Analyze the two-span continuous beam shown in Fig. 10.4.1 using the moment distribution method.

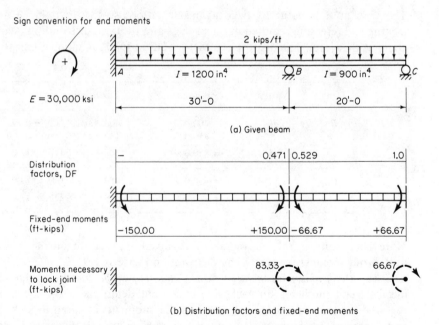

Figure 10.4.1 Beam of Example 10.4.1, showing moments to lock joints prior to applying moment distribution.

Solution. This example will be explained step by step to assist the reader in understanding the concepts.

(a) *Determine fixed-end moments and distribution factors.* At the start, all joints are considered locked such that zero slope occurs at every joint. This means that *fixed-end moments* (see Section 9.2) act on the ends of each member. For this example, referring to Fig. 9.2.4, the fixed-end moments are

$$M_{FAB} = -\frac{wL_1^2}{12} = -\frac{2(30)^2}{12} = -150.00 \text{ ft-kips}$$

$$M_{FBA} = +\frac{wL_1^2}{12} = +150.00 \text{ ft-kips}$$

$$M_{FBC} = -\frac{wL_2^2}{12} = -\frac{2(20)^2}{12} = -66.67 \text{ ft-kips}$$

$$M_{FCB} = +\frac{wL_2^2}{12} = +66.67 \text{ ft-kips}$$

Each of these is shown in Fig. 10.4.1b in its actual direction and also with a sign consistent with *clockwise direction as positive.*

Next, the stiffnesses are computed using Eq. (10.3.5),

$$S_{AB} = S_{BA} = \frac{4EI_1}{L_1} = \frac{4(30,000)(1200)}{30(12)} = 400,000 \text{ in.-kips}$$

$$S_{BC} = S_{CB} = \frac{4EI_2}{L_2} = \frac{4(30,000)(900)}{20(12)} = 450,000 \text{ in.-kips}$$

The joint at A is naturally fixed against rotation; so there would be no alternate locking and releasing required in the process and the term distribution factor does not apply at that joint. The distribution factors DF at joints B and C are next computed,

Joint	Member	Stiffness	Distribution factor DF
B	BA	400,000	400,000/850,000 = 0.471
	BC	450,000	450,000/850,000 = 0.529
		$\Sigma S = 850,000$	$\Sigma \text{DF} = 1.00$
C	CB	450,000	450,000/450,000 = 1.00
		$\Sigma S = 450,000$	$\Sigma \text{DF} = 1.00$

Note that at joint C, only one member is involved so the entire unbalance is distributed to that one member for which the distribution factor is 1.0.

(b) *Determine joint moments to maintain locked position.* The algebraic sum of the fixed-end moments on each side of a joint determines how much moment is required to maintain the locked position. The moments required to lock the joints B and C are 83.33 ft-kips clockwise at B and 66.67 ft-kips clockwise at C, as shown in Fig. 10.4.1b. Note that there is no locking moment required at A; so it is not defined at all and a dash is placed for DF at end A of member AB in Fig. 10.4.1b.

(c) *Release joint B keeping other joints locked.* The release of the locking moment at B is equivalent to applying 83.33 ft-kips (see Figs. 10.4.1b and 10.4.2) counterclockwise on joint B. This amount is distributed to BA (83.33 times 0.471 equals 39.25 ft-kips) and to BC (83.33 times 0.529 equals 44.08 ft-kips) in proportion to the distribution factors. The signs of these balancing moments are opposite to the sign of the original unbalance $(+150.00 - 66.67 = +83.33)$. In other words, after application of -83.33 ft-kips at B the joint is in equilibrium. The application of -39.25 ft-kips at end B induces -19.62 ft-kips at fixed end A; that is, the carry-over moment to A of -19.62 is obtained from -39.25 times the carry-over factor 0.5. Similarly, one-half of the -44.08 is carried over to end C as -22.04.

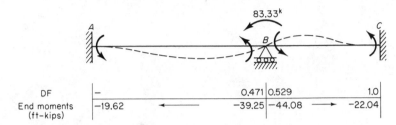

DF	–			0.471	0.529		1.0
End moments (ft–kips)	−19.62	←		−39.25	−44.08	→	−22.04

Figure 10.4.2 Effect of unlocking joint B in moment distribution sequence for Example 10.4.1.

(d) *Release joint C keeping other joints locked.* Examine Fig. 10.4.3 to see the effects of unlocking joint C. Since joint C could have been unlocked prior to joint B and since the sequence of steps should not affect the final answer, for the beginner at least it may be best to study separately the effects of releasing each joint. Release of joint C with joint B fixed means a moment of 66.67 ft-kips is applied counterclockwise to joint C to bring that joint into equilibrium. Since only member CB intersects joint C, the entire balancing moment $(-66.67$ ft-kips) is distributed (using distribution factor 1.0) to end C of member CB. This induces a carry-over moment at B of -33.33 ft-kips, equal to the applied 66.67 times the 0.5 carry-over factor. This concludes the first cycle of balancing (or in the more complete name, moment distribution) with each joint having been released toward its actual condition *once*, while simultaneously keeping all other joints locked. The summary of the situation is given in Fig. 10.4.4.

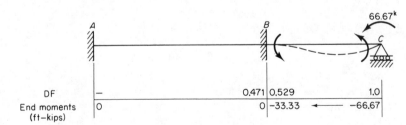

DF	–			0.471	0.529		1.0
End moments (ft–kips)	0			0	−33.33	←	−66.67

Figure 10.4.3 Effects of unlocking joint C in moment distribution sequence for Example 10.4.1.

(e) *Second cycle of moment distribution.* The situation of locked joints in Fig. 10.4.4 is similar to the locked joint situation of Fig. 10.4.1 that existed at the beginning of the process. Due to the fact that the carry-over factor is less than unity (one-half

for prismatic members), after each cycle of balancing and carrying over, the moments
necessary to lock the joints are much smaller. The second cycle of moment distribution is given in Table 10.4.1, as well as successive cycles until the amount carried over
is considered negligible.

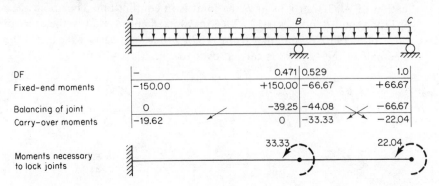

DF	–	0.471 0.529	1.0
Fixed-end moments	–150.00	+150.00 –66.67	+66.67
Balancing of joint	0	–39.25 –44.08	–66.67
Carry-over moments	–19.62	0 –33.33	–22.04

Figure 10.4.4 Summary of situation after one cycle of moment distribution in
Example 10.4.1.

(f) *Final results.* Once a sufficient number of cycles of moment distribution have
been performed, all the fixed-end moments, balancing moments, and carry-over
moments are added together at each end of every member. The results appear in the
row marked "total" in Table 10.4.1. Note that the total end moment M_{BA} (+122.86)
is equal numerically to end moment M_{BC} (−122.86), but the two moments have
opposite rotational direction; that is, for equilibrium joint B has equal and opposite
moments of 122.86 ft-kips acting on it. The end moment at C on CB is zero, as it must
be for a simple support. For design, the remainder of the structural analysis would
be to draw the shear and bending moment diagrams. To do that, the rotational sign
convention for end moments must be converted into the bending moment sign convention; that is, tension at the bottom fiber is caused by positive bending moment, and
tension at the top fiber is caused by negative bending moment. For this example, the
bending moment M_A is −163.55 ft-kips, M_B is −122.86 ft-kips, and M_C is zero.

10.5 ACCURACY OF THE METHOD

The method of moment distribution may be termed a method of successive approximations but it is not an approximate method. One may stop the process after a few
cycles and obtain an approximate answer, or a large number of cycles may be used to
compute final moments to any degree of exactness desired. When comparing various
structural analysis methods to ascertain the correctness of an answer, more exactness
may be desirable than is necessary for use in design.

For the two-span continuous beam of Example 10.4.1, the degree of exactness
is given in Table 10.5.1 for the answer after each cycle of moment distribution. Note
that after two cycles the answer is already within 3 to 4% of the exact answer, and
after 3 cycles it is within 1%. Cycles 5 through 9 really gave no practical improvement.
For design purposes, rarely should more than three cycles be used.

TABLE 10.4.1 MOMENT DISTRIBUTION FOR EXAMPLE 10.4.1

Joint	A	B		C
End moment	AB	BA	BC	CB
DF	—	0.471	0.529	1.0
FEM	−150.00	+150.00	−66.67	+66.67
BAL	0	−39.25	−44.08	−66.67
CO	−19.62	0	−33.33	−22.04
BAL	0	+15.70	+17.63	+22.04
CO	+7.85	0	+11.02	+8.82
BAL	0	−5.19	−5.83	−8.82
CO	−2.58	0	−4.41	−2.91
BAL	0	+2.08	+2.33	+2.91
CO	+1.04	0	+1.45	+1.16
BAL	0	−0.68	−0.77	−1.16
CO	−0.34	0	−0.58	−0.38
BAL	0	+0.27	+0.31	+0.38
CO	+0.13	0	+0.19	+0.15
BAL	0	−0.09	−0.10	−0.15
CO	−0.04	0	−0.07	−0.05
BAL	0	+0.03	+0.04	+0.05
CO	+0.01	0	+0.02	+0.02
BAL	0	−0.01	−0.01	−0.02
Total M	−163.55	+122.86	−122.86	0

cycle 1 (FEM, BAL)
cycle 2 (CO, BAL)
cycle 3 (CO, BAL)

TABLE 10.5.1 ACCURACY ACHIEVED AFTER VARIOUS CYCLES OF MOMENT DISTRIBUTION FOR BEAM OF EXAMPLE 10.4.1

Bending moment at:	M_A	Approx. Exact	M_B	Approx. Exact
Cycles				
1	−150.00	0.92	−110.75	0.90
2	−169.62	1.04	−126.45	1.03
3	−161.77	0.99	−121.26	0.99
4	−164.35	1.004	−123.34	1.004
5	−163.31	0.998	−122.66	0.998
6	−163.65	1.001	−122.93	1.001
7	−163.52	1.000	−122.84	1.000
8	−163.56	1.000	−122.87	1.000
9	−163.55	1.000	−122.86	1.000

10.6 USE OF MODIFIED STIFFNESS TO ACCOUNT FOR A HINGED SUPPORT

When a support is actually a simple support instead of a restrained support, the moment distribution process is greatly speeded up if the modified stiffness as expressed by Eq. (10.3.6) is used in computing the distribution factors. For instance, in Example 10.4.1 (see Fig. 10.4.2) when joint B is unlocked and the counterclockwise moment of 83.33 ft-kips is applied to balance the joint, there is no reason why joint C should have to be locked. If joint C were permitted to rotate when the releasing moment is applied at joint B, there would be no moment induced at C. The stiffness S_{BC} could be adjusted to treat C as a hinge when computing the distribution factors at B. This case has been treated in Section 10.3, resulting in Eq. (10.3.6) for the modified stiffness (which is three-fourths of the regular stiffness for a prismatic member) when the far end is hinged.

Example 10.6.1

Re-analyze the two-span continuous beam of Example 10.4.1 (shown again in Fig. 10.6.1) using the moment distribution method with stiffness modified to account for the hinge at C.

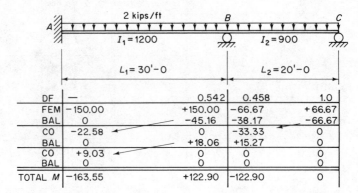

Figure 10.6.1 Moment distribution for Example 10.6.1, using modified stiffness $S_{BC} = 3EI_2/L_2$.

Solution

(a) *Fixed-end moments.* Even though the stiffness S_{BC} will be computed assuming the hinge at C it is still expeditious to begin the process exactly as in Example 10.4.1, that is, with all joints locked into the fixed position of having zero rotation. Thus, the fixed-end moments are as in Example 10.4.1, as follows:

$$M_{FAB} = -150.00 \text{ ft-kips}; \qquad M_{FBA} = +150.00 \text{ ft-kips}$$

$$M_{FBC} = -66.67 \text{ ft-kips}; \qquad M_{FCB} = +66.67 \text{ ft-kips}$$

The alternative would be to treat the span BC as a beam fixed at B but simply supported at C at the outset. Then the fixed-end moment at B would be numerically larger than if end C is also fixed. This modified fixed-end moment at end B can itself be obtained by applying the moment distribution method to the separate beam BC as shown in Fig. 10.6.2.

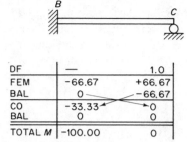

DF	—	1.0
FEM	−66.67	+66.67
BAL	0	−66.67
CO	−33.33	0
BAL	0	0
TOTAL M	−100.00	0

Figure 10.6.2 Modified fixed-end moment at B of Example 10.6.1.

(b) *Member stiffnesses.* Using Eq. (10.3.5) for S_{BA} and S_{CB}, where application of any balancing moments will be on member ends when the far end will be considered as fixed, and recognizing that only the relative stiffnesses rather than absolute stiffnesses are needed for the computation,

$$S_{BA} = \frac{4EI_1}{L_1} = \frac{4E(1200)}{30} = 160E$$

$$S_{CB} = \frac{4EI_2}{L_2} = \frac{4E(900)}{20} = 180E$$

For the modified stiffness S_{BC}, Eq. (10.3.6) applies,

$$(S_{BC})_{\text{modified}} = \frac{3EI_2}{L_2} = \frac{3E(900)}{20} = 135E$$

(c) *Distribution factors.* For joint A, the term distribution factor does not apply and for joint C the distribution factor DF is 1.0, exactly as for Example 10.4.1. For joint B,

Member	Stiffness	DF
BA	$160E$	$160/295 = 0.542$
BC (modified)	$135E$	$135/295 = 0.458$
	$\sum S = \overline{295E}$	$\sum (DF) = \overline{1.00}$

(d) *Moment distribution.* Since joint C begins as a fixed support but will be considered hinged when joint B is unlocked, joint C must be unlocked and balanced first. Thereafter, joint C remains unlocked throughout the entire process of moment distribution and it merely rotates as moments are applied at end B. Referring to Fig. 10.6.1, when joint C is balanced by application of −66.67 ft-kips, one-half of that amount is carried over to joint B. In the illustration each joint was balanced, after which the carry-over moments were considered as new unbalanced moments.

Alternatively, as has been shown in Fig. 10.6.2, after balancing joint C as the first operation, the carry-over moment of -33.33 ft-kips could be thought of as induced at BC *prior* to the first release of the locking moment at B. Referring to Fig. 10.6.3, with the modified fixed-end moment at end B of span BC of -100 ft-kips, the net moment required to lock the joint becomes equal to $+150.00 - 100.00 = +50.00$ ft-kips. Now the first release of joint B would require the balance moment of -50.00, giving -27.10 and -22.90 as the distributed moments. There will be no carry-over moment to the hinged support at C and one-half of the -27.10 will be carried over to the fixed support at A, thus completing the moment distribution using only one distribution at B, once the modified fixed-end moment has been used at end B of span BC at the outset.

	AB		BA	BC		CB
FEM	−150.00		+150.00	−66.67		+66.67
				−33.33		−66.67
MOD FEM	−150.00		+150.00	−100.00		0
BAL	0		−27.10	−22.90		
CO	−13.55		0	0		
BAL	0		0	0		
TOTAL M	−163.55		+122.90	−122.90		

Figure 10.6.3 Moment distribution for Example 10.6.1 using modified stiffness $S_{BC} = 3EI_2/L_2$ and combining carry-over moment with unbalance before unlocking joint.

The beginner will be wise to systematically balance each joint, then make all carry-overs, then rebalance each joint, make carry-overs, and so on, in the manner of Table 10.4.1 and Fig. 10.6.1 or 10.6.3. However, the unlocking and carrying over can actually be done in any sequence, with the carry-over moment combined with the already existing restraining moment before unlocking and balancing the joint. When such an irregular procedure is used, the analyst should always place a horizontal line immediately below the distributed moments as soon as a joint is balanced. There may be different numbers of distributions and balances at each joint by the time the process is completed.

10.7 CHECK ON MOMENT DISTRIBUTION

A check on moment distribution may be made by evaluating the slopes at each side of a rotated joint to show that they are the same and, if there is a fixed support, the slope there when evaluated should equal zero. The end slopes θ_A and θ_B expressed in terms of the end moments M_A and M_B (when there is no load on the member) are expressed by Eqs. (10.3.1) and (10.3.2) before they are equated to 1 and 0; or

$$\theta_A \text{ (clockwise)} = +\frac{M_A L}{3EI} - \frac{M_B L}{6EI} \tag{10.7.1}$$

$$\theta_B \text{ (clockwise)} = -\frac{M_A L}{6EI} + \frac{M_B L}{3EI} \tag{10.7.2}$$

However, if there is load on the span causing fixed-end moments M_{FAB} and M_{FBA}, then only the additional end moments, $(M_A - M_{FAB})$ and $(M_B - M_{FBA})$, should be the correct moments to maintain the slopes θ_A and θ_B. Substituting $(M_A - M_{FAB})$ and $(M_B - M_{FBA})$ for M_A and M_B in Eqs. (10.7.1) and (10.7.2), respectively,

$$\left. \begin{aligned} \theta_A &= \frac{(M_A - M_{FAB}) - 0.5(M_B - M_{FBA})}{3EI/L} \\ \theta_B &= \frac{(M_B - M_{FBA}) - 0.5(M_A - M_{FAB})}{3EI/L} \end{aligned} \right\} \quad (10.7.3)$$

Note that $(M_A - M_{FAB})$, typically, is actually the change in moment from the fixed-end moment to the final value after the moment distribution process. Thus, from the above equations the slope θ at one end equals the change in moment at that end minus one-half the change in moment at the far end, divided by $3EI/L$. The check may be made on the results of Example 10.6.1, as follows:

	AB	BA	BC	CB
FEM	−150.00	+150.00	−66.67	+66.67
Final M	−163.55	+122.90	−122.90	0
Change	−13.55	−27.10	−56.23	−66.67

$$\theta_{AB} = \frac{(-13.55) - 0.5(-27.10)}{3E(1200)/30} = 0 \quad \text{(check)}$$

$$\theta_{BA} = \frac{(-27.10) - 0.5(-13.55)}{3E(1200)/30} = \frac{-0.1694}{E}$$

$$\theta_{BC} = \frac{(-56.23) - 0.5(-66.67)}{3E(900)/20} = \frac{-0.1696}{E}$$

$$\theta_{BA} = \theta_{BC} \quad \text{(check)}$$

The number of checks obtainable on any moment distribution of continuous beams is equal to the degree of statical indeterminacy. Often these checks are called *continuity checks*. In the present problem the beam is statically indeterminate to the second degree, and the two continuity checks are: (1) the slope $\theta_A = 0$ shows that the beam is *continuous* with the fixed support at A, and (2) the fact $\theta_{BA} = \theta_{BC}$ shows that the beam is *continuous* (i.e., having a common tangent) over the support at B. As a by-product, the slope θ_{CB} may be computed:

$$\theta_{BC} = \frac{(-66.67) - 0.5(-56.23)}{3E(900)/20} = \frac{-0.2856}{E}$$

The process of making the check should be included in a tabular form at the end of the moment distribution process (as illustrated in Example 10.8.1).

10.8 EXAMPLES

The following two examples further illustrate the moment distribution procedure, as well as the check.

Example 10.8.1

Analyze the continuous beam of Fig. 10.8.1 using the moment distribution method. This is the same beam analyzed by the slope deflection method in Example 9.3.3.

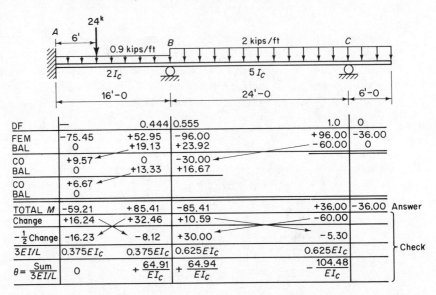

Figure 10.8.1 Moment distribution for Example 10.8.1, using $3EI/L$ as the stiffness S_{BC}.

Solution

(a) *Fixed-end moments.* The fixed-end moments are computed using Fig. 9.2.4.

$$M_{FAB} = -\frac{0.9(16)^2}{12} - \frac{24(6)(10)^2}{(16)^2} = -75.45 \text{ ft-kips}$$

$$M_{FBA} = +\frac{0.9(16)^2}{12} + \frac{24(10)(6)^2}{(16)^2} = +52.95 \text{ ft-kips}$$

$$M_{FBC} = -\frac{2.0(24)^2}{12} = -96.0 \text{ ft-kips}$$

$$M_{FCB} = +\frac{2.0(24)^2}{12} = +96.0 \text{ ft-kips}$$

$$M_{FCD} = -\frac{2.0(6)^2}{2} = -36.0 \text{ ft-kips}$$

Note that the last value is the moment M_{CD} obtainable using statics and is the same whether or not joint C is fixed. In this case M_{FCD} is negative because it acts *counterclockwise* on end C of the cantilever CD.

(b) *Distribution factors.* Since the moment at C is determinable from statics, joint C may be treated as hinged for the stiffness S_{BC}. In other words, joint C will be unlocked only once; after that it is permitted to rotate every time when joint B is released. Thus,

Joint A The term DF does not apply.
Joint B
$\quad$ BA $4E(2I_c)/(16) = 0.500EI_c$ $(DF)_{BA} = 0.444$
$\quad$ BC $3E(5I_c)/(24) = 0.625EI_c$ $(DF)_{BC} = \underline{0.555}$
$\quad\quad\quad\quad\quad \Sigma S = \overline{1.125EI_c}$ $\Sigma(DF) = 0.999$
Joint C
$\quad$ CB $\quad\quad\quad\quad\quad\quad\quad\quad\quad (DF)_{CB} = 1.0$

The summation of the distribution factors at a joint must total 1.0; however, the sum 0.999 will provide as accurate results as necessary. It is not necessary to compute distribution factors to more than three significant figures.

At joint C the distribution factor is 1.0 to the member BC since it is the only member framing into joint C. The cantilever CD does not affect the stiffness or distribution of moment at C; the load on the cantilever may be thought of as a constant moment applied at joint C, just as the fixed-end moment at C on span CB in the fixed condition.

(c) *Moment distribution.* The moment distribution is given in Fig. 10.8.1, where the values are entered into the table directly below the correct location on the beam. For continuous beams it is convenient to do it directly oriented with the structure. Note that once joint C is balanced there is a carry-over to joint B; however, because the modified stiffness $S_{BC} = 3EI/L$ was used to compute the distribution factors at joint B, there is no carry-over to joint C.

(d) *Check on moment distribution.* The check on the correctness of the distribution process, as described in Section 10.7, is made in a tabular form immediately at the conclusion of the distribution. Note that the term $(-1/2$ change) is used, because it is easier for the analyst to add the values on the two lines marked as (change) and $(-1/2$ change). This check verifies the correctness of the distribution process; however, it does not verify the correctness of the fixed-end moments and the distribution factors. If these starting values are wrong, the answer will be wrong.

(e) *Shear and bending moment diagrams.* After the moment distribution is completed, the rotational end moments obtained must be interpreted into the bending moment sign convention. When clockwise end moments (as they act on the member) are taken as positive, the end moment on the left end of a span will have the sign agreeing with the bending moment sign convention; or the end moment on the right end of a span will have the sign opposite to the bending moment sign convention. Thus, for this example the bending moments are $M_A = -59.21$, $M_B = -85.41$, and $M_C = -36$ ft-kips. The bending moment diagram was previously drawn as a part of Example 9.3.3 (Fig. 9.3.3).

Example 10.8.2

Analyze the continuous beam of Fig. 10.8.2 using the moment distribution method.

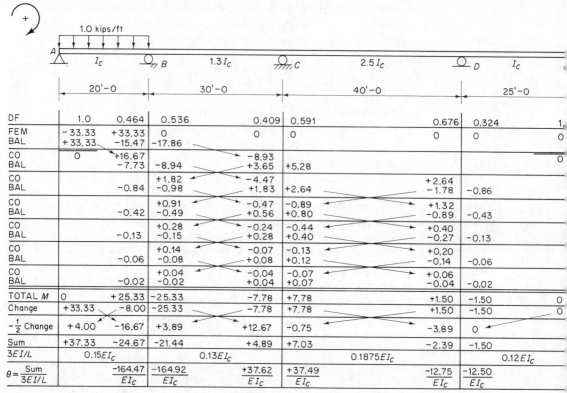

Figure 10.8.2 Moment distribution for Example 10.8.2.

Solution

(a) *Fixed-end moments.* The fixed-end moments are computed using Fig. 9.2.4,

$$M_{FAB} = -\frac{wL^2}{12} = -\frac{1.0(20)^2}{12} = -33.33 \text{ ft-kips}$$

$$M_{FBA} = +33.33 \text{ ft-kips}$$

All other fixed-end moments are zero since span *AB* is the only one loaded.

(b) *Distribution factors.*

Joint *A* $(DF)_{AB} = 1.0$

Joint *B*

BA	$3E(I_c)/20 = 0.150EI_c$	$(DF)_{BA} = 0.464$	
BC	$4E(1.3I_c)/30 = 0.173EI_c$	$(DF)_{BC} = 0.536$	
	$\Sigma S = 0.323EI_c$	$\Sigma (DF) = 1.000$	

Joint *C*

CB	$4E(1.3I_c)/30 = 0.173EI_c$	$(DF)_{CB} = 0.409$	
CD	$4E(2.5I_c)/40 = 0.250EI_c$	$(DF)_{CD} = 0.591$	
	$\Sigma S = 0.423EI_c$	$\Sigma (DF) = 1.000$	

            Moment Distribution Method—Application to Beams   Chap. 10

Joint D

DC	$4E(2.5I_c)/40 = 0.250EI_c$	$(DF)_{DC} = 0.676$
DE	$3E(I_c)/25 \quad = 0.120EI_c$	$(DF)_{DE} = 0.324$
	$\sum S = \overline{0.370EI_c}$	$\sum(DF) = \overline{1.000}$

Joint E

$$(DF)_{ED} = 1.0$$

(c) *Moment distribution.* The moment distribution is carried out in Fig. 10.8.2, where the check is also shown. Since the stiffnesses S_{BA} and S_{DE} were taken as the value $3EI/L$ as modified to account for the hinged conditions at A and E, no carry-over is made from B to A or from D to E as the distributions are performed. The reader may note that the θ values computed in the check are as close as necessary to show no error in the moment distribution process. To obtain much closer deformation checks, the distribution factors as well as the balances and carry-overs would have to have been computed to more significant figures.

(d) *Shears and bending moments.* The shears and bending moment diagrams are given in Fig. 10.8.3.

(e) *Other comments regarding results.* Note that loading on span AB affects primarily span AB and the adjacent span BC. Span CD is relatively little affected and DE even less. This is a typical result. Frequently, a long structure may be "broken apart" and analyzed by taking the span adjacent to the span of interest and considering its far end fixed. In this example if the support at C were treated as fixed, the support moment at B would have been 6% higher than its correct values and the maximum positive moment in span BC would have been 2% lower than its correct value. Frequently, such accuracy would be satisfactory for design.

Another item of interest on Fig. 10.8.3 is that points m and n where zero moment occurs in unloaded spans are "fixed points," as discussed in Section 8.4. No matter

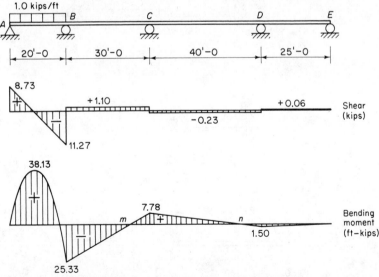

Figure 10.8.3 Shear and bending moment diagrams for four-span continuous beam of Example 10.8.2.

what loads are placed in span AB, points m and n will have zero moment; their location is a property of the structure, typically about the 0.2 to 0.25 of the span from the end (in this case, 0.24 of span BC from C and 0.16 of span CD from D).

SELECTED REFERENCES

1. Hardy Cross, "Continuity as a Factor in Reinforced Concrete Design," *Proceedings*, ACI, *25* (1929), 669–708.
2. Hardy Cross, "Simplified Rigid Frame Design," Report of Committee 301, *Proceedings*, ACI, *26* (December 1929). [*ACI Journal, 1* (1929–30), 170–183.]
3. Hardy Cross, "Analysis of Continuous Frames by Distributing Fixed-End Moments," *Proceedings ASCE, 56* (May 1930), 919–928. See also *Transactions*, ASCE, *96* (1932), 1–10; Discussion, *96* (1932), 11–156.
4. James M. Gere, *Moment Distribution*, D. Van Nostrand Company, Princeton, N.J., 1963, 378 pp.

PROBLEMS

10.1–10.14. Analyze the following continuous beams using the method of moment distribution. Wherever an exterior unrestrained support occurs, use the corresponding modified stiffness in the analysis. After completing the moment distribution, make the continuity checks to ascertain correctness. Finally, draw the shear and bending moment diagrams to scale vertically below a diagram of the given beam.

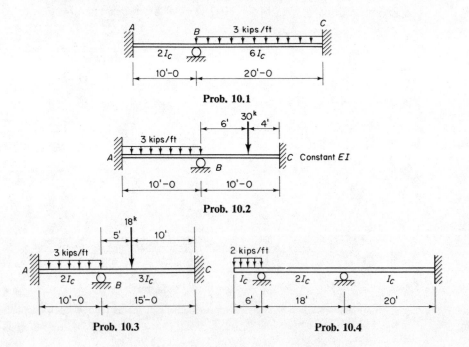

Prob. 10.1

Prob. 10.2

Prob. 10.3   Prob. 10.4

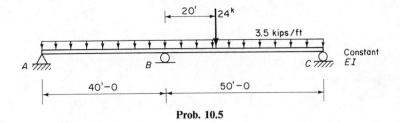

Prob. 10.5

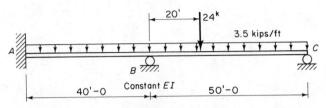

Prob. 10.6

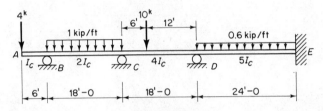

Prob. 10.7

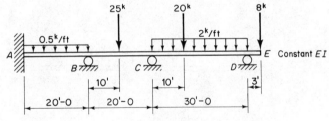

Prob. 10.8

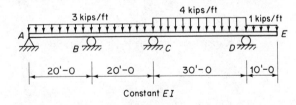

Prob. 10.9

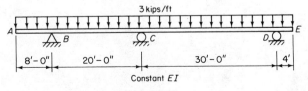

Prob. 10.10

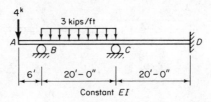

Prob. 10.11

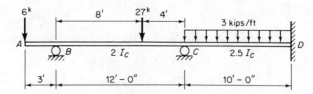

Prob. 10.12

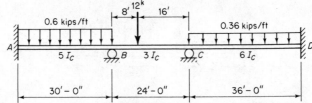

Prob. 10.13

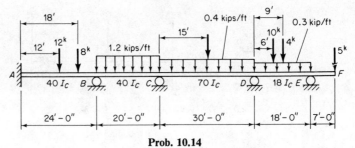

Prob. 10.14

[11]

Rigid Frames

11.1 DEFINITIONS AND ASSUMPTIONS

A rigid frame is a skeleton structure wherein the members meet at rigid joints. At a rigid joint the angle between the elastic curves of any two adjacent members remains constant during loading. For instance, two members meeting at 90° to one another must after loading remain at 90° to one another *at the joint* even though the elastic curves for both members are different. Typical of joints considered as rigid are the intersections of beams and columns in reinforced concrete construction which is cast monolithically. In steel construction, when beams are fully welded over their flanges and web to columns the frame is considered to be rigid. Some typical rigid frame joints are shown in Fig. 11.1.1.

A rigid frame structure may have rigid joints connecting members in three dimensions; or more commonly it may have rigid joints in only one plane with more flexible joints connecting the elements spanning perpendicular to the rigid frame. Only plane frames are treated in this chapter.

Just as for the connections and supports for truss members and beams, the connections in framed structures must be idealized in order to make the structural analysis. Depending on the actual arrangement of pieces and the connecting method (such as monolithic concrete, welding, or bolting) the designer must decide on the degree of rigidity before making the analysis. Few actual joints are 100% rigid, that is, during loading the joint rotates as a whole so that there is zero angle change between the members joined. In an ordinary building frame, this would be to assume that the beam elastic curve is always perpendicular to the column elastic curve at the intersection. However, when the restraining moment at the beam end is at least 90% of the amount

(a) Rigid joints at intersection of columns and beams in reinforced concrete frame.

(b) Haunched rigid joint on structural steel plate girder.

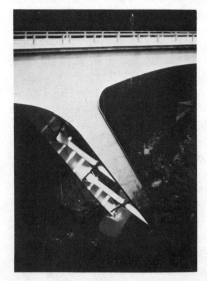

(c) Rigid joints at intersections of members in reinforced concrete frame.

Figure 11.1.1 Rigid joints. (Photos by C. G. Salmon.)

to prevent any relative angle change between the column and the beam, the joint would be idealized as rigid. At the other extreme, when the restraining moment is as low as 20 % of the amount to prevent relative angle change, the joint is usually idealized as hinged. It is possible that the joint may be considered semi-rigid; that is, it can transmit to the member end a moment between zero and that for a fully rigid connection.

The continuous beam, as treated in Chapters 8 to 10, is really a special case of the rigid frame. At an interior support the joint is considered rigid; in the usual case the angle is 180° and is maintained constant even though deflection occurs in each span.

A highway bridge may consist of two or more parallel continuous beams, connected transversely by members using simple connections (assumed hinged). Each of the continuous beams is analyzed as a planar structure. Commercial buildings frequently consist of planar rigid frames one-story high, having flat or gabled roof systems. All connections transverse to the frames are simple connections, assumed hinged to the frames and having negligible effect on the analysis of the plane frames.

Three-dimensional rigid frames, complex rigid frames involving curved or non-prismatic members, multistory rigid frames, and rigid frame assemblies for special structures, such as for aircraft, space vehicles, and ocean vessels, are outside the scope of this text.

Plane rigid frames may be either statically determinate or statically indeterminate. Although statically determinate rigid frames, as shown in Fig. 11.1.2a, b, and c, only rarely, if ever, exist as real structures, an understanding of their behavior is a necessary prerequisite to the force method of analyzing statically indeterminate rigid frames. In fact, the rigid frames in Fig. 11.1.2a, b, and c are the usual choices for the "basic" statically determinate structures (i.e., redundants removed) for those statically indeterminate structures of Fig. 11.1.2d, e, and f.

A straight member in a rigid frame, such as one of those in Fig. 11.1.2, is in general subjected to axial force, shear, and bending moment along its length. Shear deformation has been neglected in the treatment (Chapter 5) of deformation of beams.

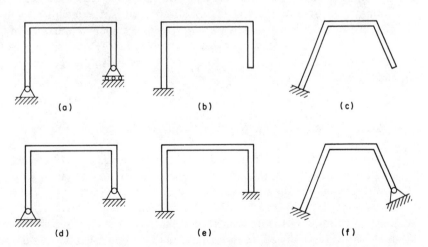

Figure 11.1.2 Statically determinate and statically indeterminate rigid frames.

Such deformation is usually small in flexural members of ordinary proportions, and is also neglected in the ordinary analysis of rigid frames. Elongation or shortening due to axial force is the only deformation that occurs in trusses having pinned joints and has been duly considered in truss analysis (Chapters 3 and 7). For most rigid frames, the effect of axial deformation on the rotations and deflections of the joints is overshadowed by the effect of deformation due to bending moment; in fact, axial deformation usually affects the magnitudes of the redundant forces in a statically indeterminate rigid frame only on the order of several percent. In this chapter it is assumed that axial deformation due to axial forces does not exist, or it is zero in all members even though there are axial forces in them. This is the practical assumption used for most rigid frames. For high-rise buildings such an assumption may not be valid.

In first-order analysis (as previously discussed in Sections 1.9 and 3.1), the transverse displacement of either end of a member in a structure is assumed not to change its length. When the member length does not change, the longitudinal components of the end displacements, such as AA_1 and BB_1 in Fig. 11.1.3, must be equal. This means that the displaced member has a length $A'B'$ which is considered equal to its longitudinal projection A_1B_1 or the original length AB.

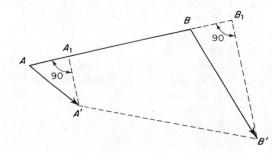

Figure 11.1.3 Plane motion of a member AB in a rigid frame to a new position $A'B'$.

While studying the analysis of rigid frames, the reader should at all times keep in mind the two assumptions just discussed; that is, the member lengths are neither changed by axial forces nor by transverse displacements. Awareness of these assumptions allows for reevaluation of them when making the analysis of high-rise buildings or other structures where axial forces become relatively large compared to bending moments.

11.2 DEGREE OF STATICAL INDETERMINACY

A loaded structure can be defined either by the magnitudes of the external reactions and the internal forces (axial force, shear, and bending moment) along the length of its members, or by the rotations, horizontal displacements, and vertical displacements of its joints. The former situation is called the *equilibrium state*; and the latter, the *compatible state*. In the force method of analysis, the equilibrium state is determined first before the compatible state; and in the displacement method, the compatible state is determined first before the equilibrium state.

The degree of statical indeterminacy of a rigid frame (or of any structure) can be ascertained either by the concept of the force method or by that of the displacement method. When the force method is to be used, the degree of statical indeterminacy is necessary to establish the number of redundant reactions or internal forces. When the displacement method is to be used, the degree of statical indeterminacy is not actually needed in the analysis procedure.

Force Method Concept

The degree of statical indeterminacy of a rigid frame can be ascertained by deriving a basic statically determinate rigid frame and counting the number of redundant reactions (including reacting moments) acting in the directions of the removed restraints and the number of internal forces (axial force, shear, and bending moment) acting at both sides of a cut within the length of a member. For example, the single-story, single-span, rigid frame of Fig. 11.2.1a can be made statically determinate by changing the hinged support at D into a roller support and replacing the horizontal restraint by the redundant reaction H_D. This "basic" structure is statically determinate because it has only three unknown external reactions, which may be determined from the three equations of statics. The rigid frame of Fig. 11.2.1b is statically indeterminate to the third degree because three redundant reactions M_D, H_D, and V_D replace the restraints originally offered by the fixed support at D. The "basic" statically determinate structure is a cantilever rigid frame with only one fixed support. The rigid frame of Fig. 11.2.1c can be made statically determinate by removing the fixed support at F together with a complete cut somewhere in the loop $ABDC$ in order that the bending moment everywhere can be obtained by statics. The six redundants in this case are the reactions M_F, H_F, and V_F and the internal forces N_1, V_1, and M_1 at the cut.

Displacement Method Concept

One must note at the outset that the degree of statical indeterminacy of a structure is an inherent property of the structure and as such it is not subject to the viewpoint of the analyst, that is, whether axial deformation is to be considered or not. In the displacement method of analysis, the rigid frame is considered to be divided into a number of members connected at the rigid joints.

When axial deformation is considered, the total number of unknown joint displacements (or the degree of freedom, or the available number of equations of equilibrium) is equal to

$$NP = 3(NJ) - 3(NFS) - 2(NHS) - (NRS) \qquad (11.2.1)$$

in which NJ is the number of joints including supports, NFS is the number of fixed supports, NHS is the number of hinged supports, and NRS is the number of roller supports. The reason for Eq. (11.2.1) is that for a typical joint, there are three unknown joint displacements; one in rotation, one in horizontal displacement, and one in vertical displacement. The number of unknown internal forces is three times the number of members, because the statics of each member is completely defined by the axial force in it and the two end moments. The external reactions are no longer independent unknowns, since they can be computed from the member-end forces of the

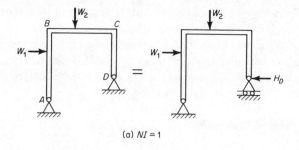

(a) $NI = 1$

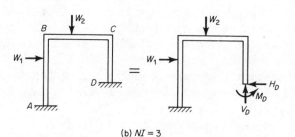

(b) $NI = 3$

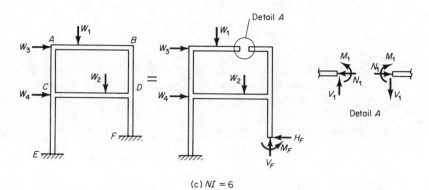

(c) $NI = 6$

Figure 11.2.1 Degrees of statical indeterminacy from the force method concept.

members entering into a support. Thus, the number of unknowns to completely define the equilibrium state is

$$NF = 3(NM) \tag{11.2.2}$$

in which NF is the total number of unknown internal forces and NM is the number of members. The degree of statical indeterminacy NI is obtained by subtracting Eq. (11.2.1) from Eq. (11.2.2), or

$$NI = 3(NM) - 3(NJ) + 3(NFS) + 2(NHS) + NRS \tag{11.2.3}$$

For the rigid frames in Fig. 11.2.1,

$$NI(\text{Fig. 11.2.1a}) = 3(3) - 3(4) + 3(0) + 2(2) + (0) = 1$$
$$NI(\text{Fig. 11.2.1b}) = 3(3) - 3(4) + 3(2) + 2(0) + (0) = 3$$
$$NI(\text{Fig. 11.2.1c}) = 3(6) - 3(6) + 3(2) + 2(0) + (0) = 6$$

When the member lengths are assumed not to change by the presence of axial forces, the degree of freedom of a rigid frame is reduced from that of Eq. (11.2.1). In such a case, it is best to observe the number of unknown joint rotations NPR and the number of joint translations NPS. Based on the two important assumptions that member lengths are neither changed by axial forces nor by transverse displacements, the only unknown joint translation in the rigid frames of Fig. 11.2.1 is the horizontal movement of the horizontal members, commonly called the *sidesway* (thus the symbol NPS in which "S" means sidesway). Because the number of equilibrium equations should correspond to the degree of freedom, it is

$$NP = NPR + NPS \tag{11.2.4}$$

Once the compatible state is defined, the axial forces can no longer be obtained from Hooke's Law because axial deformations are all zero. Thus the axial forces are no longer independent unknowns and the independent unknowns are only the moments at the ends of the members, so

$$NF = 2(NM) \tag{11.2.5}$$

Substracting Eq. (11.2.5) from Eq. (11.2.4),

$$NI = 2(NM) - (NPR + NPS) \tag{11.2.6}$$

For the rigid frames in Fig. 11.2.1,

$$NI(\text{Fig. 11.2.1a}) = 2(3) - 4 - 1 = 1$$

$$NI(\text{Fig. 11.2.1b}) = 2(3) - 2 - 1 = 3$$

$$NI(\text{Fig. 11.2.1c}) = 2(6) - 4 - 2 = 6$$

Use of Eq. (11.2.3) to obtain NI is somewhat "foolproof," although this approach will not be followed through in this elementary text until later in the full use of computer analysis. Use of Eq. (11.2.6) requires a graphical interpretation of the two assumptions involving zero effect of axial deformation and transverse displacement on member length in order to arrive at the degree of freedom in sidesway. Establishing the degree of statical indeterminacy by counting the number of redundants requires the ability of the analyst to derive a "basic" determinate structure acted upon by the known loads as well as the redundants to substitute for the original structure.

11.3 AXIAL FORCE, SHEAR, AND BENDING MOMENT IN STATICALLY DETERMINATE RIGID FRAMES

Structural analysis of a rigid frame involves determination of the shear and bending moment diagrams as well as the axial force in each member. Regarding bending moment, the sign convention for beams of using positive for tension in the bottom of a member and negative for tension in the top is not adequate for the bending moment on vertical members. However, if for the beam one draws the positive moment diagram above the beam axis, this would be drawing the diagram on the compression side of the beam. This convention can then be applied to the vertical members in frames; that is, draw the bending moment on the left or right side of a vertical line representing the member axis depending upon whether the left or right side of the member is in compression. Rather than use plus or minus, state on the diagram that bending moments are drawn on the compression side (or the tension side if the designer chooses and so states).

Example 11.3.1

Analyze the rigid frame of Fig. 11.3.1a for shears, bending moments, and axial forces.

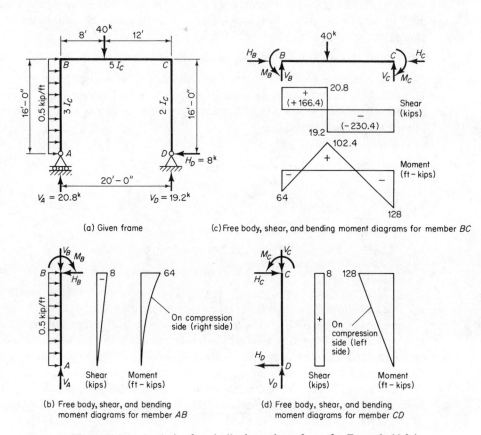

(a) Given frame

(c) Free body, shear, and bending moment diagrams for member BC

(b) Free body, shear, and bending moment diagrams for member AB

(d) Free body, shear, and bending moment diagrams for member CD

Figure 11.3.1 Analysis of statically determinate frame for Example 11.3.1.

Solution

(a) *Establish whether or not the structure is statically determinate.* The degree of freedom is established at six by noting that there are four joint rotations (A, B, C, and D) and two sideways (joint A horizontal and member BC horizontal). There are also six unknown internal forces (end moments), two on each member. Thus,

$$NI = NF - NP$$
$$= 6 - 6 = 0$$

The frame is statically determinate. Further, this frame is indeed statically determinate at a glance by the concept of the force method, that is, only three unknown reactions and no closed loop above the beam.

(b) *Compute reactions.* Taking moments about A in Fig. 11.3.1a,

$$40(8) + 0.5(16)(8) = 20V_D$$
$$V_D = 19.2 \text{ kips}$$

Taking moments about D in Fig. 11.3.1a,

$$40(12) = 0.5(16)(8) + 20V_A$$
$$V_A = 20.8 \text{ kips}$$

Check by $\sum F_y = 0$,

$$V_A + V_D = 40$$
$$19.2 + 20.8 = 40 \qquad \text{OK}$$

Equating to zero the sum of horizontal forces in Fig. 11.3.1a,

$$H_D = 8 \text{ kips}$$

(c) *Draw shear and bending moment diagrams.* From the free body of member AB (Fig. 11.3.1b),

$$H_B = 0.5(16) = 8 \text{ kips}$$
$$M_B = 0.5(16)(8) = 64 \text{ ft-kips}$$
$$V_B = V_A = 20.8 \text{ kips}$$

The bending moment diagram is drawn to the right of its zero axis, indicating compression on the right side of member AB. The sign on the shear diagram follows the convention of looking at the vertical member from the right side toward the left.

From the free body of member BC (Fig. 11.3.1c),

$$V_B = 20.8 \text{ kips}$$
$$M \text{ at 40 kips load} = -M_B + V_B(8)$$
$$= -64 + 20.8(8) = 102.4 \text{ ft-kips}$$
$$M_C = -M_B + V_B(20) - 40(12)$$
$$= -64 + 20.8(20) - 40(12) = -128 \text{ ft-kips}$$

Note that the negative sign before 64 or 128 is not used on the bending moment diagrams of Fig. 11.3.1b and d.

$$V_C = 40 - V_B = 40 - 20.8 = 19.2 \text{ kips}$$
$$H_C = H_B = 8 \text{ kips}$$

The bending moment diagram can also be obtained by use of the area of the shear diagram. Starting with 64 under the zero line, rise by the positive shear area of $+(20.8)(8) = +166.4$ to 102.4 above the zero line at B, and then drop by the negative shear area of $-(19.2)(12) = -230.4$ to 128 under the zero line at C. This procedure is no different from that done before for a span in a continuous beam. The same procedure can be applied to vertical members by looking at them from the right side.

From the free body of member CD (Fig. 11.3.1d),

$$H_D = H_C = 8 \text{ kips} \quad \text{(check)}$$

$$V_D = V_C = 19.2 \text{ kips} \quad \text{(check)}$$

$$M_D = M_C - H_C(16) = 128 - 8(16) = 0 \quad \text{(check)}$$

One may also proceed from A to B, then from D to C, and check the three equations of statics on the free-body of member BC.

11.4 DEFLECTIONS BY THE VIRTUAL WORK (UNIT LOAD) METHOD

The virtual work (unit load) method, discussed in Section 5.3 for determining deflections and slopes for points along beams, is also applicable for determining the displacements on rigid frames. Equation (5.3.6) states that

$$Y_t = \int_0^L \frac{Mm}{EI} \, dx \qquad [5.3.6]$$

For frames, the integration applies to all members.

Example 11.4.1

Determine the horizontal deflection Δ_H of the roller support at point A of the frame of Fig. 11.3.1a, using the virtual work (unit load) method.

Solution

(a) *Establish the virtual loading.* Apply a horizontal unit load at A as in Fig. 11.4.1, where the virtual loading bending moment m diagrams are also shown. The bending moment diagrams for the real loading are in Fig. 11.3.1b, c, and d.

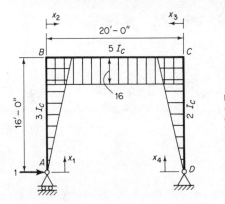

Bending moment diagrams drawn on compression side

Figure 11.4.1 Virtual loading for determination of the horizontal deflection of joint A, Example 11.4.1.

(b) *Apply Eq. (5.3.6) to members AB, BC, and CD.* For member AB, measuring x_1 from A (Fig. 11.4.1), and recognizing that both M and m are of the same sign,

$$M = \frac{1}{2}(0.5)x_1^2 = 0.25x_1^2$$

$$m = x_1$$

$$\Delta_{H1} = \int_0^{16} \frac{(0.25x_1^2)x_1 \, dx_1}{3EI_c} = \frac{0.25}{3EI_c}\left[\frac{x_1^4}{4}\right]_0^{16} = \frac{+1365.33}{EI_c}$$

For member BC, taking the segment from B to the 40-kip load,

$$M = -64 + 20.8x_2$$

$$m = -16$$

$$\Delta_{H2} = \int_0^8 \frac{(-64 + 20.8x_2)(-16) \, dx_2}{5EI_c}$$

$$= \frac{1}{5EI_c}\left[+1024x_2 - 332.8\frac{x_2^2}{2}\right]_0^8 = \frac{-491.52}{EI_c}$$

For member BC, taking the segment from C to the 40-kip load,

$$M = -128 + 19.2x_3$$

$$m = -16$$

$$\Delta_{H3} = \int_0^{12} \frac{(-128 + 19.2x_3)(-16) \, dx_3}{5EI_c}$$

$$= \frac{1}{5EI_c}\left[+2048x_3 - 307.2\frac{x_3^2}{2}\right]_0^{12} = \frac{+491.52}{EI_c}$$

For member CD, measuring x_4 from D (Fig. 11.4.1) and noting that both M and m have the same sign,

$$M = 8x_4$$

$$m = x_4$$

$$\Delta_{H4} = \int_0^{16} \frac{(8x_4)x_4 \, dx_4}{2EI_c} = \frac{4}{EI_c}\left[\frac{x_4^3}{3}\right]_0^{16} = \frac{+5461.33}{EI_c}$$

Then

$$\Delta_H = \Delta_{H1} + \Delta_{H2} + \Delta_{H3} + \Delta_{H4}$$

$$= \frac{1}{EI_c}(+1365.33 - 491.52 + 491.52 + 5461.33) = \frac{+6826.66}{EI_c}$$

The positive sign means the displacement Δ_H is in the direction of the unit load, that is, to the right.

Example 11.4.2

Determine the slope θ_B at joint B of the frame of Fig. 11.3.1a, using the virtual work (unit load) method.

Solution

(a) *Establish the virtual loading.* To obtain the slope θ_B apply a unit moment at B, as in Fig. 11.4.2, where the virtual loading bending moment m diagrams are also shown. Since the unit moment is counterclockwise, a positive answer for θ_B will mean a counterclockwise rotation of the joint. The bending moment diagrams for the real loading are in Fig. 11.3.1b, c, and d.

(b) *Apply Eq. (5.3.6) for members AB, BC, and CD.* From the m diagrams of Fig. 11.4.2, it is apparent that only member BC influences Eq. (5.3.6) since the m diagrams for AB and CD are zero. Thus, for member BC, taking the segment from B to the 40-kip load,

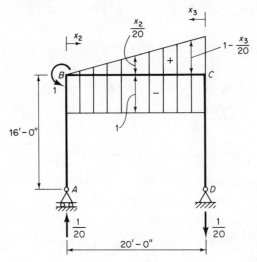

Figure 11.4.2 Virtual loading for determination of the slope at B, Example 11.4.2.

$$M = -64 + 20.8x_2$$

$$m = -1 + \frac{x_2}{20}$$

$$\theta_{B1} = \int_0^8 \frac{(-64 + 20.8x_2)(-1 + x_2/20)\, dx_2}{5EI_c}$$

$$= \frac{1}{5EI_c}\left[64x_2 - 20.8\frac{x_2^2}{2} - 3.2\frac{x_2^2}{2} + 1.04\frac{x_2^3}{3}\right]_0^8 = \frac{-15.70}{EI_c}$$

For member BC, taking the segment from C to the 40-kip load,

$$M = -128 + 19.2x_3$$

$$m = -\frac{x_3}{20}$$

$$\theta_{B2} = \int_0^{12} \frac{(-128 + 19.2x_3)(-x_3/20)\, dx_3}{5EI_c}$$

$$= \frac{1}{5EI_c}\left[+6.4\frac{x_3^2}{2} - 0.96\frac{x_3^3}{3}\right]_0^{12} = \frac{-18.43}{EI_c}$$

Then

$$\theta_B = \theta_{B1} + \theta_{B2} = \frac{-15.70 - 18.43}{EI_c} = \frac{-34.13}{EI_c}$$

Since the answer is negative, the rotation θ_B is opposite to that of the unit moment; that is, θ_B is clockwise.

Example 11.4.3

Determine the horizontal and vertical deflections of the free end A of the cantilever rigid frame $ABCD$ shown in Fig. 11.4.3, using the virtual work (unit load) method.

Solution

(a) *Horizontal deflection of point A.* The unit load is applied as the virtual loading in Fig. 11.4.3d together with the resulting bending moment m diagrams. Applying Eq. (5.3.6),

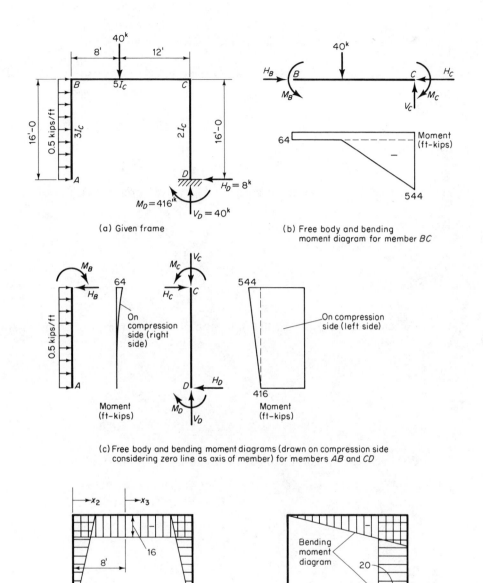

(a) Given frame

(b) Free body and bending moment diagram for member BC

(c) Free body and bending moment diagrams (drawn on compression side considering zero line as axis of member) for members AB and CD

(d) Virtual loading for horizontal deflection at A

(e) Virtual loading for vertical deflection at A

Figure 11.4.3 Cantilever rigid frame for Example 11.4.3.

$$\Delta_H \text{ of } A = \int \frac{Mm}{EI} \, dx$$

For member AB, where both M and m are of the same sign,

$$M = \frac{1}{2}(0.5)x_1^2 = 0.25x_1^2$$

$$m = x_1$$

$$\Delta_{H1} = \int_0^{16} \frac{(0.25x_1^2)x_1 \, dx_1}{3EI_c}$$

$$= \frac{0.25}{3EI_c}\left[\frac{x_1^4}{4}\right]_0^{16} = \frac{+1365.33}{EI_c}$$

For member BC, taking the constant $M_B = -64$ from B to C,

$$M = -64$$

$$m = -16$$

$$\Delta_{H2} = \int_0^{20} \frac{(-64)(-16) \, dx_2}{5EI_c}$$

$$= \frac{204.8}{EI_c}\left[x_2\right]_0^{20} = \frac{+4096.00}{EI_c}$$

For member BC from the 40-kip load to point C for the effect of the 40-kip load,

$$M = -40x_3$$

$$m = -16$$

$$\Delta_{H3} = \int_0^{12} \frac{(-40x_3)(-16) \, dx_3}{5EI_c}$$

$$= \frac{128.0}{EI_c}\left[\frac{x_3^2}{2}\right]_0^{12} = \frac{+9216.0}{EI_c}$$

For member CD, considering the constant $M = 416$ portion,

$$M = 416$$

$$m = x_4$$

$$\Delta_{H4} = \int_0^{16} \frac{(416)(x_4) \, dx_4}{2EI_c}$$

$$= \frac{208}{EI_c}\left[\frac{x_4^2}{2}\right]_0^{16} = \frac{+26{,}624.0}{EI_c}$$

For member CD, considering the triangular portion of the M diagram,

$$M = \left(\frac{544 - 416}{16}\right)x_4 = 8x_4$$

$$m = x_4$$

$$\Delta_{H5} = \int_0^{16} \frac{(8x_4)(x_4) \, dx_4}{2EI_c}$$

$$= \frac{4}{EI_c}\left[\frac{x_4^3}{3}\right]_0^{16} = \frac{+5461.33}{EI_c}$$

The total horizontal deflection Δ_H of A is

$$\Delta_H \text{ of } A = \Delta_{H1} + \Delta_{H2} + \Delta_{H3} + \Delta_{H4} + \Delta_{H5}$$

$$= \frac{1365.33 + 4906.00 + 9216.00 + 26,624.00 + 5461.33}{EI_c}$$

$$= \frac{46,762.66}{EI_c} \quad \text{(to the right)}$$

(b) *Vertical deflection of point A.* The unit load is applied as the virtual loading in Fig. 11.4.3e along with the resulting bending moment m diagrams. Applying Eq. (5.3.6),

$$\Delta_V \text{ of } A = \int \frac{Mm}{EI} \, dx$$

There is no contribution to Δ_V from member AB because m is zero. For member BC, taking the rectangular portion ($M = -64$) of the M diagram,

$$M = -64$$

$$m = -x_2$$

$$\Delta_{V1} = \int_0^{20} \frac{(-64)(-x_2) \, dx_2}{5EI_c}$$

$$= \frac{12.8}{EI_c}\left[\frac{x_2^2}{2}\right]_0^{20} = \frac{+2560}{EI_c}$$

For the triangular portion of M on member BC,

$$M = -40x_3$$

$$m = -(x_3 + 8)$$

$$\Delta_{V2} = \int_0^{12} \frac{(-40x_3)[-(x_3 + 8)] \, dx_3}{5EI_c}$$

$$= \frac{8}{EI_c}\left[+\frac{x_3^3}{3} + 8\left(\frac{x_3^2}{2}\right)\right]_0^{12} = \frac{+9216}{EI_c}$$

For the rectangular portion of M on member CD,

$$M = 416$$

$$m = 20$$

$$\Delta_{V3} = \int_0^{16} \frac{416(20) \, dx_4}{2EI_c}$$

$$= \frac{4160}{EI_c}\left[x_4\right]_0^{16} = \frac{+66,560}{EI_c}$$

For the triangular portion of M on member CD,

$$M = \left(\frac{544 - 416}{16}\right)x_4 = 8x_4$$

$$m = 20$$

$$\Delta_{V4} = \int_0^{16} \frac{(8x_4)(20)\,dx_4}{2EI_c}$$

$$= \frac{80}{EI_c}\left[\frac{x_4^2}{2}\right]_0^{16} = \frac{+10{,}240}{EI_c}$$

The total vertical deflection Δ_V (positive indicates down) of A is

$$\Delta_V = \Delta_{V1} + \Delta_{V2} + \Delta_{V3} + \Delta_{V4}$$

$$= \frac{+2560 + 9216 + 66{,}560 + 10{,}240}{EI_c} = \frac{+88{,}576}{EI_c} \quad \text{(down)}$$

11.5 DEFLECTIONS BY THE MOMENT AREA METHOD

The moment area theorems (and the supplementary conjugate beam theorems for application between two points of zero or known deflections) presented with examples for beams in Section 5.5 (and Section 5.6) are equally applicable for rigid frames. In continuous beams the two sides of a joint are 180° to one another, whereas in rigid frames the sides of a joint may be at *any* angle to one another. Frequently, they are at 90°. If a given joint rotates an amount θ, all members of the frame interesecting at the joint rotate the same amount θ. As for beams it is usually helpful to sketch the elastic curve at the start of an analysis.

Example 11.5.1

For the frame of Fig. 11.3.1a (shown again in Fig. 11.5.1) determine the horizontal deflection of joint A using the moment area method.

Solution

(a) *Sketch the estimated elastic curve, as shown in Fig.* 11.5.1b. Since members AB and CD are assumed to have no change in length, the displaced positions B' and C' of joints B and C are horizontal from their original positions and the distance $B'C'$ is the same as the distance BC. It may be unknown in which way joints B and C rotate. The initial assumption takes θ_B clockwise and θ_C counterclockwise, as in Fig. 11.5.1b.

(b) *Compute θ_B and θ_C* (refer to Fig. 11.5.1c).

$$\theta_B = \frac{\Delta_{CB}'}{L_{BC}} = \frac{\text{moment of } M/EI \text{ area between } B \text{ and } C \text{ about } C}{20}$$

$$= \frac{1}{5EI_c(20)}\left[1920\left(\frac{32}{3}\right) - 640\left(\frac{40}{3}\right) - 1280\left(\frac{20}{3}\right)\right] = \frac{34.13}{EI_c} \quad \text{(clockwise)}$$

$$\theta_C = \frac{\Delta_{BC}'}{L_{BC}} = \frac{\text{moment of } M/EI \text{ area between } B \text{ and } C \text{ about } B}{20}$$

$$= \frac{1}{5EI_c(20)}\left[1920\left(\frac{28}{3}\right) - 640\left(\frac{20}{3}\right) - 1280\left(\frac{40}{3}\right)\right] = \frac{-34.13}{EI_c}$$

The negative sign indicates that θ_C is clockwise instead of counterclockwise as assumed and shown in Fig. 11.5.1b. A revised elastic curve is shown in Fig. 11.5.1d.

Note that the *clockwise* θ_B and the *counterclockwise* θ_C could be taken as the

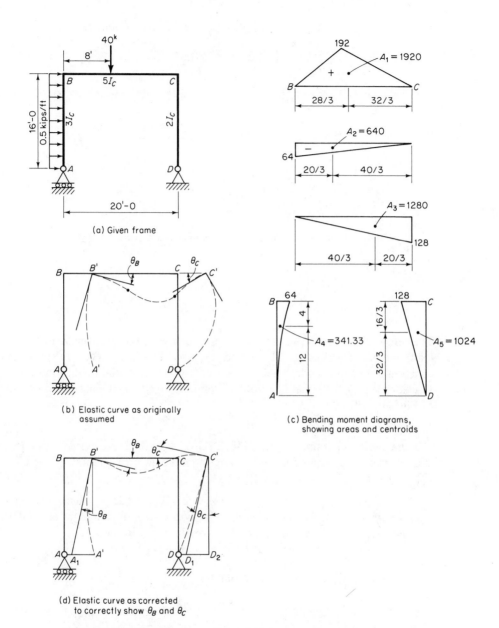

(a) Given frame

(b) Elastic curve as originally
assumed

(c) Bending moment diagrams,
showing areas and centroids

(d) Elastic curve as corrected
to correctly show θ_B and θ_C

Figure 11.5.1 Statically determinate rigid frame analyzed using the moment area method, Example 11.5.1.

upward reactions to the conjugate beam loaded by the one downward positive M/EI area and the two upward negative M/EI areas, as they are shown in Fig. 11.5.1c.

(c) *Compute DD_1 and D_1D_2 for member CD* (Fig. 11.5.1d).

$$D_1D_2 = L_{CD}\theta_C = 16\left(\frac{34.13}{EI_c}\right) = \frac{546.08}{EI_c}$$

$DD_1 = \Delta'_{DC} =$ moment of $\dfrac{M}{EI}$ area between C and D about D

$$= \frac{1}{2EI_c}\left[1024\left(\frac{32}{3}\right)\right] = \frac{5461.33}{EI_c}$$

The distance DD_2 equals the sidesway CC' of point C, which equals the sidesway BB' of point B.

$$DD_2 = CC' = BB' = \frac{546.08 + 5461.33}{EI_c} = \frac{6007.41}{EI_c}$$

(d) *Compute Δ_H of A.*

$$\Delta_H = AA' = AA_1 + A_1A'$$
$$= BB' - L_{AB}\theta_B + A_1A'$$

$A_1A' = \Delta'_{AB} =$ moment of $\dfrac{M}{EI}$ area between A and B about A

$$= \frac{1}{3EI_c}(341.33)(12) = \frac{1365.32}{EI_c}$$

$$\Delta_H = \frac{6007.41}{EI_c} - 16\left(\frac{34.13}{EI_c}\right) + \frac{1365.32}{EI_c} = \frac{6826.65}{EI_c} \quad \text{(to the right)}$$

The results above for Δ_H of A and for θ_B check with those of the virtual work (unit load) method in Example 11.4.1 and Example 11.4.2, respectively. Note that the contribution of the bending moment on BC to the horizontal deflection of A is zero, as evidenced by the equal rotations of joints B and C in the corrected elastic curve of Fig. 11.5.1d.

Example 11.5.2

For the cantilever frame of Fig. 11.4.3a (also Fig. 11.5.2a), determine the horizontal and vertical deflections of point A using the moment area method.

Solution

(a) *Sketch the elastic curve as shown in Fig. 11.5.2c to identify the geometric relationships needed for the solution.* The component bending moment diagrams, together with their areas and centroids, are given in Fig. 11.5.2b.

(b) *Apply the moment area theorems to member CD.*

$$\theta_C = \frac{M}{EI} \text{ area between } C \text{ and } D$$

$$= \frac{1}{2EI_c}(A_3 + A_4) = \frac{6656 + 1024}{2EI_c} = \frac{3840}{EI_c} \quad \text{(counterclockwise)}$$

$C'C = \Delta'_{CD} =$ moment of $\dfrac{M}{EI}$ area between C and D about C

$$= \frac{1}{2EI_c}\left[6656(8) + 1024\left(\frac{16}{3}\right)\right] = \frac{29,354.67}{EI_c} \quad \text{(to the left)}$$

Because member BC does not lengthen or shorten,

$$B_1B = C'C$$

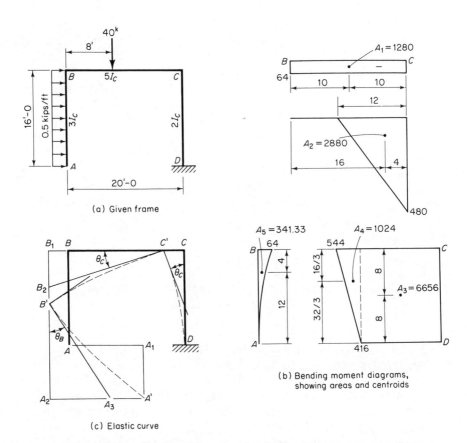

(a) Given frame

(b) Bending moment diagrams, showing areas and centroids

(c) Elastic curve

Figure 11.5.2 Cantilever rigid frame analyzed using the moment area method, Example 11.5.2.

(c) *Apply the moment area theorems to member BC.*

$$\theta_B = \theta_C + \left(\frac{M}{EI} \text{ area between } B \text{ and } C \right)$$

$$= \frac{3840}{EI_c} + \frac{1}{5EI_c}(1280 + 2880) = \frac{4672}{EI_c} \quad \text{(counterclockwise)}$$

$$B_1B' = B_1B_2 + B_2B'$$
$$= L_{BC}\theta_C + \Delta'_{BC}$$

$$\Delta'_{BC} = \text{moment of } \frac{M}{EI} \text{ area between } B \text{ and } C \text{ about } B$$

$$= \frac{1}{5EI_c}[1280(10) + 2880(16)] = \frac{11{,}776}{EI_c}$$

$$B_1B' = \frac{20(3840) + 11{,}776}{EI_c} = \frac{88{,}576}{EI_c} \quad \text{(down)}$$

Because member *AB* does not lengthen or shorten,

$$B_1B' = A_1A'$$

$$\Delta_V \text{ of } A = A_1A' = \frac{88,576}{EI_c} \quad \text{(down)}$$

(d) *Apply the moment area theorems to member AB.*

$$\Delta_H \text{ of } A = AA_1 = A_2A_3 + A_3A' - B_1B \quad \text{(from part b)}$$

$$A_2A_3 = L_{AB}\theta_B = 16\left(\frac{4672}{EI_c}\right) = \frac{74,752}{EI_c}$$

$$A_3A' = \Delta'_{AB} = \text{moment of } \frac{M}{EI} \text{ area between } A \text{ and } B \text{ about } A$$

$$= \frac{1}{3EI_c}[341.33(12)] = \frac{1365.32}{EI_c}$$

$$\Delta_H \text{ of } A = \frac{74,752}{EI_c} + \frac{1365.32}{EI_c} - \frac{29,354.67}{EI_c}$$

$$= \frac{46,762.65}{EI_c} \quad \text{(to the right)}$$

These values for Δ_V and Δ_H of point A are identical with those obtained in Example 11.4.3 by the virtual work (unit load) method.

11.6 FORCE METHOD—CONSISTENT DEFORMATION USING REACTIONS AS REDUNDANTS

A fundamental method of analyzing statically indeterminate structures using consistent deformation, as applied to beams in Section 8.2, may also be easily applied to relatively simple rigid frames. The compatibility, or consistent deformation, conditions to solve for redundant reactions are equal in number to the degree of statical indeterminacy. For the rigid frame $ABCD$ shown in Fig. 11.6.1a, using Eq. (11.2.6),

$$NI = 2(NM) - (NPR + NPS) = 6 - (4 + 1) = 1$$

This rigid frame with two hinged supports may be considered equivalent to either of the "basic" statically determinate frames of Fig. 11.6.1b or c. Whichever statically determinate frame is used, the horizontal displacement at the roller support is computed, both due to the given external loads W_1 and W_2 and due to the redundant reaction R (either R_3 for Fig. 11.6.1b or R_1 for Fig. 11.6.1c) applied as a load. The redundant reaction may be solved from the compatibility condition that the horizontal deflection of the roller support (at D in Fig. 11.6.1b or at A in Fig. 11.6.1c) must be zero.

The rigid frame of Fig. 11.6.1d is statically indeterminate to the third degree as shown by Eq. (11.2.6),

$$NI = 2(NM) - (NPR + NPS) = 6 - (2 + 1) = 3$$

Two choices for sets of redundants to be used are given in Fig. 11.6.1e and f. The "basic" statically determinate frame of Fig. 11.6.1e is acted on by the actual applied loads W_1 and W_2 as well as the redundants R_4, R_5, and R_6 acting also as applied loads. The three compatibility conditions are that under the combined action of the W

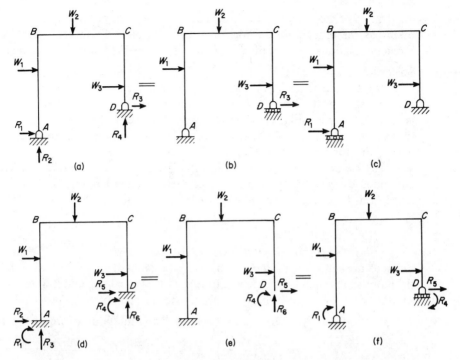

Figure 11.6.1 Force method—consistent deformation using reactions as redundants.

forces and the R forces the rotation, horizontal deflection, and vertical deflection at D must be zero. The "basic" statically determinate frame of Fig. 11.6.1f is acted on by the actual applied loads W_1 and W_2 as well as the redundants R_1, R_4, and R_5 acting also as applied loads. The three compatibility conditions are that under the combined action of all of these forces (W's and R's) the rotations at A and D must be zero, and the horizontal deflection at D must be zero.

To summarize this application of the *force method*, the analysis involves the computation of joint rotations or deflections of the "basic" statically determinate rigid frame obtained by removal of reaction restraints equal in number to the degree of statical indeterminacy. The joint rotations or deflections are obtained first due to the actual applied loads and then due to each of the redundants. These joint displacements may be obtained by any method, particularly by the virtual work (unit load) method as shown in Section 11.4 or by the moment area method as shown in Section 11.5.

Alternative to using reactions as redundants, internal restraints such as the bending moment or shear at any location could also be used as redundants. For simplicity, rarely would arbitrary internal restraints be used, although there may be situations in the advanced study of structures (such as in limit analysis and design) where member end moments might be used. In this introductory text, it is sufficient to emphasize the consistent-deformation concept in the force method by applying it to single-span, one-story rigid frames.

Example 11.6.1

Analyze the rigid frame of Fig. 11.6.2a using the method of consistent deformation taking as the redundant the horizontal reaction at A. Draw the shear and bending moment diagrams and compute the joint displacements in the final elastic curve.

Solution

(a) *Establish the "basic" statically determinate structure.* As prescribed by the problem statement, consider H_A as the redundant. The statically determinate structure is then the frame of Fig. 11.6.2b previously analyzed in Example 11.5.1. The "basic" frame has a roller support at A and is subject to the applied loads (Fig. 11.6.2b) as well as the horizontal force H_A (Fig. 11.6.2c).

(b) *Obtain the horizontal deflection of joint A due to applied loads.* This may be computed using either the moment area method as in Example 11.5.1 or the virtual work (unit load) method as in Example 11.4.1. Thus,

$$\Delta_H \text{ at } A \text{ due to loads} = \frac{+6826.66}{EI_c} \quad \text{(to the right)}$$

(c) *Obtain the horizontal deflection at A due to force H_A.* Again, either the moment area method or the virtual work (unit load) method may be used. Using the moment area method, referring to Fig. 11.6.2c, d, and e, and noting the symmetry of M/EI area *for member BC*, θ equals zero at midspan of BC. Thus,

$$\theta_B = M/EI \text{ area between } B \text{ and midpoint of } BC$$

$$= \frac{1}{5EI_c}\left(\frac{1}{2}A_2\right) = \frac{1}{5EI_c}\left(\frac{1}{2}\right)(320H_A) = \frac{32H_A}{EI_c} = \theta_C$$

$$D_2D = D_2D_1 + D_1D = L_{CD}\theta_C + \Delta'_{DC}$$

$$\Delta'_{DC} = \frac{128H_A(32/3)}{E(2I_c)} = \frac{682.67H_A}{EI_c}$$

$$D_2D = 16\left(\frac{32H_A}{EI_c}\right) + \frac{682.67H_A}{EI_c} = \frac{1194.67H_A}{EI_c}$$

$$\Delta_H \text{ at } A = AA' = AA_1 + A_1A_2 + A_2A'$$

$$AA_1 = B'B = C'C = D_2D = \frac{1194.67H_A}{EI_c}$$

$$A_1A_2 = L_{AB}\theta_B = 16\left(\frac{32H_A}{EI_c}\right) = \frac{512H_A}{EI_c}$$

$$A_2A' = \Delta'_{AB} = \frac{128H_A(32/3)}{E(3I_c)} = \frac{455.11H_A}{EI_c}$$

$$\Delta_H \text{ at } A = \frac{H_A}{EI_c}(1194.67 + 512 + 455.11) = \frac{2161.78H_A}{EI_c}$$

(d) *Apply the consistent deformation requirement.* Since the actual horizontal displacement at A is zero,

$$\Delta_H \text{ due to loads} - \Delta_H \text{ due to } H_A = 0$$

$$\frac{6826.66}{EI_c} - \frac{2161.78H_A}{EI_c} = 0$$

$$H_A = 3.158 \text{ kips} \quad \text{(to the left)}$$

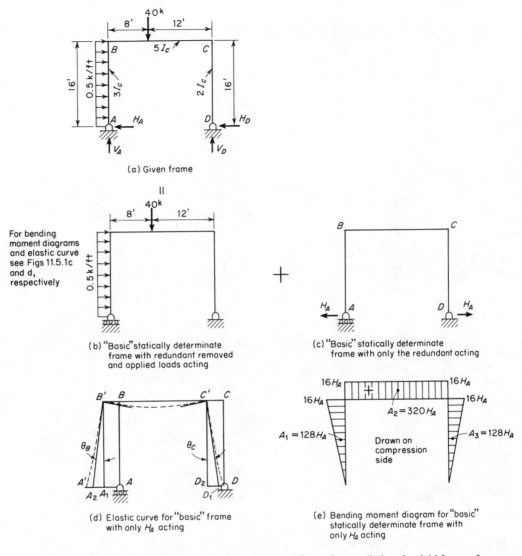

(a) Given frame

For bending moment diagrams and elastic curve see Figs 11.5.1c and d, respectively

(b) "Basic" statically determinate frame with redundant removed and applied loads acting

(c) "Basic" statically determinate frame with only the redundant acting

(d) Elastic curve for "basic" frame with only H_A acting

(e) Bending moment diagram for "basic" statically determinate frame with only H_A acting

Figure 11.6.2 Force method—consistent deformation applied to the rigid frame of Example 11.6.1.

(e) *Alternative computation to obtain the horizontal deflection due to force H_A.* The computation may be done using the virtual work (unit load) method instead of the moment area method. The virtual loading may be taken as that shown in Fig. 11.6.2c, except the multiplier H_A should be deleted. Thus,

$$\Delta_H = \int_0^L \frac{Mm\,dx}{EI}$$

where M is given as Fig. 11.6.2e and m is simply the same expression without the multiplier H_A. Evaluating, noting all terms will be positive since the M and m are of the same sign,

$$\Delta_H = \int_0^{16} \frac{(H_A x_1)(x_1)\, dx_1}{3EI_c} + \int_0^{20} \frac{(16H_A)(16)\, dx_2}{5EI_c}$$

$$+ \int_0^{16} \frac{(H_A x_4)(x_4)\, dx_4}{2EI_c}$$

$$\Delta_H = \left[\frac{H_A}{EI_c} \frac{x_1^3}{9}\right]_0^{16} + \left[\frac{H_A}{EI_c} \frac{256 x_2}{5}\right]_0^{20} + \left[\frac{H_A}{EI_c} \frac{x_4^3}{6}\right]_0^{16}$$

$$= \frac{H_A}{EI_c}(455.11 + 1024 + 682.67) = \frac{H_A(2161.78)}{EI_c}$$

The consistent deformation requirement would be

$$\Delta_H \text{ due to loads} - \Delta_H \text{ due to } H_A = 0$$

$$\frac{6826.66}{EI_c} - \frac{2161.78 H_A}{EI_c} = 0$$

exactly the same as in part (d).

(f) *Apply the laws of statics to obtain reactions, shears, and bending moments.*
From Fig. 11.6.2a,

$$H_A = 3.158 \text{ kips (determined in part d)}$$

$$H_D = 8 - 3.158 = 4.842 \text{ kips}$$

$$V_A = \frac{40(12) - 8(8)}{20} = 20.8 \text{ kips}$$

$$V_D = \frac{40(8) + 8(8)}{20} = 19.2 \text{ kips}$$

Note that in this case the two columns are of the same height so that the vertical reactions are not affected by the magnitude of the horizontal reaction. The shear and moment diagrams are shown in Fig. 11.6.3b and c.

(g) *Check compatibility of deformation and sketch the final elastic curve.* The shape of the elastic curve must match the bending moment diagram; where tension is on one side it must be shown that way on the elastic curve; where there is an inflection point (point of zero moment) the elastic curve must show it, too.

Critical values of deformations may be computed as follows using the moment area method with the M/EI areas and centroid locations given in Fig. 11.6.3e. Note that the denominator EI_c is not shown in Fig. 11.6.3e.

$$\theta_B = \frac{\Delta_{CB}}{L_{BC}} = \frac{384(32/3) - 26.94(40/3) - 154.94(20/3)}{EI_c(20)}$$

$$= \frac{135.19}{EI_c} \text{ kip-ft}^2 \quad \text{(clockwise)}$$

$$\theta_C = \frac{\Delta'_{BC}}{L_{BC}} = \frac{384(28/3) - 26.94(20/3) - 154.94(40/3)}{EI_c(20)}$$

$$= \frac{66.93}{EI_c} \text{ kip-ft}^2 \quad \text{(counterclockwise)}$$

Alternatively, using the conjugate beam concept, clockwise θ_B and counterclockwise

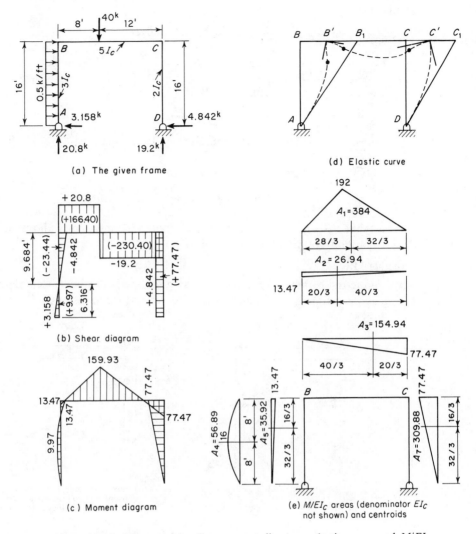

Figure 11.6.3 Shear and bending moment diagrams, elastic curve, and M/EI areas used in compatibility checks, Example 11.6.1.

θ_C are the upward reactions to the conjugate beam BC under the downward load of A_1 and the upward loads of A_2 and A_3.

$$\theta_A = \theta_B + \left(\frac{M}{EI} \text{ area between } A \text{ and } B\right)$$

Note that bending moment causing compression on the left side of vertical member AB causes the slope θ to increase in a clockwise direction going from top to bottom, Thus,

$$\theta_A = \frac{1}{EI_c}(135.19 + 56.89 - 35.92) = \frac{156.16}{EI_c} \text{ kip-ft}^2 \quad \text{(clockwise)}$$

Similarly,

$$\theta_D = \theta_C + \left(\frac{M}{EI} \text{ area between } C \text{ and } D\right)$$

$$= \frac{1}{EI_c}(-66.93 + 309.88) = \frac{242.95}{EI_c} \text{ kip-ft}^2 \quad \text{(clockwise)}$$

For the "sidesway" displacements at joints B and C,

$$BB' = BB_1 - B_1B' = L_{AB}\theta_A - \Delta'_{BA}$$

$$= \frac{1}{EI_c}\left[16(156.16) - 56.89(8) + 35.92\left(\frac{16}{3}\right)\right] = \frac{2235.0}{EI_c} \text{ kip-ft}^3$$

$$CC' = CC_1 - C_1C' = L_{CD}\theta_D - \Delta'_{CD}$$

$$= \frac{1}{EI_c}\left[16(242.95) - 309.88\left(\frac{16}{3}\right)\right] = \frac{2234.5}{EI_c} \text{ kip-ft}^3$$

The fact that BB' and CC' are equal, as computed independently above, assures that the compatibility of deformation is satisfied.

Example 11.6.2

Analyze the rigid frame of Fig. 11.6.4a using the method of consistent deformation, taking as the redundants the horizontal and vertical reactions at A. Draw the shear

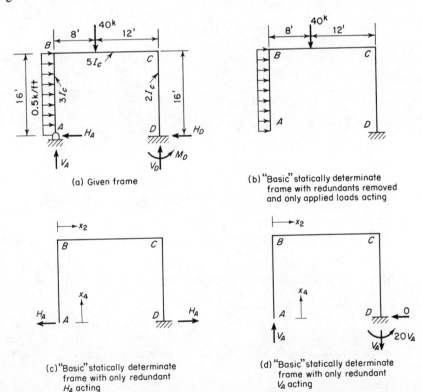

(a) Given frame

(b) "Basic" statically determinate frame with redundants removed and only applied loads acting

(c) "Basic" statically determinate frame with only redundant H_A acting

(d) "Basic" statically determinate frame with only redundant V_A acting

Figure 11.6.4 Force method—consistent deformation applied to the rigid frame of Example 11.6.2.

and bending moment diagrams and compute the joint displacements in the final elastic curve.

Solution

(a) *Establish the "basic" statically determinate structure.* In this case the redundants are prescribed by the problem statement to be H_A and V_A. The "basic" statically determinate structure is then the frame of Fig. 11.6.3b previously analyzed in Examples 11.4.3 and 11.5.2 with the applied loads acting.

(b) *Horizontal deflection Δ_H of joint A due to applied loads only* (Fig. 11.6.4b). This computation has been previously shown using the virtual work (unit load) method in Example 11.4.3 and the moment area method in Example 11.5.2,

$$\Delta_H \text{ at } A \text{ due to loads} = \frac{46{,}762.66}{EI_c} \quad \text{(to the right)}$$

(c) *Vertical deflection Δ_V of joint A due to applied loads only* (Fig. 11.6.4b). This computation has been previously shown in Examples 11.4.3 and 11.5.2,

$$\Delta_V \text{ at } A \text{ due to loads} = \frac{+88{,}576}{EI_c} \quad \text{(down)}$$

(d) *Horizontal deflection Δ_H at A due to H_A only.* Either the virtual work (unit load) method or the moment area method may be used. Using the former,

$$\Delta_H = \int_0^L \frac{Mm\,dx}{EI}$$

where M is obtained using the loading of Fig. 11.6.4c, and m is the same as M without the multiplier H_A.

$$\Delta_H \text{ at } A \text{ due to } H_A = \frac{1}{3EI_c}\int_0^{16}(H_A x_1)(x_1)\,dx_1 + \frac{1}{5EI_c}\int_0^{20}16H_A(16)\,dx_2$$

$$+ \frac{1}{2EI_c}\int_0^{16}(H_A x_4)(x_4)\,dx_4$$

$$\Delta_H = \frac{2161.78 H_A}{EI_c} \text{ kip-ft}^3 \quad \text{(to the left)}$$

(e) *Vertical deflection Δ_V at A due to H_A only.* This time the M is from Fig. 11.6.4c, and m is from Fig. 11.6.4d without the multiplier V_A; a positive answer will show upward deflection.

$$\Delta_V \text{ at } A \text{ due to } H_A = \frac{1}{5EI_c}\int_0^{20}(16H_A)(x_2)\,dx_2 + \frac{1}{2EI_c}\int_0^{16}(H_A x_4)(20)\,dx_4$$

$$= \frac{1920 H_A}{EI_c} \text{ kip-ft}^3 \quad \text{(up)}$$

(f) *Horizontal deflection Δ_H due to V_A only.* Taking M from Fig. 11.6.4d and m from Fig. 11.6.4c without the multiplier H_A,

$$\Delta_H \text{ at } A \text{ due to } V_A = \frac{1}{5EI_c}\int_0^{20}(V_A x_2)(16)\,dx_2 + \frac{1}{2EI_c}\int_0^{16}(20V_A)(x_4)\,dx_4$$

$$= \frac{1920 V_A}{EI_c} \text{ kip-ft}^3 \quad \text{(to the left)}$$

(g) *Vertical deflection Δ_V due to V_A only.* Using M from Fig. 11.6.4d and m the same as M except omitting the multiplier V_A,

$$\Delta_V \text{ at } A \text{ due to } V_A = \frac{1}{5EI_c} \int_0^{20} (V_A x_2)(x_2)\, dx_2 + \frac{1}{2EI_c} \int_0^{16} (20V_A)(20)\, dx_4$$

$$= \frac{3733.33 V_A}{EI_c} \text{ kip-ft}^3 \quad (\text{up})$$

(h) *Comments relating to Maxwell's reciprocal theorem.* Note is made that the deflections due to the action of the redundants H_A and V_A as computed in parts (d) to (g) may each be thought of as the deflections δ due to a unit load times H_A or V_A. Thus, one may state, using the terminology introduced in Section 5.4, that

$$\Delta_H \text{ at } A(\text{to left}) \text{ due to } H_A(\text{to left}) = H_A \delta_{11}$$
$$\Delta_H \text{ at } A(\text{left}) \text{ due to } V_A(\text{up}) \quad = V_A \delta_{12}$$
$$\Delta_V \text{ at } A(\text{left}) \text{ due to } H_A(\text{left}) \quad = H_A \delta_{21}$$
$$\Delta_V \text{ at } A(\text{up}) \text{ due to } V_A(\text{up}) \quad = V_A \delta_{22}$$

In the equations above, the first subscript on δ indicates the place where the deflection is measured (1 means horizontal to left at A, 2 means vertically up at A) and the second subscript indicates the place where the unit load is applied (1 means applied horizontally to left at A, 2 means applied vertically upward at A).

Maxwell's theorem of reciprocal deflections states

$$\delta_{12} = \delta_{21}$$

and this was verified in parts (e) and (f) of this example.

(i) *Apply the consistent deformation requirements.* The consistent deformation (compatibility) conditions are that the horizontal and vertical deflections of point A due to applied loads along with H_A and V_A must be zero (the sum of loadings in Fig. 11.6.4b, c, and d).

$$\Delta_H = \frac{46,762.66}{EI_c} - \frac{2161.78H_A}{EI_c} - \frac{1920V_A}{EI_c} = 0$$

$$\Delta_V = \frac{88,576}{EI_c} - \frac{1920H_A}{EI_c} - \frac{3733.33V_A}{EI_c} = 0$$

Solving the two equations above,

$$H_A = +1.030 \text{ kips} \quad (\text{to the left})$$
$$V_A = +23.196 \text{ kips} \quad (\text{up})$$

(j) *Shear and bending moment diagrams.* Applying the laws of statics to the given frame of Fig. 11.6.4a,

$$H_D = 8 - 1.030 = 6.970 \text{ kips} \quad (\text{to the left})$$
$$V_D = 40 - 23.196 = 16.804 \text{ kips} \quad (\text{up})$$
$$M_D = 40(8) + 8(8) - 16.804(20) = 47.92 \text{ ft-kips} \quad (\text{counterclockwise})$$

The complete shear and bending moment diagrams are shown in Fig. 11.6.5b and c. Note that in the process of determining the critical values on the moment diagram by the method of adding shear areas, a check is always obtained by showing that the moment at the terminal end of a member is in fact equal to the moment already indicated there on the free-body diagram.

(k) *Elastic curve and compatibility checks.* The number of compatibility checks on the elastic curve is always equal to the degree of statical indeterminacy: two in

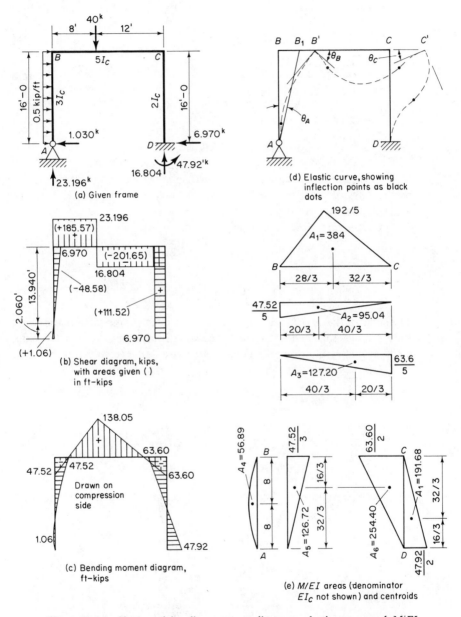

Figure 11.6.5 Shear and bending moment diagrams, elastic curve, and M/EI areas used to make compatibility of deformation checks, Example 11.6.2.

this problem. One can see that the slope at C can be independently obtained using member BC and then member DC, and, the sidesway of joint B and of joint C can be independently obtained using member AB and then member DC. Note that the elastic curve in Fig. 11.6.5d is sketched to match the moment diagram of Fig. 11.6.5c, that curvature agrees with the sign of the bending moment area, and that inflection points on the elastic curve are at the zero values on the bending moment diagram.

Using moment area method on member BC and referring to Fig. 11.6.5e,

$$\theta_B = \frac{\Delta'_{CB}}{L_{BC}} = \frac{A_1(32/3) - A_2(40/3) - A_3(20/3)}{20EI_c}$$

$$= \frac{384(32/3) - 95.04(40/3) - 127.20(20/3)}{20EI_c} = \frac{99.04}{EI_c} \text{ kip-ft}^2 \quad \text{(clockwise)}$$

$$\theta_C = \frac{\Delta'_{BC}}{L_{BC}} = \frac{384(28/3) - 95.04(20/3) - 127.29(40/3)}{20EI_c}$$

$$= \frac{62.72}{EI_c} \text{ kip-ft}^2 \quad \text{(counterclockwise)}$$

If the conjugate-beam concept were used, clockwise θ_B and counterclockwise θ_C would be the upward reactions to the conjugate beam BC loaded with A_1 downward, A_2 upward, and A_3 upward.

On member BA,

$$\theta_A = \theta_B + \left(\frac{M}{EI} \text{ area between } A \text{ and } B\right)$$

$$= \frac{1}{EI_c}(99.04 + 56.89 - 126.72) = \frac{29.21}{EI_c} \text{ kip-ft}^2 \quad \text{(clockwise)}$$

The signs of the M/EI areas A_4 and A_5 used in the computation above are determined from the fact that bending moment causing compression on the left side of a vertical member makes the slope θ increase clockwise going from top to bottom.

On member CD,

$$\theta_C = \theta_D + \frac{A_6}{EI_c} - \frac{A_7}{EI_c}$$

$$= 0 + \frac{254.40}{EI_c} - \frac{191.68}{EI_c} = \frac{62.72}{EI_c} \text{ kip-ft}^2 \quad \text{(counterclockwise)}$$

Thus, one of the compatibility checks is satisfied; that is, θ_C computed from $B'C'$ is identical to that computed from DC'.

Next, compute the "sidesway" displacement BB' using member AB':

$$BB' = BB_1 + B_1B'$$

$$= L_{AB}\theta_A + \Delta'_{BA}$$

$$= 16\left(\frac{29.21}{EI_c}\right) + \frac{1}{EI_c}\left[126.72\left(\frac{16}{3}\right) - 56.89(8)\right]$$

$$= \frac{688.80}{EI_c} \text{ kip-ft}^3 \quad \text{(to the right)}$$

Now compute the "sidesway" displacement CC' using member DC',

Rigid Frames   Chap. 11

$$CC' = \Delta'_{CD}$$

$$= \frac{1}{EI_c}\left[191.68\left(\frac{32}{3}\right) - 254.40\left(\frac{16}{3}\right)\right] = \frac{687.79}{EI_c} \text{ kip-ft}^3 \quad \text{(to the right)}$$

Thus, the second compatibility check is satisfied; that is, CC' computed from DC' equals BB' computed from AB'. As said earlier, the number of compatibility checks must equal the number of degrees of statical indeterminacy.

11.7 SLOPE DEFLECTION METHOD

The slope deflection method, a displacement rather than a force method of structural analysis, was first introduced in Chapter 9 and applied to beams. In Section 9.2 the basic equations were derived for the end moments on *any* segment AB of a structure, and are restated here, as follows:

$$\left.\begin{array}{l} M_A = M_{FAB} + M'_A \\ M_B = M_{FBA} + M'_B \end{array}\right\} \qquad [9.2.1]$$

or

$$\left.\begin{array}{l} M_A = M_{FAB} + \dfrac{4EI}{L}(\theta_A - R_{AB}) + \dfrac{2EI}{L}(\theta_B - R_{AB}) \\[2mm] M_B = M_{FBA} + \dfrac{2EI}{L}(\theta_A - R_{AB}) + \dfrac{4EI}{L}(\theta_B - R_{AB}) \end{array}\right\} \qquad [9.2.7]$$

The definition of terms is given in Fig. 9.2.1, and the sign convention is that clockwise moments and rotations are positive.

The slope deflection method has the advantage of being little different in set up for frame analysis than for beam analysis. Equations (9.2.7) are for *any segment* of a structure, whether it be a full beam span, a partial beam span as discussed in Section 9.4, or a frame element, as will be shown by examples in this section.

Example 11.7.1

Analyze the rigid frame of Fig. 11.7.1a by the slope deflection method.

Solution

(a) *Establish the displacement unknowns to be used in the solution.* Unless the high-speed digital computer is to be used, it is usually desirable to use the minimum number of displacement unknowns since that number equals the number of simultaneous equations that must be solved. To begin, the displacements relating to the statically determinate portions AB and DE of the structure need not be used as unknowns. The internal forces in members BC, CD, BF, and CG can only be determined if the deflection and slope are known at each joint. For the given structure, assumed to have no axial deformation, no joints can deflect (i.e., change coordinate location) but joints B, C, and D adjoining the statically indeterminate members do rotate. Thus, in Fig. 11.7.1b the necessary unknown displacements, θ_B, θ_C, and θ_D are shown. These displacements may be termed "degrees of freedom." The degrees of freedom are the joint displacements (either deflection or rotation), the magnitude of which will permit determination of the internal forces. The minimum number of

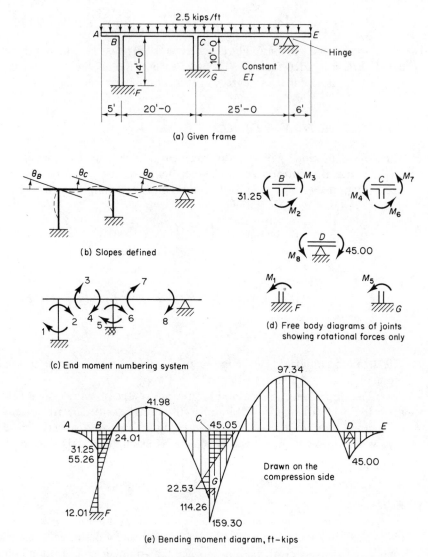

(a) Given frame

(b) Slopes defined

(c) End moment numbering system

(d) Free body diagrams of joints showing rotational forces only

(e) Bending moment diagram, ft–kips

Figure 11.7.1 Frame analyzed by the slope deflection method, Example 11.7.1.

degrees of freedom to solve this problem is three. More degrees of freedom are involved if more segments are used in making the analysis. For instance, if segment *AB* were to be used, there would be two additional degrees of freedom (deflection and rotation of joint *A*) and there would be two additional end moments, M_{AB} and M_{BA}, involved in the solution. In the use of the minimum number of displacement unknowns, the end moments involved are numbered and shown in Fig. 11.7.1c. The unknown moments are shown clockwise on the ends of the members involved and are numbered in any desired sequence for later reference.

Rigid Frames   Chap. 11

(b) *Compute fixed-end moments.* These moments required to maintain zero slope and deflection at member ends may be obtained by using the formulas in Fig. 9.2.4.

$$M_{F3} = M_{FBC} = \frac{-wL^2}{12} = -\frac{2.5(20)^2}{12} = -83.33 \text{ ft-kips}$$

$$M_{F4} = M_{FCB} = +83.33 \text{ ft-kips}$$

$$M_{F7} = M_{FCD} = -\frac{2.5(25)^2}{12} = -130.21 \text{ ft-kips}$$

$$M_{F8} = M_{FDC} = +130.21 \text{ ft-kips}$$

Also, the cantilever moments, M_{BA} and M_{DE} will be required for the joint equilibrium equations, and are computed here,

$$M_{BA} = +\frac{wL^2}{2} = +\frac{2.5(5)^2}{2} = +31.25 \text{ ft-kips}$$

$$M_{DE} = -\frac{2.5(6)^2}{2} = -45.00 \text{ ft-kips}$$

(c) *Write the slope deflection equations.* Using Eqs. (9.2.7) with R equal to zero because the member axes do not rotate, and noting that θ_F and θ_G are equal to zero,

$$M_1 = \frac{2EI}{14}\theta_B$$

$$M_2 = \frac{4EI}{14}\theta_B$$

$$M_3 = -83.33 + \frac{4EI}{20}\theta_B + \frac{2EI}{20}\theta_C$$

$$M_4 = +83.33 + \frac{2EI}{20}\theta_B + \frac{4EI}{20}\theta_C$$

$$M_5 = \frac{2EI}{10}\theta_C$$

$$M_6 = \frac{4EI}{10}\theta_C$$

$$M_7 = -130.21 + \frac{4EI}{25}\theta_C + \frac{2EI}{25}\theta_D$$

$$M_8 = +130.21 + \frac{2EI}{25}\theta_C + \frac{4EI}{25}\theta_D$$

(d) *Joint equilibrium equations.* The three equilibrium equations corresponding to the degrees of freedom are, from Fig. 11.7.1d,

$$M_2 + M_3 + 31.25 = 0$$

$$M_4 + M_6 + M_7 = 0$$

$$M_8 - 45.00 = 0$$

Substitution of the equations for the M's from part (c) gives

$$EI\theta_B\left(\frac{4}{14} + \frac{4}{20}\right) + EI\theta_C\left(\frac{2}{20}\right) \qquad\qquad = +52.08$$

$$EI\theta_B\left(\frac{2}{20}\right) \quad + EI\theta_C\left(\frac{4}{20} + \frac{4}{10} + \frac{4}{25}\right) + EI\theta_D\left(\frac{2}{25}\right) = +46.88$$

$$EI\theta_C\left(\frac{2}{25}\right) \qquad\qquad + EI\theta_D\left(\frac{4}{25}\right) = -85.21$$

Note that the matrix of coefficients on the left side of the equal sign has the expected symmetry with respect to the diagonal. The proof for this symmetry is in Chapter 17.

Solving for θ_B, θ_C, and θ_D gives

$$EI\theta_B = +84.04 \text{ ft-kips}$$
$$EI\theta_C = +112.63 \text{ ft-kips}$$
$$EI\theta_D = -588.86 \text{ ft-kips}$$

(e) *Final end moments.* Substitution of the displacements θ_B, θ_C, and θ_D in the end moment equations of part (c) gives

$$M_1 = \frac{2}{14}(84.04) = +12.01 \text{ ft-kips}$$

$$M_2 = \frac{4}{14}(84.04) = +24.01 \text{ ft-kips}$$

$$M_3 = -83.33 + \frac{4}{20}(84.04) + \frac{2}{20}(112.63) = -83.33 + 28.07$$
$$= -55.26 \text{ ft-kips}$$

$$M_4 = +83.33 + \frac{2}{20}(84.04) + \frac{4}{20}(112.63) = +83.33 + 30.93$$
$$= +114.26 \text{ ft-kips}$$

$$M_5 = \frac{2}{10}(112.63) = +22.53 \text{ ft-kips}$$

$$M_6 = \frac{4}{10}(112.63) = +45.05 \text{ ft-kips}$$

$$M_7 = -130.21 + \frac{4}{25}(112.63) + \frac{2}{25}(-588.86) = -130.21 - 29.09$$
$$= -159.30 \text{ ft-kips}$$

$$M_8 = +130.21 + \frac{2}{25}(112.63) + \frac{4}{25}(-588.86) = +130.21 - 85.21$$
$$= +45.00 \text{ ft-kips}$$

(f) *Shear and bending moment diagrams.* As in any analysis the shears on the ends of the members are computed, and utilizing the shear diagram areas between the support and the point of zero shear as the change in bending moment, the maximum positive bending moments are computed. For purposes of this example, the shear diagram has been omitted and only the bending moment diagram is shown in Fig. 11.7.1e.

(g) *Check for correctness.* When the results of a structural analysis satisfy both equilibrium and compatibility, their correctness is assured. The number of equilibrium checks is equal to the degree of freedom. In this problem the facts that (1) $M_2 + M_3 + 31.25 = 0$, (2) $M_4 + M_6 + M_7 = 0$, and (3) $M_8 - 45.00 = 0$ are obvious once the displacements θ_B, θ_C, and θ_D are back-substituted into the slope deflection equations to obtain the values of M_1 to M_8. The number of compatibility checks is equal to the degree of statical indeterminacy; in this case $NI = NF - NP = 8 - 3 = 5$. Although not shown, critical values on the final elastic curve should be evaluated by the moment area method; the facts that (1) θ_B from member $BC = \theta_B$ from member FB, (2) θ_C from member $CB = \theta_C$ from member CD, (3) θ_C from member $CB = \theta_C$ from member GC, (4) Δ_H of B from member $FB = 0$, and (5) Δ_H of C from member $GC = 0$, constitute the compatibility checks.

Example 11.7.2

Analyze the frame of Fig. 11.7.1a if the hinge at D is made a roller support.

Solution

(a) *Determine the horizontal reaction at D.* As a preliminary to analyzing a frame having a roller at D (which precludes there being any horizontal reaction), examine the statics of member $ABCDE$ as shown in Fig. 11.7.2. The horizontal shears acting at the top of member BF and CG are

$$H_{BF} = \frac{M_1 + M_2}{14} = \frac{12.01 + 24.01}{14} = 2.573 \text{ kips} \quad \text{(to the left)}$$

$$H_{CG} = \frac{M_5 + M_6}{10} = \frac{22.53 + 45.05}{10} = 6.758 \text{ kips} \quad \text{(to the left)}$$

These shears act to the right on member $ABCDE$; thus, for equilibrium the horizontal reaction at D must be

$$H_D = H_{BF} + H_{CG} = 2.573 + 6.758 = 9.331 \text{ kips} \quad \text{(to the left)}$$

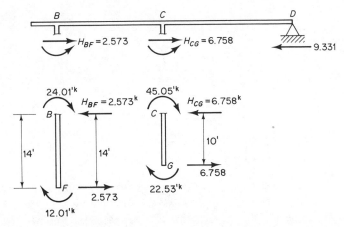

Figure 11.7.2 Determination of horizontal reaction at the hinge at D, Example 11.7.2.

The replacement of the hinge by a roller support would be equivalent to applying 9.331 kips horizontally acting along line $ABCDE$ to the right to make the net reaction $H_D = 0$. The solution for the structure having the roller at D could be obtained by superposition of the structure in Example 11.7.1 with a structure having a horizontal degree of freedom along $ABCDE$ and a horizontal load of 9.331 kips applied at that level acting to the right.

Alternatively, the entire problem can be solved as in the following parts of this example using four displacements, $\theta_B, \theta_C, \theta_D$, and Δ (the sidesway displacement of $ABCDE$).

(b) *Establish the displacement unknowns to be used in the solution.* The analysis of the frame in Fig. 11.7.3a requires the use of four degrees of freedom (four displacement unknowns) as a minimum. These are defined in Fig. 11.7.3b.

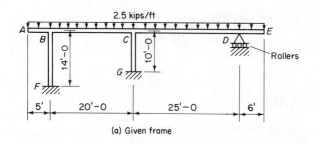

(a) Given frame

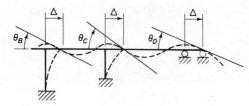

(b) Displacements defined

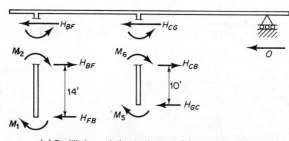

(c) Equilibrium relating to degree of freedom in sidesway

Figure 11.7.3 Frame having one degree of freedom in sidesway, Example 11.7.2.

(c) *Compute fixed-end moments.* These were computed in Example 11.7.1(b).

(d) *Write the slope deflection equations.* Using Eqs. (9.2.7) and the numbering shown in Fig. 11.7.1c, the equations for M_3, M_4, M_7, and M_8 are identical to those in Example 11.7.1(c). The equations for M_1, M_2, M_5 and M_6 in this example are different because they now involve Δ (Fig. 11.7.3). Thus, those slope deflection equations become

$$M_1 = \frac{4EI}{14}\left(-\frac{\Delta}{14}\right) + \frac{2EI}{14}\left(\theta_B - \frac{\Delta}{14}\right)$$

$$M_2 = \frac{2EI}{14}\left(-\frac{\Delta}{14}\right) + \frac{4EI}{14}\left(\theta_B - \frac{\Delta}{14}\right)$$

$$M_5 = \frac{4EI}{10}\left(-\frac{\Delta}{10}\right) + \frac{2EI}{10}\left(\theta_C - \frac{\Delta}{10}\right)$$

$$M_6 = \frac{2EI}{10}\left(-\frac{\Delta}{10}\right) + \frac{4EI}{10}\left(\theta_C - \frac{\Delta}{10}\right)$$

The $R_{AB} = \Delta/L$ in the equations above is put into the slope deflection equations as a positive quantity because the sidesway Δ is taken to the right such as to give a clockwise rotation to the axes of members BF and CG.

(e) *Joint equilibrium equations.* The three rotational equilibrium equations used in Example 11.7.1(d) (from free bodies in Fig. 11.7.1d) are applicable here (summation of *counterclockwise* moments acting on joints B, C, or D equals zero),

$$M_2 + M_3 + 31.25 = 0 \qquad \text{(a)}$$

$$M_4 + M_6 + M_7 \quad = 0 \qquad \text{(b)}$$

$$M_8 - 45.00 \quad\quad = 0 \qquad \text{(c)}$$

The fourth equilibrium equation is the horizontal force equilibrium of member $ABCDE$, shown in Fig. 11.7.3c (summation of forces acting on $ABCDE$ to the *left* equals zero),

$$H_{BF} + H_{CG} = 0 \qquad \text{(d)}$$

where

$$H_{BF} = -\frac{M_1 + M_2}{14}$$

$$H_{CG} = -\frac{M_5 + M_6}{10}$$

Thus, Eq. (d) becomes

$$-\frac{M_1}{14} - \frac{M_2}{14} - \frac{M_5}{10} - \frac{M_6}{10} = 0 \qquad \text{(e)}$$

Note that in order to observe a symmetrical matrix of coefficients in the left side of the simultaneous equations involving the unknown displacements, Eqs. (a), (b), and (c) above are taking the summation of *counterclockwise* moments on joints B, C, and D, while Eq. (d) is taking the summation of the forces acting on $ABCDE$ to the *left*; in each case the summation is taken in a direction *opposite* to the positive direction of the degree of freedom. The reason for holding this rigorous requirement is that the relationship between the slope deflection method and the matrix displacement method will be described in subsequent chapters.

Substitution of M_1, M_2, M_5, and M_6 in Eq. (e) gives

$$EI\theta_B\left(-\frac{6}{196}\right) + EI\theta_C\left(-\frac{6}{100}\right) + EI\Delta\left[\frac{12}{14}\left(\frac{\Delta}{14}\right)\left(\frac{1}{14}\right) + \frac{12}{10}\left(\frac{\Delta}{10}\right)\left(\frac{1}{10}\right)\right] = 0$$

Writing the equation above together with Eqs. (a) to (c) shows the symmetry of the coefficients on the left side of the equal sign; thus,

$$EI\theta_B\left(\frac{4}{14}+\frac{4}{20}\right)+EI\theta_C\left(\frac{2}{20}\right) \qquad\qquad\qquad +EI\Delta\left(\frac{-6}{196}\right) \quad=+52.08$$

$$EI\theta_B\left(\frac{2}{20}\right) \qquad +EI\theta_C\left(\frac{4}{20}+\frac{4}{10}+\frac{4}{25}\right)+EI\theta_D\left(\frac{2}{25}\right)+EI\Delta\left(\frac{-6}{100}\right) \quad=+46.88$$

$$+EI\theta_C\left(\frac{2}{25}\right) \qquad\qquad +EI\theta_D\left(\frac{4}{25}\right) \qquad\qquad =-85.21$$

$$EI\theta_B\left(-\frac{6}{196}\right) +EI\theta_C\left(-\frac{6}{100}\right) \qquad\qquad +EI\Delta\left(\frac{44.928}{2744}\right)=0$$

(f) *Remainder of solution.* From this stage in the solution, the four equations above must be solved for θ_B, θ_C, θ_D, and Δ. These quantities are then substituted into the equations for M_1 through M_8, following which the four equilibrium checks and the four compatibility checks should be made. The compatibility checks would be (1) $\theta_{BF}=\theta_{BC}$; (2) $\theta_{CB}=\theta_{CG}$; (3) $\theta_{CG}=\theta_{CD}$; and (4) that Δ computed from BF equals Δ computed from CG. The completion of this solution illustrates no additional concepts and is therefore not shown.

Example 11.7.3

Analyze the rigid frame having two hinged supports previously analyzed in Example 11.6.1 (shown again in Fig. 11.7.4a). Use the slope deflection method.

Solution

(a) *Establish the displacement unknowns to be used in the solution.* Assuming that the minimum number of displacements is to be used, it is noted that joint rotations occur at all four joints, and in addition member BC may displace to the side. Thus, the displacement method must involve as a minimum five displacements (four angular and one linear). The displacement unknowns are shown in Fig. 11.7.4b.

(b) *Compute fixed-end moments.* These moments required to maintain zero slope and deflection at member ends may be obtained by using the formulas in Fig. 9.2.4 (clockwise moments positive); thus,

$$M_{F1}=M_{FAB}=-\frac{wL^2}{12}=-\frac{0.5(16)^2}{12}=-10.67\text{ ft-kips}$$

$$M_{F2}=M_{FBA}=+10.67\text{ ft-kips}$$

$$M_{F3}=M_{FBC}=-\frac{Wab^2}{L^2}=-\frac{40(8)(12)^2}{(20)^2}=-115.20\text{ ft-kips}$$

$$M_{F4}=M_{FCB}=+\frac{Wa^2b}{L^2}=+\frac{40(8)^2(12)}{(20)^2}=+76.80\text{ ft-kips}$$

(c) *Write the slope deflection equations.* Using Eqs. (9.2.7),

$$M_1=-10.67+\frac{4E(3I_c)}{16}\left(\theta_A-\frac{\Delta}{16}\right)+\frac{2E(3I_c)}{16}\left(\theta_B-\frac{\Delta}{16}\right)$$

$$M_2=+10.67+\frac{2E(3I_c)}{16}\left(\theta_A-\frac{\Delta}{16}\right)+\frac{4E(3I_c)}{16}\left(\theta_B-\frac{\Delta}{16}\right)$$

$$M_3=-115.20+\frac{4E(5I_c)}{20}\theta_B+\frac{2E(5I_c)}{20}\theta_C$$

$$M_4 = +76.80 + \frac{2E(5I_c)}{20}\theta_B + \frac{4E(5I_c)}{20}\theta_C$$

$$M_5 = \frac{4E(2I_c)}{16}\left(\theta_C - \frac{\Delta}{16}\right) + \frac{2E(2I_c)}{16}\left(\theta_D - \frac{\Delta}{16}\right)$$

$$M_6 = \frac{2E(2I_c)}{16}\left(\theta_C - \frac{\Delta}{16}\right) + \frac{4E(2I_c)}{16}\left(\theta_D - \frac{\Delta}{16}\right)$$

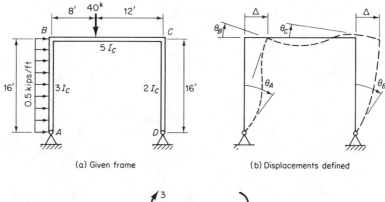

(a) Given frame

(b) Displacements defined

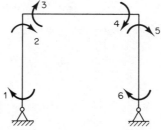

(c) End moment numbering system

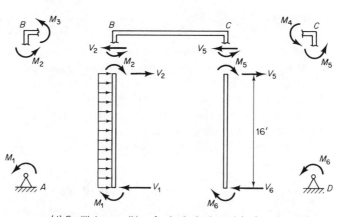

(d) Equilibrium conditions for θ_A, θ_B, θ_C, θ_D, and Δ of member BC

Figure 11.7.4 Rigid frame analyzed by the slope deflection method, Example 11.7.3.

(d) *Equilibrium equations.* The rotational equilibrium at joints A, B, C, and D, as shown in Fig. 11.7.4d, is satisfied by

$$M_1 = 0 \qquad \text{(a)}$$

$$M_2 + M_3 = 0 \qquad \text{(b)}$$

$$M_4 + M_5 = 0 \qquad \text{(c)}$$

$$M_6 = 0 \qquad \text{(d)}$$

For the "sidesway" the horizontal equilibrium of member BC may be used (member BC is presumed not to change in length). Horizontal force equilibrium of member BC requires (Fig. 11.7.4d)

$$V_2 + V_5 = 0 \qquad \text{(e)}$$

where

$$V_2 = -\frac{M_1 + M_2}{16} - \frac{0.5(16)}{2}$$

$$V_5 = -\frac{M_5 + M_6}{16}$$

which gives for Eq. (e),

$$-\frac{M_1}{16} - \frac{M_2}{16} - \frac{M_5}{16} - \frac{M_6}{16} = 4.0 \qquad \text{(f)}$$

Substitution of the end moment equations of part (c) into Eqs. (a) to (e) gives

$$\frac{3}{4}\,EI_c\theta_A + \frac{3}{8}\,EI_c\theta_B \qquad\qquad\qquad -\frac{9}{128}\,EI_c\Delta = \;+10.67$$

$$\frac{3}{8}\,EI_c\theta_A + \frac{7}{4}\,EI_c\theta_B + \frac{1}{2}EI_c\theta_C \qquad -\frac{9}{128}\,EI_c\Delta = +104.53$$

$$+\frac{1}{2}\,EI_c\theta_B + \frac{3}{2}EI_c\theta_C + \frac{1}{4}EI_c\theta_D - \frac{3}{64}\,EI_c\Delta = \;-76.80$$

$$+\frac{1}{4}EI_c\theta_C + \frac{1}{2}EI_c\theta_D - \frac{3}{64}\,EI_c\Delta = \quad 0.00$$

$$-\frac{9}{128}EI_c\theta_A - \frac{9}{128}EI_c\theta_B - \frac{3}{64}EI_c\theta_C - \frac{3}{64}EI_c\theta_D + \frac{15}{1024}EI_c\Delta = \;+4.00$$

Notice that the numerical coefficients on the left side of the equals signs form an array (matrix) symmetrical with respect to the diagonal, as expected for all structures. As will be discussed later in the chapters treating matrix methods, the symmetrical matrix above is known as the "global" *stiffness matrix* (the word "global" is inserted here in line with the popular use of the word in computer analysis, it means the stiffness matrix of the whole structure).

The solution of the foregoing set of equations for the displacements gives

$$EI_c\theta_A = +156.16 \text{ kip-ft}^2$$

$$EI_c\theta_B = +135.19 \text{ kip-ft}^2$$

$$EI_c\theta_C = -66.93 \text{ kip-ft}^2$$

$$EI_c\theta_D = +242.95 \text{ kip-ft}^2$$

$$EI_c\Delta = +2234.8 \text{ kip-ft}^3$$

(e) *Final end moments.* Substitution of the displacements back into the slope deflection equations of part (c) gives the end moments (answers in ft-kips),

$$M_1 = -10.67 + \frac{12}{16}\left(+156.16 - \frac{2234.8}{16}\right) + \frac{6}{16}\left(+135.19 - \frac{2234.8}{16}\right) = +0.01 \approx 0$$

$$M_2 = +10.67 + \frac{6}{16}\left(+156.16 - \frac{2234.8}{16}\right) + \frac{12}{16}\left(+135.19 - \frac{2234.8}{16}\right) = +13.48$$

$$M_3 = -115.20 + \frac{20}{20}(+135.19) + \frac{10}{20}(-66.93) \qquad\qquad = -13.47$$

$$M_4 = +76.80 + \frac{10}{20}(+135.19) + \frac{20}{20}(-66.93) \qquad\qquad = +77.47$$

$$M_5 = \frac{8}{16}\left(-66.93 - \frac{2234.8}{16}\right) + \frac{4}{16}\left(+242.95 - \frac{2234.8}{16}\right) \qquad = -77.48$$

$$M_6 = \frac{4}{16}\left(-66.93 - \frac{2234.8}{16}\right) + \frac{8}{16}\left(+242.95 - \frac{2234.8}{16}\right) \qquad = -0.01 \approx 0$$

These answers agree with the analysis results for this problem given in Fig. 11.6.3.

(f) *Concluding comments.* The reader is reminded that, as discussed in Section 9.2, the total end moment at any location consists of the portion M_F arising from holding all "joint" displacements to zero (there is still an elastic curve for each member) and the portion M' arising from displacement θ at the two ends and the relative transverse deflection Δ. In the equation of part (e), the fixed-end moments, if any, are combined with M' due to joint displacements.

11.8 MOMENT DISTRIBUTION METHOD

The moment distribution method as applied to beams has been presented in Chapter 10. No new concepts are required to apply the method to frames. The following example illustrates the basic application to a rigid frame having *no* unknown joint deflection.

Example 11.8.1
Analyze the rigid frame of Fig. 11.7.1a (also shown as Fig. 11.8.1) by the moment distribution method.

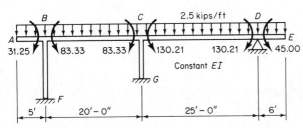

Fixed end moments on vertical members *BF* and *CG* are zero

Figure 11.8.1 Fixed end moments (ft-kips) for the rigid frame of Example 11.8.1 (clockwise moments taken as positive).

Solution

(a) *Fixed-end moments.* Since the moment distribution process always begins with the condition of having all displacements (slopes and deflections) zero, compute the fixed-end moments, as follows (taking clockwise end moments as positive):

$$M_{FBA} = +\frac{2.5(5)^2}{2} = +31.25 \text{ ft-kips}$$

$$M_{FBC} = -\frac{2.5(20)^2}{12} = -83.33 \text{ ft-kips}$$

$$M_{FCB} = +\frac{2.5(20)^2}{12} = +83.33 \text{ ft-kips}$$

$$M_{FCD} = -\frac{2.5(25)^2}{12} = -130.21 \text{ ft-kips}$$

$$M_{FDC} = +\frac{2.5(25)^2}{12} = +130.21 \text{ ft-kips}$$

$$M_{FDE} = -\frac{2.5(6)^2}{2} = -45.00 \text{ ft-kips}$$

$$M_{FBF} = F_{FFB} = M_{FCG} = M_{FGC} = 0$$

(b) *Distribution factors.* The distribution factors DF at joints B, C, and D are computed using Eq. (10.3.14) and the modified stiffness S_{CD} is used in accordance with Section 10.6. Whenever the moment at one end of a member is statically determinate, the modified stiffness should be used for the far end. Thus,

Joint	Member	Stiffness S		DF
B	BF	$4EI/L = 4EI/14 = 0.286EI$		0.588
	BC	$4EI/L = 4EI/20 = 0.200EI$		0.412
		$\sum S = 0.486EI$	$\sum DF = 1.000$	
C	CB	$4EI/L = 4EI/20 = 0.200EI$		0.278
	CG	$4EI/L = 4EI/10 = 0.400EI$		0.555
	CD	$3EI/L = 3EI/25 = 0.120EI$		0.167
		$\sum S = 0.720EI$	$\sum DF = 1.000$	
D	DC	$4EI/L = 4EI/25 = 0.160EI$		1.000
		$\sum S = 0.160EI$	$\sum DF = 1.000$	

(c) *Moment distribution process.* The operations of the moment distribution procedure appear in Table 11.8.1. Note that the carry-over factor used for all members is 0.5, the value for prismatic members.

(d) *Check on moment distribution.* Note that the joint rotations are obtained in the check process as by-products of the moment distribution, whereas in the slope deflection method these joint rotations are determined from solving the simultaneous equations before the final end moments are obtained.

TABLE 11.8.1 MOMENT DISTRIBUTION FOR EXAMPLE 11.8.1

Joint	B			C			D		F	G
End moment	BA	BF	BC	CB	CG	CD	DC	DE	FB	GC
DF	0	0.588	0.412	0.278	0.555	0.167	1.0	0	—	—
FEM BAL	+31.25 / 0	0. / +30.62	−83.33 / +21.46	+83.33 / +13.03	0 / +26.02	−130.21 / +7.83	+130.21 / −85.21	−45.00 / 0.	— / 0.	— / 0.
CO BAL		−3.83	+6.52 / −2.69	+10.73 / +8.86	+17.69	−42.60 / +5.32			+15.31	+13.01
CO BAL		−2.60	+4.43 / −1.83	−1.34 / +0.37	+0.75	+0.22			−1.91	+8.84
CO BAL		−0.11	+0.18 / −0.07	−0.91 / +0.25	+0.51	+0.15			−1.30	+0.37
CO BAL		−0.07	+0.12 / −0.05	−0.03 / +0.01	+0.02	0.			−0.05	+0.25
final carry-over to fixed supports →									−0.03	+0.01
Total M	+31.25	+24.01	−55.26	+114.30	+44.99	−159.29	+45.00	−45.00	+12.02	+22.48
Change	+24.01	+24.01	+28.07	+30.97	+44.99	−29.08	−85.21		+12.02	+22.48
−½ change	−6.01	−6.01	−15.48	−14.03	−11.24	+42.60	+14.54		−12.00	−22.50
Sum	+18.00	+18.00	+12.59	+16.94	+33.75	+13.52	−70.67		≈0.	≈0.
3EI/L	0.215EI	0.215EI	0.15EI	0.15EI	0.30EI	0.12EI	0.12EI			
$\theta = \dfrac{\text{sum}}{3EI/L}$	$+\dfrac{83.7}{EI}$	$+\dfrac{83.7}{EI}$	$+\dfrac{83.9}{EI}$	$+\dfrac{112.9}{EI}$	$+\dfrac{112.5}{EI}$	$+\dfrac{112.7}{EI}$	$-\dfrac{588.9}{EI}$			

(e) *Bending moment and shear diagrams*. The bending moment diagrams have been drawn in Fig. 11.7.1. The shear diagrams are omitted here, but would be needed in a practical situation for the design of the frame.

(f) *Horizontal reaction at support D*. Although there are no horizontal loads on this frame, there are horizontal reactions developed at supports D, F, and G. Horizontal reactions will always develop in a rigid frame. When the structure and the loading are both symmetrical, the sum of the horizontal reactions will be zero; otherwise, the frame will either have a side motion, usually referred to as *sidesway*, or must have a horizontal restraint to such motion. Each of the frames of Examples 11.7.2 and 11.7.3 has sidesway. The horizontal reaction at support D for this example was computed in Example 11.7.2 to be 9.331 kips directed to the left, thus preventing sidesway. The treatment of sidesway using moment distribution is given in the next section.

11.9 SIDESWAY ANALYSIS IN MOMENT DISTRIBUTION

The analysis of frames with sidesway using the procedure of moment distribution involves the superposition of two or more parts. The structure is first analyzed assuming that no sidesway occurs; that is, using one or more restraints to prevent any joint from change in position. Then successive analyses are made applying releasing forces in the direction opposite to the restraints (i.e., the restraints are removed). The following example illustrates the procedure for a structure having one degree of freedom in sidesway.

Example 11.9.1

Analyze the frame of Fig. 11.9.1a using the method of moment distribution.

Solution

(a) *Determine the degrees of freedom in sidesway*. On the basis of first-order assumption and zero axial deformation, it is apparent that the only freedom that the joints can change position is the shift of horizontal member $ABCD$ in the horizontal direction. Thus, the degree of freedom in sidesway is one and it may be represented by the horizontal movement Δ of member $ABCDE$.

(b) *General moment distribution procedure*. Prevent sidesway by imagining a horizontal restraint at the level of member $ABCDE$. This would be accomplished if there were a hinge at D instead of a roller. The analysis is made for the "no sidesway" case. This has been done in Example 11.8.1 with the results appearing in Table 11.8.1. Next, the magnitude of the horizontal restraint is determined by using statics, as shown in Fig. 11.7.2, where the horizontal restraint was determined to be 9.331 kips to the left. Now the 9.331 kips is applied horizontally to the right at the level of member $ABCDE$ to the frame of Fig. 11.9.1a (having the roller at D). The results of the two moment distribution analyses are added together. The details of this analysis for sidesway loading are given in the succeeding parts of this example.

(c) *Fixed-end moments*. For the analysis due to a horizontal force applied along member $ABCDE$ there are no vertical loads acting. The fixed-end moments are those due to Δ only with all θ's equal to zero, as shown in Fig. 9.2.3d, where the fixed-end

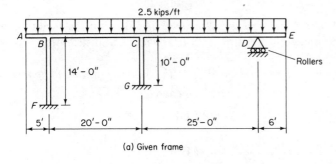

(a) Given frame

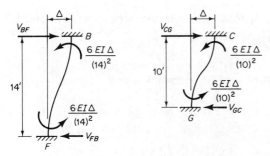

(b) Fixed end moments from sidesway Δ

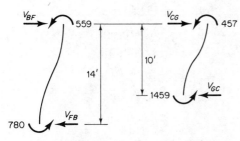

(c) Statics of members *BF* and *CG*

Figure 11.9.1 Rigid frame having one degree of freedom in sidesway, Example 11.9.1.

moments are given (clockwise positive) by

$$M_F = \frac{6EI\Delta}{L^2}$$

For a given Δ to the right on the given frame, the fixed-end moments are as shown in Fig. 11.9.1b. *Since there is no way of knowing what Δ corresponds to the applied horizontal force of 9.331 kips*, the sidesway problem is first solved by using an arbitrary Δ. In fact, the solution is usually simplest if an arbitrary fixed-end moment (say 10, 100, or 1000 ft-kips) is chosen for one of the members. In this case, *choose* for member *BF*,

$$M_{FBF} = M_{FFB} = -1000 \text{ ft-kips}$$

Then, because member *CG* is shorter, the fixed-end moments on member *CG* will be

$$M_{FCG} = M_{FGC} = -1000\left(\frac{14}{10}\right)^2 = -1960 \text{ ft-kips}$$

The 1000 and 1960 are relative values for which the analysis will be made.

(d) *Moment distribution for the sidesway loading.* The distribution factors and carry-over factors were computed in Example 11.8.1. The moment distribution procedure is shown in Table 11.9.1.

(e) *Determine horizontal force corresponding to the arbitrary fixed-end moments from Fig. 11.9.1c.*

$$V_{BF} = \frac{559 + 780}{14} = 95.6 \text{ kips}$$

$$V_{CG} = \frac{957 + 1459}{10} = 241.6 \text{ kips}$$

The horizontal force acting to the right for the moment distribution in Table 11.9.1 is

$$H = 95.6 + 241.6 = 337.2 \text{ kips}$$

The solution desired is for $H = 9.331$ kips. Thus, a proportion (9.331/337.2) of the end moments in Table 11.9.1 must be obtained and added to the no-sidesway analysis of Table 11.8.1.

(f) *Final solution for end moments.* Taking the no-sidesway results from Table 11.8.1 plus 0.02767 (i.e., 9.331/337.2) times the results from Table 11.9.1 gives

$$M_{BF} = +24.01 - 15.47 = +8.54 \text{ ft-kips}$$

$$M_{BC} = -55.26 + 15.50 = -39.76 \text{ ft-kips}$$

$$M_{CB} = +114.30 + 18.18 = +132.48 \text{ ft-kips}$$

$$M_{CG} = +44.98 - 26.48 = +18.50 \text{ ft-kips}$$

$$M_{CD} = -159.29 + 8.33 = -150.96 \text{ ft-kips}$$

$$M_{DC} = +45.00 + 0 = +45.00 \text{ ft-kips}$$

$$M_{FB} = +12.02 - 21.58 = -9.56 \text{ ft-kips}$$

$$M_{GC} = +22.48 - 40.37 = -17.89 \text{ ft-kips}$$

(g) *Joint rotations and sidesway.* The joint rotations θ_B, θ_C, and θ_D, as well as the sidesway Δ may be obtained by superposition in the same way as for the final end moments, that is, by adding the joint rotations in the check procedure of Table 11.8.1 and 0.02767 times those in the check procedure of Table 11.9.1. Moreover, the sidesway may be computed by applying the formula

$$M_F = \frac{-6EI\Delta}{L^2}$$

to either column BF or CG, wherein M_F is 0.02767 times the fixed-end moments used in the beginning of the moment distribution in Table 11.9.1. Thus, for column BF,

$$0.02767(-1000) = -\frac{6EI(\text{actual } \Delta)}{(14)^2}$$

$$\text{Actual } \Delta = +\frac{903.9}{EI} \quad \text{(to the right)}$$

For column CG,

TABLE 11.9.1 MOMENT DISTRIBUTION FOR EXAMPLE 11.9.1

Joint	B			C			D		F	G
End moment	BA	BF	BC	CB	CG	CD	DC	DE	FB	GC
DF	0	0.588	0.412	0.278	0.555	0.167	1.0	0	—	—
FEM	0	−1000	0	0	−1960	0	0	0	−1000	−1960
BAL		+588	+412	+545	+1088	+327	0	0		
CO			+272	+206					+294	+544
BAL		−160	−112	−58	−115	−34				
CO			−29	−56					−80	−57
BAL		+17	+12	+16	+31	+9				
CO			+8	+6					+8	+15
BAL		−5	−3	−2	−3	−1				
CO			−1	−1					−2	−1
BAL		+1	0	0	+1	0				
Total M	0	−559	+559	+656	−957	+301	0	0	−780	−1459
Change		+441	+559	+656	+1003	+301	0	0	+220	+501
$-\tfrac{1}{2}$ change		−110	−328	−280	−250	0	−150		−220	−502
Sum		+331	+231	+376	+753	+301	−150			
$3EI/L$		0.214EI	0.15EI	0.15EI	0.30EI	0.12EI	0.12EI			
$\theta = \dfrac{\text{sum}}{3EI/L}$		$\dfrac{+1547}{EI}$	$\dfrac{+1540}{EI}$	$\dfrac{+2507}{EI}$	$\dfrac{+2510}{EI}$	$\dfrac{+2508}{EI}$	$\dfrac{-1250}{EI}$		0	0

$$0.02767(-1960) = -\frac{6EI(\text{actual } \Delta)}{10^2}$$

$$\text{Actual } \Delta = \frac{+903.9}{EI} \text{ to the right} \quad \text{(check)}$$

Applying superposition to the joint rotations,

$$\theta_B = +\frac{83.8}{EI} + 0.02767\left(\frac{+1543}{EI}\right) = +\frac{126.5}{EI}$$

$$\theta_C = +\frac{112.7}{EI} + 0.02767\left(\frac{+2508}{EI}\right) = +\frac{182.1}{EI}$$

$$\theta_D = -\frac{588.9}{EI} + 0.02767\left(\frac{-1250}{EI}\right) = -\frac{623.5}{EI}$$

(h) *Correlation with the slope deflection method.* This problem with sidesway has been partially solved by the slope deflection method in Example 11.7.2 up to the establishment of the four simultaneous equations in part (e) of the solution for that example. Substituting the results for θ_B, θ_C, θ_D, and Δ, as by-products of the moment distribution method into those simultaneous equations gives

$$+61.44 + 18.21 \ + \ 0 \ \ - 27.67 \overset{?}{=} +52.08$$

$$+12.65 + 138.40 - 49.88 - 54.23 \overset{?}{=} +46.88$$

$$0 \ \ + 14.57 \ - 99.76 + \ 0 \ \ \overset{?}{=} -85.21$$

$$-3.87 - 10.93 \ + \ 0 \ \ + 14.80 \overset{?}{=} \ \ 0$$

Considering the three significant figures used for the distribution factors in the moment distribution method, the four equations above are indeed satisfied. Thus, the moment distribution method can be looked on as a numerical relaxation procedure of solving the simultaneous equations in the slope deflection method.

When the degree of freedom NPS in sidesway is more than one as in Fig. 11.9.2a, a systematic approach must be taken to perform the moment distribution in $(1 + \text{NPS})$ stages. In the first stage, all degrees of freedom in joint deflection are restrained by providing fictitious supports such as those at B and E in Fig. 11.9.2b. The moment distribution for the "no-sidesway" case is carried out, and the restraining forces R_{01} and R_{02} are obtained. In the second stage, one obtains a set of consistent fixed-end moments owing to an arbitrary joint deflection Δ_1, carries out the moment distribution, and computes the joint forces P_{11} and P_{21}, as shown in Fig. 11.9.2c, from the statics of the balanced member end moments. The third stage begins with a set of fixed-end moments consistent with an arbitrary joint deflection Δ_2 and the corresponding joint forces P_{12} and P_{22}, as shown in Fig. 11.9.2d, are computed from the statics of the member end moments after balancing by the moment distribution procedure. If there were more than two degrees of freedom in sidesway, there would be an additional stage similar to stages two and three for each additional sidesway degree of freedom. Each required member end moment in the given rigid frame of Fig. 11.9.2a equals the the sum of its value in Fig. 11.9.2b, an unknown ratio k_1 times its value in Fig. 11.9.2c, and an unknown ratio k_2 times its value in Fig. 11.9.2d. The unknown ratios k_1 and k_2 are determined from the conditions

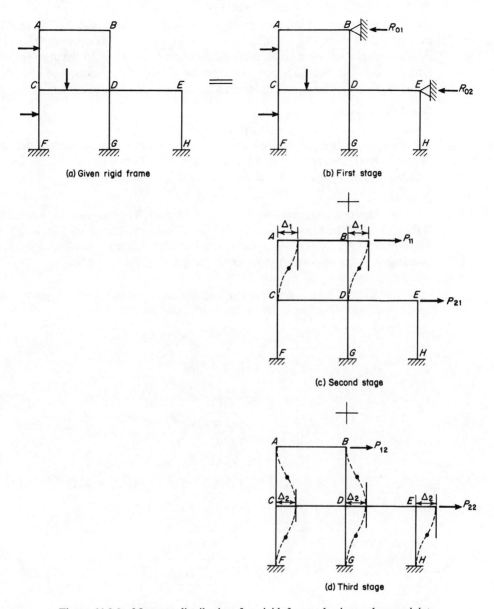

(a) Given rigid frame

(b) First stage

$+$

(c) Second stage

$+$

(d) Third stage

Figure 11.9.2 Moment distribution for rigid frames having unknown joint deflections.

$$k_1 P_{11} + k_2 P_{12} = R_{01}$$
$$k_1 P_{21} + k_2 P_{22} = R_{02}$$

The actual amounts of Δ_1 and Δ_2, if desired, may be obtained from the actual fixed-end moments in Fig. 11.9.2c and d, which are k_1 and k_2 times those fixed-end moments used at the beginning of moment distribution in the second and third stages, respectively. Similarly, each joint rotation in the given frame of Fig. 11.9.2a may be obtained as the sum of its value in Fig. 11.9.2b, k_1 times its value in Fig. 11.9.2c, and k_2 times its value in Fig. 11.9.2d, as these values are conveniently made available in the check procedure of each moment distribution table. Thus, in the moment distribution processes as outlined, not only the member end moments, but also the joint displacements (including rotations and deflections) may be obtained.

Example 11.9.2

Analyze by the moment distribution method, the rigid frame of Fig. 11.9.3a having two hinged supports, previously analyzed by the method of consistent deformation in Example 11.6.1 and by the slope deflection method in Example 11.7.3. Obtain the member end moments as well as the joint rotations and deflections.

Solution

(a) *Determine the degree of freedom in sideway.* The only possible joint deflection is the shift of the horizontal member BC in the horizontal direction, following the basic assumptions of first-order analysis and zero axial deformation. Thus, there will be two stages in the moment distribution solution: one for "no sideway" situation and one for the sideway of member BC.

(b) *Stage 1 (sideway prevented)—fixed-end moments.* For the given load and zero slope and deflection at the ends of each member,

$$M_{FAB} = -\frac{0.5(16)^2}{12} = -10.67 \text{ ft-kips}$$

$$M_{FBA} = +10.67 \text{ ft-kips}$$

$$M_{FBC} = -\frac{40(8)(12)^2}{(20)^2} = -115.20 \text{ ft-kips}$$

$$M_{FCB} = +\frac{40(8)^2(12)}{(20)^2} = +76.80 \text{ ft-kips}$$

$$M_{FCD} = M_{FDC} = 0$$

(c) *Distribution factors.* Adjusting for the hinges at A and D by using $3EI/L$ for S_{BA} and S_{CD},

Joint	Member	Stiffness, S	DF
A	AB	$4EI/L = 4E(3I_c)/16 = 0.75EI_c$	1.00
		$\Sigma S = \overline{0.75EI_c}$	$\Sigma DF = \overline{1.00}$
B	BA	$3EI/L = 3E(3I_c)/16 = 0.5625EI_c$	0.360
	BC	$4EI/L = 4E(5I_c)/20 = 1.000\ EI_c$	0.640
		$\Sigma S = \overline{1.5625EI_c}$	$\Sigma DF = \overline{1.000}$
C	CB	$4EI/L = 4E(5I_c)/20 = 1.000EI_c$	0.7273
	CD	$3EI/L = 3E(2I_c)/16 = 0.375EI_c$	0.2727
		$\Sigma S = \overline{1.375EI_c}$	$\Sigma DF = \overline{1.000}$
D	DC	$4EI/L = 4E(2I_c)/16 = 0.500EI_c$	1.00
		$\Sigma S = \overline{0.500EI_c}$	$\Sigma DF = \overline{1.00}$

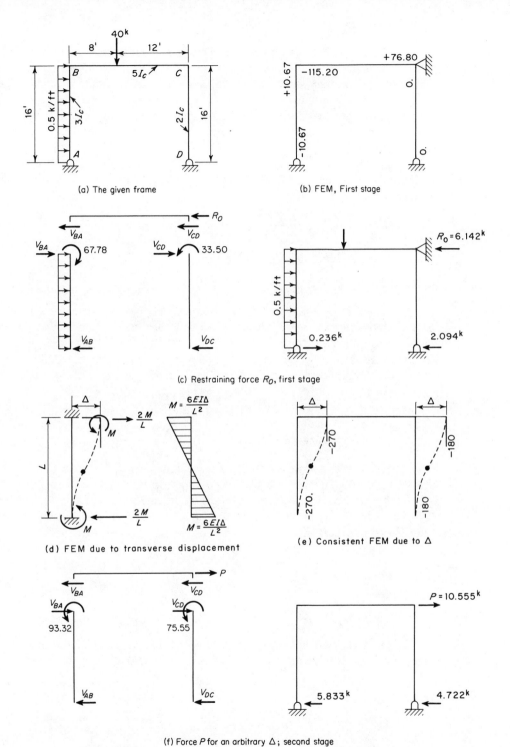

(a) The given frame

(b) FEM, First stage

(c) Restraining force R_O, first stage

(d) FEM due to transverse displacement

(e) Consistent FEM due to Δ

(f) Force P for an arbitrary Δ; second stage

Figure 11.9.3 The moment distribution method, Example 11.9.2.

377

(d) *Stage 1—moment distribution.* The moment distribution is given in Table 11.9.2. In this process joint A is balanced at the outset and $+5.33$ is carried over to joint B as a preliminary step to obtain the adjusted fixed-end moments under BA and CD. The process is speeded up by doing this prior to any balancing at joint B. Note that an accuracy of two decimal places is decided on for Table 11.9.2 for purpose of illustration for the learner; more accuracy than is usually necessary for practical design purposes.

Using the results of the moment distribution the restraining force at the top of the frame is determined by statics. The shears V_{BA} and V_{CD} (Fig. 11.9.3b) at the top of the vertical members are

$$V_{BA} = -\frac{0.5(16)}{2} - \frac{67.78}{16} = -8.236 \text{ kips}$$

$$V_{CD} = +\frac{33.50}{16} = +2.094 \text{ kips}$$

$$R_0 = -V_{BA} - V_{CD} = +8.236 - 2.094 = +6.142 \text{ kips} \quad \text{(to the left)}$$

(e) *Stage 2—sidesway moment distribution.* In order to remove the restraining force a relationship is needed between the end moments in the members and a horizontal force applied horizontally along member BC. The so-called "sidesway" moment distribution is begun by taking an arbitrary Δ and computing fixed-end moments arising therefrom according to the relationship shown in Fig. 11.9.3d. Thus,

$$M_{FAB} = M_{FBA} = \frac{-6E(3I_c)\Delta}{(16)^2} = -\frac{18EI_c\Delta}{(16)^2}$$

$$M_{FCD} = M_{FDC} = \frac{-6E(2I_c)\Delta}{(16)^2} = -\frac{12EI_c\Delta}{(16)^2}$$

From the above it is apparent that fixed-end moments in the ratio of 1.5 to 1 for members AB and DC, respectively, will be appropriate. Arbitrary values of -270 and -180 were chosen, as shown in Fig. 11.9.3e, and used in Table 11.9.3.

The horizontal force P corresponding to the arbitrary Δ is next computed. Referring to Fig. 11.9.3f,

$$V_{BA} = +\frac{93.32}{16} = +5.833 \text{ kips}$$

$$V_{CD} = +\frac{75.55}{16} = +4.722 \text{ kips}$$

$$P = V_{BA} + V_{CD} = 5.833 + 4.722 = 10.555 \text{ kips} \quad \text{(to the right)}$$

Since the restraining force R_0 to be removed is 6.142 kips, the moments from the stage 2 moment distribution must be such as to satisfy that

$$R_0 = kP$$

or

$$k = \frac{R_0}{P} = \frac{6.142}{10.555} = 0.5819$$

In other words, the moments from stage 2 must be multiplied by k to make the horizontal force at C equal to 6.142 kips.

TABLE 11.9.2 MOMENT DISTRIBUTION (STAGE 1), EXAMPLE 11.9.2

Joint	A	B		C		D
End moment	AB	BA	BC	CB	CD	DC
DF	1.0	0.360	0.640	0.7273	0.2727	1.0
FEM BAL	−10.67 +10.67	+10.67 +5.33	−115.20	+76.80	0. 0.	0. 0.
Modified FEM BAL	0.	+16.00 +35.71	−115.20 +63.49	+76.80 −55.86	0. −20.94	0.
CO BAL		+10.05	−27.93 +17.88	+31.74 −23.08	−8.66	
CO BAL		+4.15	−11.54 +7.39	+8.94 −6.50	−2.44	
CO BAL		+1.17	−3.25 +2.08	+3.70 −2.69	−1.01	
CO BAL		+0.48	−1.34 +0.86	+1.04 −0.76	−0.28	
CO BAL		+0.14	−0.38 +0.24	+0.43 −0.31	−0.12	
CO BAL		+0.06	−0.16 +0.10	+0.12 −0.09	−0.03	
CO BAL		+0.01	−0.04 +0.03	+0.05 −0.04	−0.01	
CO BAL		+0.01	−0.02 +0.01	+0.02 −0.01	−0.01	
Total M	0.	+67.78	−67.78	+33.50	−33.50	0.
Change	+10.67	+57.11	+47.42	−43.30	−33.50	0.
$-\frac{1}{2}$ change	−28.56	−5.33	+21.65	−23.71	0	+16.75
Sum	−17.89	+51.78	+69.07	−67.01	−33.50	+16.75
3EI/L	$0.5625EI_c$		$0.75EI_c$		$0.375EI_c$	
$\theta = \dfrac{\text{sum}}{3EI/L}$	$\dfrac{-31.80}{EI_c}$	$\dfrac{+92.05}{EI_c}$	$\dfrac{+92.09}{EI_c}$	$\dfrac{-89.35}{EI_c}$	$\dfrac{-89.33}{EI_c}$	$\dfrac{+44.67}{EI_c}$

TABLE 11.9.3 MOMENT DISTRIBUTION, ARBITRARY SIDESWAY (STAGE 2), EXAMPLE 11.9.2

Joint	A	B		C		D
End moment	AB	BA	BC	CB	CD	DC
DF	1.0	0.360	0.640	0.7273	0.2727	1.0
FEM BAL	−270 +270 ────►	−270 +135	0.	0.	−180 +90 ◄────	−180 +180
Modified FEM BAL	0.	−135 +48.60	0. +86.40	0. +65.46	−90 +24.54	.0
CO BAL		−11.78	+32.73 −20.95	+43.20 −31.42	−11.78	
CO BAL		+5.66	−15.71 +10.05	−10.47 +7.61	+2.86	
CO BAL		−1.37	+3.80 −2.43	+5.02 −3.65	−1.37	
CO BAL		+0.66	−1.82 +1.16	−1.21 +0.88	+0.33	
CO BAL		−0.16	+0.44 −0.28	+0.58 −0.42	−0.16	
CO BAL		+0.08	−0.21 +0.13	−0.14 +0.10	+0.04	
CO BAL		−0.02	+0.05 −0.03	+0.06 −0.04	−0.02	
CO BAL		+0.01	−0.02 +0.01	−0.02 +0.01	+0.01	
Total M	0	−93.32	+93.32	+75.55	−75.55	
Change	+270.00	+176.68	+93.32	+75.55	+104.45	+180.00
½ change	−88.34	−135.00	−37.78	−46.66	−90.00	−52.23
Sum	+181.66	+41.68	+55.54	+28.89	+14.45	+127.77
$3EI/L$	$0.5625EI_c$		$0.75EI_c$		$0.375EI_c$	
$\theta = \dfrac{\text{sum}}{3EI/L}$	$+\dfrac{323.0}{EI_c}$	$+\dfrac{74.10}{EI_c}$	$+\dfrac{74.05}{EI_c}$	$+\dfrac{38.52}{EI_c}$	$+\dfrac{38.53}{EI_c}$	$+\dfrac{340.7}{EI_c}$

(f) *Final moments for given frame* (Fig. 11.9.3a). The final end moments are obtained by adding k times the values for stage 2 to the values from stage 1, as follows:

End moment	AB	BA	BC	CB	CD	DC
Stage 1 (Table 11.9.2)	0	+67.78	−67.78	+33.50	−33.50	0
Stage 2 (0.5819 × Table 11.9.3)	0	−54.30	+54.30	+43.96	−43.96	0
Total *M* (ft-kips)	0	+13.48	−13.48	+77.46	−77.46	0

(g) *Final displacements.* The actual value of Δ may be computed from the fixed-end moment formula, using either the left or right column, as given in part (e). Thus,

$$\frac{18EI_c\Delta}{(16)^2} = 0.5819(270) \quad \text{or} \quad \frac{12EI_c\Delta}{(16)^2} = 0.5819(180)$$

from which

$$\Delta = \frac{2234.5}{EI_c} \text{ kip-ft}^3$$

The final joint rotations θ are the sums of those from stage 1 and 0.5819 times those of stage 2, as follows:

End rotation	AB	BA	BC	CB	CD	DC
Stage 1 (Table 11.9.2) (times $1/EI_c$)	−31.80	+92.05	+92.09	−89.35	−89.33	+44.67
Stage 2 (0.5819 × Table 11.9.3) (times $1/EI_c$)	+187.95	+43.12	+43.09	+22.41	+22.42	+198.25
Total θ (times $1/EI_c$)	+156.15	+135.17	+135.18	−66.94	−66.91	+242.92

The results check well also with the results obtained in Examples 11.6.1 and 11.7.3.

11.10 CHOICE OF METHOD

The choice of method depends on the calculating equipment that is available, the degree of simplicity or complexity of the frame, the purpose of the analysis as to whether it is a prelude to a trial design or a review of the final design, and the familiarity of the analyst with the chosen method.

Some pocket calculators may have provisions for solving up to 10 or 20 simultaneous equations and some of them have built-in "chips" for moment distribution processes. Thus, for a frame, say, up to a degree of freedom of five, ten, or fifteen, either the slope deflection method or the moment distribution method may be used, with the moment distribution method having some advantage because of the many sets of handwritten slope deflection equations to be processed in the slope deflection method.

When the degree of freedom reaches 30, 40, 50, or more, such as in the review of the final design of a multistory building subject to lateral loading, the high-speed digital computer should be used. In this case the slope deflection method can be processed in matrix notation so that the complete operations become automatic within a general computer program. The matrix formulation of the slope deflection method is now called the matrix displacement method, which is treated in later chapters. However, the emphasis is on the introduction of the basic concepts rather than detailed description of the computer algorithm or programming.

11.11 INFLUENCE LINES

The loading of multistory frames for maximum and minimum moments, shears, and axial forces, involves the use of influence lines, utilizing the concepts developed for statically indeterminate beams in Sections 8.4 and 8.5. Of course, in the rare case of a statically determinate frame the influence lines will be straight-line segments as discussed in Chapter 6. The need for bending moment and shear envelopes to obtain the forces for which to design is equally important for rigid frames. Most practical multistory rigid frames are "regular" in the sense of being comprised of beams and columns horizontally and vertically oriented, as shown in Fig. 11.11.1.

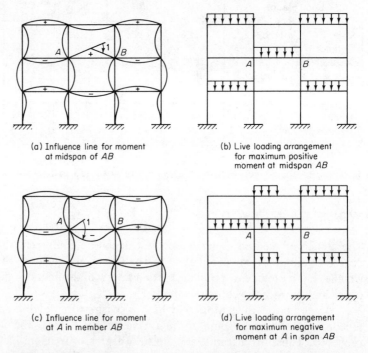

(a) Influence line for moment at midspan of *AB*

(b) Live loading arrangement for maximum positive moment at midspan *AB*

(c) Influence line for moment at *A* in member *AB*

(d) Live loading arrangement for maximum negative moment at *A* in span *AB*

Figure 11.11.1 Influence lines and loading for maximum moment in frames.

The qualitative influence lines for rigid frames are most conveniently obtained by using the Müller-Breslau principle as described in Section 8.5. The influence line for bending moment at midspan of member AB of Fig. 11.11.1a is obtained by imagining a hinge inserted at the midspan location and giving the two sides of the hinge a unit relative rotation and graphically sketching the deflection diagram that results. This influence line indicates that live loading should be placed as shown in Fig. 11.11.1b, the so-called "checkerboard" loading. The influence line for the bending moment at the left end of member AB is as shown in Fig. 11.11.1c. The partial span loadings indicated would be neglected in practical analysis situations for multistory frames, as would any partial span loadings indicated for maximum moment effects at locations between "fixed points"* and the ends of members such as AB. In obtaining the bending moment envelope for the practical design of members in multistory frames, one would rarely conform strictly to the pattern loading indicated by the influence line for more than the story above and the story below the floor in question. As discussed in Section 8.4, the effect of loads on spans more than two spans from the one in question may usually be considered negligible. The term "span" refers either to a horizontal or vertical member.

PROBLEMS

Statically Determinate Rigid Frames

11.1. For the statically determinate rigid frame of the accompanying figure, derive a formula for the horizontal deflection at the roller support by using the moment area method.

11.2. Solve Prob. 11.1 using the virtual work (unit load) method.

11.3. Determine (a) the horizontal deflection at the roller support, (b) the slope at joint B, and (c) the slope at A of the rigid frame of the accompanying figure. Use the moment area method.

11.4. Solve Prob. 11.3 using the virtual work (unit load) method.

11.5. For the statically determinate rigid frame of the accompanying figure: (a) sketch the bending moment diagram, (b) sketch the elastic curve, and (c) compute the horizontal deflections of joints C and D and the slope at A by the moment area method.

11.6. Solve Prob. 11.5 using the virtual work (unit load) method.

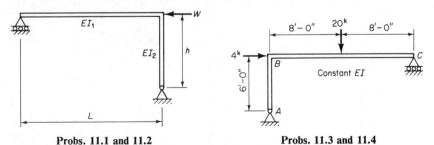

Probs. 11.1 and 11.2   Probs. 11.3 and 11.4

*See Section 8.4.

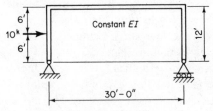

Probs. 11.5 and 11.6

11.7–11.10. For the rigid frame of the accompanying figure, compute the horizontal deflection at D by the moment area method. Sketch the elastic curve and label the displacement components.

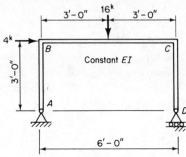

Probs. 11.7 and 11.11

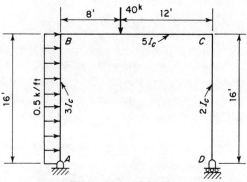

Probs. 11.8 and 11.12

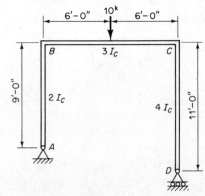

Prob. 11.9

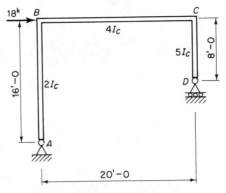

Prob. 11.10

11.11. Solve Prob. 11.7 using the virtual work (unit load) method.

11.12. Solve Prob. 11.8 by the virtual work (unit load) method.

11.13. For the frame of the accompanying figure, obtain the slope and vertical deflection at C using the virtual work (unit load) method.

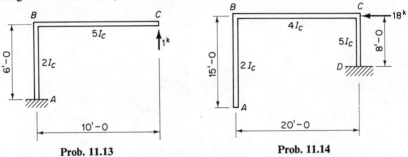

Prob. 11.13   **Prob. 11.14**

11.14. For the frame of the accompanying figure, compute the slope and horizontal deflection of the free end A using the virtual work (unit load) method. State whether the deflection is to the left or right and whether the slope is clockwise or counterclockwise.

11.15 and 11.16. For the frame of the accompanying figure, determine the rotation, horizontal deflection, and the vertical deflection of the free end D. Solve by the virtual work (unit load) method.

11.17. Solve Prob. 11.15 using the moment area method.

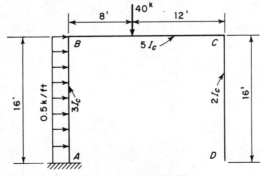

Probs. 11.15 and 11.17

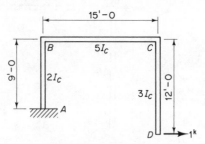

Probs. 11.16 and 11.18

11.18. Solve Prob. 11.16 using the moment area method.

Statically Indeterminate Rigid Frames

11.19. Analyze the rigid frame of the accompanying figure having two hinged supports by using the method of consistent deformation with the horizontal reaction at D as the redundant. Refer to Prob. 11.8 for the horizontal deflection at D for the "basic" statically determinate structure with the applied loads acting but with the redundant removed. Refer to Example 11.6.1 (Fig. 11.6.3) for the bending moment diagram and the elastic curve.

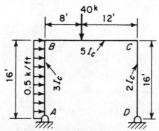

Prob. 11.19

11.20. Analyze the rigid frame of the accompanying figure having one fixed support and one hinged support using the method of consistent deformation with the horizontal reaction at A and the moment reaction at D as the redundants. Refer to Example 11.5.1 for the joint rotations and deflections of the "basic" statically determinate structure with the applied loads acting but with the redundants removed. Refer to Example 11.6.2 for the moment diagram and the elastic curve.

11.21. Analyze the rigid frame of the accompanying figure having two fixed supports using the method of consistent deformation with the horizontal, vertical, and moment reactions

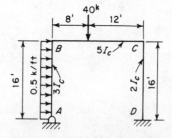

Probs. 11.20, 11.22, and 11.26

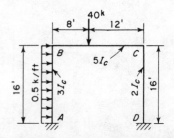

Probs. 11.21, 11.23, and 11.27

at *D* as the redundants. Draw the shear and bending moment diagrams, compute the joint displacements, and draw the final elastic curve.

11.22. Analyze the frame of Prob. 11.20 having one support hinged and one fixed using the slope deflection method.

11.23. Analyze the frame of Prob. 11.21 having two fixed supports using the slope deflection method.

11.24. Analyze the frame of the accompanying figure using the slope deflection method. Draw the final bending moment and shear diagrams for the structure, as well as the elastic curve.

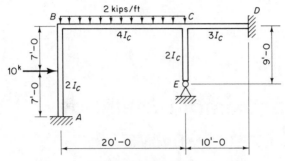

Probs. 11.24 and 11.25

11.25. Analyze the rigid frame of Prob. 11.24 using the method of moment distribution.

11.26. Analyze the rigid frame of Prob. 11.20 using the method of moment distribution.

11.27. Analyze the rigid frame of Prob. 11.21 using the method of moment distribution.

11.28. Analyze the rigid frame of the accompanying figure using the method of moment distribution. Solve the problem (a) neglecting any degree of freedom in sidesway; (b) considering such a degree of freedom.

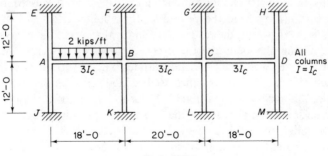

Prob. 11.28

[12]

Approximate Methods of Frame Analysis

12.1 CONCEPT OF APPROXIMATE ANALYSIS FOR STATICALLY INDETERMINATE STRUCTURES

Having the high-speed large-capacity digital computer available, the structural analyst may conclude that no use exists for approximate analysis. The classical methods discussed in earlier chapters and the matrix methods discussed in later chapters of this text certainly provide the procedure for obtaining fast "exact" solutions from the computer. However, the reader is reminded that structural analysis involves idealizing, or modeling, the loads and the structure on which they act. The loading may be estimated or code-prescribed in magnitude, but its pattern of distribution is set by the structural analyst. The structure itself is modeled such that the joints are taken to be "rigid," "hinged," or "semi-rigid." The material behavior must be idealized. The degree of precision (such as whether the effect of the product of axial force and deflection on the bending moment in a beam should be included) must be decided upon. Thus, even though the computer may be made to do thousands of calculations to arrive at an "exact" solution consistent with all input data, the results cannot be considered exact with respect to what the actual structure will experience.

Many procedures of approximate analysis have been developed over the years, originally because without the present-day computers such approximate procedures were a necessity. However, a general knowledge of certain approximate analysis procedures is not only useful in preliminary design, but also helps the analyst or designer to develop an "intuitive feeling" for how the structure will respond under a given load.

From the discussion of statically indeterminate structures in Chapters 7 to 11, the reader is reminded that such an analysis *cannot* be made without having the relative stiffnesses of the members. Thus, in the process of structural design the first step is to estimate the member sizes or at least the relative stiffnesses. The structural analysis that follows may be "exact"; however, the member sizes may change from the original estimate as a result of the ensuing design, leaving the analyst with an "exact" analysis for the *wrong* structure. The process is iterative involving successive revisions of member sizes followed by structural analysis. Certainly, approximate analysis may be sufficient or even desirable at least for preliminary design.

In this chapter, several useful approximate methods of continuous beam and plane frame analysis are described.

12.2 APPROXIMATE ANALYSIS OF CONTINUOUS BEAMS

In many framed structures a portion containing several beam spans of a story level may be idealized as a continuous beam for approximate analysis. Structures are frequently dissected into idealized portions for which the analysis can be made. After deciding that the model will be a continuous beam, the analyst may next realize that any loaded span within the continuous beam should have end moment restraint more than the zero value of a simple support but less than the value for full fixity, as shown in Fig. 12.2.1. The moment diagrams for continuous beams up to five equal spans under uniform loading are shown in Fig. 12.2.2. Note that for two and three spans the support moment is larger than that for a beam with two fixed ends, understandable if one realizes that such higher value at one support is the result of zero end restraint at the opposite support. For a typical interior span having restraint at both ends, the

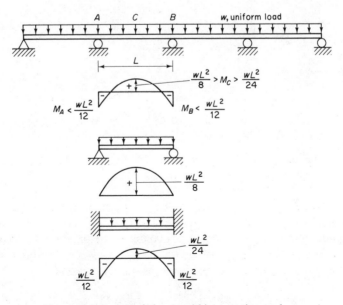

Figure 12.2.1 A loaded span within a continuous beam.

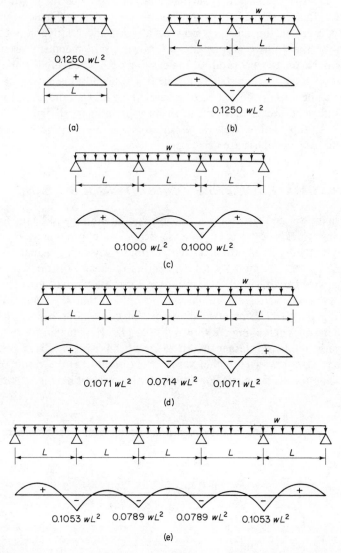

Figure 12.2.2 Moment diagrams for uniformly loaded continuous beams of equal spans.

use of an end moment equal to $\frac{1}{11}wL^2$ is generally valid considering the effect of various live load patterns. A moment of $\frac{1}{10}wL^2$ is generally used for the first interior support. Use of such coefficients is common for preliminary design and is explicitly permitted for final design in certain situations by some design codes* for "buildings of usual types of construction, spans, and story heights."

*See *Building Code Requirements for Reinforced Concrete (ACI 318–83)*, American Concrete Institute, Detroit, Mich., 1983 (specifically, Section 8.3).

Even when the idealized portion of the structure includes the upper and lower columns monolithically attached to the floor beams in question, as shown in Fig. 12.2.3, the values of end moments on the beam spans are not drastically affected and coefficients described earlier for continuous beams over knife-edge supports may commonly be used for approximate analysis.

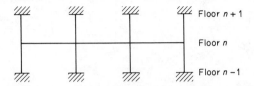

Figure 12.2.3 Idealization for analysis of a floor in a multistory building frame.

12.3 APPROXIMATE ANALYSIS OF ONE-STORY RIGID FRAMES UNDER VERTICAL LOAD

The one-story rigid frame is a structure that could without too much difficulty be analyzed using the methods discussed in Chapter 11. Again, however, the behavior of one-story frames provides insight into the behavior of multistory frames, whose analysis requires considerably more effort. The approximate analysis of one-story frames is treated by Salvadori and Levy [1].

Consider the two one-story frames of Fig. 12.3.1, one with hinged bases and the other with fixed bases. The stiffness of the beam can be expressed by $K_B = I_B/L_B$ and that of the column, by $K_C = I_C/L_C$. Let γ be the ratio of the beam stiffness to the column stiffness so that

$$\gamma = \frac{K_B}{K_C} \qquad (12.3.1)$$

If the beam is very stiff compared with the column, then each column gives only a small restraint at the beam end, and the end moment is close to zero. On the other hand, if the column is very stiff compared with the beam, the beam has almost fixed ends, and the end moment should be near the fixed-end moment.

An analysis by the slope deflection method described in Chapter 11 gives the following coefficients in terms of wL_B^2 for the magnitude of the negative bending moment at the beam ends:

$\gamma = K_B/K_C$	0	1	2	3	4	∞
Beam end moment (hinged base)	0.0833	0.0500	0.0357	0.0278	0.0227	0
Beam end moment (fixed base)	0.0833	0.0556	0.0417	0.0333	0.0278	0

From the tabulation above, one can also see that the columns with fixed bases provide more restraint to the beam ends than the columns with hinged bases.

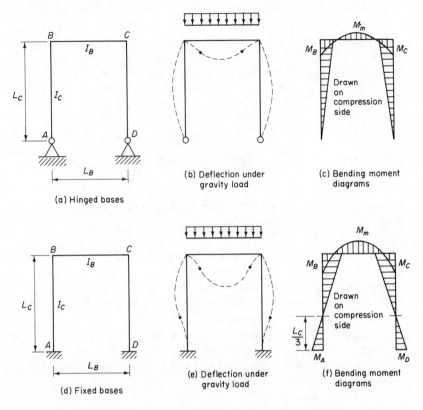

(a) Hinged bases

(b) Deflection under gravity load

(c) Bending moment diagrams

(d) Fixed bases

(e) Deflection under gravity load

(f) Bending moment diagrams

Figure 12.3.1 Hinged base and fixed base one-story rigid frames.

It is also seen from Fig. 12.3.1f that the inflection point in the column lies at the lower one-third point of the column for any combination of relative stiffnesses. Approximate analysis for one-story frames usually involves assumptions regarding the location of the inflection point. Locating the inflection point in each column as well as in the beam makes the structure statically determinate.

When multibay single-story frames are to be analyzed for gravity loads, one may assume [1] that for all practical purposes the tops of the intermediate columns *do not rotate* and *do not move laterally*. Thus, as shown in Fig. 12.3.2, sidesway may be neglected in this gravity load situation.

The student of approximate analysis of frames under gravity loading is referred to Johnson, Bryan, and Turneaure [2, pp. 530–545], where the slope deflection method is used to develop many useful formulas. This was a necessity before the use of either the digital computer or even the method of moment distribution. Today (1983) such approaches to approximate analysis are still valid as tools for preliminary design and to prevent obscuring the fact that in most structural frames the effect of a load at any location is a local one affecting only the two, or at most three, adjacent spans significantly.

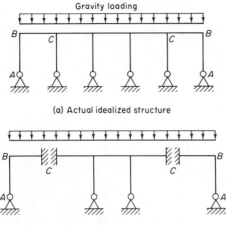

(a) Actual idealized structure

(b) Simplified structure for approximate analysis

Figure 12.3.2 Approximate analysis of symmetrical one-story frames under gravity loading.

12.4 APPROXIMATE ANALYSIS OF SIMPLE PORTAL FRAMES

Building frames must support, in addition to the usual gravity dead and live loads, the horizontal loads caused by wind (or the statical equivalent of earthquake forces). The idealizing, or modeling, of the loads was discussed in Sections 1.4 and 1.5. In the design of framed structures, either as rigid frames with monolithic joints or as beam and column structures interconnected with simple or semi-rigid connections, the lateral loads of wind or earthquake may be taken by (1) diagonal bracing members whose only purpose is to accommodate such loads, or (2) by the interaction in flexure of the beams and columns rigidly or semi-rigidly connected. Many structures have the beams and columns designed for gravity loads only and the bracing elements separately added for lateral load.

One of the earliest "rigid frames" was the bridge "portal" or the opening between the two structures through which the vehicles pass (see Fig. 6.10.1). On the early bridges there were trusses on each side as the main structures of the bridge. These two trusses were linked at the bottom by floor beams and the roadway. At the top, above the traffic, was a truss system to tie the two vertical trusses together and to resist the incidental forces (wind or traffic) that might act perpendicular to the bridge. In the inclined plane containing the two end diagonals on both sides of the traffic, a shallow truss-like assembly was usually provided, as in Fig. 12.4.1. This portal truss acts like a stiff girder so that the interaction of the end posts (acting as columns) of the side trusses with the transverse truss (acting as the girder) should be analyzed as a rigid frame (commonly called a portal frame), or at least as a "composite" structure where the end posts are subject to bending moments but all the portal truss members are considered to be carrying only axial forces. A similar situation occurs frequently in an industrial building (so-called mill building) where a truss supported on columns is used to support the long-span roof. In order to provide stiffness to resist lateral load,

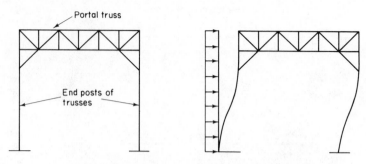

Figure 12.4.1 Portal frame and lateral loading.

the truss must be attached to stiff columns to form a "mill bent" that will act like a rigid frame.

When the portal frame of a bridge or a mill bent of an industrial building is to be analyzed, the restraint at the base of the columns is usually assumed to be fixed (and also designed to be as nearly fixed as practical). If the bases of the columns are assumed fixed, the structure will be statically indeterminate to the third degree. If the bases are assumed hinged, the structure will be statically indeterminate to the first degree. Any approximate method requires assumptions equal in number to the degree of statically indeterminacy in order to eliminate the statical indeterminacy. One of the earliest treatments of single- and multiple single-story portal frames is that of Ketchum [3].

For single-story portal frames or mill bents the approximate analysis for lateral load involves the following assumptions or observations:

1. The columns at the windward and leeward side are identical and deflect identically.
2. The position of the inflection point in the columns is estimated somewhere between midheight (for fixed base columns) and the bottom (for hinged base columns). Note that the theoretical location of the inflection point *is* at the midheight of the column if the girder is infinitely stiff.
3. The girder (or truss) deforms such that there is a point of inflection at midspan. Note that this location of the inflection point in the girder is theoretically correct as well.
4. The lateral force is divided at the inflection point between the two columns equally for a single bay bent supported by equal-size columns.

Since the single-story portal frame has column base restraint that is never precisely determinable in actual structures, no analysis can be "exact" even if performed by a computer; thus, an approximate analysis is in fact satisfactory for design purposes.

Example 12.4.1

Determine the shear and bending moment diagrams for the portal frame of Fig. 12.4.2a by an approximation analysis.

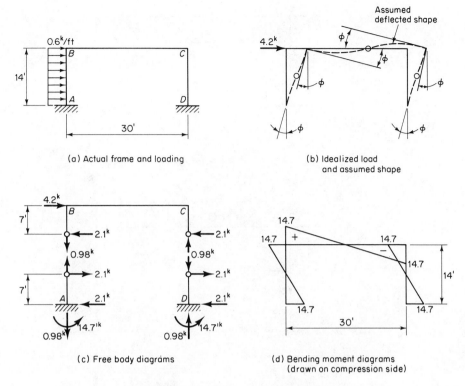

Figure 12.4.2 Portal frame of Example 12.4.1.

Solution

(a) *Idealized load and assumed deflected shape.* In order to convert the statically indeterminate frame into a statically determinate one, the columns are assumed to deform identically with inflection points at midheight, as in Fig. 12.4.2b. This inflection point location will be correct as long as the base restraint is such that the slope ϕ at the base equals the joint rotation ϕ at the top of columns. There would also be an inflection point at the midspan of *BC*. In portal analysis the uniformly distributed (or other distribution) wind load is usually taken as concentrated loads acting at the story levels; in this case one-half (4.2 kips) of the lateral load is assumed to act at *B*.

(b) *Approximate analysis for free-body, shear, and bending moment diagrams.* The free-body diagrams are given in Fig. 12.4.2c. The shear in the columns is constant at 2.1 kips in each column, while in the beam the shear is a constant 0.98 kip. The bending moment diagrams are plotted on the compression side in Fig. 12.4.2d. It is entirely feasible to treat the column as acted upon by uniform loading, although such treatment does not affect the maximum moment or the maximum shear.

Example 12.4.2

Determine the reactions for the mill bent given in Fig. 12.4.3a by an approximate analysis. Also show the general procedure for finding the forces in the truss members.

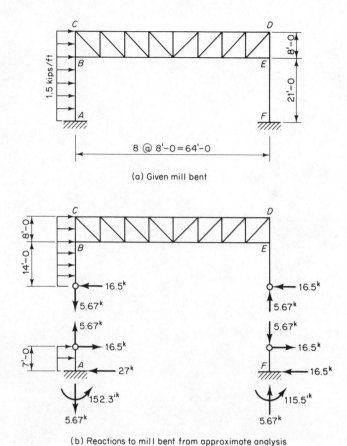

(a) Given mill bent

(b) Reactions to mill bent from approximate analysis

Figure 12.4.3 Approximate analysis of a mill bent, Example 12.4.2.

Solution

(a) *Assumptions to obtain statically determinate structure.* With a relatively stiff truss as the horizontal system, little rotation is expected at *B* (Fig. 12.4.3a), where the lower chord is assumed to be hinged to the column. If points *B* and *C* are deflected laterally approximately equal amounts because of the stiffness of the truss, segment *BC* remains vertical (i.e., no joint rotation occurs at *B* or *C*). At the base, full fixity is not attainable in practice. Thus, for this mill bent the point of inflection in the column will be closer to the base than midheight. A common assumption is to locate the inflection point at one-third the distance from the base to the lower chord of the truss. The other assumption of dividing equally between the columns the total lateral force up to the inflection point level is used in part (b).

(b) *Determine the base reactions.* The total horizontal shear above the column points of contraflexure is divided equally between the columns. The total shear at the inflection points is

$$\text{Shear} = 1.5(14 + 8) = 33 \text{ kips}$$

The horizontal reaction at A is

$$H_A = \frac{33}{2} + 7(1.5) = 27 \text{ kips}$$

The uniform loading on the left column below the inflection point makes the horizontal reactions at A and F unequal. At point F,

$$H_F = \frac{33}{2} = 16.5 \text{ kips}$$

The base moments are

$$M_A = 16.5(7) + \frac{1}{2}(1.5)(7)^2 = 152.3 \text{ ft-kips}$$

$$M_F = 16.5(7) = 115.5 \text{ ft-kips}$$

The axial force reactions may be obtained by taking the whole frame as free body, or by taking the portion of the frame above the inflection points as free body; thus, using the former,

$$V_F = \frac{1.5(29)(14.5) - 152.3 - 115.5}{64} = 5.67 \text{ kips} \quad \text{(compression)}$$

$$V_A = 5.67 \text{ kips} \quad \text{(tension)}$$

The reactions are shown in Fig. 12.4.3b.

(c) *Determine the force in the top chord member U_0U_1 in the first panel of the truss.* The free-body diagram is given in Fig. 12.4.4. Taking summation of moments about L_1 gives

$$U_0U_1(8) + 1.5(8)(4) - 1.5(14)(7) + 16.5(14) - 5.67(8) = 0$$

$$U_0U_1 = -10.83 \text{ kips} \quad \text{(compression)}$$

The force in diagonal U_0L_1 is obtained using vertical force equilibrium,

$$0.707U_0L_1 + 5.67 = 0$$

$$U_0L_1 = -8.02 \text{ kips} \quad \text{(compression)}$$

Note that when the horizontal wind force is from the opposite direction the diagonal U_0L_1 will be in tension (8.02 kips); thus, it must be designed to accommodate either force.

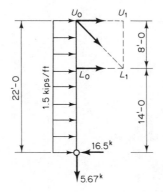

Figure 12.4.4 Free-body diagram of column and truss segment in the mill bent for Example 12.4.2.

12.5 APPROXIMATE ANALYSIS OF MULTISTORY FRAMES BY THE PORTAL METHOD

The general approach used in the approximate analysis of simple portal frames due to lateral loading can be extended to multistory frames. This method, generally known as the *portal method*, has been extremely useful either for preliminary design only or for both preliminary and final design. The method originated with Albert Smith [4] in 1915, and has been treated in detail by Sutherland and Bowman [5].

The assumptions for the portal method as applied to multistory multibay frames of equal span are as follows:

1. The position of the point of inflection is at midheight of each column.
2. The girders deform such that there is a point of inflection at midspan of each girder.
3. The total horizontal shear is proportioned among the columns so that each interior column carries twice the shear of an exterior one.

Assumption 3 arises from considering multibay frame behavior as the sum of those from the individual bays; thus, an interior column gets a shear contribution from each adjacent bay while an exterior column only gets a shear contribution from one bay.

The usefulness of the method has been considered satisfactory for most common multistory frames up to, say, 25 stories in height [6]. Prior to the availability of the computer to analyze structures having a large number of degrees of freedom the basic portal method or one of its modifications was the only available procedure for such analysis and has been widely used. The alternative *cantilever method*, discussed in the next section, has certain advantages for high, narrow buildings.

Example 12.5.1

Evaluate the internal forces (moment, shear, and axial force) in the frame of Fig. 12.5.1 using the portal method.

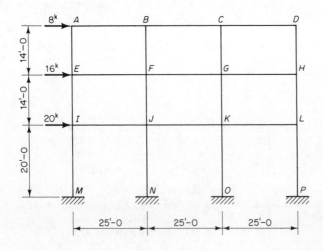

Figure 12.5.1 Multistory frame subject to lateral loading, Example 12.5.1.

Solution

(a) *Forces in the top story*. Assuming inflection points (idealized as hinges) at the midheight of the upper story, the free body of that story is given in Fig. 12.5.2. First the total shear of 8 kips is proportioned among the columns with an interior column taking twice that at an exterior column. Then one may note that equilibrium at a joint requires the sum of the column end moments in one direction to equal the sum of the beam end moments in the opposite direction.

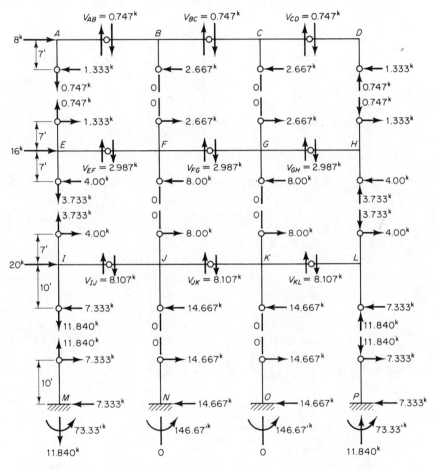

Figure 12.5.2 Results of approximate analysis for frame of Example 12.5.1; portal method.

At joint A,

$$V_{AB}(12.5) = 1.333(7); \qquad V_{AB} = 0.747 \text{ kip}$$

At joint B,

$$V_{BC}(12.5) + 0.747(12.5) = 2.667(7); \qquad V_{BC} = 0.747 \text{ kip}$$

It is obvious that the axial forces in all interior columns are zero so that the axial force in either exterior column can be obtained by taking the whole upper 7-ft section of the frame as the free body; thus,

$$\text{Axial force in exterior column} = \frac{8(7)}{75} = 0.747 \text{ kip}$$

(b) *Forces at story level EFGH.* The total shear at the assumed inflection points below this story is 16 kips plus the 8 kips acting on the story above. Again, assuming that the interior columns take twice as much shear as exterior columns, the column shears are 4 kips, 8 kips, 8 kips, and 4 kips, respectively, on columns EI, FJ, GK, and HL.

To determine the beam shear in member EF, examine the equilibrium of joint E,

$$4(7) + 1.333(7) = V_{EF}(12.5); \qquad V_{EF} = 2.987 \text{ kips}$$

At joint F,

$$8(7) + 2.667(7) = 2.987(12.5) + V_{FG}(12.5); \qquad V_{FG} = 2.987 \text{ kips}$$

Again, the axial forces in the interior columns are zero and the axial force in either exterior column can be found by taking the whole 21-ft section above the inflection points as the free body. Thus,

$$\text{Axial force in exterior column} = \frac{8(21) + 16(7)}{75} = 3.733 \text{ kips}$$

(c) *Girder moments.* The girders are assumed to deform in double curvature with maximum bending moments at the girder ends, equal in magnitude to the shear times the distance from midspan to the column.

(d) *Remainder of forces.* The entire structure may be considered separated into free bodies at the assumed hinges. All the hinge shears and column axial forces, as well as base end moments are given in Fig. 12.5.2.

(e) *Check for correctness of computation.* Note that in the process of moving down the frame from the top, there is a numerical check in that the axial force in the exterior column (compression in the leeward side and tension in the windward side) can be obtained (1) first by using $\Sigma F_y = 0$ for the free body containing half of the girder span and the exterior column between the two points of inflection, and (2) second by using $\Sigma M = 0$ for the free body containing the entire frame above each level of inflection points. Thus for the lowest story, referring to Fig. 12.5.2,

$$\text{Axial force in exterior column} = 3.733 + 8.107 = 11.840 \text{ kips}$$

Also, referring to Fig. 12.5.1,

$$\text{Axial force in exterior column} = \frac{8(14 + 14 + 10) + 16(14 + 10) + 20(10)}{75}$$

$$= 11.840 \text{ kips} \quad \text{(check)}$$

(f) *Discussion.* If the girder spans are unequal, the total shear should be divided in the ratio of $L_1/2$, $(L_1 + L_2)/2$, ... to $L_n/2$. In this way, the axial forces in the interior columns will still be zero and the entire bending moment to each inflection point level is resisted by the pair of axial forces in the windward and leeward columns.

12.6 APPROXIMATE ANALYSIS OF MULTISTORY FRAMES BY THE CANTILEVER METHOD

The cantilever method is an appropriate approximation for tall buildings of relatively small lateral dimension. Essentially, the method assumes the building to be a vertically positioned cantilever beam fixed at the base. The horizontal loads are assumed to cause compression or tension in the vertical members in proportion to their distance from the center of gravity of all the columns. First presented by A. C. Wilson [7] in 1908, the method provides a satisfactory approximation for narrow buildings of 25 to 35 stories or less.

The assumptions of the cantilever method are as follows:

1. The position of the point of inflection is at the midheight of each column.
2. The girders deform such that there is a point of inflection at midspan of each girder.
3. The compressive or tensile unit stresses in the columns vary as the distances of the columns from the center of gravity of the bent. Since it is usually assumed that all columns in a story are of equal area, it means that the total column axial forces will vary as the distances from the center of gravity of the entire group of columns.

It is noted that the first two assumptions above are identical to those for the portal method. The last one involves an assumed distribution of column axial forces, whereas the portal method involves zero axial forces in the interior columns.

Example 12.6.1

Evaluate the internal forces (moment, shear, and axial force) in the frame of Fig. 12.5.1 using the cantilever method.

Solution

(a) *Axial forces in the top-story columns.* The whole frame from the top to the assumed point of inflection in the top-story columns is examined as a free body in Fig. 12.6.1. The columns are assumed to be the same size; thus, if the magnitude of the tensile and compressive axial forces in the two interior columns is called by the unknown X, then the magnitude of the axial forces in the exterior columns is equal to $37.5/12.5$ times X, or $3X$. The exterior columns are 37.5 ft from the center of grav-

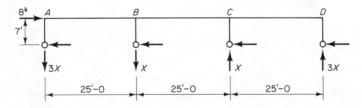

Figure 12.6.1 Free-body diagram of top story of frame for Example 12.6.1; cantilever method.

ity, whereas the interior columns are only 12.5 ft from it. Next, write the moment equilibrium equation with respect to the assumed hinge location in the left exterior column AE. Thus,

$$8(7) + X(25) - X(50) - 3X(75) = 0$$

$$X = 0.224 \text{ kip}$$

(b) *Shears in the top-story beams and columns.* Utilizing the second assumption of the method, the inflection points in the beams are at midspan. For the free body of the joint at A (Fig. 12.6.2) bounded by the assumed inflection points (hinges) 12.5 from A in AB and at midheight of column AE, moment equilibrium about the hinge in AB gives

$$V_{AE}(7) = 0.672(12.5); \qquad V_{AE} = 1.20 \text{ kips}$$

and from vertical force equilibrium,

$$V_{AB} = 0.672 \text{ kip}$$

For the free body of the joint at B (Fig. 12.6.2) bounded by the assumed inflection points (hinges) in spans AB, BC, and BF, moment equilibrium about the hinge in BC gives

$$V_{BF}(7) - 0.224(12.5) - 0.672(25) = 0; \qquad V_{BF} = 2.80 \text{ kips}$$

and from vertical force equilibrium,

$$V_{BC} = 0.672 + 0.224 = 0.895 \text{ kip}$$

Joints C and D would be treated in a similar manner to obtain the other shears in the top-story members. However, because of the symmetry of this frame, the shears in columns CG and DH are 2.80 kips and 1.20 kips, respectively.

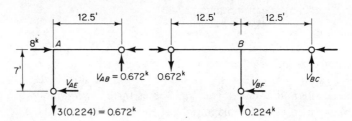

Figure 12.6.2 Free body diagrams of two segments of top story, Example 12.6.1; cantilever method.

(c) *Other forces in the frame.* Each story in succession is treated in a manner the same as the top story. The resulting forces are shown in Fig. 12.6.3 for the entire multistory frame.

(d) *Check for correctness of computation.* Note that in the process of moving down the frame from the top, there is a numerical check in that the horizontal shears in all the columns in a particular story should tally with the total horizontal force acting on the frame between the top and that story. For instance, taking vertical equilibrium of joint I in Fig. 12.6.3 as a free body,

$$\text{Shear in girder } IJ = 10.656 - 3.360 = 7.296 \text{ kips}$$

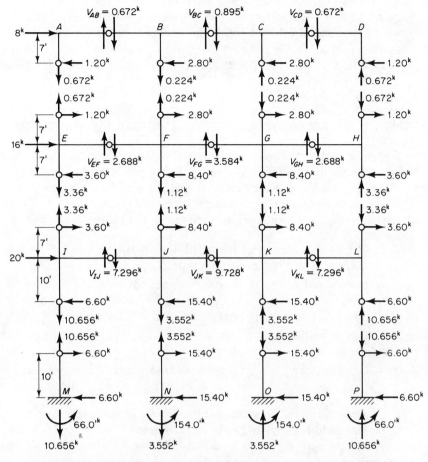

Figure 12.6.3 Results of approximate analysis for frame of Example 12.6.1; cantilever method.

Then applying moment equilibrium to this free body gives

$$\text{Shear in column below joint } I = \frac{7.296(12.5) - 3.60(7)}{10} = 6.60 \text{ kips}$$

Similarly, using $\sum F_y = 0$ for joint J gives

$$\text{Shear in girder } JK = 7.296 + 3.552 - 1.12 = 9.728 \text{ kips}$$

and, using $\sum M = 0$ for joint J gives

$$\text{Shear in column below joint } J = \frac{9.728(12.5) + 7.296(12.5) - 8.40(7)}{10}$$

$$= 15.40 \text{ kips}$$

Then the sum of the shears in the four columns below joints I, J, K, and L is $2(6.60 + 15.40)$ or 44 kips, exactly the total horizontal force above the columns at this story.

(e) *Discussion.* If the girder spans are unequal, or if the column areas are un-equal, then the centroid of the column areas will have first to be located. Then the axial forces in the columns will be proportional to the products of the column area and the distance to the centroid. A moment equation applied to the whole frame up to each level of inflection points is used to determine the axial forces in the columns at each level.

12.7 CHOICE OF METHOD AND INACCURACIES OF METHODS

Either the cantilever or the portal method is adequate for preliminary design of multi-story frames up to about 25 stories, with the portal method preferable for the wide multipanel frames and the cantilever method preferable for the tall narrow frames. These methods quite likely involve no greater approximation than that arising from the inadequacy of any analysis method to account for the following:

1. Part of the wind shear is carried directly by the building walls and partitions rather than all by the skeleton frame.
2. Floor construction is usually rigid enough so that it prevents a given frame from acting independently of adjacent parallel frames; in other words, a given frame may not merely carry the wind load tributary to it.
3. Unsymmetrical wind forces on symmetrical and unsymmetrical frames may give rise to torsional effects that are neglected.
4. Connections are neither simple (i.e., transmit zero moment) nor rigid, thus giving rise to end moments that are different from those obtained in an analysis.

Since the digital computer is relatively inexpensive to use, there is little interest in making modifications to the basic cantilever or portal methods. Use the simplest approximate method; then if a more detailed computation seems desirable, make the analysis using matrix methods with the computer. Over the years some modifications in these approximate analysis methods have been proposed, such as by Witmer [8] and Sutherland and Bowman [5]. These are largely only of historical interest at the present time.

For very tall buildings (say above 35 stories), the stiffness of the members has a significant effect and these approximate methods should not be relied upon. Further treatment of multistory building analysis is to be found in the matrix method approach of Clough, King, and Wilson [9], and the approximate analysis state-of-the-art work of Goldberg [10].

SELECTED REFERENCES

1. Mario Salvadori and Matthys Levy, *Structural Design in Architecture*, Prentice-Hall, Inc., Englewood Cliffs, N.J., 1967, Chap. 7.
2. J. B. Johnson, C. W. Bryan, and F. E. Turneaure, *The Theory and Practice of Modern Framed Structures*, 10th ed., John Wiley & Sons, Inc., New York, 1929.

3. Milo S. Ketchum, *Design of Steel Mill Buildings*, 2nd ed., Engineering News Publishing Co., New York, 1907, pp. 84–93, 109–119.

4. Albert Smith, "Wind Stresses in the Frames of Office Buildings," *Journal of the Western Society of Engineers*, April 1915, p. 341.

5. Hale Sutherland and Harry Lake Bowman, *Structural Theory*, 4th ed., John Wiley & Sons, Inc., New York, 1950, pp. 295–301.

6. ASCE Committee on Steel, Subcommittee No. 31, "Wind Bracing in Steel Buildings," *Transactions*, ASCE, *105* (1940), 1713–1739. [Also, *Proceedings*, ASCE, *65* (1939), p. 966.]

7. A. C. Wilson, "Wind Bracing with Knee Braces or Gusset Plates," *Engineering Record*, September 5, 1908, 272–274.

8. F. P. Witmer, "Wind Stress Analysis by the *K*-Percentage Method," *Proceedings*, ASCE, *67* (1941), 961–974. [Also, *Transactions*, ASCE, *107* (1942), 925–938; Discussion, 939–954.]

9. Ray W. Clough, Ian P. King, and Edward L. Wilson, "Structural Analysis of Multistory Buildings," *Journal of the Structural Division*, ASCE, *90*, ST3 (June 1964), 19–34.

10. John E. Goldberg, "Approximate Elastic Analysis, State-of-Art Report 2," *Proceedings of the International Conference on Planning and Design of Tall Buildings*, August 21–26, 1972, Vol. II, 14–169.

PROBLEMS

12.1–12.4. For the accompanying figures, compute moments, shears, and axial forces in all members using (a) the portal method; (b) the cantilever method.

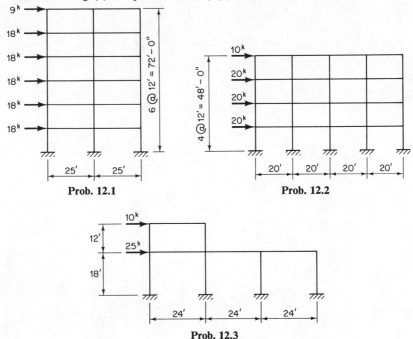

Prob. 12.1   Prob. 12.2

Prob. 12.3

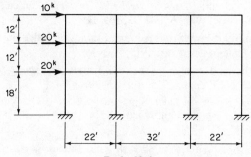

Prob. 12.4

12.5. Use the portal method to determine the bending moments, shears, and axial forces in all members of the rigid frame (so-called Vierendeel truss) in the accompanying figure.

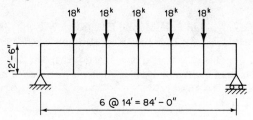

Prob. 12.5

[13]

Matrix Operations

___

13.1 INTRODUCTION

The practical solution of structural analysis problems on the high-speed digital computer involves the use of matrix methods by which systems of linear equations must be operated according to the laws of matrix algebra. It is expected that many students would have had a course in matrix or linear algebra before this textbook is studied. However, the barest minimum of matrix definitions, matrix multiplication, and matrix inversion is introduced here in the eventuality that the reader as yet has had no background in matrices. Only the essentials needed for the study of the next several chapters are included. For the reader already familiar with matrix operations, Chapter 13 may be omitted.

13.2 MATRIX FORM OF A SYSTEM OF LINEAR EQUATIONS

Suppose that the relationships between three values of x and four values of y are expressible by the following three linear equations:

$$\left.\begin{array}{l} x_1 = 2y_1 + 3y_2 + 4y_3 + 5y_4 \\ x_2 = 6y_1 + 7y_2 + 8y_3 + 9y_4 \\ x_3 = 10y_1 + 11y_2 + 12y_3 + 13y_4 \end{array}\right\} \quad (13.2.1)$$

The 12 constants (i.e., the numerical values 2 through 13) which appear in Eqs. (13.2.1) may be arranged in the form of a *matrix*, which is a rectangular array (or block) of numbers. Such a matrix may be written as

$$[A]_{3 \times 4} = \begin{bmatrix} +2 & +3 & +4 & +5 \\ +6 & +7 & +8 & +9 \\ +10 & +11 & +12 & +13 \end{bmatrix} \qquad (13.2.2)$$

A *rectangular matrix* is denoted by enclosing a capital letter inside square brackets, where the first and second subscripts indicate the number of rows and columns, respectively. Similarly, the x or y values of Eqs. (13.2.1) may be arranged in the form of a *column matrix*, which is the special name of a rectangular matrix with only one column; thus,

$$\{X\}_{3 \times 1} = \begin{bmatrix} x_1 \\ x_2 \\ x_3 \end{bmatrix} \qquad \{Y\}_{4 \times 1} = \begin{bmatrix} y_1 \\ y_2 \\ y_3 \\ y_4 \end{bmatrix} \qquad (13.2.3)$$

A column matrix is denoted by enclosing a capital letter inside curved brackets, called braces.

The matrix form of a system of linear equations like Eqs. (13.2.1) is simply

$$\{X\}_{3 \times 1} = [A]_{3 \times 4} \{Y\}_{4 \times 1}$$

where $\{X\}$, $[A]$, and $\{Y\}$ are as defined in Eqs. (13.2.2) and (13.2.3). In general terms, the matrix form of a system of linear equations

$$x_1 = a_{11} y_1 + a_{12} y_2 + \cdots a_{1j} y_j + \cdots + a_{1M} y_M$$
$$x_2 = a_{21} y_1 + a_{22} y_2 + \cdots a_{2j} y_j + \cdots + a_{2M} y_M$$
$$\cdot$$
$$\cdot$$
$$\cdot$$
$$x_i = a_{i1} y_1 + a_{i2} y_2 + \cdots a_{ij} y_j + \cdots + a_{iM} y_M \qquad (13.2.4)$$
$$\cdot$$
$$\cdot$$
$$\cdot$$
$$x_L = a_{L1} y_1 + a_{L2} y_2 + \cdots a_{Lj} y_j + \cdots + a_{LM} y_M$$

is as follows

$$\{X\}_{L \times 1} = [A]_{L \times M} \{Y\}_{M \times 1} \qquad (13.2.5)$$

where

$$\{X\}_{L \times 1} = \begin{bmatrix} x_1 \\ x_2 \\ \cdot \\ \cdot \\ \cdot \\ x_L \end{bmatrix}; \qquad [A]_{L \times M} = \begin{bmatrix} a_{11} & a_{12} & \cdots & a_{1M} \\ a_{21} & a_{22} & \cdots & a_{2M} \\ \cdot & & & \\ \cdot & & & \\ \cdot & & & \\ a_{L1} & a_{L2} & \cdots & a_{LM} \end{bmatrix}; \qquad \{Y\}_{M \times 1} = \begin{bmatrix} y_1 \\ y_2 \\ \cdot \\ \cdot \\ \cdot \\ y_M \end{bmatrix}$$

One may note that a specific element a_{ij} of the $[A]$ matrix is that element at the ith row and jth column.

13.3 MATRIX MULTIPLICATION

Given the following two systems of linear equations:

$$\left.\begin{aligned} x_1 &= 2y_1 + 3y_2 + 4y_3 + 5y_4 \\ x_2 &= 6y_1 + 7y_2 + 8y_3 + 9y_4 \\ x_3 &= 10y_1 + 11y_2 + 12y_3 + 13y_4 \end{aligned}\right\} \quad (13.3.1)$$

and

$$\left.\begin{aligned} y_1 &= 14z_1 + 15z_2 \\ y_2 &= 16z_1 + 17z_2 \\ y_3 &= 18z_1 + 19z_2 \\ y_4 &= 20z_1 + 21z_2 \end{aligned}\right\} \quad (13.3.2)$$

From sets of Eqs. (13.3.1) and (13.3.2) a third system of linear equations may be derived in which the three values of x are expressed directly by the two values of z. This is accomplished by substituting Eqs. (13.3.2) into Eqs. (13.3.1); thus,

$$\begin{aligned} x_1 &= 2(14z_1 + 15z_2) + 3(16z_1 + 17z_2) + 4(18z_1 + 19z_2) + 5(20z_1 + 21z_2) \\ &= 248z_1 + 262z_2 \\ x_2 &= 6(14z_1 + 15z_2) + 7(16z_1 + 17z_2) + 8(18z_1 + 19z_2) + 9(20z_1 + 21z_2) \quad (13.3.3) \\ &= 520z_1 + 550z_2 \\ x_3 &= 10(14z_1 + 15z_2) + 11(16z_1 + 17z_2) + 12(18z_1 + 19z_2) + 13(20z_1 + 21z_2) \\ &= 792z_1 + 838z_2 \end{aligned}$$

The three systems of equations, Eqs. (13.3.1) to (13.3.3) may be written in matrix form as

$$\{X\}_{3 \times 1} = [A]_{3 \times 4}\{Y\}_{4 \times 1} \quad (13.3.4)$$

$$\{Y\}_{4 \times 1} = [B]_{4 \times 2}\{Z\}_{2 \times 1} \quad (13.3.5)$$

$$\{X\}_{3 \times 1} = [C]_{3 \times 2}\{Z\}_{2 \times 1} \quad (13.3.6)$$

Note that standard symbols are the curved brackets for the column matrices and square brackets for a rectangular matrix. If Eq. (13.3.5) is substituted into Eq. (13.3.4), the following is obtained:

$$\{X\} = [A]\{Y\} = [A][B]\{Z\} \quad (13.3.7)$$

Comparing Eq. (13.3.7) with Eq. (13.3.6), it becomes apparent that

$$[C]_{3 \times 2} = [A]_{3 \times 4}[B]_{4 \times 2} \quad (13.3.8)$$

The $[C]$ matrix is thus defined as the product of the matrices $[A]$ and $[B]$. It is important to note that the number of columns (in this case, four) in the premultiplier matrix $[A]$ must equal the number of rows in the postmultiplier matrix $[B]$.

The arithmetic involved in the matrix multiplication of premultiplier matrix $[A]$ by postmultiplier matrix $[B]$ is tabulated as shown:

$$[B]_{4 \times 2} = \begin{array}{|c|c|} \hline 14 & 15 \\ \hline 16 & 17 \\ \hline 18 & 19 \\ \hline 20 & 21 \\ \hline \end{array}$$

$$[A]_{3 \times 4} = \begin{array}{|c|c|c|c|} \hline 2 & 3 & 4 & 5 \\ \hline 6 & 7 & 8 & 9 \\ \hline 10 & 11 & 12 & 13 \\ \hline \end{array}$$

$$[C]_{3 \times 2} = \begin{array}{|c|c|} \hline
\begin{aligned} 2 \times 14 &= 28 \\ 3 \times 16 &= 48 \\ 4 \times 18 &= 72 \\ 5 \times 20 &= \underline{100} \\ & 248 \end{aligned} &
\begin{aligned} 2 \times 15 &= 30 \\ 3 \times 17 &= 51 \\ 4 \times 19 &= 76 \\ 5 \times 21 &= \underline{105} \\ & 262 \end{aligned} \\ \hline
\begin{aligned} 6 \times 14 &= 84 \\ 7 \times 16 &= 112 \\ 8 \times 18 &= 144 \\ 9 \times 20 &= \underline{180} \\ & 520 \end{aligned} &
\begin{aligned} 6 \times 15 &= 90 \\ 7 \times 17 &= 119 \\ 8 \times 19 &= 152 \\ 9 \times 21 &= \underline{189} \\ & 550 \end{aligned} \\ \hline
\begin{aligned} 10 \times 14 &= 140 \\ 11 \times 16 &= 176 \\ 12 \times 18 &= 216 \\ 13 \times 20 &= \underline{260} \\ & 792 \end{aligned} &
\begin{aligned} 10 \times 15 &= 150 \\ 11 \times 17 &= 187 \\ 12 \times 19 &= 228 \\ 13 \times 21 &= \underline{273} \\ & 838 \end{aligned} \\ \hline
\end{array}$$

where c_{ij} is shown to be the sum of the products of a_{ik} and b_{kj} as k increases from 1 to the number of columns in the premultiplier or the number of rows in the post-multiplier. Thus, the rule for multiplying $[A]_{L \times M}$ and $[B]_{M \times N}$ is

$$C(i, j) = \sum_{k=1}^{k=M} [A(i, k) * B(k, j)] \qquad (13.3.9)$$

One may note that the tabulation of matrix multiplication shown above is exactly identical to the operations involved in simplifying Eqs. (13.3.3). For instance, in the first equation of Eqs. (13.3.3), the coefficient 248 is the sum of products 2(14), 3(16), 4(18), and 5(20). This is done in the upper left corner of the tabulation, wherein the 2, 3, 4, and 5 in the first row of $[A]$ and the 14, 16, 18, and 20 in the first column of $[B]$, both move *inward* and converge at the first row and the first column of $[C]$; thus, the name *inner product* rule for matrix multiplication has been used. Consequently, as another example, to obtain the inner product at the third row and second column of $[C]$, one will draw visually a horizontal line through 10, 11, 12, and 13 of $[A]$ and a vertical line through 15, 17, 19, and 21 of [B] until they intersect at location C(3, 2), where the numbers are paired, multiplied, and added.

13.4 MATRIX INVERSION

Suppose that three values of x are expressible in terms of three values of y by three linear equations as follows:

$$
\left.
\begin{aligned}
x_1 &= 2y_1 + 3y_2 + 4y_3 \\
x_2 &= 5y_1 + 6y_2 + 7y_3 \\
x_3 &= 8y_1 + 9y_2 + 6y_3
\end{aligned}
\right\} \tag{13.4.1}
$$

The matrix form of Eqs. (13.4.1) is

$$\{X\}_{3 \times 1} = [A]_{3 \times 3}\{Y\}_{3 \times 1} \tag{13.4.2}$$

in which

$$
[A]_{3 \times 3} = \begin{bmatrix} +2 & +3 & +4 \\ +5 & +6 & +7 \\ +8 & +9 & +6 \end{bmatrix} \tag{13.4.3}
$$

As long as each of Eqs. (13.4.1) is independent of the other two, it is possible to express the three values of y in terms of the three values of x. Such a result may be expressed,

$$
\left.
\begin{aligned}
y_1 &= b_{11}x_1 + b_{12}x_2 + b_{13}x_3 \\
y_2 &= b_{21}x_1 + b_{22}x_2 + b_{23}x_3 \\
y_3 &= b_{31}x_1 + b_{32}x_2 + b_{33}x_3
\end{aligned}
\right\} \tag{13.4.4}
$$

which in matrix form becomes

$$\{Y\}_{3 \times 1} = [B]_{3 \times 3}\{X\}_{3 \times 1} \tag{13.4.5}$$

The matrix [B] containing the elements b_{ij} is

$$
[B]_{3 \times 3} = \begin{bmatrix} b_{11} & b_{12} & b_{13} \\ b_{21} & b_{22} & b_{23} \\ b_{31} & b_{32} & b_{33} \end{bmatrix} \tag{13.4.6}
$$

The matrix [B] is defined as the *inverse* of the matrix [A]; the arithmetic operation of converting from matrix [A] to matrix [B] is known as *matrix inversion*. In symbolic form the relationship is expressed

$$[B] = [A]^{-1} = [A^{-1}] \tag{13.4.7}$$

Note that the inverse of matrix [A], designated $[A]^{-1}$, is itself a matrix; thus, the symbol $[A^{-1}]$, not $1/[A]$, has to be used.

Equation (13.4.5) may now be written as

$$\{Y\} = [A^{-1}]\{X\} \tag{13.4.8}$$

Substitution of Eq. (13.4.8) into Eq. (13.4.2) gives

$$
\begin{aligned}
\{X\} = [A]\{Y\} &= [A][A^{-1}]\{X\} \\
&= [AA^{-1}]\{X\}
\end{aligned}
\tag{13.4.9}
$$

From the fact of having the column matrix $\{X\}$ on both sides of the equation, the matrix $[AA^{-1}]$ must have no effect on the equation. Thus, in conventional linear equation format, Eq. (13.4.9) must be

$$\left.\begin{array}{l} x_1 = 1.0x_1 + 0 \ x_2 + 0 \ x_3 \\ x_2 = 0 \ x_1 + 1.0x_2 + 0 \ x_3 \\ x_3 = 0 \ x_1 + 0 \ x_2 + 1.0x_3 \end{array}\right\} \qquad (13.4.10)$$

which means that the matrix $[AA^{-1}]$ must be

$$[AA^{-1}]_{3\times3} = \begin{bmatrix} 1 & 0 & 0 \\ 0 & 1 & 0 \\ 0 & 0 & 1 \end{bmatrix} = [I]_{3\times3} \qquad (13.4.11)$$

The symbol $[I]$ is to represent a *unit matrix* or an *identity matrix*, which is a square matrix having the value of 1 on the main diagonal and the value of 0 elsewhere. Thus, the product of a matrix and its own inverse, which exists whenever the original matrix arises from *independent* linear equations, is a unit matrix.

If a square matrix contains linear equations that are not independent, such a matrix is called a *singular matrix*. A singular matrix will have no inverse, because if there are not enough independent conditions expressing, say, the x's in terms of the y's, the y's cannot be solved in terms of the x's. This happens, for instance, when the third row in the matrix $[A]$ of Eq. (13.4.3) can be obtained by adding a constant k_1 times the first row and a constant k_2 times the second row.

One of the ways to find the inverse of a matrix is to solve the system of linear equations simultaneously by first interchanging the left and right sides of Eqs. (13.4.1); thus,

$$\left.\begin{array}{l} 2y_1 + 3y_2 + 4y_3 = x_1 \\ 5y_1 + 6y_2 + 7y_3 = x_2 \\ 8y_1 + 9y_2 + 6y_3 = x_3 \end{array}\right\} \qquad (13.4.12)$$

Solving Eqs. (13.4.12) for the three values of y gives

$$\left.\begin{array}{l} y_1 = -\dfrac{9}{4}x_1 + \dfrac{3}{2}x_2 - \dfrac{1}{4}x_3 \\[2mm] y_2 = +\dfrac{13}{6}x_1 - \dfrac{5}{3}x_2 + \dfrac{1}{2}x_3 \\[2mm] y_3 = -\dfrac{1}{4}x_1 + \dfrac{1}{2}x_2 - \dfrac{1}{4}x_3 \end{array}\right\} \qquad (13.4.13)$$

The solution given by Eqs. (13.4.13) is best obtained by letting successively ($x_1 = +1$, $x_2 = 0$, $x_3 = 0$), ($x_1 = 0$, $x_2 = +1$, $x_3 = 0$), and ($x_1 = 0$, $x_2 = 0$, $x_3 = +1$) in order that the coefficients of x_1, x_2, and x_3, respectively, of Eqs. (13.4.13) may be obtained separately. These coefficients of x_1, x_2, and x_3 should appear, respectively, at the first, second, and third columns in the inverse matrix $[A^{-1}]$. Thus, the inverse of the matrix $[A]$ of Eq. (13.4.3) is

$$[A]^{-1} = \begin{bmatrix} +2 & +3 & +4 \\ +5 & +6 & +7 \\ +8 & +9 & +6 \end{bmatrix}^{-1} = \begin{bmatrix} -\dfrac{9}{4} & +\dfrac{3}{2} & -\dfrac{1}{4} \\ +\dfrac{13}{6} & -\dfrac{5}{3} & +\dfrac{1}{2} \\ -\dfrac{1}{4} & +\dfrac{1}{2} & -\dfrac{1}{4} \end{bmatrix} \qquad (13.4.14)$$

Once the inverse is determined, a check should be made showing that the product of the premultiplier matrix $[A]$ times the postmultiplier matrix $[A^{-1}]$ equals the unit matrix or the identity matrix $[I]$; thus,

$$[AA^{-1}] = \begin{bmatrix} +2 & +3 & +4 \\ +5 & +6 & +7 \\ +8 & +9 & +6 \end{bmatrix}\begin{bmatrix} -\dfrac{9}{4} & +\dfrac{3}{2} & -\dfrac{1}{4} \\ +\dfrac{13}{6} & -\dfrac{5}{3} & +\dfrac{1}{2} \\ -\dfrac{1}{4} & +\dfrac{1}{2} & -\dfrac{1}{4} \end{bmatrix} = \begin{bmatrix} 1 & 0 & 0 \\ 0 & 1 & 0 \\ 0 & 0 & 1 \end{bmatrix} = [I] \qquad (13.14.15)$$

13.5 SOLUTION OF LINEAR SIMULTANEOUS EQUATIONS

The most common method of solving a set of linear simultaneous equations is by the process of successively eliminating unknowns, referred to as the *Gauss–Jordan elimination method*. In this method the solution of a system of equations such as

$$\left.\begin{array}{l} a_{11}x_1 + a_{12}x_2 + a_{13}x_3 = a_{14} \\ a_{21}x_1 + a_{22}x_2 + a_{23}x_3 = a_{24} \\ a_{31}x_1 + a_{32}x_2 + a_{33}x_3 = a_{34} \end{array}\right\} \qquad (13.5.1)$$

is obtained by operating on the original matrix of coefficients three times to reach the final matrix. The first elimination would be the following:

$$\begin{bmatrix} a_{11} & a_{12} & a_{13} & a_{14} \\ a_{21} & a_{22} & a_{23} & a_{24} \\ a_{31} & a_{32} & a_{33} & a_{34} \end{bmatrix} \longrightarrow \begin{bmatrix} 1 & a'_{12} & a'_{13} & a'_{14} \\ 0 & a'_{22} & a'_{23} & a'_{24} \\ 0 & a'_{32} & a'_{33} & a'_{34} \end{bmatrix} \qquad (13.5.2)$$

(1st elimination)

For the first elimination, row 1 is divided by a_{11} to obtain row 1 of the new matrix. This new row 1 is then multiplied by a_{21} and subtracted from original row 2 to obtain new row 2. Then new row 1 is multiplied by a_{31} and subtracted from original row 3 to obtain new row 3.

The second elimination is as follows:

$$\begin{bmatrix} 1 & a'_{12} & a'_{13} & a'_{14} \\ 0 & a'_{22} & a'_{23} & a'_{24} \\ 0 & a'_{32} & a'_{33} & a'_{34} \end{bmatrix} \longrightarrow \begin{bmatrix} 1 & 0 & a''_{13} & a''_{14} \\ 0 & 1 & a''_{23} & a''_{24} \\ 0 & 0 & a''_{33} & a''_{34} \end{bmatrix} \qquad (13.5.3)$$

(2nd elimination)

For the second elimination row 2 is divided by a'_{22} to obtain row 2 of the new matrix. This new row 2 is then multiplied by a'_{12} and subtracted from original row 1 to obtain new row 1. Then new row 2 is multiplied by a'_{32} and subtracted from original row 3 to obtain new row 3.

The third elimination is as follows:

$$\begin{bmatrix} 1 & 0 & a''_{13} & a''_{14} \\ 0 & 1 & a''_{23} & a''_{24} \\ 0 & 0 & a''_{33} & a''_{34} \end{bmatrix} \longrightarrow \begin{bmatrix} 1 & 0 & 0 & a'''_{14} \\ 0 & 1 & 0 & a'''_{24} \\ 0 & 0 & 1 & a'''_{34} \end{bmatrix} \tag{13.5.4}$$

(3rd elimination)

For the third elimination row 3 is divided by a''_{33} to obtain row 3 of the new matrix. This new row 3 is then multiplied by a''_{13} and subtracted from original row 1 to obtain new row 1. Then new row 3 is multiplied by a''_{23} and subtracted from original row 2 to obtain new row 2.

The final matrix gives the solution

$$\{X\} = \begin{bmatrix} x_1 \\ x_2 \\ x_3 \end{bmatrix} = \begin{bmatrix} a'''_{14} \\ a'''_{24} \\ a'''_{34} \end{bmatrix} \tag{13.5.5}$$

To summarize the process, the ith row at the end of the ith elimination is called the *pivotal row* whose elements are obtained by dividing the ith row in the previous matrix by a_{ii} (the pivot) in that matrix. The rows other than the ith row at the end of the ith elimination are obtained by subtracting, from the corresponding row in the previous matrix, the product of a constant and the pivotal row (the ith row) with the purpose of attaining the value of zero directly above or below the value of 1 in the pivotal row.

Example 13.5.1

Determine the inverse of the given matrix $[A]$ by Gauss–Jordan elimination.

$$[A] = \begin{bmatrix} +2 & +3 & +4 \\ +5 & +6 & +7 \\ +8 & +9 & +6 \end{bmatrix}$$

Solution. The required operation is equivalent to solving Eqs. (13.4.12) for three different sets of x values on the right-hand side of the equations, as follows:

$$2y_1 + 3y_2 + 4y_3 = 1, \quad 0, \quad 0$$
$$5y_1 + 6y_2 + 7y_3 = 0, \quad 1, \quad 0$$
$$8y_1 + 9y_2 + 6y_3 = 0, \quad 0, \quad 1$$

(a) *First Gauss–Jordan elimination.* The first elimination makes column 1 of the new matrix, $+1, 0, 0,$

$$
\begin{bmatrix}
+2 & +3 & +4 & +1 & 0 & 0 \\
+5 & +6 & +7 & 0 & +1 & 0 \\
+8 & +9 & +6 & 0 & 0 & +1
\end{bmatrix}
\longrightarrow
\begin{bmatrix}
+1 & +\dfrac{3}{2} & +2 & +\dfrac{1}{2} & 0 & 0 \\
0 & -\dfrac{3}{2} & -3 & -\dfrac{5}{2} & +1 & 0 \\
0 & -3 & -10 & -4 & 0 & +1
\end{bmatrix}
$$

(1st elimination)

For example, the second row at first elimination $(0, -\frac{3}{2}, -3, -\frac{5}{2}, +1, 0)$ is obtained by subtracting $+5$ times the pivotal row $(+1, +\frac{3}{2}, +2, +\frac{1}{2}, 0, 0)$ from the second row in the original matrix $(+5, +6, +7, 0, +1, 0)$. Likewise, the third row $(0, -3, -10, -4, 0, +1)$ is obtained by subtracting $+8$ times the pivotal row $(+1, +\frac{3}{2}, +2, +\frac{1}{2}, 0, 0)$ from the third row in the original matrix $(+8, +9, +6, 0, 0, +1)$. For this elimination row 1 is called the pivotal row because it is the one containing the $+1$ along the diagonal (i.e., the ith row of the ith elimination where i is 1).

(b) *Second elimination.* The second elimination is as follows:

$$
\begin{bmatrix}
+1 & +\dfrac{3}{2} & +2 & +\dfrac{1}{2} & 0 & 0 \\
0 & -\dfrac{3}{2} & -3 & -\dfrac{5}{2} & +1 & 0 \\
0 & -3 & -10 & -4 & 0 & +1
\end{bmatrix}
\longrightarrow
\begin{bmatrix}
+1 & 0 & -1 & -2 & +1 & 0 \\
0 & +1 & +2 & +\dfrac{5}{3} & -\dfrac{2}{3} & 0 \\
0 & 0 & -4 & +1 & -2 & +1
\end{bmatrix}
$$

(c) *Third elimination.* The third elimination is as follows:

$$
\begin{bmatrix}
+1 & 0 & -1 & -2 & +1 & 0 \\
0 & +1 & +2 & +\dfrac{5}{3} & -\dfrac{2}{3} & 0 \\
0 & 0 & -4 & +1 & -2 & +1
\end{bmatrix}
\longrightarrow
\begin{bmatrix}
+1 & 0 & 0 & -\dfrac{9}{4} & +\dfrac{3}{2} & -\dfrac{1}{4} \\
0 & +1 & 0 & +\dfrac{13}{6} & -\dfrac{5}{3} & +\dfrac{1}{2} \\
0 & 0 & +1 & -\dfrac{1}{4} & +\dfrac{1}{2} & -\dfrac{1}{4}
\end{bmatrix}
$$

(d) *Final matrix.* The solution for the inverse of matrix $[A]$ is

$$
[A]^{-1} = [A^{-1}]
$$

$$
\begin{bmatrix}
+2 & +3 & +4 \\
+5 & +6 & +7 \\
+8 & +9 & +6
\end{bmatrix}^{-1}
=
\begin{bmatrix}
-\dfrac{9}{4} & +\dfrac{3}{2} & -\dfrac{1}{4} \\
+\dfrac{13}{6} & -\dfrac{5}{3} & +\dfrac{1}{2} \\
-\dfrac{1}{4} & +\dfrac{1}{2} & -\dfrac{1}{4}
\end{bmatrix}
$$

For the solution in Example 13.5.1 containing three rows, the pivot values were taken successively from a_{11}, to a'_{22}, to a''_{33}, using symbols of Eqs. (13.5.2) to (13.5.4). There are many variations that may be used to improve accuracy of the solution, including interchanging of rows. The reader is referred to texts [1–5] on numerical analysis for a more complete treatment of these other methods.

Example 13.5.2

Determine the inverse of the given matrix $[A]$ by the Gauss–Jordan elimination method.

$$[A] = \begin{bmatrix} +2 & +3 & +4 \\ +5 & +6 & +7 \\ +8 & +9 & +10 \end{bmatrix}$$

Solution

(a) *First elimination.* The first elimination in the procedure is as follows:

$$\begin{bmatrix} +2 & +3 & +4 & +1 & 0 & 0 \\ +5 & +6 & +7 & 0 & +1 & 0 \\ +8 & +9 & +10 & 0 & 0 & +1 \end{bmatrix} \longrightarrow \begin{bmatrix} +1 & +\frac{3}{2} & +2 & +\frac{1}{2} & 0 & 0 \\ 0 & -\frac{3}{2} & -3 & -\frac{5}{2} & +1 & 0 \\ 0 & -3 & -6 & -4 & 0 & +1 \end{bmatrix}$$

(1st elimination)

(b) *Second elimination.* The second elimination is as follows:

$$\begin{bmatrix} +1 & +\frac{3}{2} & +2 & +\frac{1}{2} & 0 & 0 \\ 0 & -\frac{3}{2} & -3 & -\frac{5}{2} & +1 & 0 \\ 0 & -3 & -6 & -4 & 0 & +1 \end{bmatrix} \longrightarrow \begin{bmatrix} +1 & 0 & -1 & -2 & +1 & 0 \\ 0 & +1 & +2 & +\frac{5}{3} & -\frac{2}{3} & 0 \\ 0 & 0 & 0 & +1 & -2 & +1 \end{bmatrix}$$

(2nd elimination)

At the end of the second elimination, the third row is

$$0y_1 + 0y_2 + 0y_3 = +1, -2, +1$$

which is impossible. This indicates the matrix $[A]$ is *singular*; that is, the three equations are not independent. As mentioned in Section 13.4, a square matrix having no inverse is called a singular matrix. In this case the dependency between equations should be obvious because the difference in values between row 1 and row 2 is equal to the difference in values between row 2 and row 3. In other words, row 3 in the original matrix can be obtained by adding $+2$ times row 2 and -1 times row 1.

An alternative indication of singularity for a matrix is that the determinant of a singular matrix is zero. In this example,

$$\text{Determinant} \begin{vmatrix} +2 & +3 & +4 \\ +5 & +6 & +7 \\ +8 & +9 & +10 \end{vmatrix}$$

$$= (+2)(+6)(+10) + (+5)(+9)(+4) + (+8)(+3)(+7)$$
$$- (+8)(+6)(+4) - (+5)(+3)(+10) - (+2)(+9)(+7)$$
$$= 120 + 180 + 168 - 192 - 150 - 126 = 0$$

SELECTED REFERENCES

1. C. K. Wang, *Matrix Methods of Structural Analysis*, 2nd ed., American Publishing Company, Madison, Wis., 1970, Appendix B.

2. Charles H. Norris, John B. Wilbur, and Senol Utku, *Elementary Structural Analysis*, 3rd ed., McGraw-Hill Book Company, New York, 1976, pp. 547–556.

3. Merlin L. James, Gerald M. Smith, and James C. Wolford, *Applied Numerical Methods*, International Textbook Company, Scranton, Pa., 1967, Chap. 4.

4. Mario G. Salvadori and Melvin L. Baron, *Numerical Methods in Engineering*, 2nd ed., Prentice-Hall, Inc., Englewood Cliffs, N.J., 1961, pp. 21–53.

5. Ralph G. Stanton, *Numerical Methods for Science and Engineering*, Prentice-Hall, Inc., Englewood Cliffs, N.J., 1961, Chap. 8.

PROBLEMS

13.1. Given: $[A]_{1 \times 2} = [+3 \quad -2]$; $[B]_{2 \times 1} = \begin{bmatrix} -5 \\ -7 \end{bmatrix}$.

Find: $[AB]_{1 \times 1}$.

13.2. Compute the product of the following two matrices:

$$\begin{bmatrix} 4 & 2 \\ 3 & 5 \end{bmatrix} \begin{bmatrix} 7 \\ 6 \end{bmatrix}$$

13.3. Given:

$$[A] = \begin{bmatrix} -1 & +2 & -3 \\ +4 & -5 & -6 \\ +7 & -8 & +9 \end{bmatrix} \quad \text{and} \quad [B] = \begin{bmatrix} +10 & -11 \\ -12 & +13 \\ -14 & +15 \end{bmatrix}$$

(a) Compute $[C] = [A][B]$; (b) compute $\{X\}$ from $\{X\} = [C]\{Z\}$ directly if $z_1 = +3$ and $z_2 = -2$; and then (c) compute $\{X\}$ from $\{X\} = [A]\{Y\}$ after obtaining $\{Y\}$ from $\{Y\} = [B]\{Z\}$ when $z_1 = +3$ and $z_2 = -2$.

13.4. Given:

$$[A] = \begin{bmatrix} a & 0 & 0 \\ 0 & b & 0 \\ 0 & 0 & c \end{bmatrix} \quad \text{and} \quad [B] = \begin{bmatrix} e & f \\ g & h \\ i & j \end{bmatrix}$$

Find: $[AB]$. This problem is intended to illustrate the effect of a *diagonal matrix* (a square matrix in which the only nonzero elements are on the diagonal) when used as a premultiplier.

13.5. Given:

$$[A] = \begin{bmatrix} e & f & g \\ h & i & j \end{bmatrix} \quad \text{and} \quad [B] = \begin{bmatrix} a & 0 & 0 \\ 0 & b & 0 \\ 0 & 0 & c \end{bmatrix}$$

Find: $[AB]$. This problem is intended to illustrate the effect of a *diagonal matrix* (see Prob. 13.4) when used as a postmultiplier.

13.6. Solve the two given simultaneous equations using the Gauss–Jordan elimination method,

$$+10x_1 - 3x_2 = -5$$
$$-3x_1 + 20x_2 = -94$$

13.7. Find the inverse of $\begin{bmatrix} +2 & 0 \\ 0 & -5 \end{bmatrix}$

13.8. Solve the two given simultaneous equations using the Gauss–Jordan elimination method,

$$+4x_1 + 3x_2 = +18$$
$$+2x_1 + 5x_2 = +16$$

13.9. By using the Gauss–Jordan elimination method, determine the inverse of the matrix

$$\begin{bmatrix} +0.6 & -0.8 \\ +0.8 & +0.6 \end{bmatrix}$$

13.10. Solve the given two simultaneous equations by the Gauss–Jordan elimination method,

$$+5x_1 - 2x_2 = -14$$
$$+7x_1 + 3x_2 = -37$$

13.11. By using the Gauss–Jordan elimination method, determine the inverse of the matrix

$$\begin{bmatrix} +0.8 & -0.2 \\ -0.2 & +1.3 \end{bmatrix}$$

13.12. By the Gauss–Jordan elimination method, obtain the inverse of the matrix

$$\begin{bmatrix} -\dfrac{9}{4} & +\dfrac{3}{2} & -\dfrac{1}{4} \\[2ex] +\dfrac{13}{6} & -\dfrac{5}{3} & +\dfrac{1}{2} \\[2ex] -\dfrac{1}{4} & +\dfrac{1}{2} & -\dfrac{1}{4} \end{bmatrix}$$

Note that the answer is the original matrix of Eq. (13.4.14).

13.13. By the Gauss–Jordan elimination method, obtain the inverse of the matrix

$$\begin{bmatrix} +a & +b \\ +b & +c \end{bmatrix}$$

Take special note of the result because it will be useful later in the treatment of the stiffness and flexibility matrices of a member in bending.

[14]

General Concepts of Flexibility and Stiffness Methods of Analysis

---

14.1 TRANSITION FROM CLASSICAL TO MATRIX METHODS

One might call the treatment in the first 12 chapters of this text the classical methods of structural analysis. The term "classical methods" is used because these methods have been in use long before the advent of the electronic computer for which the classical theory is put in matrix notation. Having had described in Chapter 13 the basics of matrix multiplication and inversion, it may be useful at this time to recapitulate some of the common concepts and theorems in the classical approach in terms of the matrix notations, to provide an introduction to the matrix methods in Chapters 15 to 20.

In the force method of structural analysis, the forces (reactions and axial forces in trusses; shears and moments in beams; axial forces, shears, and moments in frames) are determined first before the displacements. If the structure is statically determinate, these forces can be obtained by the laws of statics alone, without need of having the section properties of the individual members. If the structure is statically indeterminate, the redundant forces (beyond and above those required for statical determinacy) must be determined from the physical compatibility (consistent deformation) conditions; only then the remaining unknown forces are obtained by the laws of statics. If the deformed shape of the structure under load is also required, it is ascertained after the equilibrium state (free-body diagram) of the structure has been completely defined.

In the displacement method of structural analysis, the deformed shape of the structure under load in terms of the joint displacements is determined first before the forces. These unknown joint displacements are found by solving an equal number of simultaneous equations, each of which is an equilibrium condition in the direction of the unknown displacement. Once these joint displacements are obtained, the free-body diagram of each individual member becomes solely dependent on the geometrical deformation of its two ends. Whether the structure is statically determinate or indeterminate is not relevant in the displacement method; the section properties of the individual members are always needed in the analysis procedure.

Historically, trusses have been analyzed by the force method, because using the displacement method would involve the solution of a large number of simultaneous equations even for the ordinary statically determinate six-panel or eight-panel trusses. For continuous beams and rigid frames, the slope deflection method (a form of displacement method) is preferable to the redundant reaction method, although the three-moment equation method (a form of force method) works as well for continuous beams. One could argue that the moment distribution method is a force method, because the end moments are obtained first before the end rotations, which are computed as the by-product in the check procedure. Nevertheless, the moment distribution method is in fact a relaxation procedure of solving the simultaneous equations in the slope deflection method, only the real unknowns (the joint rotations) are hidden. The moment distribution method should be classified as a displacement method.

One will find that in terms of matrix notations the displacement method is far superior to the force method, because the solution of a large number of simultaneous equations is no longer a problem for the electronic computer. Thus, Chapters 16 to 20 are entirely devoted to the analysis of different types of structures by the matrix displacement method.

14.2 TRUSS ANALYSIS BY THE MATRIX FORCE METHOD USING FLEXIBILITY COEFFICIENTS

Consider the truss shown in Fig. 14.2.1a, which is statically indeterminate to the second degree. This truss has been analyzed by Castigliano's second theorem in Example 7.7.2 using R_1, the axial force in $L_1 U_2$, and R_2, the reaction at the interior roller support, as the redundants. Rather than removing the bar $L_1 U_2$ as in Example 7.7.2, the basic determinate truss of Fig. 14.2.1b only has the bar $L_1 U_2$ cut at some point between the joints L_1 and U_2 where the pair of R_1 forces is applied.

Adopt the notation $\delta(i, j)$ as the deflection in the ith direction due to a unit force applied in the jth direction to the basic statically determinate truss. For solving the redundants R_1 and R_2, two physical compatibility (consistent deformation) conditions are required; they are: (1) the overlap at the cut ends of the redundant bar is zero, and (2) the upward vertical deflection at joint L_2 is zero. Thus,

$$\delta(R_1, R_1)R_1 + \delta(R_1, R_2)R_2 + \Delta_1 = 0 \qquad (14.2.1)$$

$$\delta(R_2, R_1)R_1 + \delta(R_2, R_2)R_2 + \Delta_2 = 0 \qquad (14.2.2)$$

in which Δ_1 and Δ_2 are the deflections in the R_1 and R_2 directions due to the applied loads. Equations (14.2.1) and (14.2.2) can be written in matrix form as

$$[\delta]_{2 \times 2}\{R\}_{2 \times 1} + \{\Delta\}_{2 \times 1} = 0 \qquad (14.2.3)$$

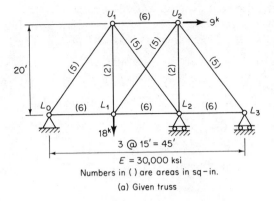

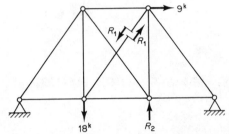

(b) Basic determinate truss acted upon
by applied loads and redundants
R_1 and R_2

Figure 14.2.1 Truss analysis by the matrix force method.

wherein

$$[\delta]_{2 \times 2} = \begin{bmatrix} \delta(R_1, R_1) & \delta(R_1, R_2) \\ \delta(R_2, R_1) & \delta(R_2, R_2) \end{bmatrix} \qquad (14.2.4)$$

and

$$\{R\}_{2 \times 1} = \begin{Bmatrix} R_1 \\ R_2 \end{Bmatrix}; \qquad \{\Delta\}_{2 \times 1} = \begin{Bmatrix} \Delta_1 \\ \Delta_2 \end{Bmatrix} \qquad (14.2.5)$$

The $[\delta]$ matrix of Eq. (14.2.4) is called the flexibility matrix which contains the flexibility coefficients $\delta(i, j)$ for $i = 1$ to NI and $j = 1$ to NI; NI is the degree of statical indeterminacy. Of course, for any statically indeterminate structure, there is a different flexibility matrix for a different set of redundants. By virtue of the Maxwell's theorem of reciprocal deflection, the flexibility matrix is always a symmetric matrix.

The method of solution indicated by Eq. (14.2.3) can be called the force method, the Maxwell–Mohr method, the consistent deformation method, or the flexibility method.

Example 14.2.1

Referring to the solution in Example 7.7.2 for the statically indeterminate truss of Fig. 14.2.1, write out numerically the matrix equation $[\delta]\{R\} + \{\Delta\} = 0$.

Solution. In part (c) of the solution in Example 7.7.2, bar 6 (the redundant bar, member $L_1 U_2$) is removed; therefore, the horizontal line along bar 6 is left blank in the table. Had this bar only been cut with the redundant forces applied at the cut ends as shown in Fig. 14.2.1b, the value of u_{j1} for this bar would be $+1$ and the only addition would be for $u_{j1}^2/S_j = (+1)^2/500 = 2000 \times 10^{-6}$ in./kip. Thus,

$$\delta(R_1, R_1) = (7840 + 2000) \times 10^{-6} = +9840 \times 10^{-6} \text{ in./kip}$$

$$\delta(R_1, R_2) = \delta(R_2, R_1) = (1150 + 0) \times 10^{-6} = +1150 \times 10^{-6} \text{ in./kip}$$

$$\delta(R_2, R_2) = (4486.1 + 0) \times 10^{-6} = +4486.1 \times 10^{-6} \text{ in./kip}$$

Also from part (c) of the solution,

$$\Delta_1 = -124,500 \times 10^{-6} \text{ in.}$$

$$\Delta_2 = -55,916 \times 10^{-6} \text{ in.}$$

The matrix equation is

$$\begin{bmatrix} +9840 & +1150 \\ +1150 & +4486.1 \end{bmatrix} \begin{Bmatrix} R_1 \\ R_2 \end{Bmatrix} + \begin{Bmatrix} -124,500 \\ -55,916 \end{Bmatrix} = 0$$

14.3 BEAM ANALYSIS BY THE MATRIX FORCE METHOD USING FLEXIBILITY COEFFICIENTS

Consider the fixed-ended beam shown in Fig. 14.3.1a, which is statically indeterminate to the second degree. This beam has been analyzed by the force method (consistent deformation method) in Example 8.2.2, using first the two end moments as the redundants and then the moment and vertical reaction at the right end as the redundants. These two solutions can be put into matrix form, as shown by the following two examples.

Example 14.3.1

Using the matrix force method, obtain formulas for the redundants R_1 and R_2 as shown in Fig. 14.3.1 (positive if both act clockwise on member ends, different from the ordinary bending moment sign convention) for the fixed-ended beam subject to a concentrated load.

Solution

(a) *The matrix* $[\delta]$. The positive directions for R_1 and R_2 have been established as clockwise acting on member ends; so are the positive directions for the rotations at the ends of the basic statically determinate beam. Thus, $\delta(R_1, R_1)$ and $\delta(R_2, R_1)$ are the clockwise rotations at the left and right ends of the beam in Fig. 14.3.1c; or

$$\delta(R_1, R_1) = +\frac{BB_1}{L} = +\frac{1}{2}\frac{1.0}{EI}(L)\left(\frac{2}{3}L\right)\bigg/L = +\frac{L}{3EI}$$

$$\delta(R_2, R_1) = -\frac{AA_1}{L} = -\frac{1}{2}\frac{1.0}{EI}(L)\left(\frac{1}{3}L\right)\bigg/L = -\frac{L}{6EI}$$

Similarly, $\delta(R_1, R_2)$ and $\delta(R_2, R_2)$ are the clockwise rotations at the left and right ends of the beam in Fig. 14.3.1d; or

$$\delta(R_1, R_2) = -\frac{BB_1}{L} = -\frac{1}{2}\frac{1.0}{EI}(L)\left(\frac{1}{3}L\right)\bigg/L = -\frac{L}{6EI}$$

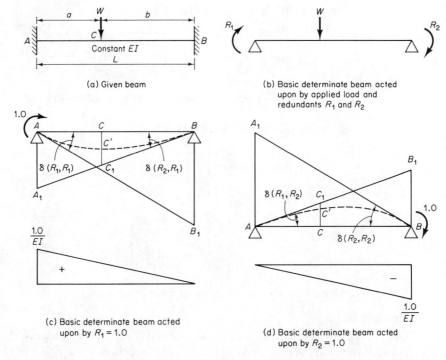

(a) Given beam

(b) Basic determinate beam acted upon by applied load and redundants R_1 and R_2

(c) Basic determinate beam acted upon by $R_1 = 1.0$

(d) Basic determinate beam acted upon by $R_2 = 1.0$

Figure 14.3.1 Matrix analysis of a fixed-ended beam using end moments as redundants.

$$\delta(R_2, R_2) = +\frac{AA_1}{L} = +\frac{1}{2}\frac{1.0}{EI}(L)\left(\frac{2}{3}L\right)\Big/L = +\frac{L}{3EI}$$

Therefore,

$$[\delta] = \begin{bmatrix} +\dfrac{L}{3EI} & -\dfrac{L}{6EI} \\[2ex] -\dfrac{L}{6EI} & +\dfrac{L}{3EI} \end{bmatrix}$$

(b) *The matrix* [Δ]. The elements in this matrix are the clockwise rotations at the ends of the simple beam AB due to the action of the concentrated load W. These are, from Example 8.2.2(a),

$$\Delta_1 = +\frac{Wab}{6EI}\left(\frac{L+b}{L}\right)$$

$$\Delta_2 = -\frac{Wab}{6EI}\left(\frac{L+a}{L}\right)$$

Alternatively, using Maxwell's theorem of reciprocal deflections,

$$\Delta_1 = +W \text{ times } CC' \text{ in Fig. 14.3.1c}$$
$$= +W(CC_1 - C'C_1)$$
$$= +W\left[b\left(\frac{L}{6EI}\right) - \frac{1}{2}\frac{b^2}{LEI}\left(\frac{b}{3}\right)\right]$$
$$= +\frac{Wab(L+b)}{6LEI} \quad \text{(positive means clockwise)}$$

$$\Delta_2 = -W \text{ times } CC' \text{ in Fig. 14.3.1d}$$

$$= -W(CC_1 - C'C_1)$$

$$= -W\left[a\left(\frac{L}{6EI}\right) - \frac{1}{2}\frac{a^2}{LEI}\left(\frac{a}{3}\right)\right]$$

$$= -\frac{Wab(L + a)}{6LEI} \quad \text{(negative means counterclockwise)}$$

(c) *The matrix equation* $[\delta]\{R\} + \{\Delta\} = 0$

$$\begin{bmatrix} +\dfrac{L}{3EI} & -\dfrac{L}{6EI} \\[2mm] -\dfrac{L}{6EI} & +\dfrac{L}{3EI} \end{bmatrix}\begin{Bmatrix} R_1 \\[2mm] R_2 \end{Bmatrix} + \begin{Bmatrix} +\dfrac{Wab(L+b)}{6LEI} \\[2mm] -\dfrac{Wab(L+a)}{6LEI} \end{Bmatrix} = 0$$

$$\begin{Bmatrix} R_1 \\[2mm] R_2 \end{Bmatrix} = \begin{Bmatrix} -\dfrac{Wab^2}{L^2} \\[2mm] +\dfrac{Wba^2}{L^2} \end{Bmatrix}$$

The results above are identical to those of Example 8.2.2.

Example 14.3.2

Using the matrix force method, obtain formulas for the redundants R_1 (clockwise moment acting at the right end) and R_2 (upward reaction at the right end) as shown in Fig. 14.3.2 for the fixed-ended beam subject to a concentrated load.

Solution

(a) *The matrix* $[\delta]$. Referring to Figs. 14.3.2c and 14.3.2d and using the moment area method,

$$\delta(R_1, R_1) = +L\left(\frac{1}{EI}\right) = +\frac{L}{EI}$$

$$\delta(R_2, R_1) = -\left(\frac{L}{EI}\right)\left(\frac{L}{2}\right) = -\frac{L^2}{2EI}$$

$$\delta(R_1, R_2) = -\frac{1}{2}\left(\frac{L}{EI}\right)(L) = -\frac{L^2}{2EI}$$

$$\delta(R_2, R_2) = +\frac{1}{2}\left(\frac{L^2}{EI}\right)\left(\frac{2}{3}L\right) = +\frac{L^3}{3EI}$$

Note that $\delta(R_2, R_1)$ is the deflection in the positive R_2 direction (i.e., upward) due to a unit moment in the positive R_1 direction (i.e., clockwise). Since $\delta(R_2, R_1)$ is actually downward, it is given the negative sign. Similarly, $\delta(R_1, R_2)$ is the rotation in the positive R_1 direction (i.e., clockwise) due to a unit force in the positive R_2 direction (i.e., upward). Since $\delta(R_1, R_2)$ is actually counterclockwise, it is given the negative sign. The fact that $\delta(R_2, R_1)$ is equal to $\delta(R_1, R_2)$ is a natural consequence of Maxwell's theorem of reciprocal deflections, even though in this case $\delta(R_1, R_2)$ is a rotation and $\delta(R_2, R_1)$ is a linear deflection. Thus,

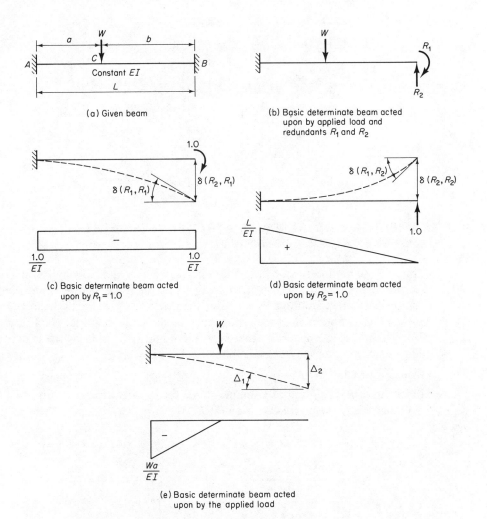

Figure 14.3.2 Matrix analysis of a fixed-ended beam using reacting moment and force at one end as redundants.

$$[\delta] = \begin{bmatrix} +\dfrac{L}{EI} & -\dfrac{L^2}{2EI} \\[2ex] -\dfrac{L^2}{2EI} & +\dfrac{L^3}{3EI} \end{bmatrix}$$

(b) *The matrix* $\{\Delta\}$. Referring to Fig. 14.3.2e and using the moment area method,

$$\Delta_1 = +\frac{1}{2}\frac{Wa}{EI}(a) = +\frac{Wa^2}{2EI}$$

$$\Delta_2 = -\frac{1}{2}\frac{Wa^2}{EI}\left(b + \frac{2}{3}a\right) = -\frac{Wa^2}{6EI}(2a + 3b)$$

(c) *The matrix equation* $[\delta]\{R\} + \{\Delta\} = 0$

$$\begin{bmatrix} +\dfrac{L}{EI} & -\dfrac{L^2}{2EI} \\[2ex] -\dfrac{L^2}{2EI} & +\dfrac{L^3}{3EI} \end{bmatrix} \begin{Bmatrix} R_1 \\[2ex] R_2 \end{Bmatrix} + \begin{Bmatrix} +\dfrac{Wa^2}{2EI} \\[2ex] -\dfrac{Wa^2}{6EI}(2a + 3b) \end{Bmatrix} = 0$$

$$\begin{Bmatrix} R_1 \\[2ex] R_2 \end{Bmatrix} = \begin{Bmatrix} +\dfrac{Wba^2}{L^2} \\[2ex] +\dfrac{Wa^2}{L^3}(a + 3b) \end{Bmatrix}$$

The results above are identical to those of Example 8.2.2.

14.4 FRAME ANALYSIS BY THE MATRIX FORCE METHOD USING FLEXIBILITY COEFFICIENTS

Consider the rigid frame shown in Fig. 14.4.1a, which is statically indeterminate to the second degree. This frame has been analyzed by the force method (consistent deformation method) in Example 11.6.2, using the reaction H_A (to the left) and V_A (upward) at the hinged support A as the redundants, as shown in Fig. 14.4.1b. The procedure followed in that example can again be put into matrix form using the matrix of flexibility coefficients as shown in the following example.

Example 14.4.1

Referring to the solution in Example 11.6.2 for the rigid frame of Fig. 14.4.1, write out numerically the matrix equation $[\delta]\{R\} + \{\Delta\} = 0$.

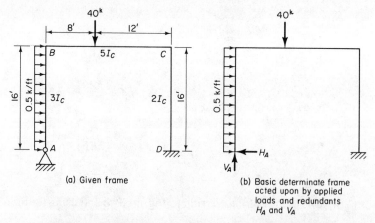

(a) Given frame

(b) Basic determinate frame acted upon by applied loads and redundants H_A and V_A

Figure 14.4.1 Frame analysis by the matrix force method.

Solution. From parts (a) and (c) of the solution to Example 11.6.2, the horizontal deflection at A due to the applied loads acting on the basic statically determinate cantilever structure is $46,762.66/EI_c$ to the right, which makes Δ_1 negative as the actual deflection is in the opposite direction to H_A. Similarly, the vertical deflection Δ_2 is

$-88{,}576/EI_c$ because it is actually downward, opposite to the positive direction for V_A. Thus,

$$\{\Delta\} = \left\{ \begin{array}{c} -\dfrac{46{,}762.66}{EI_c} \\[2ex] -\dfrac{88{,}576}{EI_c} \end{array} \right\}$$

From parts (d) to (f) of the solution to Example 11.6.2,

$$[\delta] = \left[\begin{array}{cc} +\dfrac{2161.78}{EI_c} & +\dfrac{1920}{EI_c} \\[2ex] +\dfrac{1920}{EI_c} & +\dfrac{3733.33}{EI_c} \end{array} \right]$$

The matrix equation is, canceling the denominator EI_c in every element,

$$\left[\begin{array}{cc} +2161.78 & +1920 \\ +1920 & +3733.33 \end{array} \right] \left\{ \begin{array}{c} H_A \\ V_A \end{array} \right\} + \left\{ \begin{array}{c} -44{,}762.66 \\ -88{,}576 \end{array} \right\} = 0$$

$$\left\{ \begin{array}{c} H_A \\ V_A \end{array} \right\} = \left\{ \begin{array}{c} +1.030 \text{ kips} \\ +23.196 \text{ kips} \end{array} \right\}$$

14.5 USE OF CASTIGLIANO'S THEOREM, PART II, IN THE MATRIX FORCE METHOD

The importance of the Castigliano's theorem, parts I and II, in structural analysis, conceptually speaking, cannot be overemphasized. It is to be shown that Castigliano's theorem, part II, as has been described in Sections 7.7 and 7.8, provides the ideological basis for evaluating the matrices $[\delta]$ and $\{\Delta\}$ in the equation

$$[\delta]\{R\} + \{\Delta\} = 0$$

The element δ_{ij} in the $[\delta]$ matrix is simply the displacement in the direction of the ith redundant due to a unit force applied to the basic statically determinate structure in the direction of the jth redundant. The element Δ_i in the $\{\Delta\}$ matrix is simply the displacement in the direction of the ith redundant due to the applied loads on the basic statically determinate structure. All quantities δ_{ij} or Δ_i can be determined either by the physical geometric method (e.g., Williot–Mohr graphical method, moment area method, etc.), the virtual work (unit load) method, or Castigliano's theorem, part II. The virtual work (unit load) method is, of course, just the variation obtained after the partial differentiation is done in the application of Castigliano's theorem.

In the following examples, it is shown that each row in the $[\delta]$ matrix can be obtained in one operation by applying Castigliano's theorem, part II. While this alternate method does not seem to be any more convenient than the geometric method, in cases where the element or the structure is a curved bar, a plate, a shell, or even an elastic solid, the use of Castigliano's theorem may be the only viable approach.

Example 14.5.1

Obtain the $[\delta]$ matrix in Example 14.3.1 by using Castigliano's theorem, part II.

Solution. Applying Castigliano's theorem, part II, Eq. (8.7.1), to the simple beam element of Fig. 14.5.1,

$$\frac{\partial W_i}{\partial M_i} = \phi_i \quad \text{and} \quad \frac{\partial W_i}{\partial M_j} = \phi_j$$

Then, using Eq. (8.6.6),

$$W_i = \int \frac{M^2\,dx}{2EI} = \int_0^L \frac{\left(M_i - \dfrac{M_i + M_j}{L}x\right)^2 dx}{2EI}$$

$$\phi_i = \frac{\partial W_i}{\partial M_i} = \int_0^L \frac{\left(M_i - \dfrac{M_i + M_j}{L}x\right)\left(1 - \dfrac{x}{L}\right)dx}{EI} = +\left(\frac{L}{3EI}\right)M_i - \left(\frac{L}{6EI}\right)M_j$$

$$\phi_j = \frac{\partial W_i}{\partial M_j} = \int_0^L \frac{\left(M_i - \dfrac{M_i + M_j}{L}x\right)\left(-\dfrac{x}{L}\right)dx}{EI} = -\left(\frac{L}{6EI}\right)M_i + \left(\frac{L}{3EI}\right)M_j$$

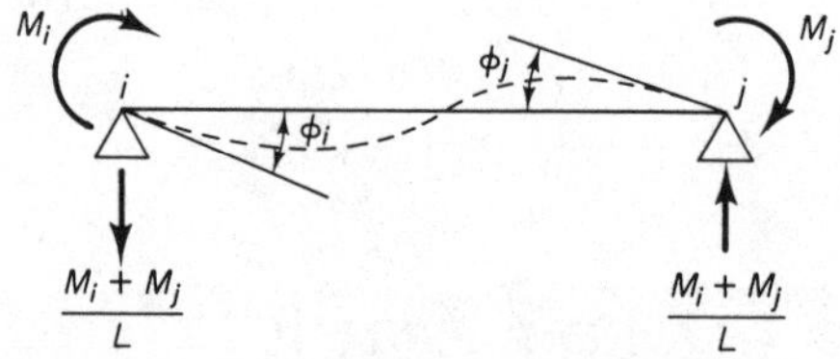

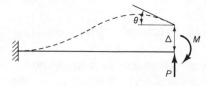

Figure 14.5.1 Flexibility matrix of a single beam element.

The two equations above are exactly identical to the first and second rows in the $[\delta]$ matrix of Example 14.3.1.

Example 14.5.2

Obtain the $[\delta]$ matrix in Example 14.3.2 by using Castigliano's theorem, part II.

Solution. Applying Castigliano's theorem, part II, Eq. (8.7.1), to the cantilever beam element of Fig. 14.5.2,

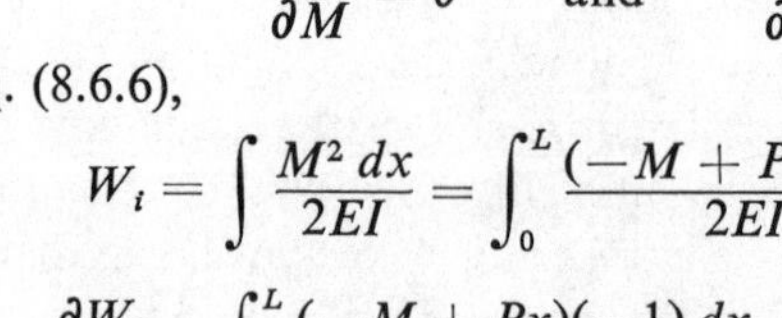

Figure 14.5.2 Flexibility matrix of a cantilever beam element.

$$\frac{\partial W_i}{\partial M} = \theta \quad \text{and} \quad \frac{\partial W_i}{\partial P} = \Delta$$

Then using Eq. (8.6.6),

$$W_i = \int \frac{M^2\,dx}{2EI} = \int_0^L \frac{(-M + Px)^2\,dx}{2EI}$$

$$\theta = \frac{\partial W_i}{\partial M} = \int_0^L \frac{(-M + Px)(-1)\,dx}{EI} = +\left(\frac{L}{EI}\right)M - \left(\frac{L^2}{2EI}\right)P$$

$$\Delta = \frac{\partial W_i}{\partial P} = \int_0^L \frac{(-M + Px)(x)\,dx}{EI} = -\left(\frac{L^2}{2EI}\right)M + \left(\frac{L^3}{3EI}\right)P$$

The two equations above are exactly identical to the first and second rows in the $[\delta]$ matrix of Example 14.3.2.

14.6 TRUSS ANALYSIS BY THE MATRIX DISPLACEMENT METHOD USING STIFFNESS COEFFICIENTS

The basic concept in the displacement method of truss analysis has been described in Section 7.4 by means of the three-bar truss shown in Fig. 7.4.1. Consider now the truss of Fig. 14.6.1. Even though this truss is statically indeterminate to the second degree from the viewpoint of the force method, in terms of the displacement method there are always as many equations of statics as there are unknown joint displacements. For the present problem, the unknown deflections are designated X_1 to X_8, as shown in Fig. 14.6.1b. The P-forces in the directions of the joint displacements, as shown in Fig. 14.6.1c, can be expressed in terms of the unknown X-values through the stiffness coefficients K_{ij} for $i = 1$ to 8 and $j = 1$ to 8; thus

$$\left.\begin{array}{l} P_1 = K_{11}X_1 + K_{12}X_2 + \ldots + K_{18}X_8 \\ P_2 = K_{21}X_1 + K_{22}X_2 + \ldots + K_{28}X_8 \\ \quad \cdot \\ \quad \cdot \\ \quad \cdot \\ P_8 = K_{81}X_1 + K_{82}X_2 + \ldots + K_{88}X_8 \end{array}\right\} \quad (14.6.1)$$

Or, in matrix notation,

$$\{P\}_{8 \times 1} = [K]_{8 \times 8}\{X\}_{8 \times 1} \quad (14.6.2)$$

The $[K]$ matrix in Eq. (14.6.2) is the stiffness matrix of the truss as a whole.

The stiffness coefficients K_{ij} can be determined by a physically comprehensible procedure as follows:

1. Express the elongations e_1 to e_{10} in each of the 10 bars as linear functions of the joint displacements X_1 to X_8, by means of the joint displacement equation (3.4.1).
2. Express the forces F_1 to F_{10} in each of the 10 bars in terms of the bar elongations e_1 to e_{10} by multiplying each elongation by the respective stiffness (value of EA/L) of the bar.
3. Express the joint forces P_1 to P_8 as linear functions of the bar forces F_1 to F_{10} by applying the equation $\sum F_x = 0$ or $\sum F_y = 0$ to each joint in the P-directions.
4. Express the bar forces F_1 to F_{10} as linear functions of the joint displacements X_1 to X_8 by substituting the results of step 1 in step 2.
5. Express the joint forces P_1 to P_8 as linear functions of the joint displacements X_1 to X_8 by substituting the results of step 4 in step 3.

The procedure described above can be placed in matrix form and is applied to a number of problems in Chapter 16.

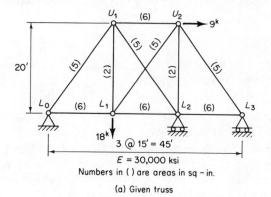

(a) Given truss

E = 30,000 ksi
Numbers in ( ) are areas in sq – in.

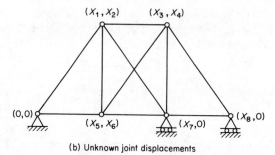

(b) Unknown joint displacements

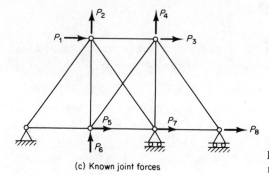

(c) Known joint forces

Figure 14.6.1 Truss analysis by the matrix displacement method.

As an alternative, one can first imagine that there is only one typical bar in a truss, as shown in Fig. 14.6.2. The forces P_1 to P_4 can be expressed in terms of the displacements X_1 to X_4 by doing exactly the same five steps as stated previously for the entire truss of Fig. 14.6.1. For the one-bar truss, the stiffness matrix will then be

$$\{P\}_{4 \times 1} = [K]_{4 \times 4}\{X\}_{4 \times 1} \tag{14.6.3}$$

Then again, if the truss has more than one bar, such as the 10 bars contained in the truss of Fig. 14.6.1, it is convenient to just accumulate the $[K]$ matrix of each bar as expressed by Eq. (14.6.3) into the "appropriate" slots in the $[K]$ matrix of the truss as a whole, by means of a "proper" algorithm in a computer program. This approach is described in more detail in Chapters 16 to 20.

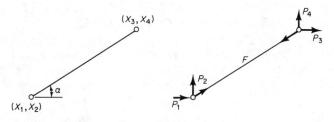

Figure 14.6.2 Stiffness matrix of a single bar in a truss.

Later, in Section 14.9, it is shown that, either the [K] matrix of the truss as a whole, or the [K] matrix of a single bar, can be established by using Castigliano's theorem, part I. For a skeleton structure, the five physical steps described at the beginning of this section are not only more easily understandable but also take fewer algebraic operations. But, again, if the element or the structure is a curved bar, a plate, a shell, or even an elastic solid, the use of Castigliano's theorem may be the only feasible approach.

Example 14.6.1

Using the physical approach, derive the 4 by 4 stiffness matrix of a single bar in a truss, inclined at angle α with the positive x-axis toward the right.

Solution. Referring to Fig. 14.6.2, and using the joint displacement equation (3.4.1),

$$e = (X_3 - X_1) \cos \alpha + (X_4 - X_2) \sin \alpha$$

Using the stiffness $S = EA/L$ of the bar,

$$F = Se = S(X_3 - X_1) \cos \alpha + S(X_4 - X_2) \sin \alpha$$

Writing out the equations of equilibrium at the lower and upper ends of the bar,

$P_1 = -F \cos \alpha$

$$= +S \cos^2 \alpha\, X_1 + S \sin \alpha \cos \alpha\, X_2 - S \cos^2 \alpha\, X_3 - S \sin \alpha \cos \alpha\, X_4$$

$P_2 = -F \sin \alpha$

$$= +S \sin \alpha \cos \alpha\, X_1 + S \sin^2 \alpha\, X_2 - S \sin \alpha \cos \alpha\, X_3 - S \sin^2 \alpha\, X_4$$

$P_3 = +F \cos \alpha$

$$= -S \cos^2 \alpha\, X_1 - S \sin \alpha \cos \alpha\, X_2 + S \cos^2 \alpha\, X_3 + S \sin \alpha \cos \alpha\, X_4$$

$P_4 = +F \sin \alpha$

$$= -S \sin \alpha \cos \alpha\, X_1 - S \sin^2 \alpha\, X_2 + S \sin \alpha \cos \alpha\, X_3 + S \sin^2 \alpha\, X_4$$

It can be observed from the four equations above that the [K] matrix of a single bar in a truss is

$$[K] = \begin{bmatrix} +T1 & +T2 & -T1 & -T2 \\ +T2 & +T3 & -T2 & -T3 \\ -T1 & -T2 & +T1 & +T2 \\ -T2 & -T3 & +T2 & +T3 \end{bmatrix} \tag{14.6.4}$$

in which $T1 = S \cos^2 \alpha$; $T2 = S \sin \alpha \cos \alpha$; $T3 = S \sin^2 \alpha$.

14.7 BEAM ANALYSIS BY THE MATRIX DISPLACEMENT METHOD USING STIFFNESS COEFFICIENTS

The analysis of beams, continuous beams in particular, by the matrix displacement method using stiffness coefficients is nothing more than placing the slope deflection method as described in Chapter 9 in matrix notation. Referring to the continuous beam of Fig. 14.7.1a, if it is to be analyzed by the slope deflection method using three segments and four joints, it will have four unknown displacements $\theta_B, \theta_C, \theta_D,$ and Δ_D, as shown in Fig. 14.7.2b. These unknown displacements are due to the action of the reversals of the fixed-end moments and reactions shown in Fig. 14.7.1c (identical to Fig. 9.3.3b). These reversals are the P-forces of Fig. 14.7.1d, actively causing the joint displacements X_1 to X_4. They are, in this case,

$$P_1 = +43.05 \text{ ft-kips}; \quad P_2 = -90.00 \text{ ft-kips};$$
$$P_3 = -6.00 \text{ ft-kips}; \quad P_4 = +6.00 \text{ kips}$$

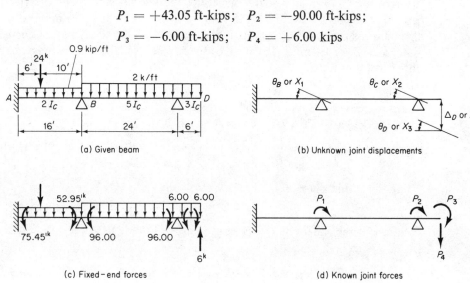

(a) Given beam

(b) Unknown joint displacements

(c) Fixed–end forces

(d) Known joint forces

Figure 14.7.1 Beam analysis by matrix displacement method.

The problem is now analogous to truss analysis, wherein the forces are all applied at the joints only. It is necessary to establish a stiffness matrix $[K]$ of the beam as a whole, by calling $\theta_B - \theta_C - \theta_D - \Delta_D$ as $X_1 - X_2 - X_3 - X_4$. In symbolic form,

$$\left.\begin{aligned}
P_1 &= K_{11}X_1 + K_{12}X_2 + K_{13}X_3 + K_{14}X_4 \\
P_2 &= K_{21}X_1 + K_{22}X_2 + K_{23}X_3 + K_{24}X_4 \\
P_3 &= K_{31}X_1 + K_{32}X_2 + K_{33}X_3 + K_{34}X_4 \\
P_4 &= K_{41}X_1 + K_{42}X_2 + K_{43}X_3 + K_{44}X_4
\end{aligned}\right\} \quad (14.7.1)$$

The stiffness coefficients K_{ij} for $i = 1$ to 4 and $j = 1$ to 4 can be obtained by the same five steps as described in Section 14.6:

1. Express the member-end rotations ϕ_1 to ϕ_6 (clockwise rotations from the member axis to the tangent to the elastic curve; see Fig. 9.2.1c or 14.7.2b) as linear functions of X_1 to X_4 by simply observing the geometry at the member ends. In this

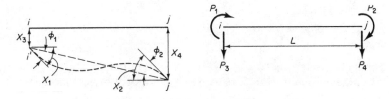

Figure 14.7.2 Stiffness matrix of a single beam element.

case,

$$\phi_1 = 0; \quad \phi_2 = X_1; \quad \phi_3 = X_1; \quad \phi_4 = X_2;$$

$$\phi_5 = X_2 - \frac{X_4}{L_{5-6}}; \quad \phi_6 = X_3 - \frac{X_4}{L_{5-6}}$$

2. Express the end moments M_1 to M_6 (exclusive of the fixed-end moments) as linear functions of member-end rotations ϕ_1 to ϕ_6 by means of Eqs. (9.2.5), which are, typically,

$$\begin{Bmatrix} M_i \\ M_j \end{Bmatrix} = \begin{bmatrix} \dfrac{4EI}{L} & \dfrac{2EI}{L} \\ \dfrac{2EI}{L} & \dfrac{4EI}{L} \end{bmatrix} \begin{Bmatrix} \phi_i \\ \phi_j \end{Bmatrix} \qquad \text{for member } i\text{-}j$$

3. Express the joint forces P_1 to P_4 as linear functions of the end moments M_1 to M_6 by applying the equations $\sum M = 0$ and $\sum F_y = 0$ to each joint in the P-directions.

4. Express the end moments M_1 to M_6 as linear functions of the joint displacements X_1 to X_4 by substituting the results of step 1 in step 2.

5. Express the joint forces P_1 to P_4 as linear functions of the joint displacements X_1 to X_4 by substituting the results of step 4 in step 3.

The procedure as described above can be placed in matrix form and is applied to a number of problems in Chapter 17.

As an alternative, one can first consider that there is only one typical segment in a beam, as shown in Fig. 14.7.2. The forces P_1 to P_4 can be expressed in terms of the joint displacements X_1 to X_4 in exactly the same five steps described previously for the entire beam of Fig. 14.7.1. For the single segment, the stiffness matrix will be

$$\{P\}_{4\times 1} = [K]_{4\times 4}\{X\}_{4\times 1} \tag{14.7.2}$$

Thus, if the entire beam has more than one segment, such as the three segments contained in the beam of Fig. 14.7.1, it is convenient to accumulate the $[K]$ matrix of each segment as expressed by Eq. (14.7.2) into the "appropriate" slots in the $[K]$ matrix of the beam as a whole, by means of a "proper" algorithm in a computer program. This approach is described in more detail in Chapters 16 to 20.

As discussed in Section 14.6, either the $[K]$ matrix of the beam as a whole, or the $[K]$ matrix of a single beam segment, can be established by using Castigliano's theorem, part II. The former is rather cumbersome, but the latter will be demonstrated in Section 14.9.

Example 14.7.1

Using the physical approach, derive the 4 by 4 stiffness matrix of a single beam segment.

Solution. Referring to Fig. 14.7.2, and using the slope deflection equations (9.2.6), which have been derived in Chapter 9 using the physical moment area method,

$$P_1 = \frac{4EI}{L}\left(X_1 - \frac{X_4 - X_3}{L}\right) + \frac{2EI}{L}\left(X_2 - \frac{X_4 - X_3}{L}\right) \tag{a}$$

$$P_2 = \frac{2EI}{L}\left(X_1 - \frac{X_4 - X_3}{L}\right) + \frac{4EI}{L}\left(X_2 - \frac{X_4 - X_3}{L}\right) \tag{b}$$

and, using statics,

$$P_3 = +\frac{P_1 + P_2}{L} \tag{c}$$

$$P_4 = -\frac{P_1 + P_2}{L} \tag{d}$$

Rearranging Eqs. (a) to (d) so as to express P_1 to P_4 in terms of X_1 to X_4, the $[K]$ matrix of a single beam segment is

$$[K] = \begin{bmatrix} +\dfrac{4EI}{L} & +\dfrac{2EI}{L} & +\dfrac{6EI}{L^2} & -\dfrac{6EI}{L^2} \\[2mm] +\dfrac{2EI}{L} & +\dfrac{4EI}{L} & +\dfrac{6EI}{L^2} & -\dfrac{6EI}{L^2} \\[2mm] +\dfrac{6EI}{L^2} & +\dfrac{6EI}{L^2} & +\dfrac{12EI}{L^3} & -\dfrac{12EI}{L^3} \\[2mm] -\dfrac{6EI}{L^2} & -\dfrac{6EI}{L^2} & -\dfrac{12EI}{L^3} & +\dfrac{12EI}{L^3} \end{bmatrix}$$

14.8 FRAME ANALYSIS BY THE MATRIX DISPLACEMENT METHOD USING STIFFNESS COEFFICIENTS

Again the analysis of rigid frames by the matrix displacement method using stiffness coefficients is simply to put the slope deflection method described in Section 11.7 in matrix notation.

The discussion in Section 14.7 relating to beams applies equally well here. The detailed treatment of rigid frame analysis by the matrix displacement method wherein axial deformation is not considered is in Chapters 18 and 19. When both axial and bending deformations are to be considered, the stiffness matrix of a single segment is a 6 by 6 matrix, combining the stiffness matrices of a truss element and a beam element. The treatment for this situation is in Chapter 20.

14.9 USE OF CASTIGLIANO'S THEOREM, PART I, IN THE MATRIX DISPLACEMENT METHOD

The Castigliano's theorem, part I, elegantly forms the basis of the displacement method (or stiffness method) of structural analysis. Its application to the displacement method of truss analysis is fully explained in Section 7.6 and illustrated in Example 7.6.1. In that example, the degree of freedom of the two-bar truss is two, so that the determination of the 2 by 2 stiffness matrix of the truss as a whole is relatively simple. As the degree of freedom becomes larger, the amount of repeated writing in setting $\partial W/\partial X_i$ (for $i = 1$ to NP, where NP is the degree of freedom) to equal P_i for each degree of freedom becomes cumbersome. Thus, the practice of accumulating the $[K]$ matrix of each single bar in a truss into the appropriate locations in the $[K]$ matrix of the truss as a whole becomes appealing. In Example 14.9.1, the $[K]$ matrix of a single bar in a truss will be derived by using Castigliano's theorem, part I.

For determining the $[K]$ matrix of a beam (containing many segments) as a whole, the elastic curve of each segment will have to be expressed in terms of the unknown joint displacements at its ends. Then the total strain energy in the entire beam is written in terms of all the unknown joint displacements by means of Eq. (8.6.7), which is

$$W_t = \int_0^L \frac{EI}{2} \left(\frac{d^2 y}{dx^2}\right)^2 dx \qquad [8.6.7]$$

Only then, the partial derivative of W_t with respect to each X_i is set to equal P_i. Again, as is more commonly done in computer programming, it is more convenient to use the compilation method. In Example 14.9.2, the $[K]$ matrix of a single beam segment is derived by using Castigliano's theorem, part I.

Example 14.9.1

Using Castigliano's theorem, part I, derive the 4 by 4 stiffness matrix of a single bar in a truss, inclined at angle α with the positive x-axis toward the right.

Solution.   The total strain energy in the bar shown in Fig. 14.6.2 is equal to one-half of the product of the force in it and its elongation. The elongation e is, by Eq. (3.4.1),

$$e = (X_3 - X_1) \cos \alpha + (X_4 - X_2) \sin \alpha$$

and the force is, for $S = EA/L$,

$$F = Se = S(X_3 - X_1) \cos \alpha + S(X_4 - X_2) \sin \alpha$$

The strain energy W_t is

$$W_t = \frac{1}{2} Fe = \frac{1}{2} S[(X_3 - X_1) \cos \alpha + (X_4 - X_2) \sin \alpha]^2$$

Differentiating the expression above with respect to X_1, X_2, X_3, and X_4,

$$P_1 = \frac{\partial W_t}{\partial X_1} = +S \cos^2 \alpha\, X_1 + S \sin \alpha \cos \alpha\, X_2 - S \cos^2 \alpha\, X_3 - S \sin \alpha \cos \alpha\, X_4$$

$$P_2 = \frac{\partial W_t}{\partial X_2} = +S \sin \alpha \cos \alpha\, X_1 + S \sin^2 \alpha\, X_2 - S \sin \alpha \cos \alpha\, X_3 - S \sin^2 \alpha\, X_4$$

$$P_3 = \frac{\partial W_t}{\partial X_3} = -S \cos^2 \alpha\, X_1 - S \sin \alpha \cos \alpha\, X_2 + S \cos^2 \alpha\, X_3 + S \sin \alpha \cos \alpha\, X_4$$

$$P_4 = \frac{\partial W_t}{\partial X_4} = -S \sin \alpha \cos \alpha\, X_1 - S \sin^2 \alpha\, X_2 + S \sin \alpha \cos \alpha\, X_3 + S \sin^2 \alpha\, X_4$$

The results above are identical to those of Example 14.6.1, where P_1 to P_4 are solved by using $\sum F_x = 0$ and $\sum F_y = 0$ at the lower and upper ends of the bar.

Example 14.9.2

Using Castigliano's theorem, part I, derive the 4 by 4 stiffness matrix of a single beam element.

Solution. Referring to Fig. 5.1.2, the differential equation of the elastic curve of an unloaded beam is

$$\frac{d^4y}{dx^4} = 0$$

which means that the equation of the elastic curve is a cubic equation, or

$$y = a_0 + a_1x + a_2x^2 + a_3x^3$$

From the boundary conditions shown in Fig. 14.7.2:

1. When $x = 0$, $\frac{dy}{dx} = +X_1$

$$a_1 = +X_1 \tag{a}$$

2. When $x = L$, $\frac{dy}{dx} = +X_2$

$$a_1 + 2a_2L + 3a_3L^2 = +X_2 \tag{b}$$

3. When $x = 0$, $y = +X_3$

$$a_0 = +X_3 \tag{c}$$

4. When $x = L$, $y = +X_4$

$$a_0 + a_1L + a_2L^2 + a_3L^3 = +X_4 \tag{d}$$

Solving Eqs. (a) to (d) simultaneously for a_0, a_1, a_2, and a_3,

$$a_0 = +X_3$$
$$a_1 = +X_1$$
$$a_2 = -\frac{2X_1}{L} - \frac{X_2}{L} - \frac{3X_3}{L^2} + \frac{3X_4}{L^2}$$
$$a_3 = +\frac{X_1}{L^2} + \frac{X_2}{L^2} + \frac{2X_3}{L^3} - \frac{2X_4}{L^3}$$

Then

$$\frac{d^2y}{dx^2} = 2a_2 + 6a_3x$$

$$= \left(-\frac{4X_1}{L} - \frac{2X_2}{L} - \frac{6X_3}{L^2} + \frac{6X_4}{L^2}\right) + \left(\frac{6X_1}{L^2} + \frac{6X_2}{L^2} + \frac{12X_3}{L^3} - \frac{12X_4}{L^3}\right)x$$

Using Eqs. (8.6.7),

$$W_i = \int_0^L \frac{EI}{2}\left(\frac{d^2y}{dx^2}\right)^2 dx = \frac{EI}{2}\int_0^L (2a_2 + 6a_3x)^2 \, dx$$

Differentiating the expression above with respect to X_1, X_2, X_3, and X_4,

$$P_1 = \frac{\partial W_i}{\partial X_1} = EI \int_0^L (2a_2 + 6a_3 x)\left(-\frac{4}{L} + \frac{6x}{L^2}\right) dx = EI(-2a_2)$$

$$= \frac{4EI}{L} X_1 + \frac{2EI}{L} X_2 + \frac{6EI}{L^2} X_3 - \frac{6EI}{L^2} X_4$$

$$P_2 = \frac{\partial W_i}{\partial X_2} = EI \int_0^L (2a_2 + 6a_3 x)\left(-\frac{2}{L} + \frac{6x}{L^2}\right) dx = EI(2a_2 + 6a_3 L)$$

$$= \frac{2EI}{L} X_1 + \frac{4EI}{L} X_2 + \frac{6EI}{L^2} X_3 - \frac{6EI}{L^2} X_4$$

$$P_3 = \frac{\partial W_i}{\partial X_3} = EI \int_0^L (2a_3 + 6a_3 x)\left(-\frac{6}{L^2} + \frac{12x}{L^3}\right) dx = 6a_3$$

$$= +\frac{6EI}{L^2} X_1 + \frac{6EI}{L^2} X_2 + \frac{12EI}{L^3} X_3 - \frac{12EI}{L^3} X_4$$

$$P_4 = \frac{\partial W_i}{\partial X_4} = EI \int_0^L (2a_2 + 6a_3 x)\left(+\frac{6}{L^2} - \frac{12x}{L^3}\right) dx = -6a_3$$

$$= -\frac{6EI}{L^2} X_1 - \frac{6EI}{L^2} X_2 - \frac{12EI}{L^3} X_3 + \frac{12EI}{L^3} X_4$$

The results above are identical to those of Example 14.7.1, where P_1 and P_2 are obtained by applying the slope deflection equations (9.2.6), and P_3 and P_4 are simply the resisting shear force equal to $(P_1 + P_2)/L$.

14.10 COMPARISON OF THE MATRIX FORCE METHOD WITH THE MATRIX DISPLACEMENT METHOD

Before the advent of the electronic computer, the force method (the flexibility method, or the consistent deformation method) had been used almost exclusively for the analysis of statically indeterminate trusses, because using the displacement method (the stiffness method) would always require the solution of a large number of simultaneous equations. For continuous beams, both the three-moment equation method and the slope deflection method involve a fair number of simultaneous equations; the former is a force method, while the latter is a displacement method. For rigid frames, the consistent deformation method is not favored, except possibly for single-span, single-story frames. For multispan, multistory rigid frames, the slope deflection method or the moment distribution method has been used.

In the early stages of computer availability, say in the late 1950s and early 1960s, the matrix force method had been more used than the matrix displacement method, even in the structural analysis of aircraft structures. Because of the limited size and speed of the computer, only the compatibility equations in the force method were solved as simultaneous equations, whereas the elements in the flexibility matrix $[\delta]$ and the deflection matrix $\{\Delta\}$, as shown in Eq. (14.2.3), were computed in subroutines by programming the method of computing deflections in statically determinate structures. Later, when computers could accommodate larger problems and became faster, the statics of the basic statically determinate structures themselves were solved by inverting the statics matrix $[A]$ (see Chapter 15). It then seemed that there was little purpose of inverting the $[A]$ matrix, when the $[K]$ matrix of the entire structure is about the same size.

In recent years, computers have rapidly become cheaper, faster, and able to accommodate larger problems. As a consequence, the matrix displacement method is universally used for structural analysis, largely because of its simple formulation. As described in Sections 14.7 to 14.9, the stiffness matrix $[K]$ of the entire structure (called the global stiffness matrix in Chapters 16 to 20) can be built up as such or by compiling the stiffness matrix $[K]$ of each single element (called the local stiffness matrix in Chapters 16 to 20). Furthermore, each stiffness matrix can be obtained either by the physical approach (using the five steps described in Sections 14.6 and 14.7) or by applying Castigliano's theorem, part I.

While the compilation method (called the direct element method, or the direct stiffness method) is suitable for a computer algorithm, it is not as amenable to physical comprehension, at least for the beginner. Consequently, in Chapters 16 to 20, the emphasis is on building up the stiffness matrix of the structure as a whole by the physical approach and on making the statics and deformation checks for the resulting internal forces and joint displacements. Indeed, structural analysis can all be based on equilibrium, Hooke's law, and compatibility on one hand, and Castigliano's theorem, parts I and II, on the other hand.

Once more it should be noted that as the student advances to the study of the finite element method of stress analysis (plane stress, plates, shells, solids, etc.), the energy theorems (Castigliano's theorems, Betti's reciprocal energy theorems, etc.) will play a more important role than the physical and geometric approach.

SELECTED REFERENCES

1. J. S. Archer, "Digital Computation for Stiffness Matrix Analysis," *Journal of the Structural Division*, ASCE, *84*, ST6 (October 1958), Paper 1814, 1–16.

2. J. H. Argyris, "Energy Theorems and Structural Analysis, Part I—General Theory," *Aircraft Engineering*, *26* (October 1954); *26* (November 1954); *27* (February 1955); *27* (March 1955); *27* (April 1955); *27* (May 1955).

3. Stanley U. Benscoter, "Matrix Analysis of Continuous Beams," *Transactions*, ASCE, *112* (1947), 1109–1140, Discussion, 1141–1172.

4. Pei-Ping Chen, "Matrix Analysis of Pin-Connected Structures," *Transactions*, ASCE, *114* (1949), 181–192.

5. R. W. Clough, "Matrix Analysis of Beams," *Journal of Engineering Mechanics Division*, ASCE, *84*, EM1 (January 1958), Paper 1494, 1–24.

6. P. H. Denke, "A General Digital Computer Analysis of Statically Indeterminate Structures," NASA, TN D-1666, 1962.

7. P. H. Denke and C. K. Wang, "Analysis of Indeterminate Structures by the Displacement Method," *Journal of Applied Mechanics*, ASME, *26*, Series E, 3 (September 1959), 455–456.

8. Harold C. Martin, "Truss Analysis by Stiffness Considerations," *Transactions*, ASCE, *123* (1958), 1182–1194.

9. Robert J. Melosh, "Matrix Methods of Structural Analysis," *Journal of the Aeronautical Sciences*, *29*, 3 (March 1962), 365–366.

10. Ming L. Pei, "Stiffness Method of Rigid Frame Analysis," *Proceedings of the Second Conference on Electronic Computation*, ASCE, 1960, 225–248.

11. J. S. Przemieniecki, "Matrix Structural Analysis of Substructure," *AIAA Journal, 1*, 1 (January 1963), 138–147.

12. M. J. Turner, R. W. Clough, M. C. Martin, and L. J. Topp, "Stiffness and Deflection Analysis of Complex Structures," *Journal of the Aeronautical Sciences, 23*, 9 (September 1956), 805–823.

13. Chu-Kia Wang, "Matrix Formulation of Slope-Deflection Equations," *Journal of Structural Division*, ASCE, *84*, ST6 (October 1958), Paper 1819, 1–19.

14. Chu-Kia Wang, "Matrix Analysis of Statically Indeterminate Trusses," *Journal of Structural Division*, ASCE, *85*, ST4 (April 1959), 23–36.

[15]

Matrix Formulation
—Method of Joints

15.1 DEFINITION OF STATICS MATRIX FOR A STATICALLY DETERMINATE TRUSS

In Section 2.1 it was shown that when a truss is statically determinate, the total number NP of possible external forces that may act at the joints must equal the number NF of internal axial forces. At each joint, two equations of statics may be applied in any two orthogonal directions.

Refer to the truss in Fig. 15.1.1, where the internal forces (and therefore the members) are identified by number and all possible external forces are shown and numbered. For clarity, the P-numbers and F-numbers are shown in two separate diagrams. Note that components in any two directions (chosen here horizontal and vertical for simplicity) completely define a force. If this truss were to be analyzed algebraically (as in Example 2.2.1) by the method of joints, only the nonzero P-forces would actually appear in the statics equations for the joints. The matrix formulation, on the other hand, is general and includes all *possible* P-forces.

For joint U_1 of Fig. 15.1.1, the free-body diagram of which is shown in Fig. 15.1.2, the two statics equations are

$$\left. \begin{array}{l} P_1 = 0.6F_1 - F_2 - 0.6F_8 \\ P_2 = 0.8F_1 + F_7 + 0.8F_8 \end{array} \right\} \quad (15.1.1)$$

These two equations may be directly written from the facts that P_1, the force acting to the right on joint U_1, is resisted positively by 60% of F_1, negatively by 100% of F_2, and negatively by 60% of F_8, and that P_2, the force acting upward on joint U_1, is resisted positively by 80% of F_1, positively by 100% of F_7, and positively by 80%

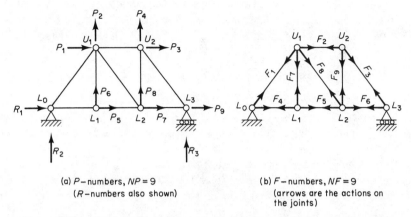

(a) P–numbers, $NP = 9$
(R–numbers also shown)

(b) F–numbers, $NF = 9$
(arrows are the actions on the joints)

Figure 15.1.1 Statically determinate truss showing numbering of all possible external forces and numbering of member axial forces.

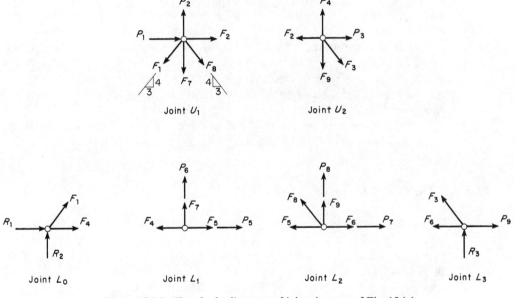

Figure 15.1.2 Free-body diagrams of joints in truss of Fig. 15.1.1.

of F_8. In a similar fashion, the equations of statics may be written for all other joints. In other words, there will be one equation for each P, that is, NP equations. The right-hand sides of these equations involve the internal forces F, that is, NF terms. Of course, not every F appears in every equation.

These statics equations relating the P external forces to the F internal forces may be written in matrix form as

$$\{P\}_{\text{NP}\times 1} = [A]_{\text{NP}\times\text{NF}}\{F\}_{\text{NF}\times 1} \tag{15.1.2}$$

The $[A]$ matrix as defined by Eq. (15.1.2) is called the *statics matrix* (sometimes the *equilibrium matrix*) because its rows are the equations of statics (or of equilibrium) for the joints. The coefficients of the F's in Eqs. (15.1.1) give the nonzero terms in the first two rows of $[A]$. Fully written out, Eqs. (15.1.1) would read

$$P_1 = 0.6F_1 - F_2 + 0F_3 + 0F_4 + 0F_5 + 0F_6 + 0F_7 - 0.6F_8 + 0F_9$$

$$P_2 = 0.8F_1 + 0F_2 + 0F_3 + 0F_4 + 0F_5 + 0F_6 + F_7 + 0.8F_8 + 0F_9$$

Note that the above two equations for equilibrium at joint U_1 are valid *whatever the values* of P_1 and P_2. The statics matrix is thus a property of the structure and not dependent on the specific loading (i.e., the specific set of values for the NP external forces).

The $[A]$ matrix may be either a square or a rectangular matrix. When the structure is statically determinate, NF equals NP and the $[A]$ matrix is square. A statically indeterminate structure will have NF greater than NP and the $[A]$ matrix will be rectangular.

For the statically determinate truss where $NF = NP$, Eq. (15.1.2) may be solved for the internal forces F; thus,

$$\{F\}_{NP \times 1} = [A^{-1}]_{NF \times NP} \{P\}_{NP \times 1} \tag{15.1.3}$$

This equation may be programmed for a computer with the external force matrix $\{P\}$ and the statics matrix $[A]$ as input, and the internal force matrix $\{F\}$ as output.

The physical meaning of the $[A^{-1}]$ matrix is that each of its columns gives the set of internal axial forces caused by one of the external forces having a unit magnitude. For sake of an academic exercise, using its physical meaning, the $[A^{-1}]$ matrix may be determined directly by the conventional method of joints and sections, and its correctness verified by confirming that $[A][A^{-1}] = [I]$.

If there exists an inverse of the statics matrix $[A]$, the truss is statically stable, but if the statics matrix $[A]$ is singular, the truss is statically unstable.

15.2 MATRIX METHOD OF JOINTS

The matrix method of joints may be applied only to a statically determinate truss, that is, where NF equals NP. The statics matrix $[A]$ is established by rows in the same manner as for an algebraic solution. Then the $[A]$ matrix is inverted. Finally, the set of F-values are obtained by postmultiplying the inverse static matrix $[A^{-1}]$ by the external force matrix $\{P\}$. If the solution is required for more than one set of external forces, the external force matrix will be rectangular, having as many columns as there are loading cases NLC. The final internal force matrix will also have NLC columns.

Summarizing the matrix method of joints for statically determinate trusses in one all-inclusive formula,

$$[F]_{NF \times NLC} = [A^{-1}]_{NF \times NP} [P]_{NP \times NLC} \tag{15.2.1}$$

wherein $NF = NP$.

Example 15.2.1

For the statically determinate truss of Example 2.2.1 (Fig. 2.2.1), shown again as Fig. 15.2.1, solve for the internal forces using the matrix method of joints.

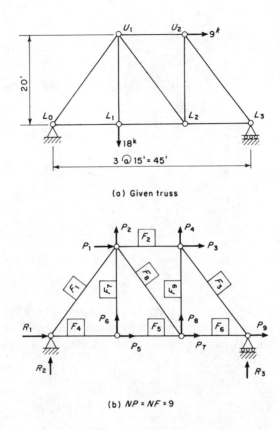

(a) Given truss

(b) $NP = NF = 9$

Figure 15.2.1 Truss for Example 15.2.1.

Solution

(a) *Establish the statics matrix* $[A]$. This matrix is a compilation of the equations of statics at each joint. For instance, using $\sum F_x = 0$ and $\sum F_y = 0$ for the joint U_1 of Fig. 15.1.2, Eqs. (15.1.1) were obtained. Those equations comprise the first two rows of the statics matrix. Using the free-body diagrams of the joints in Fig. 15.1.2 (which are the joints for the truss of this example), the remainder of the statics matrix is obtained as shown on page 444.

With a little practice, the reader will find it relatively easy to fill out the nonzero spots in the $[A]$ matrix by simply observing the two truss diagrams showing the P-numbers and the F-numbers, as in Fig. 15.1.1. For instance, to figure out the nonzero values in the 7th row, one would observe that P_7 acting to the right on joint L_2 of Fig. 15.1.1b will be resisted positively by 100% of F_5, negatively by 100% of F_6, and positively by 60% of F_8. Note that the arrows drawn as pulling forces on the joints of Fig. 15.1.1b will aid in this observation, and that for P_7 acting to the *right* on the joint, all resistances offered by the F-forces to the *left* are helpful, therefore positive.

(b) *Obtain the inverse of matrix* $[A]$. This inversion will normally be done by a computer subroutine on matrix inversion. Just for the purpose of understanding the

P / F	1	2	3	4	5	6	7	8	9
1	+0.6	−1.0						−0.6	
2	+0.8						+1.0	+0.8	
3		+1.0	−0.6						
4			+0.8						+1.0
5				+1.0	−1.0				
6							−1.0		
7					+1.0	−1.0		+0.6	
8								−0.8	−1.0
9			+0.6			+1.0			

$[A]_{9 \times 9} =$ (matrix above)

physical meaning of the $[A^{-1}]$ matrix, the algebraic method of joints and sections is used to determine the contents of this matrix. Since each column of the $[A^{-1}]$ matrix is actually the set of bar forces due to each unit value of P, the matrix $[A^{-1}]$ is obtained as follows:

F / P	1	2	3	4	5	6	7	8	9
1	$+\frac{5}{9}$	$+\frac{5}{6}$	$+\frac{5}{9}$	$+\frac{5}{12}$	0	$+\frac{5}{6}$	0	$+\frac{5}{12}$	0
2	$-\frac{1}{3}$	$+\frac{1}{4}$	$+\frac{2}{3}$	$+\frac{1}{2}$	0	$+\frac{1}{4}$	0	$+\frac{1}{2}$	0
3	$-\frac{5}{9}$	$+\frac{5}{12}$	$-\frac{5}{9}$	$+\frac{5}{6}$	0	$+\frac{5}{12}$	0	$+\frac{5}{6}$	0
4	$+\frac{2}{3}$	$-\frac{1}{2}$	$+\frac{2}{3}$	$-\frac{1}{4}$	$+1$	$-\frac{1}{2}$	$+1$	$-\frac{1}{4}$	$+1$
5	$+\frac{2}{3}$	$-\frac{1}{2}$	$+\frac{2}{3}$	$-\frac{1}{4}$	0	$-\frac{1}{2}$	$+1$	$-\frac{1}{4}$	$+1$
6	$+\frac{1}{3}$	$-\frac{1}{4}$	$+\frac{1}{3}$	$-\frac{1}{2}$	0	$-\frac{1}{4}$	0	$-\frac{1}{2}$	$+1$
7	0	0	0	0	0	-1	0	0	0
8	$-\frac{5}{9}$	$+\frac{5}{12}$	$-\frac{5}{9}$	$-\frac{5}{12}$	0	$+\frac{5}{12}$	0	$-\frac{5}{12}$	0
9	$+\frac{4}{9}$	$-\frac{1}{3}$	$+\frac{4}{9}$	$+\frac{1}{3}$	0	$-\frac{1}{3}$	0	$-\frac{2}{3}$	0

$[A^{-1}]_{9 \times 9} =$ (matrix above)

Note that the second column in the $[A^{-1}]$ matrix could be obtained by applying a unit load P_2 and then determining the internal forces, as shown in Fig. 15.2.2a. The third column could be obtained using the internal forces resulting from application of a unit load P_3, as in Fig. 15.2.2b. If both the $[A]$ and $[A^{-1}]$ matrices are determined independently, the result may be verified by the matrix multiplication $[A][A^{-1}]$ to show that the product is a unit (identity) matrix $[I]$.

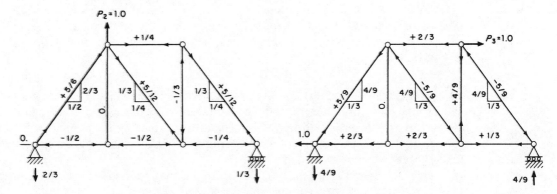

Figure 15.2.2 Axial forces due to unit external forces in Example 15.2.1.

(c) *Determine internal force matrix* $\{F\}$. Once the $[A^{-1}]$ matrix is obtained by matrix inversion, matrix multiplication is used, as follows:

$$\{F\}_{9\times1} = [A^{-1}]_{9\times9}\{P\}_{9\times1}$$

where the $\{P\}$ matrix of the loads and the final $\{F\}$ matrix of internal forces are shown, as follows:

$\{P\}_{9\times1} =$			$\{F\}_{9\times1} = [A^{-1}]\{P\} =$	
	1	0.0	1	−10.0
	2	0.0	2	+ 1.5
	3	+9.0	3	−12.5
	4	0.0	4	+15.0
	5	0.0	5	+15.0
	6	−18.0	6	+ 7.5
	7	0.0	7	+18.0
	8	0.0	8	−12.5
	9	0.0	9	+10.0

(d) *Discussion*. In large-scale application of this method on the computer, it is conceivable that one need only give the computer the shape of the truss and the loads and then ask for the bar forces. The shape of the truss may be defined by the locations of all the joints properly numbered, and the starting and terminal joints of each bar.

The loads may be defined by the horizontal and vertical forces acting on each joint. The computer program would consist primarily of getting the computer to construct its own [A] matrix from the directions of the bars entering into each degree of freedom.

15.3 MATRIX TRANSPOSITION

In Chapter 3 the deflection of trusses (i.e., the displacement of truss joints) was treated; whereas in Sections 3.1 and 3.4 the relationship between the member elongation e and the joint displacements X was derived. The joint displacement equation, Eq. (3.4.1), is as follows:

$$e = (X_3 - X_1) \cos \alpha + (X_4 - X_2) \sin \alpha \qquad [3.4.1]$$

where e is the member elongation, X_1 and X_2 are the horizontal and vertical coordinates of the joint displacement at the near end of the member where the angle α is measured counterclockwise from the horizontal line extending to the right, and X_3 and X_4 are the horizontal and vertical coordinates of the joint displacement at the far end of the member (see also Fig. 3.4.1).

One may note that an equation like Eq. (3.4.1) may be written for each member of a truss. Thus, in matrix formulation, Eq. (3.4.1) becomes

$$\{e\}_{NF \times 1} = [B]_{NF \times NP} \{X\}_{NP \times 1} \qquad (15.3.1)$$

Since there are NF members in a truss, there will be NF elongations e; similarly, since there are NP possible joint external forces, there will be NP deflection components. The matrix $[B]$ comprises the coefficients by which the X's are multiplied, such as the right side of Eq. (3.4.1).

The matrix $[B]$ as defined in Eq. (15.3.1) gives the member elongations in terms of the joint displacements; it is a deformation relationship and as such it may be called the *deformation matrix*. Sometimes it is also called the *compatibility matrix* because it expresses the internal member elongations compatible with the external joint displacements.

It may now be stated that in first-order analysis the deformation matrix $[B]$ is the *transpose* of the statics matrix $[A]$. Although the proof for this relationship will be delayed until Chapter 16, it is expedient to learn the definition of matrix transposition at this time. If one rectangular matrix is the *transpose* of the other, the ith row of one is the ith column of the other. For instance, if

$$[A] = \begin{bmatrix} 2 & 3 & 4 \\ 5 & 6 & 7 \end{bmatrix} \quad \text{and} \quad [B] = \begin{bmatrix} 2 & 5 \\ 3 & 6 \\ 4 & 7 \end{bmatrix}$$

then $[B]$ is the transpose of $[A]$, and vice versa. Expressed in matrix notation,

$$[A] = [B^T] \quad \text{and} \quad [B] = [A^T] \qquad (15.3.2)$$

When one matrix is square, its transpose is still a square matrix; otherwise, the number of rows in one is equal to the number of columns in the other.

Since, for a statically determinate truss $NF = NP$, the $[B]$ matrix of Eq. (15.3.1) will be a square matrix, then

$$\{X\}_{NP \times 1} = [B^{-1}]_{NP \times NF} \{e\}_{NF \times 1} \qquad (15.3.3)$$

If matrix $[B]$ is the transpose of matrix $[A]$, then the inverse matrix $[B^{-1}]$ will be the transpose of inverse matrix $[A^{-1}]$. Thus, Eq. (15.3.3) may be written

$$\{X\}_{\text{NP}\times 1} = [A^{-1}]_{\text{NP}\times\text{NF}}^T \{e\}_{\text{NF}\times 1} \tag{15.3.4}$$

In the following section of this chapter, the transpose matrix $[A^{-1}]^T$ is examined from the viewpoint of the virtual work (unit load) method, as first presented in Section 3.3. Further discussion of the deformation matrix $[B]$ in Eq. (15.3.1) will be made in Section 16.3. The proof for the transposition relationship between the matrices $[A]$ and $[B]$ is presented in Section 16.4.

15.4 DEFLECTIONS OF TRUSSES BY THE MATRIX VIRTUAL WORK (UNIT LOAD) METHOD

According to the virtual work (unit load) method as developed in Section 3.3, a truss joint displacement X may be obtained by applying Eq. (3.2.4),

$$X = \sum_{i=1}^{i=\text{NF}} u_i F_i \frac{L_i}{E_i A_i} \tag{3.2.4}$$

Since $e = FL/EA$, Eq. (3.2.4) may also be written

$$X = \sum_{i=1}^{i=\text{NF}} u_i e_i \tag{15.4.1}$$

where u_1 through u_{NF} are the internal forces in the members of the truss due to a *unit load* applied at the joint and in the direction where X is measured. The elongations e_1 through e_{NF} are the actual elongations of the truss members caused by the actual loading.

Equation (15.4.1) gives one joint displacement. If all *NP* joint displacements are desired, there will be *NP* equations. Thus, according to Eq. (15.4.1),

$$\begin{Bmatrix} X_1 \\ X_2 \\ \cdot \\ \cdot \\ \cdot \\ X_{\text{NP}} \end{Bmatrix} = \begin{Bmatrix} \sum_{i=1}^{i=\text{NF}} (u_{i,1} e_i) \\ \sum_{i=1}^{i=\text{NF}} (u_{i,2} e_i) \\ \cdot \\ \cdot \\ \cdot \\ \sum_{i=1}^{i=\text{NF}} (u_{i,\text{NP}} e_i) \end{Bmatrix} \tag{15.4.2}$$

in which $u_{i,1}, u_{i,2}, \ldots, u_{i,\text{NP}}$ are the forces in the ith member due to a unit load applied in the first, second, $\ldots$, NPth direction, respectively, while $X_1, X_2, \ldots, X_{\text{NP}}$ are the actual displacements in the first, second, $\ldots$, NPth directions, respectively. Writing out Eq. (15.4.2) gives

$$\left.\begin{aligned} X_1 &= u_{1,1}e_1 + u_{2,1}e_2 + u_{3,1}e_3 + \cdots + u_{\text{NF},1}e_{\text{NF}} \\ X_2 &= u_{1,2}e_1 + u_{2,2}e_2 + u_{3,2}e_3 + \cdots + u_{\text{NF},2}e_{\text{NF}} \\ X_3 &= u_{1,3}e_1 + u_{2,3}e_2 + u_{3,3}e_3 + \cdots + u_{\text{NF},3}e_{\text{NF}} \\ &\ \ \vdots \qquad \vdots \qquad \vdots \qquad \quad \vdots \\ X_{\text{NP}} &= u_{1,\text{NP}}e_1 + u_{2,\text{NP}}e_2 + u_{3,\text{NP}}e_3 + \cdots + u_{\text{NF},\text{NP}}e_{\text{NF}} \end{aligned}\right\} \tag{15.4.3}$$

Using parentheses to enclose subscripts as is usually done in computer programming, Eq. (15.4.3) may be written in expanded matrix form, as follows:

$$\begin{Bmatrix} X_1 \\ X_2 \\ \cdot \\ \cdot \\ \cdot \\ X_{NP} \end{Bmatrix} = \begin{bmatrix} u(1,1) & u(2,1) & \cdots & u(NF,1) \\ u(1,2) & u(2,2) & \cdots & u(NF,2) \\ \cdot & \cdot & & \cdot \\ \cdot & \cdot & & \cdot \\ \cdot & \cdot & & \cdot \\ u(1,NP) & u(2,NP) & \cdots & u(NF,NP) \end{bmatrix} \begin{Bmatrix} e_1 \\ e_2 \\ \cdot \\ \cdot \\ \cdot \\ e_{NF} \end{Bmatrix} \tag{15.4.4}$$

A study of the meaning of Eq. (15.4.4) shows that any one row in the $[u]$ matrix gives the forces in all the truss members due to a unit load at the location and in the direction of the X identified by the second subscript of u. For example, the first row, $u(1,1), u(2,1), \ldots, u(NF,1)$, gives the forces in the NF members due to a unit load applied at the location and in the direction of X_1. Since there is a one-to-one correlation between possible external forces P and joint displacements X, the first row of u values in Eq. (15.4.4) are the forces due to P_1 equal to a unit value. In the inverse statics matrix $[A^{-1}]$ defined by Eq. (15.1.3), each *column* gives the forces in the NF members due to a unit P; for instance, the first column of $[A^{-1}]$ of Eq. (15.1.3) gives the forces in the members due to P_1 equal to a unit value. Since the rows of $[u]$ correspond to the columns of $[A^{-1}]$, the $[u]$ matrix is the transpose of the $[A^{-1}]$ matrix. Thus,

$$[u] = [A^{-1}]^T \tag{15.4.5}$$

Also,

$$[u] = [B^{-1}] \tag{15.4.6}$$

Thus, Eq. (15.4.4) becomes

$$\{X\}_{NP \times 1} = [A^{-1}]^T_{NP \times NF} \{e\}_{NF \times 1} \tag{15.4.7}$$

One may note that Eqs. (15.3.4) and (15.4.6), although given two different equation numbers, are in fact identical, stating that the joint displacement matrix $\{X\}$ is the product of the premultiplier matrix $[A^{-1}]^T$ and the postmultiplier matrix $\{e\}$. If Eq. (15.3.4) is proved first, as will be done in Section 16.3, from the fact that $[B] = [A^T]$, then the virtual work (unit load) method for finding truss deflections follows. On the other hand, if Eq. (15.4.6) is accepted by quoting the virtual work (unit load) method proved earlier in Chapter 3, then the matrix method as shown by Eqs. (15.3.4) or (15.4.6) is simply a compilation of the procedure to find all values of X.

Example 15.4.1

For the statically determinate truss of Fig. 15.4.1a (previously treated in Section 3.4), obtain the set of joint displacements due to the set of bar elongations given in Fig. 15.4.1b by means of the matrix virtual work (unit load) method.

Solution

(a) *Obtain the $[A^{-1}]$ matrix.* Ordinarily, when the computer is to be used, the statics matrix $[A]$ would be used as the input and then inverted to obtain $[A^{-1}]$. Of course, this operation is possible only for statically determinate and stable structures where the $[A]$ matrix is square and nonsingular. It is suggested here that the reader verify the $[A^{-1}]$ matrix shown below by computing it directly according to its physical

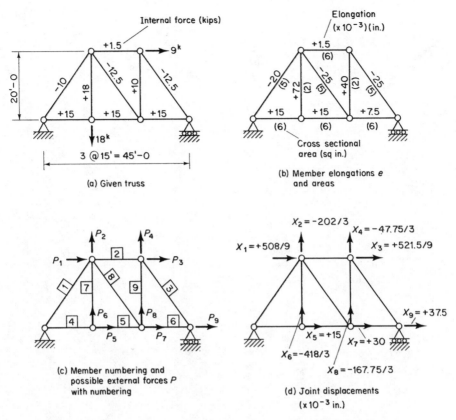

(a) Given truss

(b) Member elongations e and areas

(c) Member numbering and possible external forces P with numbering

(d) Joint displacements (x 10⁻³ in.)

Figure 15.4.1 Joint displacements by the matrix unit load method, Example 15.4.1.

meaning using the conventional method of joints (see Section 2.2). For this truss it has already been so computed in Example 15.2.1b.

(b) *Obtain the transpose* $[A^{-1}]^T$ *matrix.* The $[A^{-1}]^T$ matrix is obtained by interchanging the rows of the $[A^{-1}]$ matrix to become columns of the $[A^{-1}]^T$ matrix as shown at the top of page 450. Note that the $[A^{-1}]$ matrix expresses the F-values in terms of the P-values while the $[A^{-1}]^T$ matrix expresses the X-values in terms of the e-values. The former is a force-to-force relationship and the latter is a deformation-to-deformation relationship.

(c) *Obtain the elongation matrix* $\{e\}$. The elongations of the members and their number designation are given in Fig. 15.4.1b and c. The matrix $\{e\}$, a column matrix, may be written as at the lower left of page 450.

(d) *Determine the* $\{X\}$ *matrix.* Using Eq. (15.4.6),

$$\{X\} = [A^{-1}]^T\{e\} \qquad [15.4.7]$$

where the premultiplier matrix of part (b) is multiplied by the postmultiplier column matrix of part (c), to give as at the lower right of page 450. The results are shown in Fig. 15.4.1d.

Sec. 15.4   Deflections of Trusses by the Matrix Virtual Work (Unit Load) Method   **449**

$$[A^{-1}]^T_{9\times9} =$$

e \ x	1	2	3	4	5	6	7	8	9
1	$+\frac{5}{9}$	$-\frac{1}{3}$	$-\frac{5}{9}$	$+\frac{2}{3}$	$+\frac{2}{3}$	$+\frac{1}{3}$	0	$-\frac{5}{9}$	$+\frac{4}{9}$
2	$+\frac{5}{6}$	$+\frac{1}{4}$	$+\frac{5}{12}$	$-\frac{1}{2}$	$-\frac{1}{2}$	$-\frac{1}{4}$	0	$+\frac{5}{12}$	$-\frac{1}{3}$
3	$+\frac{5}{9}$	$+\frac{2}{3}$	$-\frac{5}{9}$	$+\frac{2}{3}$	$+\frac{2}{3}$	$-\frac{1}{3}$	0	$-\frac{5}{9}$	$+\frac{4}{9}$
4	$+\frac{5}{12}$	$+\frac{1}{2}$	$+\frac{5}{6}$	$-\frac{1}{4}$	$-\frac{1}{4}$	$-\frac{1}{2}$	0	$-\frac{5}{12}$	$+\frac{1}{3}$
5	0	0	0	$+1$	0	0	0	0	0
6	$+\frac{5}{6}$	$+\frac{1}{4}$	$+\frac{5}{12}$	$-\frac{1}{2}$	$-\frac{1}{2}$	$-\frac{1}{4}$	-1	$+\frac{5}{12}$	$-\frac{1}{3}$
7	0	0	0	$+1$	$+1$	0	0	0	0
8	$+\frac{5}{12}$	$+\frac{1}{2}$	$+\frac{5}{6}$	$-\frac{1}{4}$	$-\frac{1}{4}$	$-\frac{1}{2}$	0	$-\frac{5}{12}$	$-\frac{2}{3}$
9	0	0	0	$+1$	$+1$	$+1$	0	0	0

$$\{e\}_{9\times1} = \begin{bmatrix} -20 \\ +1.5 \\ -25 \\ +15 \\ +15 \\ +7.5 \\ +72 \\ -25 \\ +40 \end{bmatrix}$$

$$\{X\}_{NF\times1} = \begin{bmatrix} +\frac{508}{9} \\ -\frac{202}{3} \\ +\frac{521.5}{9} \\ -\frac{47.75}{3} \\ +15 \\ -\frac{418}{3} \\ +30 \\ -\frac{167.75}{3} \\ +37.5 \end{bmatrix} \times 10^{-3} \text{ in.}$$

PROBLEMS

15.1. For the statically determinate truss of Prob. 2.5, establish the statics matrix $[A]$.

15.2. For the statically determinate truss of Prob. 2.6, establish the statics matrix $[A]$.

15.3–15.8. For the statically determinate truss of each accompanying figure, establish the statics matrix $[A]$.

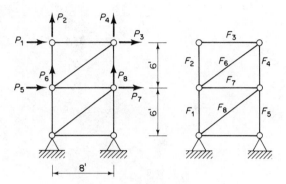

Prob. 15.3

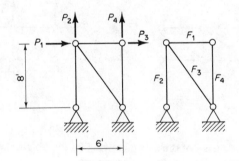

Prob. 15.4

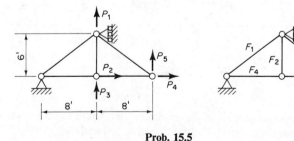

Prob. 15.5

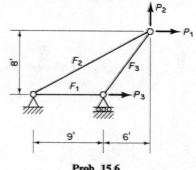

Prob. 15.6

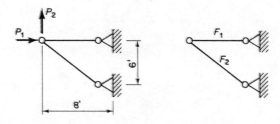

Prob. 15.7

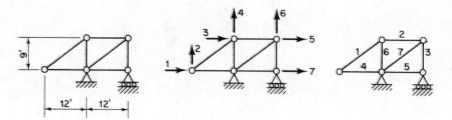

Prob. 15.8

15.9. For the truss of the accompanying figure, obtain the third column of the $[A^{-1}]$ matrix directly using the physical meaning.

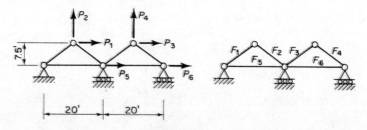

Prob. 15.9

15.10. Assuming the truss of the accompanying figure to be analyzed by the matrix method of joints, determine the statics matrix $[A]$ and the joint force matrix $\{P\}$.

Matrix Formulation—Method of Joints   Chap. 15

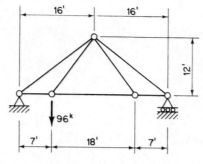

Prob. 15.10

15.11 and 15.12. For the truss of each accompanying figure, obtain the statics matrix $[A]$ and its inverse $[A^{-1}]$ independently using the physical definition relating to each. Show that $[A][A^{-1}] = [I]$.

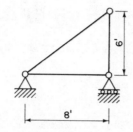

Prob. 15.11

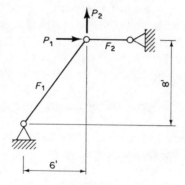

Prob. 15.12

[16]

Matrix Displacement Method Applied to Trusses

16.1 DEGREE OF FREEDOM VS. DEGREE OF STATICAL INDETERMINACY

The terms *degree of freedom* and *degree of statical indeterminacy* with regard to trusses were defined and discussed in Sections 1.14, 2.1, and 7.1. Furthermore, in Section 15.1 use was made of NP external forces applied at truss joints. The symbol NP was used to denote the number of *possible* forces, defined by joint location and horizontal or vertical direction, that might act on a truss. Also in Section 15.4, note was made that there will be NP joint displacements necessary to completely describe the deformed position of a truss. Thus, there is a one-to-one correlation between possible joint forces and actual joint displacements. In fact, since all the NP displacements are necessary, and some of them might be zero only by chance, whereas some or even many of the possible joint forces may be zero, it is easier and more appropriate to refer to a structure in terms of how it may deform rather than how it may be loaded. Thus, the term *degree of freedom* is used to identify the number of joint displacements required to represent the ways in which a framed structure may *freely* respond to any external disturbance. The actual number of loads is not a property of the structure, whereas the actual number of displacements is such a property.

Regarding the *degree of statical indeterminacy*, when the number of internal forces NF exceeds the number of equations of statics that may be written at the joints, the excess is the degree of statical indeterminacy. In Section 7.1 it was shown that NP, the number of possible joint forces, or the number of actual joint displacements, represents the number of statics equations that are available. In other words, the degree of freedom NP equals the number of available statics equations. Thus, as stated by Eq. (7.1.4), the degree of NI of statical indeterminacy equals $NF - NP$.

16.2 DEFINITION OF STATICS MATRIX

The statics matrix [A] has been previously discussed in the context of a statically determinate truss in Section 15.1. However, statics must be satisfied whether or not the truss is statically determinate. Thus, the statics matrix of a statically indeterminate truss is defined in the same way as that of a statically determinate truss; that is, it contains NP rows and NF columns.

Referring to Fig. 16.2.1a, the degree of freedom is noted to be four; two components X_1 and X_2 are needed to define the displacement of the top chord joint, and the bottom chord joints because of the roller supports have only horizontal displacements X_3 and X_4. Each joint displacement (degree of freedom) also identifies the location of a possible external force. Thus, in the numbering system used, such as P_1-X_1, the same vector (arrow) may designate either the force or the displacement.

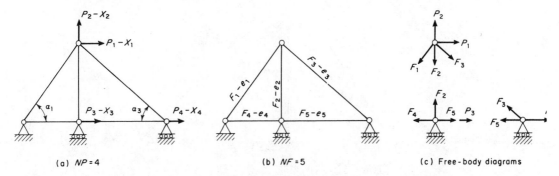

Figure 16.2.1 Numbering system and free-body diagrams for obtaining statics matrix.

In the member identification system shown in Fig. 16.2.1b, designations such as F_1-e_1 are used, so that the same numbering by subscript may be used to refer to either the internal force or the elongation in the member.

The joint free-body diagrams in Fig. 16.2.1c are used to compile the equations of statics in accordance with Eq. (15.1.2). The statics equations for the particular structure can be expressed in matrix form as

$$\{P\}_{4\times 1} = [A]_{4\times 5}\{F_{5\times 1}\} \tag{16.2.1}$$

in which the 4 by 5 *statics matrix* is

$$[A]_{4\times 5} =$$

P＼F	1	2	3	4	5
1	$+\cos \alpha_1$	0	$-\cos \alpha_3$	0	0
2	$+\sin \alpha_1$	$+1$	$+\sin \alpha_3$	0	0
3	0	0	0	$+1$	-1
4	0	0	$+\cos \alpha_3$	0	$+1$

Unlike the statics matrices encountered in Chapter 15, which are square matrices, the statics matrix here is rectangular; that is, NF exceeds NP by one (the degree of statical indeterminacy). A solution of Eq. (16.2.1) for the F-values in terms of the P-values is impossible because the inverse matrix $[A^{-1}]$ is obtainable only for a square (and nonsingular) matrix. As should be expected for a statically indeterminate structure, compatibility of deformation (treated in the next section) has to be satisfied in addition to equilibrium (statics).

16.3 DEFINITION OF DEFORMATION MATRIX

The relationship provided by the deformation matrix is that between the member elongations (internal) and the joint displacements (external). On the basis of first-order analysis (see Section 3.1), only the component of the joint displacement along the bar length changes its length. Referring to Fig. 16.3.1, the elongation e_1 is directly

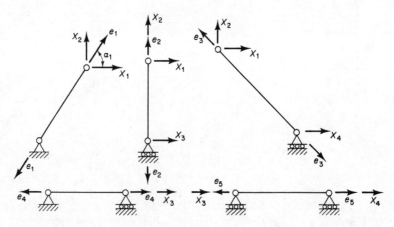

Figure 16.3.1 Compatibility between elongations and joint displacements for the truss of Fig. 16.2.1a.

dependent on joint displacements X_1 and X_2; in other words, e_1 equals the components of X_1 and X_2 parallel to and extending the length of bar 1. Similarly, the other member elongations may be expressed in terms of joint displacements, giving rise to the following set of equations:

$$\left.\begin{aligned}
e_1 &= +X_1 \cos \alpha_1 + X_2 \sin \alpha_1 \\
e_2 &= +X_2 \\
e_3 &= -X_1 \cos \alpha_3 + X_2 \sin \alpha_3 + X_4 \cos \alpha_3 \\
e_4 &= +X_3 \\
e_5 &= -X_3 + X_4
\end{aligned}\right\} \quad (16.3.1)$$

Note that e_1 is dependent on the displacements at one end of the member only because the other end is a hinged support having zero displacement. In general, e will be a function of the displacements at both ends. In effect, Eqs. (16.3.1) are the set of equations obtained by applying the joint displacement equation, Eq. (3.4.1), to each member of the truss. Expressed in matrix form Eqs. (16.3.1) become

$$\{e\}_{5 \times 1} = [B]_{5 \times 4}\{X\}_{4 \times 1} \qquad (16.3.2)$$

in which the 5 by 4 *deformation* matrix is

$$[B]_{5 \times 4} =$$

e \ X	1	2	3	4
1	$+\cos \alpha_1$	$+\sin \alpha_1$	0	0
2	0	$+1$	0	0
3	$+\cos \alpha_3$	$+\sin \alpha_3$	0	$+\cos \alpha_3$
4	0	0	$+1$	0
5	0	0	-1	$+1$

Note that joint displacements cannot be obtained by directly solving Eq. (16.3.2) since the 5 by 4 deformation matrix cannot be inverted.

16.4 TRANSPOSITION RELATIONSHIP BETWEEN STATICS MATRIX AND DEFORMATION MATRIX

Comparison of the elements in the statics matrix $[A]$ of Eq. (16.2.1) with the elements in the deformation matrix $[B]$ of Eq. (16.3.2) shows that one is the transpose of the other; that is,

$$[B] = [A^T] \qquad \text{or} \qquad [A] = [B^T] \qquad (16.4.1)$$

Proof for the transposition relationship may be made by use of the principle of virtual work (see Section 3.2), which states that the external and internal work, each done by the forces in the equilibrium state of one system in going through the deformations of a compatible state of the same or different system, must be equal. The general relationship was shown to be

$$\sum Q_i X_i = \sum u_i e_i \qquad [3.2.1]$$

Take the equilibrium state as in Fig. 16.4.1a, where all internal forces are zero except F_3, which means that the Q-values are $P_1 = A_{13}F_3, P_2 = A_{23}F_3, P_3 = A_{33}F_3$, and $P_4 = A_{43}F_3$, and the u-values are zero except for F_3. Take the compatible state as in Fig. 16.4.1b, where all joint displacements are zero except X_2, which means that all elongations are functions of X_2, or $e_1 = B_{12}X_2, e_2 = B_{22}X_2, e_3 = B_{32}X_2, e_4 = B_{42}X_2$, and $e_5 = B_{52}X_2$. Thus, applying Eq. (3.2.1) gives

$$P_1(0) + P_2 X_2 + P_3(0) + P_4(0) = 0(e_1) + 0(e_2) + F_3 e_3 + 0(e_4) + 0(e_5)$$

$$P_2 X_2 = F_3 e_3$$

$$(A_{23}F_3)X_2 = F_3(B_{32}X_2)$$

$$A_{23} = B_{32}$$

In general,

$$A_{ij} = B_{ji} \qquad (16.4.2)$$

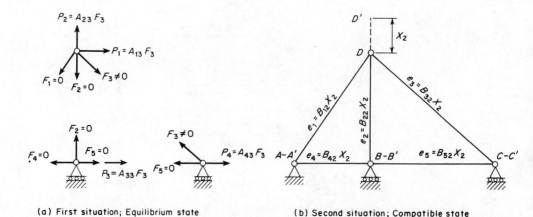

(a) First situation; Equilibrium state
$(F_1 = F_2 = F_4 = F_5 = 0; F_3 \neq 0)$

(b) Second situation; Compatible state
$(X_1 = X_3 = X_4 = 0; X_2 \neq 0)$

Figure 16.4.1 Proof for transposition relationship between [A] and [B] by the principle of virtual work.

Thus, Eq. (16.4.1) is verified. The statics matrix [A] and the deformation matrix [B] are the transpose of one another; that is, the ith row of one is the ith column of the other.

16.5 MEMBER STIFFNESS MATRIX

The relationship between member internal force and member elongation is given for elastic (linear) systems by Hooke's Law, which states that stress is proportional to strain,

$$\sigma_i = E\epsilon_i \tag{16.5.1}$$

where σ_i is the unit stress (ksi or MPa, for example) in the member, ϵ_i is the unit strain (in./in. or mm/mm, for example), and E is the modulus of elasticity of the material. Defining A_i and L_i as the cross-sectional area and length of the member, respectively, and noting that

$$\sigma_i = \frac{F_i}{A_i} \quad \text{and} \quad \epsilon_i = \frac{e_i}{L_i} \tag{16.5.2}$$

permits substitution of σ_i and ϵ_i into Eq. (16.5.1) to give

$$\frac{F_i}{A_i} = \frac{Ee_i}{L_i}$$

or

$$F_i = \frac{EA_i}{L_i}e_i = S_i e_i \tag{16.5.3}$$

There will be an Eq. (16.5.3) for each member of the truss, relating the internal force F_i to the member elongation e_i by its *stiffness* S_i. For a truss having five members, such as in Fig. 16.2.1, the equations relating F to e may be expressed in matrix form as

$$\{F\}_{5\times1} = [S]_{5\times5}\{e\}_{5\times1} \tag{16.5.4}$$

in which the *member stiffness matrix* $[S]$ is

$$[S]_{5\times5} =$$

F \ e	1	2	3	4	5
1	$\dfrac{EA_1}{L_1}$				
2		$\dfrac{EA_2}{L_2}$			
3			$\dfrac{EA_3}{L_3}$		
4				$\dfrac{EA_4}{L_4}$	
5					$\dfrac{EA_5}{L_5}$

The member stiffness matrix $[S]$ contains the respective forces required to elongate the bars by a unit length. The member stiffness matrix will be a square matrix and for trusses will have all elements except those on the principal diagonal equal to zero (such a matrix is commonly called a diagonal matrix).

16.6 GLOBAL STIFFNESS MATRIX $[K]$

The relationship between the possible external forces P and the joint displacements X as expressed in matrix form is defined here as the *global stiffness matrix*. The term "global" is used because it is a relationship involving the entire structure rather than just one member. In matrix terminology,

$$\{P\} = [K]\{X\} \tag{16.6.1}$$

The global stiffness matrix $[K]$ may be obtained by matrix operations using the statics matrix $[A]$, the deformation matrix $[A^T]$, and the member stiffness matrix $[S]$; or it can be established directly by its own physical meaning.

Using matrix operations, starting with the equilibrium equation, involving the statics matrix $[A]$, Eq. (16.2.1),

$$\{P\}_{NP\times1} = [A]_{NP\times NF}\{F\}_{NF\times1} \tag{16.2.1}$$

Next, the Hooke's Law equation involving the member stiffness matrix $[S]$, Eq. (16.5.3), is used,

$$\{F\}_{NF\times1} = [S]_{NF\times NF}\{e\}_{NF\times1} \tag{16.5.4}$$

Then, the compatibility equation, Eq. (16.3.2), involving the deformation matrix $[B]$ is used,

$$\{e\}_{NF\times1} = [B]_{NF\times NP}\{X\}_{NP\times1} \tag{16.3.2}$$

Now, substituting Eq. (16.3.2) into Eq. (16.5.4) gives

$$\{F\}_{NF \times 1} = [SB]_{NF \times NP}\{X\}_{NP \times 1} \tag{16.6.2}$$

and since $[B] = [A^T]$,

$$\{F\}_{NF \times 1} = [SA^T]_{NF \times NP}\{X\}_{NP \times 1} \tag{16.6.3}$$

Finally, substituting Eq. (16.6.3) into Eq. (16.2.1) gives

$$\{P\}_{NP \times 1} = [ASA^T]_{NP \times NP}\{X\}_{NP \times 1} \tag{16.6.4}$$

Thus, comparing Eqs. (16.6.4) and (16.6.1), the reader may see that the global stiffness matrix $[K]$ equals

$$[K]_{NP \times NP} = [ASA^T]_{NP \times NP} \tag{16.6.5}$$

The global stiffness matrix $[K]$ can be shown to be a symmetrical matrix; that is, the original matrix is identical to its transpose (or any ith row is identical to the ith column) by the following:

Given: $[A]$, $[S] = [S^T]$, $[K] = [ASA^T]$

To Prove: $[K] = [K^T]$

Proof: Right side $= [K^T] = [ASA^T]^T = [SA^T]^T[A]^T = [A][S]^T[A^T]$

                 $= [A][S][A^T] = [K] = $ left side

Example 16.6.1

Establish the statics matrix $[A]$, the member stiffness matrix $[S]$, and the deformation matrix $[B]$ for the truss of Fig. 16.6.1. Verify the relationship $[B] = [A^T]$. Finally, compute the global stiffness matrix $[K]$.

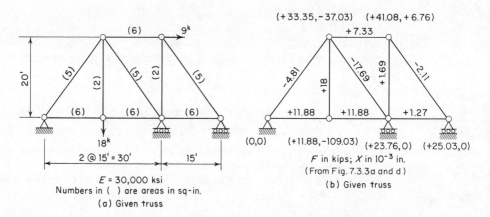

Figure 16.6.1   Statically indeterminate truss of Example 16.6.1.

Solution

(a) *Establish the numbering system for the degrees of freedom and the members.* The *P-X* and *F-e* diagrams are superimposed on Fig. 16.6.2a. The student may prefer to use two separate diagrams for ease of following.

(b) *Statics matrix* $[A]$. The statics matrix is established by rows and is given in Fig. 16.6.3. As explained in Section 15.2, the student should practice filling in the

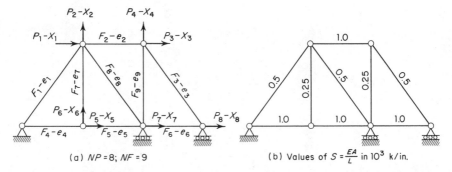

Figure 16.6.2 Numbering system and member stiffnesses for the truss of Example 16.6.1.

(a) $NP = 8$; $NF = 9$

(b) Values of $S = \frac{EA}{L}$ in 10^3 k/in.

nonzero values in this matrix by simply looking at Fig. 16.6.2a. For instance, an external force P_3 is resisted positively by 100% of F_2 and negatively by 60% of F_3.

(c) *Deformation matrix* [B]. The deformation matrix [B] is established by columns and is given in Fig. 16.6.3. Again, this matrix may be filled out by inspection of Fig. 16.6.2a. For instance, a unit X_1 displacement will extend member 1 by $+0.6X_1$, member 2 by $-1.0X_1$, and member 8 by $-0.6X_1$. A check is made to be certain that $[B] = [A^T]$.

(d) *Member stiffness matrix* [S]. The member stiffnesses EI/L are computed as labeled on Fig. 16.6.2b. The member stiffness matrix [S] is shown in Fig. 16.6.3.

(e) *Matrix operations.* For the purpose of obtaining the [K] matrix, the matrix multiplication may be done by obtaining first [AS] and then [AS][B], or by first getting [SB] and then getting [A][SB]. In the next section it will be shown that the intermediate matrix $[SB] = [SA^T]$ is useful as well as the matrix [K], whereas the [AS] matrix is not of special interest. Thus, in Fig. 16.6.3 the matrix [SB] is computed first, orienting the premultiplier matrix [S] to the left of, and the postmultiplier matrix [B] above, the product [SB]. Then for the final multiplication, the premultiplier matrix [A] is oriented to the left of, and the postmultiplier matrix [SB] above, the product matrix [ASB]. This orientation allows matrix multiplication to be performed systematically by applying the inner product rule described in Section 13.3.

16.7 MATRIX DISPLACEMENT METHOD

Now that the individual matrices and their concepts have been treated separately in the preceding sections, the matrix displacement method may be summarized. The three basic relationships are the equilibrium requirement, Eq. (16.2.1), the Hooke's Law relationship, Eq. (16.5.3), and the compatibility of deformation requirement, Eq. (16.3.2). These are repeated here, using the symbol *NLC* to represent "number of loading conditions," in addition to symbols previously used.

Equilibrium:

$$[P]_{\text{NP}\times\text{NLC}} = [A]_{\text{NP}\times\text{NF}}[F]_{\text{NF}\times\text{NLC}} \tag{16.7.1}$$

$[B] =$

e \ x	1	2	3	4	5	6	7	8
1	+0.6	+0.8						
2	-1.		+1.					
3			-0.6	+0.8				0.6
4					+1.			
5					-1.		+1.	
6							-1.	+1.
7		+1.				-1.		
8	-0.6	+0.8					+0.6	
9				+1.				

$[S] = 10^3\ \text{k/in.}$

F \ e	1	2	3	4	5	6	7	8	9
1	0.5								
2		1.0							
3			0.5						
4				1.0					
5					1.0				
6						1.0			
7							0.25		
8								0.5	
9									0.25

$[SB] = 10^3\ \text{k/in.}$

F \ x	1	2	3	4	5	6	7	8
1	+0.3	+0.4						
2	-1.0		+1.0					
3			-0.3	+0.4				+0.3
4					+1.0			
5					-1.0		+1.0	
6							-1.0	+1.0
7		+0.25				-0.25		
8	-0.3	+0.4					+0.3	
9				+0.25				

$[A] =$

p \ F	1	2	3	4	5	6	7	8	9
1	+0.6	-1.0						-0.6	
2	+0.8						+1.0	+0.8	
3		+1.0	-0.6						
4			+0.8						+1.0
5				+1.0	-1.0				
6							-1.0		
7					+1.0	-1.0		+0.6	
8			+0.6			+1.0			

$[K] = [ASB] = 10^3\ \text{k/in.}$

p \ x	1	2	3	4	5	6	7	8
1	+1.36	0.0	-1.0	0.0	0.0	0.0	-0.18	0.0
2	0.0	+0.89	0.0	0.0	0.0	-0.25	+0.24	0.0
3	-1.0	0.0	+1.18	-0.24	0.0	0.0	0.0	-0.18
4	0.0	0.0	-0.24	+0.59	0.0	0.0	0.0	+0.24
5	0.0	0.0	0.0	0.0	+2.0	0.0	-1.0	0.0
6	0.0	-0.25	0.0	0.0	0.0	+0.25	0.0	0.0
7	-0.18	+0.24	0.0	0.0	-1.0	0.0	+2.18	-1.0
8	0.0	0.0	-0.18	+0.24	0.0	0.0	-1.0	+1.18

Figure 16.6.3 Matrices and solution for Example 16.6.1; truss of Fig. 16.6.1.

Compatibility:

$$[e]_{\text{NF} \times \text{NLC}} = [B]_{\text{NF} \times \text{NP}}[X]_{\text{NP} \times \text{NLC}} = [A^T]_{\text{NF} \times \text{NP}}[X]_{\text{NP} \times \text{NLC}} \tag{16.7.2}$$

Hooke's Law:

$$[F]_{\text{NF} \times \text{NLC}} = [S]_{\text{NF} \times \text{NF}}[e]_{\text{NF} \times \text{NLC}} \tag{16.7.3}$$

In the general formulation above, recognition is made that the analysis of a given structure may be desired for more than one loading condition; thus, the symbol *NLC* is used instead of 1 as in the earlier part of the chapter. When the number of loading conditions is larger than one, the matrices [F] and [P] will be rectangular instead of column matrices.

The analysis for the internal forces F is usually the primary objective starting with the external forces P. The internal force matrix $[F]$ can be obtained directly from Eq. (16.7.1) *only* when matrix $[A]$ is square (and nonsingular) and thus may be inverted, which situation will arise only for a statically determinate truss, as discussed in Chapter 15. For the statically indeterminate truss, the internal force matrix $[F]$ may be obtained in terms of the joint displacement matrix $[X]$ by substituting Eq. (16.7.2) into Eq. (16.7.3); thus,

$$[F]_{\text{NF}\times\text{NLC}} = [S]_{\text{NF}\times\text{NF}}[A^T]_{\text{NF}\times\text{NP}}[X]_{\text{NP}\times\text{NLC}}$$

$$= [SA^T]_{\text{NF}\times\text{NP}}[X]_{\text{NP}\times\text{NLC}} \tag{16.7.4}$$

Substitution of Eq. (16.7.4) into Eq. (16.7.1) gives

$$[P]_{\text{NP}\times\text{NLC}} = [A]_{\text{NP}\times\text{NF}}[SA^T]_{\text{NF}\times\text{NP}}[X]_{\text{NP}\times\text{NLC}}$$

$$= [ASA^T]_{\text{NP}\times\text{NP}}[X]_{\text{NP}\times\text{NLC}} \tag{16.7.5}$$

Thus, the global stiffness matrix $[ASA^T] = [K]$ is used to relate the external loads to the joint displacements. The matrix $[ASA^T]$ is a square matrix that may be inverted, allowing the determination of the joint displacements; thus,

$$[X]_{\text{NP}\times\text{NLC}} = [ASA^T]^{-1}_{\text{NP}\times\text{NP}}[P]_{\text{NP}\times\text{NLC}} \tag{16.7.6}$$

Now that the displacements are known, Eq. (16.7.6) may be substituted into Eq. (16.7.4) to obtain the internal forces; thus,

$$[F]_{\text{NF}\times\text{NLC}} = [SA^T]_{\text{NF}\times\text{NP}}[ASA^T]^{-1}_{\text{NP}\times\text{NP}}[X]_{\text{NP}\times\text{NLC}} \tag{16.7.7}$$

Because the displacements are determined as the primary unknowns the method is referred to as the *displacement method*, or here with matrices as the *matrix displacement method*. Alternatively, because the global stiffness matrix $[K]$ (as well as the member stiffness matrix $[S]$) is necessary in order to determine first the displacements and then the internal forces, the method has frequently been categorized as the *stiffness method*.

Example 16.7.1

Using the matrix displacement method, analyze the truss of Fig. 16.7.1.

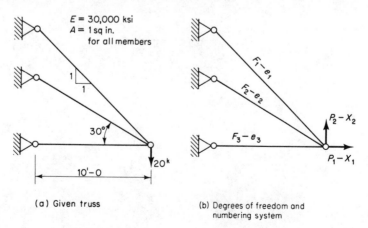

(a) Given truss

(b) Degrees of freedom and numbering system

Figure 16.7.1 Truss for Example 16.7.1.

Solution

(a) *Establish the statics matrix* [A]. From equilibrium of the joints (in this case there is only one),

$$P_1 = 0.707F_1 + 0.866F_2 + 1.0F_3$$
$$P_2 = 0.707F_1 + 0.500F_2$$

or in matrix form,

Statics matrix

$$[A] = \begin{array}{c|c|c|c} P \diagdown F & 1 & 2 & 3 \\ \hline 1 & +0.707 & +0.866 & +1.0 \\ \hline 2 & +0.707 & +0.50 & 0. \end{array}$$

(b) *Establish the deformation matrix* [B]. This matrix can be established either by columns or by rows. In the former, the effects of each X on all the e's are examined. In the latter, how each e is affected by all the X's is examined. Using the latter and examining the geometry shown in Fig. 16.7.2,

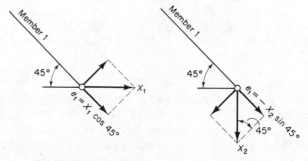

(a) Elongation of member 1 due to X_1 and X_2

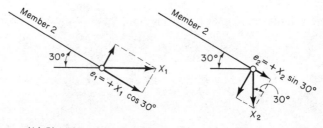

(b) Elongtion of member 2 due to X_1 and X_2

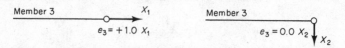

(c) Elongtion of member 3 due to X_1 and X_2

Figure 16.7.2 Elongations due to joint displacements for Example 16.7.1.

$$e_1 = +0.707X_1 + 0.707X_2$$

$$e_2 = +0.866X_1 + 0.500X_2$$

$$e_3 = +1.0X_1$$

Thus, the matrix $[B]$ is

Deformation matrix

	e \\ X	1	2
$[B] =$	1	+0.707	+0.707
	2	+0.866	+0.50
	3	+1.0	0.

After establishing matrix $[B]$ independently from matrix $[A]$ check that one is the transpose of the other.

(c) *Establish the member stiffness matrix* $[S]$. This is the Hooke's Law relationship,

$$\{F\} = [S]\{e\}$$

where $S_i = EA_i/L_i$ for each member. The matrix $[S]$ is then

Member stiffness matrix

	F \\ e	1	2	3
$[S] =$	1	+2121.6		
	2		+2599.8	
	3			+3000.0

$$S_1 = \frac{30,000(1)}{14.14} = 2121.6$$

$$S_2 = \frac{30,000(1)}{10/0.866} = 2599.8$$

$$S_3 = \frac{30,000(1)}{10} = 3000.0$$

(values in kips/ft)

(d) *Establish the load (same as external force) matrix* $\{P\}$. The load matrix links the actual loads for which analysis is to be made with the possible loads identified on the *P-X* diagram (Fig. 16.7.1b). In matrix form for the one loading condition $[NLC = 1]$, the $\{P\}$ matrix is as follows:

Load matrix

	P \\ LC	1
$\{P\} =$	1	0.
	2	+20.0

(e) *Obtain the* $[SA^T]$ *matrix.* Matrix multiplication of the premultiplier matrix $[S]$ and the postmultiplier matrix $[B]$ (i.e., $[A^T]$) gives the $[SA^T]$ matrix. For example, element (1, 1) is the sum of the inner products of row 1 of $[S]$ and column 1 of $[B]$, as follows:

$$+2121.6(+0.707) + 0(+0.866) + 0(+1.0) = +1500.0 \text{ kips/ft}$$

The complete matrix $[SB] = [SA^T]$ is as follows:

F \\ X	1	2
1	+1500.0	+1500.0
2	+2251.4	+1299.9
3	+3000.0	0.

$[SB] = [SA^T] =$ (kips/ft)

$\{F\} = [S]\{e\}$
$\{e\} = [A^T]\{X\}$
$\{F\} = [SA]^T\{X\}$

(f) *Obtain global stiffness matrix* $[K]$. Matrix multiplication of the premultiplier matrix $[A]$ and the postmultiplier matrix $[SA^T]$ gives the matrix $[K]$ (i.e., $[ASA^T]$). For example, element K_{11} is the sum of the inner product of row 1 of $[A]$ and column 1 of $[SA^T]$, as follows:

$$K_{11} = +0.707(+1500.0) + 0.866(+2251.4) + 1.0(+3000.0)$$
$$= +6010.2 \text{ kips/ft}$$

The complete matrix $[K]$ is as follows:

$[K] =$ (kips/ft)

P \\ X	1	2
1	+6010.2	+2186.2
2	+2186.2	+1710.4

$\{P\} = [A]\{F\}$
$\{F\} = [SA^T]\{X\}$
$\{P\} = [ASA^T]\{X\}$

Note that $[K]$ is a symmetric matrix, as it must be.

(g) *Obtain the inverse of the matrix* $[K]$. By Gauss–Jordan elimination as discussed in Section 13.4, the $[K]^{-1}$ matrix is obtained as follows:

$[K]^{-1} =$ (ft/kip)

X \\ P	1	2
1	+310.95	−397.44
2	−397.44	+1092.64

$\times 10^{-6}$ $\{X\} = [K]^{-1}\{P\}$

(h) *Obtain the joint displacement matrix* $\{X\}$. The joint displacement matrix $\{X\}$ is obtained by multiplying the premultiplier matrix $[K]^{-1}$ by the postmultiplier matrix $\{P\}$. The matrix operation gives

$\{X\} =$ (ft)

X \\ LC	1
1	-794.9×10^{-5}
2	$+2185.3 \times 10^{-5}$

$\{X\} = [K]^{-1}\{P\}$

(i) *Obtain the internal force matrix* $\{F\}$. This matrix is obtained by multiplying the premultiplier matrix $[SA^T]$ by the postmultiplier matrix $\{X\}$. The matrix operation gives

F \ LC	1
1	$+20.86$
2	$+10.51$
3	-23.85

$$\{F\} = [SA^T]\{X\}$$

$\{F\} =$ (kips)

(j) *Checking results.* Since output for $\{X\}$ and $\{F\}$ has been obtained, usually from the computer, the results *must* be checked. The checks are discussed in detail in Sections 16.9 and 16.10.

Example 16.7.2

Solve for all the internal forces for the truss given in Fig. 16.6.1, using the matrix displacement method.

Solution

(a) *Degrees of freedom, numbering system, statics matrix* $[A]$, *member stiffness matrix* $[S]$, *deformation matrix* $[B]$, *product matrices* $[SA^T]$ *and* $[ASA^T]$. All of this has been illustrated in Example 16.6.1: the truss numbering in Fig. 16.6.2 and the matrices in Fig. 16.6.3.

(b) *Load matrix* $[P]$. To make the analysis of the truss for only the loads in Fig. 16.6.1, constituting one loading condition ($NLC = 1$), the $\{P\}$ matrix will be the column matrix

$\{P\} =$

1	0
2	0
3	$+9$
4	0
5	0
6	-18
7	0
8	0

The elements in $\{P\}$ are obtained by comparing the actual loads with the possible P's identified on Fig. 16.6.2a.

(c) *Solution for* $\{X\}$ *and* $\{F\}$ *matrices.* To actually solve the problem requires performing a matrix inversion on the $[K]$ matrix of Fig. 16.6.3 (or a solution of the simultaneous equations $[K]\{X\} = \{P\}$), a process impractical without the digital computer. In lieu of actually showing the resulting inverted matrix $[K]^{-1}$, a check of Eqs. (16.7.4) and (16.7.5) may be made using the solution for $\{X\}$ previously determined in Example 7.3.1 (see Fig. 16.6.1b for results) by the geometric method after the internal forces had been found by the classical consistent deformation method. The product of matrix $[SA^T]$ and matrix $\{X\}$ is shown as follows to give the $\{F\}$ matrix (values in Fig. 16.6.1b) obtained by the previous solution:

$$\{X\} = \begin{array}{|c|c|} \hline 1 & +33.35 \\ \hline 2 & -37.03 \\ \hline 3 & +41.08 \\ \hline 4 & +6.76 \\ \hline 5 & +11.88 \\ \hline 6 & -109.03 \\ \hline 7 & +23.76 \\ \hline 8 & +25.03 \\ \hline \end{array}$$

$[SA^T] =$

F \ X	1	2	3	4	5	6	7	8
1	+0.3	+0.4	0	0	0	0	0	0
2	−1.0	0	+1.0	0	0	0	0	0
3	0	0	−0.3	+0.4	0	0	0	+0.3
4	0	0	0	0	+1.0	0	0	0
5	0	0	0	0	−1.0	0	+1.0	0
6	0	0	0	0	0	0	−1.0	+1.0
7	0	+0.25	0	0	0	−0.25	0	0
8	−0.3	+0.4	0	0	0	0	+0.3	0
9	0	0	0	+0.25	0	0	0	0

$$\{F\} = \begin{array}{|c|c|} \hline 1 & -4.81 \\ \hline 2 & +7.73 \\ \hline 3 & -2.11 \\ \hline 4 & +11.88 \\ \hline 5 & +11.88 \\ \hline 6 & +1.27 \\ \hline 7 & +18.00 \\ \hline 8 & -17.69 \\ \hline 9 & +1.69 \\ \hline \end{array}$$

The product of matrix $[K]$ and matrix $\{X\}$ is shown as follows to give the $\{P\}$ matrix (see part b of this example):

$$\{X\} = \begin{array}{|c|c|} \hline 1 & +33.35 \\ \hline 2 & -37.03 \\ \hline 3 & +41.08 \\ \hline 4 & +6.76 \\ \hline 5 & +11.88 \\ \hline 6 & -109.03 \\ \hline 7 & +23.76 \\ \hline 8 & +25.03 \\ \hline \end{array}$$

$[K] =$

P\X	1	2	3	4	5	6	7	8
1	+1.36	0	−1.00	0	0	0	−0.18	0
2	0	+0.89	0	0	0	−0.25	+0.24	0
3	−1	0	+1.18	−0.24	0	0	0	−0.18
4	0	0	−0.24	+0.59	0	0	0	+0.24
5	0	0	0	0	+2.00	0	−1.00	0
6	0	−0.25	0	0	0	+0.25	0	0
7	−0.18	+0.24	0	0	−1.00	0	+2.18	−1.00
8	0	0	−0.18	+0.24	0	0	−1.00	+1.18

$$[K]\{X\} = \{P\} = \begin{array}{|c|c|} \hline 1 & 0 \\ \hline 2 & 0 \\ \hline 3 & +9 \\ \hline 4 & 0 \\ \hline 5 & 0 \\ \hline 6 & -18 \\ \hline 7 & 0 \\ \hline 8 & 0 \\ \hline \end{array}$$

16.8 DIRECT METHOD OF OBTAINING GLOBAL STIFFNESS MATRIX $[K]$

The global stiffness matrix $[K]$ relates external forces P to joint displacements X as follows:

$$\{P\}_{NP \times 1} = [K]_{NP \times NP}\{X\}_{NP \times 1} \qquad [16.7.5]$$

The stiffness coefficient K_{ij} is the force at location i due to a unit displacement at location j. Written out, the equation for P_1 (involving the first row of the $[K]$ matrix) of the truss in Fig. 16.8.1a is

$$P_1 = K_{11}X_1 + K_{12}X_2 + K_{13}X_3 + K_{14}X_4$$

In terms of the matrix $[K]$ the first column of that matrix is obtained if $X_1 = 1$ and all other X's are set to zero. The external forces required for that displacement are the K_{i1} values, where i goes from 1 to NP.

For the truss of Fig. 16.8.1a, the degree of freedom NP is four, and the direction of possible joint forces P_1 to P_4 and the corresponding unknown joint displacements X_1 to X_4 are numbered in Fig. 16.8.1a. The internal force–elongation numbering system is given in Fig. 16.8.1b.

The thought process involved to obtain the K-values is to successively apply the joint displacements one at a time with the others remaining zero. In Fig. 16.8.1c, $X_1 = 1$ and X_2, X_3, and X_4 are held zero. The joint displacement $X_1 = 1$ causes internal forces F, as shown. Similarly, for equilibrium at each joint, four external forces, the four K values, are required to balance the F-forces arising from the assumed single unit displacement. Figure 16.8.1c, d, e, and f can be observed to give the four sets of joint forces K which are needed for this structure; each set of K-forces corresponds to one unit displacement with the remainder held to zero.

First the unit displacement is applied and the resulting internal forces F are determined, as labeled on the truss diagrams of Fig. 16.8.1c, d, e, and f. Next the

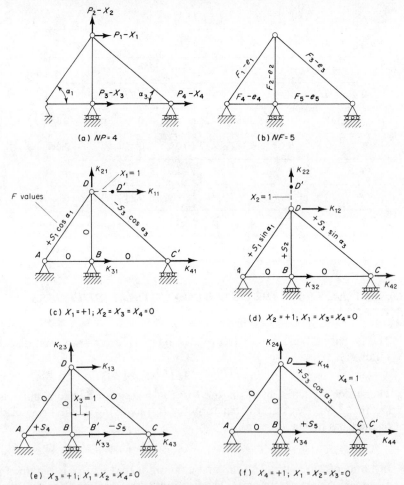

Figure 16.8.1 Direct method of obtaining the global stiffness matrix $[K]$. (S values are EA/L for the particular member.)

equilibrium of each joint is satisfied, relating the external forces K to the internal forces F.

To summarize, the relationship between internal forces F and joint displacements X for the truss of Fig. 16.8.1 can be obtained by visual inspection of Fig. 16.8.1c, d, e and f. Although labeled as $[SA^T]$ matrix, its first column is established by referring to Fig. 16.8.1c, second column to Fig. 16.8.1d, third column to Fig. 16.8.1e, and fourth column to Fig. 16.8.1f. Thus,

$[SA^T]_{5 \times 4} =$

F \ X	1	2	3	4
1	$+S_1 \cos \alpha_1$	$+S_1 \sin \alpha_1$	0	0
2	0	$+S_2$	0	0
3	$-S_3 \cos \alpha_3$	$+S_3 \sin \alpha_3$	0	$+S_3 \cos \alpha_3$
4	0	0	$+S_4$	0
5	0	0	$-S_5$	$+S_5$

in which $S_1 = EA_1/L_1$ to $S_5 = EA_5/L_5$ are the individual member stiffnesses. Note that only the component of the joint displacement in the bar direction causes a change in the length of the bar.

With the internal forces known, the statics of each joint must be satisfied, giving the external forces K. For the truss of Fig. 16.8.1, the $[K]$ matrix relating external forces P and joint displacements X is

$[K]_{4 \times 4} =$

P \ X	1	2	3	4
1	$+S_1 \cos^2 \alpha_1$ $+S_3 \cos^2 \alpha_3$	$+S_1 \sin \alpha_1 \cos \alpha_1$ $-S_3 \sin \alpha_3 \cos \alpha_3$	0	$-S_3 \cos^2 \alpha_3$
2	$+S_1 \sin \alpha_1 \cos \alpha_1$ $-S_3 \sin \alpha_3 \cos \alpha_3$	$S_1 \sin^2 \alpha_1 + S_2$ $+S_3 \sin^2 \alpha_3$	0	$+S_3 \sin \alpha_3 \cos \alpha_3$
3	0	0	$+S_4 + S_5$	$-S_5$
4	$-S_3 \cos^2 \alpha_3$	$+S_3 \sin \alpha_3 \cos \alpha_3$	$-S_5$	$+S_5 + S_5 \cos^2 \alpha_3$

Again, the $[K]$ matrix can be established by columns by noting that the external forces shown in Fig. 16.8.1c, d, e, and f represent, respectively, columns 1 through 4 of the matrix. It is noted again, as proved in Section 16.6, that $[K]$ is a symmetric matrix with respect to the principal diagonal.

These two matrices $[SA^T]$ and $[K]$ (which is $[ASA^T]$) give the two important relationships for the matrix displacement method described in Section 16.7; those are

$$\{P\} = [K]\{X\} \qquad\qquad [16.7.5]$$

from which the solution for $\{X\}$ can be obtained, and

$$\{F\} = [SA^T]\{X\} \qquad\qquad [16.7.4]$$

from which the solution for $\{F\}$ can be obtained.

Example 16.8.1

For the truss of Fig. 16.6.2, determine the first two columns of the internal force vs. joint displacement matrix $[SA^T]$ and the global stiffness matrix $[K]$ directly from physical concepts.

Solution

(a) *Degrees of freedom and numbering system.* These are given in Fig. 16.6.2.

(b) $[SA^T]$ *matrix.* Columns 1 and 2 of this matrix represent physically the internal forces arising when X_1 and X_2, respectively, equal unity while all other X's are zero. Figure 16.8.2a and c give the member elongations, respectively, for $X_1 = 1$, X_2 through $X_8 = 0$; and $X_2 = 1$, $X_1 = 0$, X_3 through $X_8 = 0$. The corresponding internal forces F are given in Fig. 16.8.2b and d.

(c) $[K]$ *matrix.* Using the sets of internal forces obtained by applying the sets of joint displacements, the equilibrium of the joints is used to obtain the elements of the matrix $[K]$. For the first two columns, the K-values are the external forces shown in Fig. 16.8.2b and d, respectively.

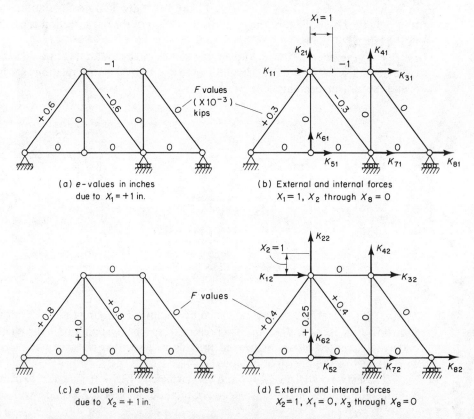

Figure 16.8.2 Elements for the $[SA^T]$ and $[K]$ matrices from the physical concept for the truss of Fig. 16.6.2, Example 16.8.2.

The results are given in Fig. 16.6.3, where the $[SA^T]$ matrix is called $[SB]$ and the $[K]$ matrix is called $[ASB]$.

16.9 STATICS CHECKS

Once the internal forces F and the joint displacements X have been obtained, usually from the computer after it has performed all the necessary matrix operations, *the results must be checked.* Two types of checks are required: statics and deformation. Since the statically indeterminate truss could not be analyzed using only statics, a check of the statics is a necessary but not sufficient check of the results. The statics check requires proving that equilibrium is satisfied at each joint. The deformation check is described in the next section.

Example 16.9.1

Illustrate the statics check for the truss analyzed in Example 16.7.1.

Solution. The internal forces obtained by the solution are entered on a diagram of the truss, as in Fig. 16.9.1, and the horizontal and vertical components are also shown. A check of statics at joint A shows:

$$\text{Horizontal forces:} \quad -23.85 + 9.10 + 14.75 = 0$$

$$\text{Vertical forces:} \quad +14.75 + 5.26 - 20.0 = +0.01 \approx 0$$

Note that the number of statics checks equals the degree of freedom, in this case, two. In fact, the joint equilibrium requirement in the direction of each reaction gives the magnitude of the reaction. For this problem, the reactions at the three hinged supports are shown in Fig. 16.9.1. Thus, in the matrix displacement method, the external reactions are last determined, after the joint displacements and the internal forces.

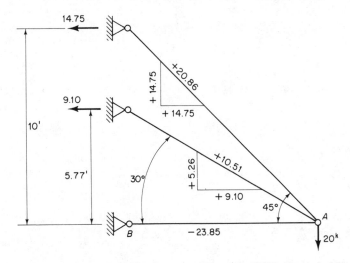

Figure 16.9.1 Statics check for the truss of Example 16.7.2; Example 16.9.1.

16.10 DEFORMATION CHECKS

In addition to the statics checks, the compatibility of deformation must be checked. For trusses, such checks consist of verifying that the elongation in a member computed from the internal force in that member must agree with the elongation computed from the joint displacements for each end of the member. For the latter, the joint displacement equation, Eq. (3.4.1), is used.

$$e = (X_3 - X_1) \cos \alpha + (X_4 - X_2) \sin \alpha \qquad [3.4.1]$$

Example 16.10.1

Illustrate the deformation checks for the truss analyzed in Example 16.7.1.

Solution

(a) *Compute the member elongations.* These are computed from the F-values given in the matrix of Example 16.7.1(i) as

$$e = \frac{FL}{EA}$$

for each truss member and entered on the diagram in Fig. 16.10.1.

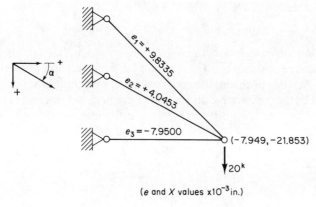

$e_1 = +9.8335$

$e_2 = +4.0453$

$e_3 = -7.9500$

$(-7.949, -21.853)$

20^k

(*e* and *X* values x10^{-3} in.)

Figure 16.10.1 Deformation check for the truss of Example 16.7.2; Example 16.10.1. (Positive X-values are to the right and upward.)

(b) *Compatibility of deformations.* For member 1, using Eq. (3.4.1),

$$e_1 = (-7.949 - 0)(+0.707) + (-21.853 - 0)(-0.707)$$
$$e_1 = +9.8301 \approx +9.8335 \text{ (Fig. 16.10.1)} \quad \text{(check)}$$

For member 2,

$$e_2 = (-7.949 - 0)(+0.866) + (-21.853 - 0)(-0.50)$$
$$e_2 = +4.0425 \approx +4.0453 \text{ (Fig. 16.10.1)} \quad \text{(check)}$$

For member 3, by inspection $e_3 = -7.9500$ equals the joint horizontal displacement of -7.949.

SELECTED REFERENCES

1. Chu-Kia Wang, *Matrix Methods of Structural Analysis*, 2nd ed., American Publishing Company, Madison, Wis., 1970.
2. M. Daniel Vanderbilt, *Matrix Structural Analysis*, Quantum Publishers, Inc., New York, 1974, Chap. 6.
3. William Weaver, Jr., and James M. Gere, *Matrix Analysis of Framed Structures*, 2nd ed., D. Van Nostrand Company, New York, 1980, Chap. 3.
4. Harold C. Martin, *Introduction to Matrix Methods of Structural Analysis*, McGraw-Hill Book Company, New York, 1966, Chaps. 2, 3.
5. Harry H. West, *Analysis of Structures*, John Wiley & Sons, Inc., New York, 1980, Chap. 14.

PROBLEMS

16.1. For the truss of the accompanying figure, determine the statics matrix $[A]$.

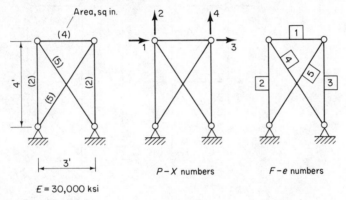

Prob. 16.1

16.2. For the truss of the accompanying figure, determine the statics matrix $[A]$.

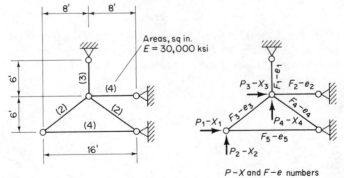

Prob. 16.2

16.3. For the truss of the accompanying figure, obtain the statics matrix $[A]$ and the deformation matrix $[B]$.

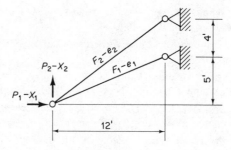

Prob. 16.3

16.4. For the truss of the accompanying figure, show numerically the sixth, seventh, and eighth columns of the deformation matrix $[B]$

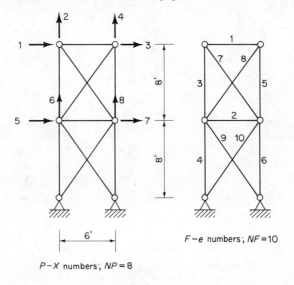

$P-X$ numbers; $NP = 8$

$F-e$ numbers; $NF = 10$

Prob. 16.4

16.5–16.12. For the truss of each accompanying figure, determine the statics matrix $[A]$, the deformation matrix $[B]$, the member stiffness matrix $[S]$, and combine those matrices to determine the global stiffness matrix $[K]$.

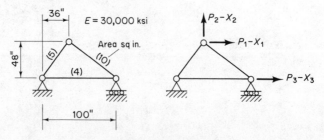

Probs. 16.5 and 16.13

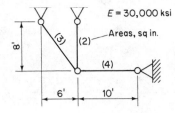
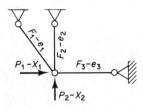

Probs. 16.6 and 16.14

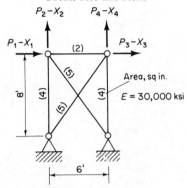

Probs. 16.7 and 16.15

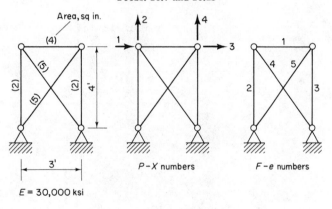

Probs. 16.8 and 16.16

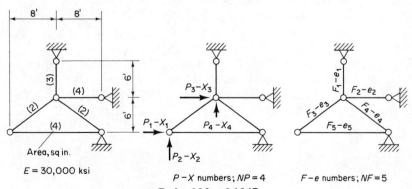

Probs. 16.9 and 16.17

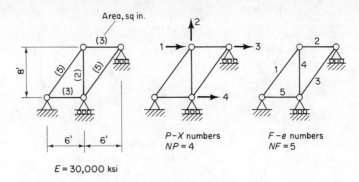

Area, sq in.

E = 30,000 ksi

P–X numbers
NP = 4

F–e numbers
NF = 5

Probs. 16.10 and 16.19

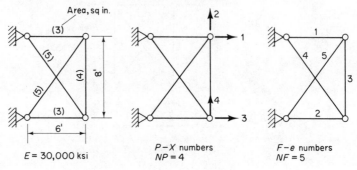

Area, sq in.

E = 30,000 ksi

P–X numbers
NP = 4

F–e numbers
NF = 5

Probs. 16.11 and 16.20

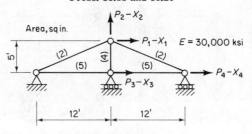

Area, sq in.

E = 30,000 ksi

Probs. 16.12 and 16.21

16.13. For the truss of Prob. 16.5, establish directly the numerical values of all elements in the second column of the $[K]$ matrix wherein $[K]$ is defined by Eq. (16.6.1). E = 30,000 ksi. (Compare with Prob. 16.5 if that problem was solved.)

16.14. For the three-bar pin-connected structure of Prob. 16.6, establish the $[SA^T]$ and $[K]$ matrices directly which should satisfy the equations $\{F\} = [SA^T]\{X\}$ and $\{P\} = [K]\{X\}$. Use P in kips and X in inches.

16.15. For the truss of Prob. 16.7, compute directly the values in the global stiffness matrix $[K]$. Compare with Prob. 16.7 if that problem was solved.

16.16. For the truss of Prob. 16.8, it is given that

$$\{X\} = \begin{Bmatrix} +2 \times 10^{-3} \text{ in.} \\ -4 \times 10^{-3} \text{ in.} \\ -5 \times 10^{-3} \text{ in.} \\ +3 \times 10^{-3} \text{ in.} \end{Bmatrix}$$

Determine the sign and magnitude of F_1 through F_5.

16.17. For the truss of Prob. 16.9, by defining $\{F\} = [SA^T]\{X\}$ and $\{P\} = [K]\{X\}$, obtain directly the third column of the $[SA^T]$ matrix and the third column of the $[K]$ matrix. Compare with Prob. 16.9 if that problem was solved.

16.18. For the truss of the accompanying figure, determine numerically (in kips per inch) the elements in the third column of the $[SA^T]$ and global stiffness $[K]$ matrices, wherein those matrices are defined by $\{F\} = [SA^T]\{X\}$ and $\{P\} = [K]\{X\}$.

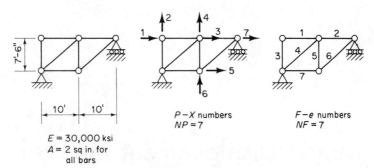

P – X numbers
NP = 7

F – e numbers
NF = 7

E = 30,000 ksi
A = 2 sq in. for
all bars

Prob. 16.18

16.19. For the truss of Prob. 16.10, compute and directly show the global stiffness matrix $[K]$. Compare with Prob. 16.10 if that problem was solved.

16.20. For the truss of Prob. 16.11, compute directly and show the global stiffness matrix $[K]$. Compare with Prob. 16.11 if that problem was solved.

16.21. For the truss of Prob. 16.12, determine numerically the elements in the second and fourth columns of the global stiffness matrix $[K]$. Compare with Prob. 16.12 if that problem was solved.

16.22. Establish the $[SA^T]$ and $[K]$ matrices directly for the truss in Example 7.3.2 (Fig. 7.3.6a). Using the numerical values of $\{P\}, \{X\}$, and $\{F\}$ in that example, verify numerically the equations $\{P\} = [K]\{X\}$ and $\{F\} = [SA^T]\{X\}$.

16.23. Establish directly the $[SA^T]$ and $[K]$ matrices of the truss in Prob. 7.4. If Prob. 7.4 was solved, use the numerical values of $\{P\}, \{X\}$, and $\{F\}$ in that problem to verify numerically the equations $\{P\} = [K]\{X\}$ and $\{F\} = [SA^T]\{X\}$. If Prob. 7.4 was not solved, verify by computing $[SA^T] = [S][B]$ and $[K] = [A][S][B]$.

16.24. Establish the $[SA^T]$ and $[K]$ matrices of the truss in Prob. 7.11. If Prob. 7.11 was solved, use the numerical values of $\{P\}, \{X\}$, and $\{F\}$ in that problem to verify numerically the equations $\{P\} = [K]\{X\}$ and $\{F\} = [SA^T]\{X\}$.

[17]

Matrix Displacement Method Applied to Beams

17.1 DEGREE OF FREEDOM VS. DEGREE OF STATICAL INDETERMINACY

The degree of statical indeterminacy as applied to beams has been defined in Section 8.1; it is equal to the number of external reactions minus two when no internal hinges are present. In general, Eq. (8.1.1) gives the degree NI of statical indeterminacy of a beam,

$$NI = NR - NIH - 2 \tag{8.1.1}$$

in which NR and NIH are the numbers of reactions and internal hinges, respectively.

The application of the computer to structural analysis requires a general treatment of the structure. The term *degree of freedom*, as it has been applied to trusses in Section 16.1, can be applied to beams in a similar manner.

A beam, whether statically determinate or indeterminate, may be separated into a finite number of *members* or *elements*, at points called *joints* or *nodes*. The "members" may be entire spans, or they may be segments of spans. The "joints" are the points connecting the "members." For the purpose of analysis, any point at which there is either an unknown reaction or an internal hinge *must* be taken as a joint. Therefore, for the beam of Fig. 17.1.1a, the minimum number of joints is four. The overhang, DE, being a statically determinate cantilever, does not have to be included in the structure; instead, the loads acting on DE can be transferred to act as a force and a moment at point D. Thus, the analysis of the beam of Fig. 17.1.1a using the displacement method (see Section 9.1) can be accomplished by obtaining the rotations of joints B, C, and D. Such a structure may be said to have three *degrees of freedom*

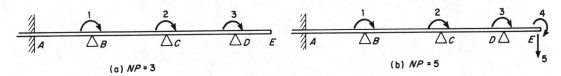

Figure 17.1.1 Degree of freedom of a beam.

(i.e., $NP = 3$). In general, the degree of freedom NP for the entire beam structure is the total number of unknown displacements at all its "joints."

In the analysis, "joints" may be taken any place, dividing the beam into as many "members" as desired. When a "joint" is created at a location other than at an external support, there will be two displacements (rotation and translation) associated with such a "joint." Thus, each added "joint" has two unknown displacements and therefore adds two degrees of freedom to the structure. For the beam of Fig. 17.1.1b, where the analyst desires to obtain the displacements at joint E as a part of the solution, there are then five degrees of freedom ($NP = 5$) for the entire beam.

In general, a "joint" is a point chosen as the boundary between two segments ("members") of a beam. The joint may be fixed against rotation, as at a built-in end (point A of Fig. 17.1.1), or it may be free to rotate under load. Further, it may be prevented from transverse displacement, as at a support (such as points B, C, and D of Fig. 17.1.1), or such a "joint" may be free to deflect under load (such as point E of Fig. 17.1.1). Thus, the degree of freedom at an arbitrary location selected as a "joint" is two (translation and rotation). If the chosen location is actually an internal hinge, then because of the two unknown rotations at either side of the hinge, the degree of freedom would be three. At exterior simple supports or at intermediate external supports, the degree of freedom reduces to one (rotation only), and at fixed supports the degree of freedom is zero.

Since the analysis by the displacement method (see Chapter 9 on slope deflection method) will involve the solving of simultaneous equations equal in number to the unknown displacements, the use of "joints" and "members" in excess of the minimum required for solution adds complexity and cost to obtaining the solution (unless a graphical output of the elastic curve is desired). The smallest degree of freedom is obtained by (a) utilizing statics to avoid treating cantilevers as "members", and (b) using as "joints" locations where the degree of freedom is zero or one (i.e., actual external supports). Thus, by excluding the cantilever DE in the structure the beam of Fig. 17.1.1a has a minimum degree of freedom of three. If, on the other hand, the portion DE is considered as a "member", the degree of freedom is five.

When a beam "member" or segment such as span AB of Fig. 17.1.1 is examined as a free body in Fig. 17.1.2, it is noted that knowledge of the end moments (internal forces using the symbol F), in addition to the fixed-end moments due to loads acting *on* the member, is sufficient to complete the analysis. Although shears are also needed for a complete analysis, it is noted that shears are determinable from the end moments. Thus, each "member" used in an analysis contains two unknown internal forces (i.e., the two end moments exclusive of fixed-end moments). Thus, the beam of Fig. 17.1.1a will have six unknown internal forces F, shown by the numbering in Fig. 17.1.2b; and the beam of Fig. 17.1.1b using four "members" will have eight internal forces F,

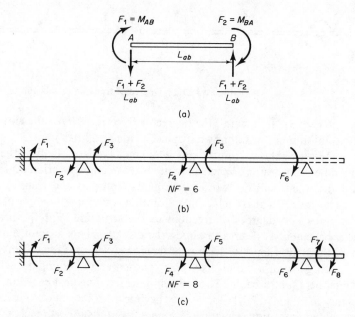

Figure 17.1.2 Internal forces used in displacement method analysis.

shown in Fig. 17.1.2c. The fact that F_8 is zero will be verified by the solution; it must still be considered as an unknown internal force in the analysis.

A major distinction between the force and displacement methods is that the degree of statical indeterminacy must be determined and used in the force method; whereas the degree of statical indeterminacy is not important in the displacement method, except perhaps to help verify the degree of freedom to be correct. If the degree of statical indeterminacy is desired, it equals (as discussed in Section 16.1) the excess of the number NF of internal forces over the number NP of degrees of freedom. Thus, the beam of Fig. 17.1.1a, which has $NP = 3$ and $NF = 6$, is statically indeterminate to the third degree. Even when an extra "member" and "joint" are used, as in Fig. 17.1.1b, the same conclusion is obtained; $NI = NF - NP = 8 - 5 = 3$. The degree of statical indeterminacy is an inherent property of the structure and cannot be affected by the analyst's choice of "members" and "joints."

17.2 DEFINITION OF STATICS MATRIX [A]

The continuous beam shown in Fig. 17.2.1a may be considered to have four joints A, B, C, and D and three members AB, BC, and CD. For matrix formulation it is convenient to label the four unknown joint displacements as X_1, X_2, X_3, and X_4, corresponding to θ_B, θ_C, θ_D, and Δ_D, the symbols used in classical methods such as the slope deflection method.

Since a continuous beam such as that of Fig. 17.2.1a subject to transverse loads will undergo rotation at joints B, C, and D and translation at joint D, the *same* displacements at the joints could be obtained by applying external loads only at the joints. In other words, a set of external loads (translational and rotational) applied at the

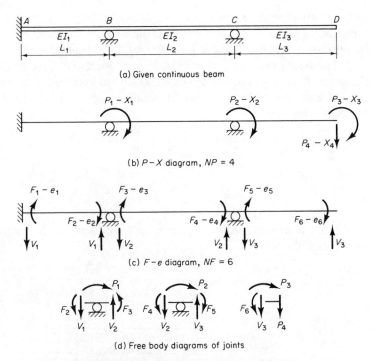

(a) Given continuous beam

(b) $P - X$ diagram, $NP = 4$

(c) $F - e$ diagram, $NF = 6$

(d) Free body diagrams of joints

Figure 17.2.1 Data for the statics matrix $[A]$.

joints in the directions of the unknown joint displacements could be the static equivalent of whatever set of actual loads might be acting. Thus, a joint force P, causing a displacement X, may be considered as existing for each degree of freedom, and for systematic treatment the force and the displacement should be numbered identically. Each degree of freedom, force P and joint displacement X, may be represented by a single rotational or translational arrow, as shown in the P-X diagram of Fig. 17.2.1b.

Once the reader understands that any external loading may be converted into statically equivalent joint forces, the next logical step is to note that the internal forces (i.e., shears and bending moments) all along a member are completely determinable from statics if the end moments on the member are known. Thus, for systematic treatment using matrix methods the unknown internal forces due to the statically equivalent joint forces are taken as the rotational moments at the two ends of a member. There is a close analogy between a continuous beam and a truss, especially when loads are applied at joints. The truss member has the equal and opposite collinear forces at the two ends of the member, whereas the beam member has a moment and a shear at each end of the member. The two end moments are independent, but the shear forces are not since they equal the sum of the end moments divided by the member length. Thus, the total number of independent internal forces (considering a moment as one type of force, sometimes called a *generalized force*) in a continuous beam is twice the number of members. The end moments and shear forces acting on the member ends are shown in their positive directions in Fig. 17.2.1c. Once F_i and F_j are designated as the end moments, the e_i and e_j are the corresponding clockwise rotations at the member ends measured from the member axis. In other words, e_i and e_j are measured to the tangent to the elastic curve at the member ends. *Whether or not the member axis is itself rotated the reference for measuring e is the member axis.*

The statics matrix $[A]$ expresses the joint forces $\{P\}$ in terms of the member end moments $\{F\}$. For example, the equations of equilibrium for the beam of Fig. 17.2.1 may be written by observing the free-body diagrams of the joints in Fig. 17.2.1d. The equations are

$$\left.\begin{array}{l} P_1 = F_2 + F_3 \\[4pt] P_2 = F_4 + F_5 \\[4pt] P_3 = F_6 \\[4pt] P_4 = -V_3 = -\dfrac{F_5 + F_6}{L_3} \end{array}\right\} \quad (17.2.1)$$

The statics matrix is the array of numbers representing the coefficients of the F's, written as follows:

$$[A]_{\text{NP} \times \text{NF}} =$$

P \ F	1	2	3	4	5	6
1	0	+1.	+1.	0	0	0
2	0	0	0	+1.	+1.	0
3	0	0	0	0	0	+1.
4	0	0	0	0	$-\dfrac{1}{L_3}$	$-\dfrac{1}{L_3}$

Note that the "rectangularity" of the statics matrix, which is $(NF - NP)$, is the degree of statical indeterminacy of the structure.

17.3 DEFINITION OF DEFORMATION MATRIX [B]

The deformation matrix $[B]$ expresses the member end rotations $\{e\}$ in terms of the joint displacement $\{X\}$. By observing the joint displacement diagrams of Fig. 17.3.1 for the beam of Fig. 17.2.1, the following relationships between e and X are obtained:

$$\left.\begin{array}{l} e_1 = 0 \\[4pt] e_2 = +X_1 \\[4pt] e_3 = +X_1 \\[4pt] e_4 = +X_2 \\[4pt] e_5 = +X_2 - \dfrac{X_4}{L_3} \\[4pt] e_6 = +X_3 - \dfrac{X_4}{L_3} \end{array}\right\} \quad (17.3.1)$$

Note that negative terms result in e_5 and e_6 because e is measured *from* the line con-

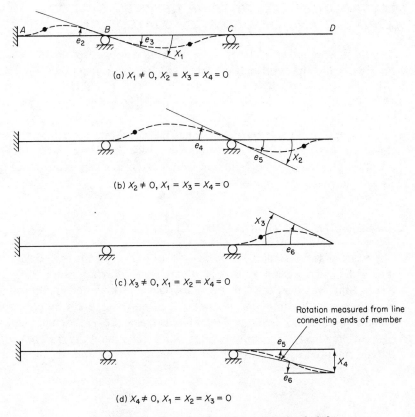

(a) $X_1 \neq 0$, $X_2 = X_3 = X_4 = 0$

(b) $X_2 \neq 0$, $X_1 = X_3 = X_4 = 0$

(c) $X_3 \neq 0$, $X_1 = X_2 = X_4 = 0$

(d) $X_4 \neq 0$, $X_1 = X_2 = X_3 = 0$

Figure 17.3.1 Data for the deformation matrix $[B]$.

necting the ends of the member to the elastic curve. The elastic curve has zero slope at C and D of Fig. 17.3.1d because X_2 and X_3 are zero. Consequently, e_5 and e_6 in Fig. 17.3.1d are counterclockwise rotations, that is, negative rotations.

The deformation matrix $[B]$ consists of the coefficients of the X's in Eqs. (17.3.1), written as follows:

$$[B]_{\text{NF} \times \text{NP}} =$$

e \ X	1	2	3	4
1	0	0	0	0
2	+1.	0	0	0
3	+1.	0	0	0
4	0	+1.	0	0
5	0	+1.	0	$-\dfrac{1}{L_3}$
6	0	0	+1.	$-\dfrac{1}{L_3}$

17.4 TRANSPOSITION RELATIONSHIP BETWEEN STATICS MATRIX [A] AND DEFORMATION MATRIX [B]

Exactly as was shown to be the case in truss analysis, so also for beams the statics matrix [A] and the deformation matrix [B] are the transpose of one another; that is, [B] = [A^T], as shown by the [A] matrix in Section 17.2 and the [B] matrix in Section 17.3. The ith column of the deformation matrix corresponds to the ith row of the statics matrix.

The proof for this relationship has been shown once before in Section 16.4 for trusses, using the principle of virtual work. The reader may write out such a similar proof for any chosen set of values of i and j in the continuous beam of Figs. 17.2.1 and 17.3.1 and that $A_{ij} = B_{ji}$.

17.5 MEMBER STIFFNESS MATRIX [S]

The member stiffness matrix consists of the coefficients of the end rotations e used to obtain the end moments F; that is, it expresses the relationship $\{F\} = [S]\{e\}$. Such a relationship was developed in the slope deflection method treatment in Chapter 9. For the present use in matrix methods, the symbols F_i and F_j are used instead of the symbols M'_A and M'_B used in Chapter 9. Also, e_i and e_j are used in place of the ϕ_A and ϕ_B used previously. Referring to Fig. 17.5.1 containing the matrix method symbols, the end rotations e_i and e_j are obtainable using, for example, the moment area method

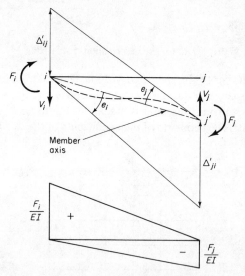

Figure 17.5.1 Member stiffness matrix [S].

(see Section 9.2 for details). The equations are

$$e_i = \frac{L}{3EI}F_i - \frac{L}{6EI}F_j$$

$$e_j = \frac{L}{6EI}F_i - \frac{L}{3EI}F_j$$

$$\left.\right\} \quad (17.5.1)$$

which are identical to Eqs. (9.2.4) obtained for use in the slope deflection method. Expressed in matrix notation, Eqs. (17.5.1) become

$$\begin{Bmatrix} e_i \\ e_j \end{Bmatrix} = [D] \begin{Bmatrix} F_i \\ F_j \end{Bmatrix} \tag{17.5.3}$$

in which

$$[D] = \begin{array}{c|c|c} \diagbox{e}{F} & F_i & F_j \\ \hline e_i & +\dfrac{L}{3EI} & -\dfrac{L}{6EI} \\ \hline e_j & -\dfrac{L}{6EI} & +\dfrac{L}{3EI} \end{array}$$

Solving Eqs. (17.5.1) for F_i and F_j gives

$$\left. \begin{aligned} F_i &= \frac{4EI}{L} e_i + \frac{2EI}{L} e_j \\ F_j &= \frac{2EI}{L} e_i + \frac{4EI}{L} e_j \end{aligned} \right\} \tag{17.5.4}$$

which are identical to Eqs. (9.2.5) obtained for use in the slope deflection method. Expressed in matrix notation, Eqs. (17.5.4) become

$$\begin{Bmatrix} F_i \\ F_j \end{Bmatrix} = [S] \begin{Bmatrix} e_i \\ e_j \end{Bmatrix} \tag{17.5.5}$$

in which

$$[S] = \begin{array}{c|c|c} \diagbox{F}{e} & e_i & e_j \\ \hline F_i & +\dfrac{4EI}{L} & +\dfrac{2EI}{L} \\ \hline F_j & +\dfrac{2EI}{L} & +\dfrac{4EI}{L} \end{array}$$

The matrix $[S]$ is the *member stiffness matrix*, and the matrix $[D]$ may be called the *member flexibility matrix*. Note that these relationships between F and e were developed for an *unloaded* beam element. Recall that in the slope deflection method the deformations and end moments relating to the unloaded element are superimposed on the fixed-end condition containing all the transverse loads (see Section 9.2 and Fig. 9.2.3). The matrix displacement method is in effect the slope deflection method in a more systematized form. The notation is changed here in order to have uniform symbols in the matrix displacement method for the truss, the continuous beam, or any other framed structure.

17.6 GLOBAL STIFFNESS MATRIX [K]

The global stiffness matrix $[K]$ is the relationship between the external forces P and the joint displacements X; thus,

$$\{P\} = [K]\{X\} \tag{17.6.1}$$

The $[K]$ matrix may be established by columns by letting one displacement equal unity at a time with all others equal zero. For example, referring to Fig. 17.6.1 (the structure

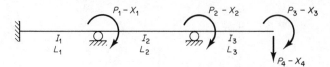

(a) Original structure showing degrees of freedom

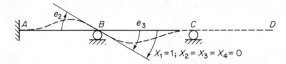

(b) Elastic curve

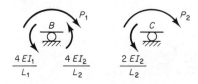

(c) End moments developed

(d) Free body diagrams of joints

Figure 17.6.1 First column of the global stiffness matrix $[K]$.

of Fig. 17.2.1), $X_1 = 1$ with X_2, X_3, and X_4 equal to zero. For member AB, with the rotation $e_2 = 1$ at B and the member fixed at A, the end moments $4EI_1/L_1$ and $2EI_1/L_1$ develop at ends B and A, respectively. For member BC, with the rotation $e_3 = 1$ at B and the slope zero at end C (i.e., $X_2 = 0$), the end moments $4EI_2/L_2$ and $2EI_2/L_2$ develop at B and C, respectively. Then examine the free-body diagrams of the joints to determine the joint forces P required for equilibrium. Thus,

$$P_1 = \frac{4EI_1}{L_1} + \frac{4EI_2}{L_2}$$

$$P_2 = \frac{2EI_2}{L_2}$$

$$P_3 = P_4 = 0$$

In a similar manner, the second and third columns of the global stiffness matrix $[K]$ are determined and values are given in Fig. 17.6.2. These should be verified by the reader. The details of column 4 are given in Fig. 17.6.3. From the equilibrium of the joints as shown in Fig. 17.6.3c, the values of P for the fourth column are

$$P_1 = 0$$

$$P_2 = -\frac{6EI_3}{L_3^2}$$

$$P_3 = -\frac{6EI_3}{L_3^2}$$

$$P_4 = +\frac{12EI_3}{L_3^3}$$

X P	1	2	3	4
1	$\dfrac{4EI_1}{L_1} + \dfrac{4EI_2}{L_2}$	$\dfrac{2EI_2}{L_2}$	0	0
2	$\dfrac{2EI_2}{L_2}$	$\dfrac{4EI_2}{L_2} + \dfrac{4EI_3}{L_3}$	$\dfrac{2EI_3}{L_3}$	$-\dfrac{6EI_3}{L_3^2}$
3	0	$\dfrac{2EI_3}{L_3}$	$\dfrac{4EI_3}{L_3}$	$-\dfrac{6EI_3}{L_3^2}$
4	0	$-\dfrac{6EI_3}{L_3^2}$	$-\dfrac{6EI_3}{L_3^2}$	$+\dfrac{12EI_3}{L_3^3}$

Figure 17.6.2 Global stiffness matrix $[K]$ for the continuous beam of Fig. 17.6.1.

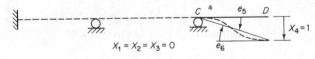

$$X_1 = X_2 = X_3 = 0$$

(a) Elastic curve

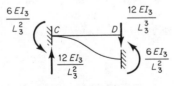

(b) End moments and shears developed

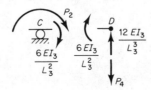

(c) Free body diagrams of joints

Figure 17.6.3 Fourth column of the global stiffness matrix $[K]$.

By the preceding illustrated process the global stiffness matrix $[K]$ of any continuous beam may be determined directly.

Alternatively, and in the authors' opinion more easily, the matrix $[K]$ may be determined by matrix operations using the statics matrix $[A]$, the deformation matrix $[A^T]$, and the member stiffness matrix $[S]$. That was first shown in Section 16.6 for trusses; however, the relationship is generally applicable for all elastic structures. The procedure is repeated here for completeness.

Using matrix operations, start with the statics matrix $[A]$, as described in Section 17.2,

$$\{P\}_{\text{NP}\times 1} = [A]_{\text{NP}\times\text{NF}}\{F\}_{\text{NP}\times 1} \tag{17.6.2}$$

Next, the Hooke's Law equation, involving the member stiffness matrix as described in Section 17.5, is used,

$$\{F\}_{\text{NF}\times 1} = [S]_{\text{NF}\times\text{NF}}\{e\}_{\text{NF}\times 1} \tag{17.6.3}$$

Then, the compatibility equation, involving the deformation matrix $[B]$ as described in Section 17.3, is used,

$$\{e\}_{\text{NF}\times 1} = [B]_{\text{NF}\times\text{NP}}\{X\}_{\text{NP}\times 1} \tag{17.6.4}$$

Now, substituting Eq. (17.6.4) in Eq. (17.6.3) gives

$$\{F\}_{\text{NF}\times 1} = [SB]_{\text{NF}\times\text{NP}}\{X\}_{\text{NP}\times 1} \tag{17.6.5}$$

and since $[B] = [A^T]$,

$$\{F\}_{\text{NF}\times 1} = [SA^T]_{\text{NF}\times\text{NP}}\{X\}_{\text{NP}\times 1} \tag{17.6.6}$$

Finally, substituting Eq. (17.6.6) into Eq. (17.6.2) gives

$$\{P\}_{\text{NP}\times 1} = [ASA^T]_{\text{NP}\times\text{NP}}\{X\}_{\text{NP}\times 1} \tag{17.6.7}$$

A comparison of Eq. (17.6.7) with Eq. (17.6.1) shows that the global stiffness matrix $[K]$ equals

$$[K]_{\text{NP}\times\text{NP}} = [ASA^T]_{\text{NP}\times\text{NP}} \tag{17.6.8}$$

As was proved in Chap. 16, and is equally applicable for flexural members, the global stiffness matrix $[K]$ is a symmetrical matrix; that is, any ith row is identical to the ith column.

As a third alternative, which is most useful in automatic computer programming, the stiffness matrix of a typical beam member (or beam element) relating to the four degrees of freedom (two at each end of the beam member) may be derived first. This matrix, called the *local stiffness matrix*, has four rows and four columns, as follows:

$$[\text{local } K] =$$

P \ X	1	2	3	4
1	$+\dfrac{4EI}{L}$	$+\dfrac{2EI}{L}$	$+\dfrac{6EI}{L^2}$	$-\dfrac{6EI}{L^2}$
2	$+\dfrac{2EI}{L}$	$+\dfrac{4EI}{L}$	$+\dfrac{6EI}{L^2}$	$-\dfrac{6EI}{L^2}$
3	$+\dfrac{6EI}{L^2}$	$+\dfrac{6EI}{L^2}$	$+\dfrac{12EI}{L^3}$	$-\dfrac{12EI}{L^3}$
4	$-\dfrac{6EI}{L^2}$	$-\dfrac{6EI}{L^2}$	$-\dfrac{12EI}{L^3}$	$+\dfrac{12EI}{L^3}$

$$\tag{17.6.9}$$

The details of derivation for the first and fourth columns of this matrix are shown in Fig. 17.6.4. In the computer program, by a single DO loop, the local stiffness matrix of each element is computed and then fed into the appropriate slots in the global stiffness matrix according to the global degree of freedom numbers existing at each end of the beam element. For any missing freedom, such as at a fixed support, a global number equal to $(NP + 1)$ may be assigned so that the $(NP + 1)$th row and $(NP + 1)$th column of the built-up global matrix may be simply discarded.

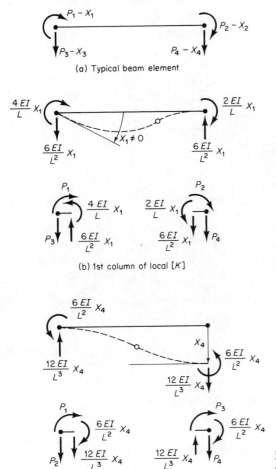

(a) Typical beam element

(b) 1st column of local $[K]$

(c) 4th column of local $[K]$

Figure 17.6.4 Derivation of the local stiffness matrix of a typical beam element.

17.7 EXTERNAL FORCE MATRIX $\{P\}$

Very rarely are the real loads on a continuous beam applied directly along the chosen degrees of freedom; in such rare cases the elements of the $\{P\}$ matrix, known as the *external force matrix,* may be assembled as in truss analysis (see Section 16.7), where the actual loads are applied along the degrees of freedom for the structure. Most often, on beams the transverse loads are applied along the members, which in turn transmit the loads to the joints.

Transmission of the beams loads to the joints is conveniently done by using the fixed condition, in which all joint displacements along the degrees of freedom are zero. As in the slope deflection method (refer to Fig. 9.2.3), the total effect on a given beam span may be treated as the sum of (1) the transverse loading effect (Fig. 9.2.3a) while maintaining zero joint displacements, and (2) the joint loadings (Fig. 9.2.3b, c, and d) causing the joint displacements. For instance, the given continuous beam of Fig. 17.7.1a may be considered to be the sum of the fixed conditioned beam of Fig. 17.7.1b and the joint-force conditioned beam of Fig. 17.7.1c.

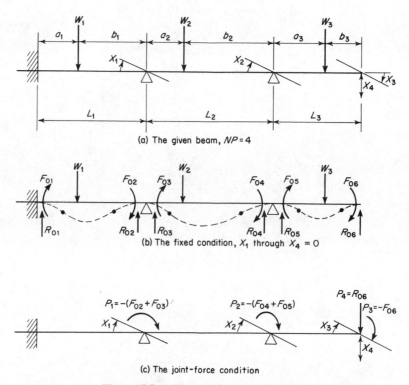

Figure 17.7.1   External force matrix $\{P\}$.

Notation for the fixed condition is changed from that used in Chapters 9 and 10, where fixed-end moments were defined by the symbols M_{F1} or M_{FAB}. Since in the matrix methods a moment is a generalized force the symbol F is used for *all* types of independent internal forces (that is, axial force, moment, and torsional moment). Since shears are *dependent* on moments, the symbol F is *not* used for them. When the internal forces are in the fixed condition, the subscript 0 precedes the internal force number. Thus, in Fig. 17.7.1b, the fixed-end moments are F_{01}, F_{02}, F_{03}, etc., and the reactions for the fixed condition are R_{01}, R_{02}, R_{03}, etc.

When the *clockwise* moments $(F_{02} + F_{03})$, $(F_{04} + F_{05})$, and F_{06}, as well as the upward force R_{06}, are applied to the beam of Fig. 17.7.1b, all joint displacements X_1 through X_4 will be zero. Then, when these four restraining forces are removed (that is, applied in the reverse direction), joint displacements X_1 to X_4 will occur as in Fig. 17.7.1c.

The actual structure of Fig. 17.7.1a is equivalent to the sum of Fig. 17.7.1b and c. Since the joint displacements X_1 to X_4 in Fig. 17.7.1a can be obtained by adding the zero joint displacements of Fig. 17.7.1b to the X_1 to X_4 displacements of Fig. 17.7.1c, Fig. 17.7.1c contains the real joint displacements. However, the internal forces (that is, the end moments) are the sum of those in the fixed condition (F_{01} to F_{06}) of Fig. 17.7.1b and those in the joint-force condition (F_1 to F_6) of Fig. 17.7.1c.

The equivalent joint forces in Fig. 17.7.1c are

$$P_1 = -(F_{02} + F_{03}) = -\left(+\frac{W_1 b_1 a_1^2}{L_1^2} - \frac{W_2 a_2 b_2^2}{L_2^2}\right)$$

$$P_2 = -(F_{04} + F_{05}) = -\left(+\frac{W_2 b_2 a_2^2}{L_2^2} - \frac{W_3 a_3 b_3^2}{L_3^2}\right)$$

$$P_3 = -F_{06} = -\left(+\frac{W_3 b_3 a_3^2}{L_3^2}\right)$$

$$P_4 = +R_{06} = +\frac{W a_3^2}{L_3^3}(3b_3 + a_3)$$

Expressions for fixed-end moments are obtained from Fig. 9.2.4. Formulas for the fixed-end reactions need not be memorized, as they can be computed by superimposing the reactions due to simple-beam action and due to the unbalanced fixed-end moments. In matrix format the equivalent joint forces become the external force matrix $\{P\}$, as follows:

$$\{P\} = $$

P \\ LC	1
1	$-\left(+\dfrac{W_1 b_1 a_1^2}{L_1^2} - \dfrac{W_2 a_2 b_2^2}{L_2^2}\right)$
2	$-\left(+\dfrac{W_2 b_2 a_2^2}{L_2^2} - \dfrac{W_3 a_3 b_3^2}{L_3^2}\right)$
3	$-\left(+\dfrac{W_3 b_3 a_3^2}{L_3^2}\right)$
4	$+\dfrac{W a_3^2}{L_3^3}(3b_3 + a_3)$

17.8 MATRIX DISPLACEMENT METHOD

The matrix displacement method for continuous beam analysis follows a procedure similar to that described in Section 16.7 for trusses. First, one must establish the degree of freedom NP to be used in the solution. Next, the method requires establishment of the statics matrix $[A]$ and the deformation matrix $[B]$, each using basic concepts, following by a visual check to assure that $[B] = [A^T]$. The member stiffness matrix $[S]$ is then established. Finally, the fixed-end moments $\{F_0\}$ are computed and the external force matrix $\{P\}$ is established; the elements in the $\{P\}$ matrix are the reversals of those forces acting on the member ends in the fixed condition.

Combining the equilibrium conditions,

$$\{P\}_{\text{NP}\times 1} = [A]_{\text{NP}\times\text{NF}}\{F\}_{\text{NF}\times 1} \tag{17.8.1}$$

the force–deformation relationships,

$$\{F\}_{\text{NF}\times 1} = [S]_{\text{NF}\times\text{NF}}\{e\}_{\text{NF}\times 1} \tag{17.8.2}$$

and the compatibility conditions,

$$\{e\}_{\text{NF}\times 1} = [B]_{\text{NF}\times\text{NP}}\{X\}_{\text{NP}\times 1} \tag{17.8.3}$$

the following two equations are obtained:

$$\{X\}_{\text{NP}\times 1} = [ASA^T]^{-1}_{\text{NP}\times\text{NP}}\{P\}_{\text{NP}\times 1} \tag{17.8.4}$$

and

$$\{F\}_{\text{NF}\times 1} = [SA^T]_{\text{NF}\times\text{NP}}\{X\}_{\text{NP}\times 1} \tag{17.8.5}$$

Then, noting that the final end moments $\{F^*\}$ are the sum of those in the fixed and joint-force conditions,

$$\{F^*\}_{\text{NF}\times 1} = \{F_0\}_{\text{NF}\times 1} + \{F\}_{\text{NF}\times 1} \tag{17.8.6}$$

The application of Eqs. (17.8.4) to (17.8.6) is illustrated in the following examples.

Example 17.8.1

Using A, B, and C as joints, analyze the continuous beam of Fig. 17.8.1a by the matrix displacement method. (This structure has previously been analyzed by the theorem of

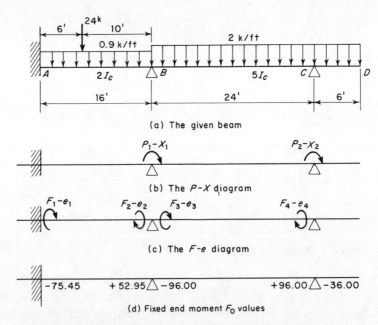

(a) The given beam

(b) The P-X diagram

(c) The F-e diagram

(d) Fixed end moment F_0 values

Figure 17.8.1 Matrix displacement method, Example 17.8.1.

Matrix Displacement Method Applied to Beams   Chap. 17

three moments, Example 8.3.1, by the slope deflection method, Example 9.3.2, and by the moment distribution method, Example 10.8.1.)

Solution

(a) *Draw the P-X and F-e diagrams* (Fig. 17.8.1b and c). The *P-X* diagram establishes the degrees of freedom that are to be used in the solution. In this problem, the minimum possible number of degrees of freedom is used. The segment *CD* does not have to be used because it is statically determinate. However, the analyst must use a minimum of two members (*AB* and *BC*) containing four *F*'s (F_1 to F_4). Thus, $(NF - NP)$ equals two, the degree of statical indeterminacy. More beam segments may be used, as will be shown in Example 17.8.2, but $(NF - NP)$ will always be two.

(b) *Establish the statics matrix* [*A*]. From equilibrium of the joints *B* and *C*, as in Fig. 17.8.2a,

(a) Equilibrium of the joints

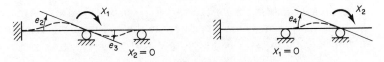

(b) Compatibility of internal deformations with joint displacements

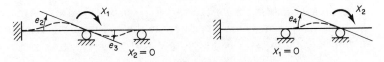

(c) Forces on joints in the fixed condition

Figure 17.8.2 Details for matrices used in Example 17.8.1.

$$P_1 = F_2 + F_3$$
$$P_2 = F_4$$

or in matrix form,

$[A]_{2 \times 4} =$	$\diagdown$ P \ F	1	2	3	4
	1		+1.0	+1.0	
	2				+1.0

Note that in this method the statically determinate moment M_{CD} on the cantilever is treated as a joint force, and thus it becomes a part of the external force matrix.

(c) *Establish the deformation matrix* [*B*]. This matrix relating the internal deformations (rotations at the ends of member segments) to the joint displacements can be established either by columns or by rows. In the former, the effects of each X on all the e's are examined. In the latter, how each e is affected by all the X's is examined. Using the former and examining the geometry shown in Fig. 17.8.2b,

$$e_2 = X_1 \quad \text{and} \quad e_3 = X_1 \qquad \text{when } X_1 \neq 0 \quad \text{and} \quad X_2 = 0$$
$$e_4 = X_2 \qquad\qquad\qquad\qquad \text{when } X_1 = 0 \quad \text{and} \quad X_2 \neq 0$$

Thus, the matrix $[B]$ is

$[B]_{4 \times 2} = $

e \ X	1	2
1		
2	+1.0	
3	+1.0	
4		+1.0

After establishing matrix $[B]$ independently from matrix $[A]$, check that one is the transpose of the other.

(d) *Establish the member stiffness matrix* $[S]$. This is the pair of equations involving EI/L, Eqs. (17.5.4), that relate the end moments F to the end rotations e. In matrix form, Eqs. (17.5.4) are expressed

$[S]_{4 \times 4} = $

F \ e	1	2	3	4
1	$\dfrac{EI_c}{2}$	$\dfrac{EI_c}{4}$		
2	$\dfrac{EI_c}{4}$	$\dfrac{EI_c}{2}$		
3			$\dfrac{5EI_c}{6}$	$\dfrac{5EI_c}{12}$
4			$\dfrac{5EI_c}{12}$	$\dfrac{5EI_c}{6}$

(e) *Establish the external force matrix* $\{P\}$. This matrix, sometimes called the load matrix, is obtained from the equilibrium of the forces acting on the joints in the fixed condition with the P-forces on the P-X diagram, as discussed in Section 17.7. For this example the fixed-end moments are computed and shown in Fig. 17.8.1d. The details of the computations are in Example 10.8.1(a). Then examining the rotational equilibrium of joints B and C in Fig. 17.8.2c, the net forces on the joints are

$$P_1 = 96.00 - 52.95 = +43.05 \text{ ft-kips}$$
$$P_2 = 36.00 - 96.00 = -60.00 \text{ ft-kips}$$

In matrix format for the one loading condition ($LC = 1$), the external force matrix is

$$\{P\}_{2\times1} =$$

P \ LC	1
1	$+43.05$
2	-60.00

LC = loading condition

(f) *Obtain the* $[SA^T]$ *matrix.* Matrix multiplication of the premultiplier matrix $[S]$ and the postmultiplier matrix $[B]$ (i.e., $[A^T]$) gives the $[SA^T]$ matrix. For example, element $(1, 1)$ is the sum of the inner products of row 1 of $[S]$ and column 1 of $[B]$, as follows:

$$\left(\frac{EI_c}{2}\right)(0) + \left(\frac{EI_c}{4}\right)(+1.0) + (0)(1.0) + (0)(0) = \frac{EI_c}{4}$$

The complete matrix $[SB] = [SA^T]$ is as follows:

$$[SA^T]_{4\times2} =$$

F \ X	1	2
1	$\dfrac{EI_c}{4}$	0
2	$\dfrac{EI_c}{2}$	0
3	$\dfrac{5EI_c}{6}$	$\dfrac{5EI_c}{12}$
4	$\dfrac{5EI_c}{12}$	$\dfrac{5EI_c}{6}$

(g) *Obtain the global stiffness matrix* $[K]$. Matrix multiplication of the premultiplier matrix $[A]$ and the postmultiplier matrix $[SA^T]$ gives the matrix $[K]$ (i.e., $[ASA^T]$). Alternatively, the matrix $[K]$ could have been determined directly as described in Section 17.6. The complete matrix $[K]$ is as follows:

$$[K] = [ASA^T]_{2\times2} =$$

P \ X	1	2
1	$\dfrac{4EI_c}{3}$	$\dfrac{5EI_c}{12}$
2	$\dfrac{5EI_c}{12}$	$\dfrac{5EI_c}{6}$

(h) *Obtain the inverse of the matrix* $[K]$. Using Eq. (17.8.4),

$$\{X\} = [ASA^T]^{-1}\{P\}$$

the $[K] = [ASA^T]$ matrix must be inverted. By Gauss–Jordan elimination as discussed in Section 13.4, or by solving the two simultaneous equations for X_1 and X_2 in terms of P_1 and P_2, the $[K]^{-1}$ matrix is obtained:

$$[K]^{-1} = \begin{array}{c|c|c|c} \diagdown\!\!\!\begin{array}{l} P \\ X \end{array} & 1 & 2 \\ \hline 1 & +\dfrac{8}{9EI_c} & -\dfrac{4}{9EI_c} \\ \hline 2 & -\dfrac{4}{9EI_c} & +\dfrac{64}{45EI_c} \end{array}$$

(i) *Obtain the joint displacement matrix* $\{X\}$. The joint displacement matrix $\{X\}$ is obtained by multiplying the premultiplier matrix $[K]^{-1}$ by the postmultiplier external force matrix $\{P\}$. The matrix operation gives

$$X_1 = +\frac{8}{9EI_c}(+43.05) - \frac{4}{9EI_c}(-60.00) = +\frac{64.93}{EI_c}$$

$$X_2 = -\frac{4}{9EI_c}(+43.05) + \frac{64}{45EI_c}(-60.00) = -\frac{104.47}{EI_c}$$

In matrix format the $\{X\}$ matrix for the one loading condition ($LC = 1$) is

$$\{X\} = \begin{array}{c|c} \diagdown\!\!\!\begin{array}{l} LC \\ X \end{array} & 1 \\ \hline 1 & +\dfrac{64.93}{EI_c} \\ \hline 2 & -\dfrac{104.47}{EI_c} \end{array}$$

(j) *Obtain the internal force matrix* $\{F\}$. Using Eq. (17.8.5), multiply the premultiplier matrix $[SA^T]$ by the postmultiplier matrix $\{X\}$.

$$\{X\} = \begin{array}{c|c} 1 & +\dfrac{64.93}{EI_c} \\ \hline 2 & -\dfrac{104.47}{EI_c} \end{array}$$

$$[SA^T] = \begin{array}{c|c|c} \diagdown\!\!\!\begin{array}{l} X \\ F \end{array} & 1 & 2 \\ \hline 1 & \dfrac{EI_c}{4} & 0 \\ \hline 2 & \dfrac{EI_c}{2} & 0 \\ \hline 3 & \dfrac{5EI_c}{6} & \dfrac{5EI_c}{12} \\ \hline 4 & \dfrac{5EI_c}{12} & \dfrac{5EI_c}{6} \end{array}$$

$$\{F\} = \begin{array}{c|c} 1 & +16.23 \\ \hline 2 & +32.47 \\ \hline 3 & +10.58 \\ \hline 4 & -60.00 \end{array}$$

(k) *Obtain the final end moments.* The total moment F^* on the end of a member is the sum of the F-value from the matrix operations and the fixed-end moment F_0.

In matrix notation,

$$\{F^*\} = \{F_0\} + \{F\}$$

For this example,

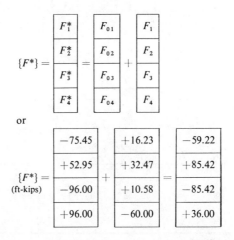

$$\{F^*\} =
\begin{array}{|c|}
\hline
F_1^* \\
\hline
F_2^* \\
\hline
F_3^* \\
\hline
F_4^* \\
\hline
\end{array}
=
\begin{array}{|c|}
\hline
F_{01} \\
\hline
F_{02} \\
\hline
F_{03} \\
\hline
F_{04} \\
\hline
\end{array}
+
\begin{array}{|c|}
\hline
F_1 \\
\hline
F_2 \\
\hline
F_3 \\
\hline
F_4 \\
\hline
\end{array}$$

or

$$\{F^*\} = \atop (\text{ft-kips})
\begin{array}{|c|}
\hline
-75.45 \\
\hline
+52.95 \\
\hline
-96.00 \\
\hline
+96.00 \\
\hline
\end{array}
+
\begin{array}{|c|}
\hline
+16.23 \\
\hline
+32.47 \\
\hline
+10.58 \\
\hline
-60.00 \\
\hline
\end{array}
=
\begin{array}{|c|}
\hline
-59.22 \\
\hline
+85.42 \\
\hline
-85.42 \\
\hline
+36.00 \\
\hline
\end{array}$$

The solution is in agreement with those obtained in Examples 8.3.1, 9.3.2, and 10.8.1.

(l) *Check.* Checks of statics and compatibility of deformations are always necessary to verify the correctness of the answer. These checks are discussed in Sections 17.9 and 17.10.

In comparing the solution for Example 17.8.1 with that of the slope deflection method in Example 8.3.1, the reader will find that the slope deflection equations, when cast in matrix form, are in fact identical to Eq. (17.8.6),

$$\{F^*\} = \{F_0\} + [SA^T]\{X\} \qquad (17.8.7)$$

and the set of equations solved simultaneously in the slope deflection method are

$$[ASA^T]\{X\} = \{P\} \qquad (17.8.8)$$

exactly the same as Eqs. (17.8.4).

The matrix displacement method, when displayed and carried out in longhand, may actually take more time than the conventional slope deflection method; however, the former method can be conveniently programmed on the computer by taking $[A]$, $[S]$, and $\{P\}$ as input, and $\{X\}$ and $\{F\}$ as output. Furthermore, the direct stiffness method of feeding the local stiffness matrix of each beam element into the proper slots of the global stiffness matrix may be used, although more logistics will be needed in the computer program. When there are several loading conditions, the matrix $\{P\}$ will simply become one with several columns.

Example 17.8.2

Analyze again the continuous beam of Example 17.8.1, except use an additional joint at D. The beam is shown again in Fig. 17.8.3a.

Solution

(a) *Draw the P-X and F-e diagrams* (Fig. 17.8.3b and c). Since the decision has been made to use D as a "joint," there are two more degrees of freedom than in Exam-

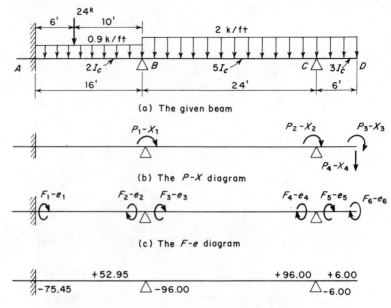

(a) The given beam

(b) The P-X diagram

(c) The F-e diagram

(d) Fixed end moment F_0 values (clockwise positive)

Figure 17.8.3 Matrix displacement method solution, Example 17.8.2.

ple 17.8.1. Joint D has a rotational and a translational degree of freedom, designated P_3-X_3 and P_4-X_4, respectively. The segment CD then brings F_5-e_5 and F_6-e_6 into the analysis. Using an additional segment in the analysis does not change the degree of statical indeterminacy (i.e., $NF - NP = 6 - 4 = 2$), since that is a property of the structure no matter how many beam segments are used in the analysis.

(b) *Establish the statics matrix* [*A*]. From equilibrium of the joints B, C, and D in Fig. 17.8.4a,

$$P_1 = F_2 + F_3$$
$$P_2 = F_4 + F_5$$
$$P_3 = F_6$$
$$P_4 = -R_{DC} = \frac{-(F_5 + F_6)}{6}$$

Note that the reaction R_{DC} on segment CD is equal to $(F_5 + F_6)/6$ acting upward on end D; thus R_{DC} acts downward on joint D.

In matrix format the statics matrix is

$[A]_{4 \times 6} =$

F \ P	1	2	3	4	5	6
1		+1.0	+1.0			
2				+1.0	+1.0	
3						+1.0
4					$-\dfrac{1}{6}$	$-\dfrac{1}{6}$

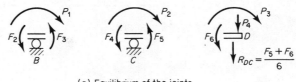

(a) Equilibrium of the joints

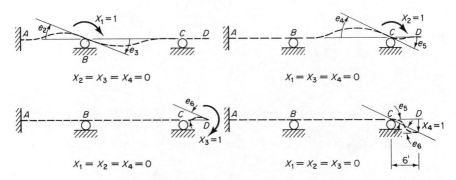

$X_2 = X_3 = X_4 = 0$ $X_1 = X_3 = X_4 = 0$

$X_1 = X_2 = X_4 = 0$ $X_1 = X_2 = X_3 = 0$

(b) Compatibility of internal deformations with joint displacements

$$R_{DC} = \frac{2(6)}{2} = 6^k$$

(c) Forces on joints in the fixed condition

Figure 17.8.4 Details for matrices used in Example 17.8.2.

(c) *Establish the deformation matrix* [B]. This matrix may be established by columns using the diagrams in Fig. 17.8.4b, as follows:

$e_2 = X_1$ and $e_3 = X_1$ when $X_1 \neq 0$ and $X_2 = X_3 = X_4 = 0$

$e_4 = X_2$ and $e_5 = X_2$ when $X_2 \neq 0$ and $X_1 = X_3 = X_4 = 0$

$e_6 = X_3$ when $X_3 \neq 0$ and $X_1 = X_2 = X_4 = 0$

$e_5 = -\dfrac{X_4}{6}$ and $e_6 = -\dfrac{X_4}{6}$ when $X_4 \neq 0$ and $X_1 = X_2 = X_3 = 0$

Since e_5 and e_6 are measured clockwise from the axis of the member to the tangent to the elastic curve, the counterclockwise angle here from the inclined member axis to the horizontal tangent will give

$$e_5 = e_6 = -\frac{X_4}{6}$$

Thus, the matrix [B] is

$[B]_{6\times4} =$

e \ X	1	2	3	4
1				
2	+1.0			
3	+1.0			
4		+1.0		
5		+1.0		$-\dfrac{1}{6}$
6			+1.0	$-\dfrac{1}{6}$

Note the transposition relationship between $[A]$ and $[B]$ again as a check.

(d) *Establish the member stiffness matrix* $[S]$. Since there are three "members" used in this example, Eqs. (17.5.4) will give a 6 by 6 matrix, as follows:

$[S]_{6\times6} = EI_c$

F \ e	1	2	3	4	5	6
1	$\dfrac{1}{2}$	$\dfrac{1}{4}$				
2	$\dfrac{1}{4}$	$\dfrac{1}{2}$				
3			$\dfrac{5}{6}$	$\dfrac{5}{12}$		
4			$\dfrac{5}{12}$	$\dfrac{5}{6}$		
5					2	1
6					1	2

(e) *Establish the external force matrix* $\{P\}$. This matrix is obtained from the forces acting on the joints in the fixed condition, as discussed in Section 17.7. For this example, the fixed-end moments (F_0-values) are shown in Fig. 17.8.3d. Examining the equilibrium of joints B, C, and D in the fixed condition of Fig. 17.8.4c, the net forces on the joints are

$$P_1 = 96.00 - 52.95 = +43.05 \text{ ft-kips}$$

$$P_2 = 6.00 - 96.00 = -90.00 \text{ ft-kips}$$

$$P_3 = -6.00 \text{ ft-kips}$$

$$P_4 = +6.00 \text{ kips}$$

Note that when segment CD is used as a "member," the fixed-end moments on CD are $wL^2/12 = 2.0(6)^2/12 = 6.00$ ft-kips, whereas in Example 17.8.1, where CD was not used as a member, the joint moment in the fixed condition was the cantilever moment

$wL^2/2 = 36.00$ ft-kips. Also in this example, the reaction R_{DC} is a force acting downward on "joint" D. Since there is a translational degree of freedom ($P_4 - X_4$) at D, the force R_{DC} corresponds to P_4 and has the ($+$) sign since it acts on the joint in the same direction as the P_4–X_4.

In matrix format for the one loading condition ($LC = 1$), the external force matrix is

$$\{P\} = $$

LC \ P	1
1	$+43.05$
2	-90.00
3	$-\ 6.00$
4	$+\ 6.00$

(f) *Obtain the matrix* $[SA^T]$. By matrix multiplication

$$[SA^T]_{6\times 4} = EI_c$$

F \ X	1	2	3	4
1	$\frac{1}{4}$			
2	$\frac{1}{2}$			
3	$\frac{5}{6}$	$\frac{5}{12}$		
4	$\frac{5}{12}$	$\frac{5}{6}$		
5		2	1	$-\frac{1}{2}$
6		1	2	$-\frac{1}{2}$

(g) *Obtain the global stiffness matrix* $[K]$. Matrix multiplication of the premultiplier matrix $[A]$ and the postmultiplier matrix $[SA^T]$ gives the matrix $[K]$ (i.e., $[ASA^T]$). The matrix is

$$[K] = [ASA^T]_{4\times 4} = EI_c$$

P \ X	1	2	3	4
1	$\frac{4}{3}$	$\frac{5}{12}$		
2	$\frac{5}{12}$	$\frac{17}{6}$	1	$-\frac{1}{2}$
3		1	2	$-\frac{1}{2}$
4		$-\frac{1}{2}$	$-\frac{1}{2}$	$+\frac{1}{6}$

Note that the $[K]$ matrix is always symmetric, as was proved in Section 16.6.

(h) *Obtain the inverse of the matrix* $[K]$. Using Eq. (17.8.4),

$$\{X\} = [ASA^T]^{-1}\{P\}$$

the inverse of the $[K] = [ASA^T]$ matrix must be computed. By Gauss–Jordan elimination as discussed in Section 13.4, the $[K]^{-1}$ matrix is obtained:

$$[K]^{-1} =$$

P / X	1	2	3	4	
1	+0.88889	−0.44444	−0.44444	−2.66667	
2	−0.44444	+1.42222	+1.42222	+8.53333	$\dfrac{1}{EI_c}$
3	−0.44444	+1.42222	+3.42222	+14.53333	
4	−2.66667	+8.53333	+14.53333	+75.20000	

(i) *Obtain the joint displacement matrix* $\{X\}$. By matrix multiplication using Eq. (17.8.4),

$$\{X\} = [K]^{-1}\{P\}$$

the $\{X\}$ matrix for the one loading condition ($LC = 1$) is

$$\{X\}_{4\times1} =$$

LC / X	1
1	$+\dfrac{64.93}{EI_c}$
2	$-\dfrac{104.47}{EI_c}$
3	$-\dfrac{80.47}{EI_c}$
4	$-\dfrac{518.8}{EI_c}$

(j) *Obtain the internal force matrix* $\{F\}$. Using Eq. (17.8.5), multiply the premultiplier matrix $[SA^T]$ by the postmultiplier matrix $\{X\}$, as shown in Fig. 17.8.5.

(k) *Obtain the final end moments.* The total moment F^* on the end of a member is the sum of the F-value that causes the joint displacements and the moment F_0 in the fixed condition that has zero joint displacements; thus,

$$\{F^*\} = \{F_0\} + \{F\}$$

In matrix format for this example,

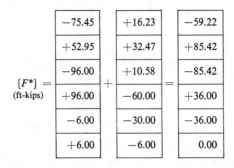

$$\{F^*\} \atop (\text{ft-kips}) = \begin{array}{|c|} -75.45 \\ +52.95 \\ -96.00 \\ +96.00 \\ -6.00 \\ +6.00 \end{array} + \begin{array}{|c|} +16.23 \\ +32.47 \\ +10.58 \\ -60.00 \\ -30.00 \\ -6.00 \end{array} = \begin{array}{|c|} -59.22 \\ +85.42 \\ -85.42 \\ +36.00 \\ -36.00 \\ 0.00 \end{array}$$

Note that the end moments on the segment CD that is statically determinate are exactly the statically determinate value of -36.0 ft-kips (negative means acting counterclockwise on CD) at end C and zero at the free end at D. The additional information gained by using the segment CD in the analysis is the slope (X_3) and the deflection (X_4) at the free end of the cantilever.

$$\{X\} = \begin{array}{c|c} \diagdown{X}{LC} & 1 \\ \hline 1 & +64.93 \\ 2 & -104.47 \\ 3 & -80.47 \\ 4 & -518.80 \end{array} \frac{1}{EI_c}$$

$$[SA^T] = \begin{array}{c|c|c|c|c} \diagdown{F}{X} & 1 & 2 & 3 & 4 \\ \hline 1 & \frac{1}{4} & & & \\ 2 & \frac{1}{2} & & & \\ 3 & \frac{5}{6} & \frac{5}{12} & & \\ 4 & \frac{5}{12} & \frac{5}{6} & & \\ 5 & & 2 & 1 & -\frac{1}{2} \\ 6 & & 1 & 2 & -\frac{1}{2} \end{array} EI_c; \quad \{F\} = \begin{array}{c|c} 1 & +16.23 \\ 2 & +32.47 \\ 3 & +10.58 \\ 4 & -60.00 \\ 5 & -30.00 \\ 6 & -6.00 \end{array}$$

Figure 17.8.5 Computation of internal force matrix $\{F\}$.

(1) *Check.* Checks of statics and compatibility of deformations are always necessary to verify the correctness of the answer. These checks are discussed in Sections 17.9 and 17.10.

17.9 STATICS CHECKS

In general, the statics checks on a continuous beam analysis are mostly by inspection to see that the clockwise moments on a joint are balanced by the counterclockwise moments. As discussed in Section 16.9, the check of statics is necessary but not sufficient to ensure correctness of results. For a statically indeterminate structure, statics alone is not sufficient to solve for the internal forces (moments), so it is also not sufficient as a check.

Example 17.9.1

Illustrate the statics check for the continuous beam analyzed in Example 17.8.2.

Solution. The solution is shown in Fig. 17.9.1, where the internal forces (F) satisfying equilibrium under the external joint forces are shown in Fig. 17.9.1b, the internal forces (F_0) in the fixed condition (all X's $= 0$) are in Fig. 17.9.1c, and the final end moments (F^*) are shown in Fig. 17.9.1d.

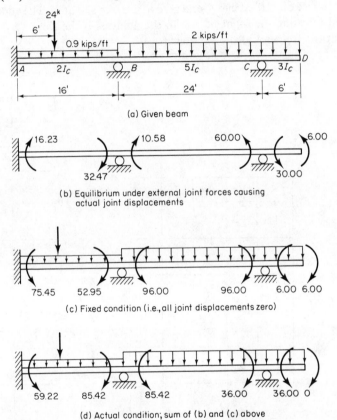

Figure 17.9.1 Final results and statics check, Examples 17.8.2 and 17.9.1.

The number of statics checks should be equal to the degree of freedom used in the solution procedure; for this example, it is four. Each check is associated with the degree of freedom in the direction of the P or X.

The check can be made either for the joint-force condition as shown in Fig. 17.9.1b, or for the actual condition as shown in Fig. 17.9.1d. The suggestion is to make use of the final actual condition in Fig. 17.9.1d.

The first three statics checks associated with X_1, X_2, and X_3 are done by inspection, in that the moments of 85.42, 36.00, and 0.00 are balanced at joints B, C, and D. The check for the fourth degree of freedom is that using CD in Fig. 17.9.1d as a free body; the force required at point D is zero for equilibrium.

17.10 DEFORMATION CHECKS

These checks are made by determining the slopes to the elastic curve at both ends of each span to verify that they are identical to the values of X from the matrix solution. When there is a joint deflection that deflection must also be computed to verify the matrix solution deflection. Note that there should be two deformation checks per member as has been used in the solution procedure.

Example 17.10.1

Illustrate the deformation checks for the continuous beam analyzed in Example 17.8.2.

Solution

(a) *Member AB.* Draw the free-body diagram of the member and the M/EI diagrams for the component loading, as in Fig. 17.10.1a. Using the moment area (or the conjugate beam) method,

$$\underset{\text{(clockwise)}}{\theta_{A1}} = \frac{\Delta'_{BA}}{L} = \frac{1}{16EI_c}\left[\frac{2}{3}(14.4)(16)(8) + \frac{1}{2}(45)(6)(12) + \frac{1}{2}(45)(10)\left(\frac{20}{3}\right)\right]$$

$$= \frac{271.80}{EI_c}$$

$$\underset{\substack{\text{(counter-}\\ \text{clockwise)}}}{\theta_{A2}} = \frac{\Delta'_{BA}}{L} = \frac{1}{16EI_c}\left[\frac{1}{2}(29.61)(16)\left(\frac{32}{3}\right) + \frac{1}{2}(42.71)(16)\left(\frac{16}{3}\right)\right]$$

$$= \frac{271.81}{EI_c}$$

$$\theta_A = \theta_{A1} - \theta_{A2} = \frac{271.80}{EI_c} - \frac{271.81}{EI_c} = 0 \quad \text{(checks for fixed end)}$$

$$\underset{\text{(clockwise)}}{\theta_{B1}} = \frac{1}{16EI_c}\left[\frac{1}{2}(29.61)(16)\left(\frac{16}{3}\right) + \frac{1}{2}(42.71)(16)\left(\frac{32}{3}\right)\right]$$

$$= \frac{306.75}{EI_c}$$

$$\underset{\substack{\text{(counter-}\\ \text{clockwise)}}}{\theta_{B2}} = \frac{1}{16EI_c}\left[\frac{2}{3}(14.4)(16)(8) + \frac{1}{2}(45)(6)(4) + \frac{1}{2}(45)(10)\left(\frac{28}{3}\right)\right]$$

$$= \frac{241.80}{EI_c}$$

$$\theta_B = X_1 = \frac{306.75 - 241.80}{EI_c} = +\frac{64.95}{EI_c} \quad \text{(checks with } +64.93/EI_c)$$

(b) *Member BC.* The M/EI diagrams of the component loadings are shown in Fig. 17.10.1b.

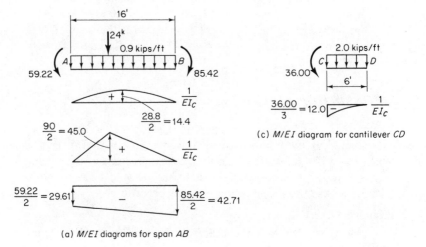

(c) M/EI diagram for cantilever CD

(a) M/EI diagrams for span AB

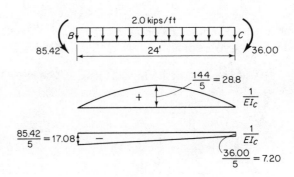

(b) M/EI diagrams for span BC

Figure 17.10.1 Deformation checks for a continuous beam, Examples 17.8.2 and 17.10.1.

$$\theta_{B1} \atop \text{(clockwise)} = \frac{1}{24EI_c}\left[\frac{2}{3}(28.8)(24)(12)\right] = \frac{230.40}{EI_c}$$

$$\theta_{B2} \atop \substack{\text{(counter-}\\\text{clockwise)}} = \frac{1}{24EI_c}\left[\frac{1}{2}(17.08)(24)(16) + \frac{1}{2}(7.20)(24)(8)\right] = \frac{165.47}{EI_c}$$

$$\theta_B = X_1 = \frac{230.40 - 165.47}{EI_c} = +\frac{64.92}{EI_c} \quad \text{(checks with } +64.93/EI_c)$$

$$\theta_{C1} \atop \text{(clockwise)} = \frac{1}{24EI_c}\left[\frac{1}{2}(17.08)(24)(8) + \frac{1}{2}(7.20)(24)(16)\right] = \frac{125.94}{EI_c}$$

$$\theta_{C2} \atop \substack{\text{(counter-}\\\text{clockwise)}} = \frac{1}{24EI_c}\left[\frac{2}{3}(28.8)(24)(12)\right] = \frac{230.40}{EI_c}$$

$$\theta_C = X_2 = \frac{125.94 - 230.40}{EI_c} = -\frac{104.46}{EI_c} \quad \text{(checks with } -104.47/EI_c)$$

(c) *Member CD.* Starting with the slope at C the slope and deflection at D may be verified.

$$\theta_D = X_3 = \theta_C + \left(\text{area of } \frac{M}{EI} \text{ between } C \text{ and } D\right)$$

$$= -\frac{104.47}{EI_c} + \frac{1}{EI_c}\left(\frac{1}{3}\right)(12)(6)$$

$$= -\frac{80.47}{EI_c} \quad (\text{checks with } -80.47/EI_c)$$

$$\Delta_D = X_4 = 6\theta_C + \left(\text{moment of } \frac{M}{EI} \text{ area between } C \text{ and } D \text{ about } D\right)$$

$$= 6\left(-\frac{104.47}{EI_c}\right) + \frac{1}{EI_c}\left(\frac{1}{3}\right)(12)(6)\left(\frac{18}{4}\right)$$

$$= -\frac{518.82}{EI_c} \quad (\text{i.e., up}) \quad (\text{checks with } -518.80/EI_c)$$

17.11 USE OF SHORT SEGMENTS

As has been illustrated in Example 17.8.2 with regard to the cantilever segment, a beam may be divided into as many elements as desired over and above the minimum for solution. This means that the segment does not have to be an entire span; "joints" may be arbitrarily taken at any place. Each additional joint used will add two degrees of freedom (one rotational and one translational) and two internal forces (end moments). A reason for using additional joints may be that one desires the deflection and slope for these additional locations (such as graphical output in computer analysis). Generally, unless such additional data are needed, one would use only the minimum number of joints to obtain the solution. Remember that the size of the matrix $[ASA^T]$ to be inverted will always be equal to the number of degrees of freedom used in the solution.

The general use of "joints" at other than external support locations is illustrated in the next example.

Example 17.11.1

Establish the matrices $[A]$, $[B]$, $[S]$, and $\{P\}$ to analyze the simply supported beam of Fig. 17.11.1a by taking "joints" at each end and one that divides the span into segments L_1 and L_2.

Solution. This problem was solved similarly by the slope deflection method in Example 9.4.1 and the procedure may be compared.

(a) *Establish the P-X and F-e diagrams.* These diagrams are shown in Fig. 17.11.1b and c. At each end where the deflection is zero there is a rotational degree of freedom, and at the intermediate "joint" there are two degrees of freedom, one rotational and one translational.

(b) *Establish the statics matrix* $[A]$. Referring to the free-body diagrams of Fig. 17.11.1c,

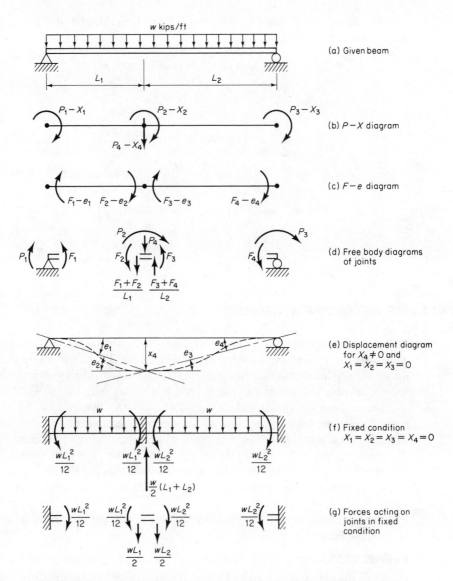

Figure 17.11.1 Matrix displacement method using an intermediate joint, Example 17.11.1.

$$P_1 = F_1$$

$$P_2 = F_2 + F_3$$

$$P_3 = F_4$$

$$P_4 = -\frac{F_1 + F_2}{L_1} + \frac{F_3 + F_4}{L_2}$$

which in matrix form is

$$[A]_{4\times4} =$$

F \\ P	1	2	3	4
1	$+1.0$			
2		$+1.0$	$+1.0$	
3				$+1.0$
4	$-\dfrac{1}{L_1}$	$-\dfrac{1}{L_1}$	$+\dfrac{1}{L_2}$	$+\dfrac{1}{L_2}$

At this stage, the matrix [A] may be inverted to solve for the F's since for a statically determinate structure the matrix [A] is a square matrix.

(c) *Establish the deformation matrix* [B]. Applying individually one displacement X at a time with all other X's being zero, the [B] matrix is obtained by columns. The effects of X_1, X_2, and X_3 are straightforward, as illustrated in Examples 17.8.1 and 17.8.2. When a translational displacement X_4 is applied, segment L_1 rotates clockwise and segment L_2 rotates counterclockwise. The angle of rotation X/L of the member axis plus the internal deformation e equals the total slope, which in this case is zero. Thus for clockwise rotation of the member axis,

$$0 = e_1 + \frac{X_4}{L_1}; \qquad e_1 = -\frac{X_4}{L_1}$$

or

$$0 = e_2 + \frac{X_4}{L_1}; \qquad e_2 = -\frac{X_4}{L_1}$$

When the member axis is rotated counterclockwise as in the case of segment L_2,

$$0 = e_3 - \frac{X_4}{L_2}; \qquad e_3 = +\frac{X_4}{L_2}$$

$$0 = e_4 - \frac{X_4}{L_2}; \qquad e_4 = +\frac{X_4}{L_2}$$

The results above may be verified graphically by looking at the displacement diagram of Fig. 17.11.1e, reading e_1 to e_4 values, by noting the clockwise angle from the inclined member axis to the horizontal tangents.

The deformation matrix [B] is

$$[B] =$$

X \\ e	1	2	3	4
1	$+1.0$			$-\dfrac{1}{L_1}$
2		$+1.0$		$-\dfrac{1}{L_1}$
3		$+1.0$		$+\dfrac{1}{L_2}$
4			$+1.0$	$+\dfrac{1}{L_2}$

The [B] matrix is verified to be the transpose of the [A] matrix.

(d) *Establish the member stiffness matrix* $[S]$. Since there are two "members" used in this analysis, Eqs. (17.5.4) will give a 4 by 4 matrix, as follows:

$$[S]_{4 \times 4} =$$

$F \backslash e$	1	2	3	4
1	$\dfrac{4EI}{L_1}$	$\dfrac{2EI}{L_1}$		
2	$\dfrac{2EI}{L_1}$	$\dfrac{4EI}{L_1}$		
3			$\dfrac{4EI}{L_2}$	$\dfrac{2EI}{L_2}$
4			$\dfrac{2EI}{L_2}$	$\dfrac{4EI}{L_2}$

(e) *Establish the external force matrix* $\{P\}$. The loads are applied to the joints in the fixed condition, as shown in Fig. 17.11.1f. The forces acting on the joints are those in Fig. 17.11.1g, as follows:

$$P_1 = +\frac{wL_1^2}{12}$$

$$P_2 = +\frac{wL_2^2}{12} - \frac{wL_1^2}{12}$$

$$P_3 = -\frac{wL_2^2}{12}$$

$$P_4 = \frac{wL_1}{2} + \frac{wL_2}{2}$$

In matrix form for the one loading condition (LC = 1),

$$\{P\}_{4 \times 1} =$$

$P \backslash LC$	1
1	$+\dfrac{wL_1^2}{12}$
2	$+\dfrac{wL_2^2}{12} - \dfrac{wL_1^2}{12}$
3	$-\dfrac{wL_2^2}{12}$
4	$+\dfrac{wL_1}{2} + \dfrac{wL_2}{2}$

Thus, the basic matrices are established for the situation of an arbitrary "joint" in a beam structure. The procedure is identical whether the structure is statically determinate or statically indeterminate. The remainder of the matrix operations and the checks are as previously shown.

SELECTED REFERENCES

The references listed for Chapter 16 include treatments of the stiffness method for either truss structures or beam structures.

PROBLEMS

17.1. Obtain numerically the statics matrix $[A]$, the deformation matrix $[B]$, and the external load matrix $\{P\}$ which are needed in the matrix displacement method analysis of the continuous beam in the accompanying figure. (a) Use the minimum number of degrees of freedom; (b) use beam segments AB, BC, and CD.

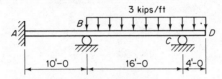

Prob. 17.1

17.2. Obtain numerically the statics matrix $[A]$, the deformation matrix $[B]$, and the external load matrix $\{P\}$ for the beam of the accompanying figure. (a) Use segments BC and CD; (b) use segments AB, BC, and CD.

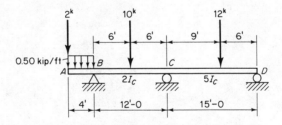

Prob. 17.2

17.3. Solve Prob. 10.1 using the matrix displacement method.

17.4. Solve Prob. 10.2 using the matrix displacement method.

17.5. Solve Prob. 10.3 using the matrix displacement method.

17.6. Solve Prob. 10.6 using the matrix displacement method.

17.7. Solve Prob. 10.12 using the matrix displacement method. Use the minimum number of beam segments.

17.8. Solve Prob. 10.13 using the matrix displacement method.

17.9. Obtain the statics matrix $[A]$ and the external force matrix $\{P\}$ required in the analysis of the continuous beam of the accompanying figure. Use (a) a two-member (e.g., members AB and BCD) approach; (b) use a three-member (e.g., members AB, BC, and CD) approach.

17.10. Write out the statics matrix $[A]$, the member stiffness matrix $[S]$, and the external force matrix $\{P\}$ for the analysis of the simple beam of the accompanying figure by the matrix displacement method. Use three segments, AB, BC, and CD.

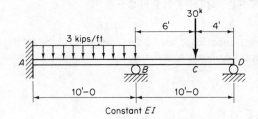

Prob. 17.9

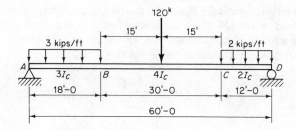

Prob. 17.10

17.11. Show the input matrices $[A]$, $[B]$, $[S]$, and $\{P\}$ for the analysis of the simple beam of the accompanying figure by the matrix displacement method. Use two segments by considering a joint at midspan.

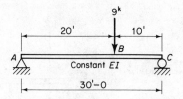

Prob. 17.11

17.12. Write numerically the $[A]$, $[B]$, $[S]$, and $\{P\}$ matrices for the analysis of the beam shown in the accompanying figure by the matrix displacement method. Use two beam segments by using a joint at the concentrated load.

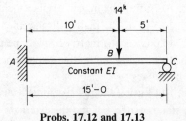

Probs. 17.12 and 17.13

17.13. Analyze the beam of Prob. 17.12, using the matrices obtained for that problem.

17.14. Write out the input matrices $[A]$, $[B]$, $[S]$, and $\{P\}$ for the analysis of the beam of the accompanying figure by the matrix displacement method. Use two segments AB and BC.

17.15. Analyze the beam of Prob. 17.14 using the matrices obtained for that problem.

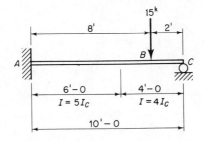

Probs. 17.14 and 17.15

17.16. Write out the input matrices $[A]$, $[B]$, $[S]$, and $\{P\}$ for the analysis of the beam shown in the accompanying figure by the matrix displacement method. Use two segments AB and BC.

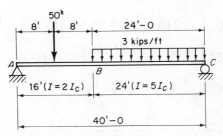

Prob. 17.16

17.17. Analyze the beam of the accompanying figure by the matrix displacement method. Use two beam segments AB and BC.

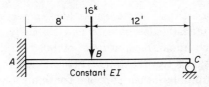

Prob. 17.17

17.18. Analyze the beam of Prob. 9.13 by the matrix displacement method. Use segments AB and BC.

17.19. Analyze the beam of Example 10.4.1 by the matrix displacement method.

17.20. Write out the input matrices $[A]$, $[B]$, $[S]$, and $\{P\}$ for the analysis of the beam shown in the accompanying figure by the matrix displacement method. Use segments AB, BC, and CD.

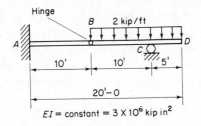

Prob. 17.20

[18]

Matrix Displacement Method Applied to Rigid Frames Without Sidesway

18.1 RIGID FRAMES WITHOUT SIDESWAY

The definition of a rigid frame was treated in Section 11.1, where in Fig. 11.1.2 are shown some statically determinate and statically indeterminate rigid frames. The members of such frames meet at rigid joints where the elastic curves of any two adjacent members have a constant angle between them during loading. In fact, in a narrow sense a joint at an interior support of a continuous beam is a rigid frame joint for two members meeting at 180°. This chapter extends the treatment of Chapter 17 to structures where each rigid joint may connect two or more members whose axes intersect at other than 180°.

The special category of rigid frames discussed in this chapter is "without sidesway." Typically, a rigid frame, such as the one-bay, one-story frame of Fig. 18.1.1a, has the freedom to have some of its joints translate, that is, have not only rotation but linear displacement as well. When joint translation is prevented, as in Fig. 18.1.1b, it is said to be "without sidesway." In Chapter 17, the treatment of a translation degree of freedom (No. 5 in Fig. 17.1.1b) for a continuous beam was presented. In those cases, the use in the solution of more joints than the minimum was optional. This chapter treats a category of rigid frames that can be analyzed by use of only rotational degrees of freedom. In other words, when the minimum number of degrees of freedom is used for solution, all the degrees of freedom will be rotational.

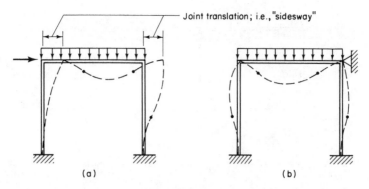

Figure 18.1.1 Rigid frames, with and without sidesway.

18.2 APPLICATIONS

All the concepts necessary to solve the rigid frame without sidesway have been treated in Chapter 17. The statics matrix, deformation matrix, member stiffness matrix, external force matrix, and the operational sequence have all been explained and illustrated in Chapter 17. The following examples show the procedure for the analysis of rigid frames without sidesway by the matrix displacement method.

Example 18.2.1

Analyze the rigid frame of Fig. 18.2.1a by the matrix displacement method. (This structure has previously been analyzed by the slope deflection method in Example 11.7.1, and by the moment distribution method in Example 11.8.1.)

Solution

(a) *Draw the P-X and F-e diagrams* (Fig. 18.2.1b and c). The *P-X* diagram establishes the degrees of freedom that are to be used in the solution. In this problem, the minimum possible number of degrees of freedom is used. The segments *AB* and *DE* do not have to be used because they are statically determinate. The analyst must use all other members (*BF, BC, CG,* and *CD*) containing eight *F*'s (F_1 to F_8). Thus, ($NF - NP$) equals five, which is the degree of statical indeterminacy. More segments may be used (in which case the rigid frame will become one with sidesway), as discussed for beams in Section 17.11, but ($NF - NP$) will always equal five for this structure.

(b) *Establish the statics matrix* [*A*]. From equilibrium of the joints *B, C,* and *D,* as in Fig. 18.2.2a,

$$P_1 = F_2 + F_3$$
$$P_2 = F_4 + F_6 + F_7$$
$$P_3 = F_8$$

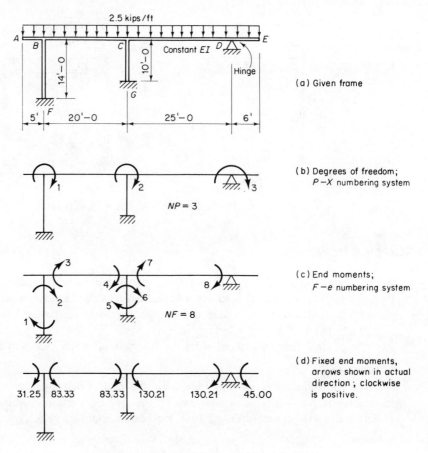

Figure 18.2.1 Rigid frame for Example 18.2.1.

or in matrix form,

$$[A]_{3\times8} =$$

P \ F	1	2	3	4	5	6	7	8
1		+1.0	+1.0					
2				+1.0		+1.0	+1.0	
3								+1.0

Note that the statically determinate moments M_{BA} and M_{DE} on the cantilevers are treated as joint forces and thus appear as part of the external force matrix.

(c) *Establish the deformation matrix* [B]. This matrix relating internal deformations (rotations at ends of member segments) to the joint displacements can be established either by columns or by rows. Doing it by columns the effects of each X on all the e's are examined, as in Fig. 18.2.2b,

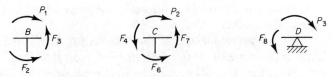

(a) Rotational equilibrium of the joints (for statics matrix $[A]$)

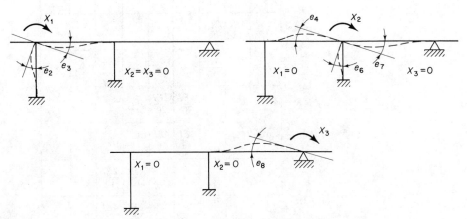

(b) Compatibility of internal deformation with joint displacements

$31.25 \quad \overline{} \quad 83.33 \qquad 83.33 \quad \overline{} \quad 130.21 \qquad 130.21 \quad \overline{} \quad 45.00$
$\qquad 0 \qquad\qquad\qquad 0$

(c) Rotational forces on joint in the fixed condition (moments are given in ft−kips)

Figure 18.2.2 Data for statics matrix, deformation matrix, and external force matrix; Example 18.2.1.

$$[B]_{8 \times 3} =$$

e \ X	1	2	3
1			
2	+1.0		
3	+1.0		
4		+1.0	
5			
6		+1.0	
7		+1.0	
8			+1.0

After establishing matrix $[B]$ independently from matrix $[A]$, check that one is the transpose of the other.

(d) *Establish the member stiffness matrix* $[S]$. This is the pair of equations, Eqs. (17.5.4), involving EI/L that relate the end moments F to the end rotations e. In matrix form, Eqs. (17.5.4) become for this problem,

$$[S]_{8 \times 8} =$$

F \ e	1	2	3	4	5	6	7	8
1	$\dfrac{4EI}{14}$	$\dfrac{2EI}{14}$						
2	$\dfrac{2EI}{14}$	$\dfrac{4EI}{14}$						
3			$\dfrac{4EI}{20}$	$\dfrac{2EI}{20}$				
4			$\dfrac{2EI}{20}$	$\dfrac{4EI}{20}$				
5					$\dfrac{4EI}{10}$	$\dfrac{2EI}{10}$		
6					$\dfrac{2EI}{10}$	$\dfrac{4EI}{10}$		
7							$\dfrac{4EI}{25}$	$\dfrac{2EI}{25}$
8							$\dfrac{2EI}{25}$	$\dfrac{4EI}{25}$

(e) *Establish the external force matrix* $\{P\}$. This matrix consists of the net forces acting on the joints when the fixed condition is released, as discussed in Section 17.7. For this example the fixed-end moments are computed and shown in Fig. 18.2.1d. The release of joints B, C, and D, as shown in Fig. 18.2.2c, gives the net clockwise forces P on the joints as

$$P_1 = 83.33 - 31.25 = +52.08 \text{ ft-kips}$$
$$P_2 = 130.21 - 83.33 = +46.88 \text{ ft-kips}$$
$$P_3 = 45.00 - 130.21 = -85.21 \text{ ft-kips}$$

Note that the release of a fixed-end moment is identical to the application of a moment in the reverse direction. In matrix form for the one loading condition (LC = 1), the external force matrix is

$$\{P\}_{3 \times 1} =$$

P \ LC	1
1	$+52.08$
2	$+46.88$
3	-85.21

(f) *Obtain the* $[SA^T]$ *matrix.* Matrix multiplication gives

$$[SA^T]_{8\times 3} =$$

F \ X	1	2	3
1	$\dfrac{2EI}{14}$		
2	$\dfrac{4EI}{14}$		
3	$\dfrac{4EI}{20}$	$\dfrac{2EI}{20}$	
4	$\dfrac{2EI}{20}$	$\dfrac{4EI}{20}$	
5		$\dfrac{2EI}{10}$	
6		$\dfrac{4EI}{10}$	
7		$\dfrac{4EI}{25}$	$\dfrac{2EI}{25}$
8		$\dfrac{2EI}{25}$	$\dfrac{4EI}{25}$

(g) *Obtain the global stiffness matrix* $[K]$. Using matrix multiplication of the premultiplier matrix $[A]$ and the postmultiplier matrix $[SA^T]$ gives

$$[K] = [ASA^T]_{3\times 3} =$$

P \ X	1	2	3
1	$\dfrac{17EI}{35}$	$\dfrac{EI}{10}$	
2	$\dfrac{EI}{10}$	$\dfrac{19EI}{25}$	$\dfrac{2EI}{25}$
3		$\dfrac{2EI}{25}$	$\dfrac{4EI}{25}$

(h) *Obtain the inverse of matrix* $[K]$. Using Eq. (17.8.4),

$$\{X\} = [ASA^T]^{-1}\{P\}$$

the $[K] = [ASA^T]$ matrix must be inverted. By Gauss–Jordan elimination as discussed in Section 13.4, or by any other method, the following $[K]^{-1}$ matrix is obtained:

$$[K]^{-1} = \frac{1}{EI}$$

X \ P	1	2	3
1	+2.119428	−0.294365	+0.147183
2	−0.294365	+1.429773	−0.714886
3	+0.147183	−0.714886	+6.607443

(i) *Obtain the joint displacement matrix* $\{X\}$. The joint displacement matrix $\{X\}$ is obtained by multiplying the premultiplier matrix $[K]^{-1}$ by the postmultiplier external force matrix $\{P\}$, in accordance with Eq. (17.8.4),

$$X_1 = \frac{1}{EI}[+2.119428(+52.08) - 0.294365(+46.88) + 0.147183(-85.21)]$$

$$= \frac{+84.04}{EI}$$

$$X_2 = \frac{1}{EI}[-0.294365(+52.08) + 1.429773(+46.88) - 0.714886(-85.21)]$$

$$= \frac{+112.61}{EI}$$

$$X_3 = \frac{1}{EI}[+0.147183(+52.08) - 0.714886(+46.88) + 6.607443(-85.21)]$$

$$= \frac{-588.87}{EI}$$

In matrix format the $\{X\}$ matrix for the one loading condition (LC = 1) is

$$\{X\} = \begin{array}{c|c} \diagdown\!\!\!\!\!\begin{smallmatrix}LC\\X\end{smallmatrix} & 1 \\ \hline 1 & +84.04/EI \\ \hline 2 & +112.61/EI \\ \hline 3 & -588.87/EI \end{array}$$

(j) *Obtain the internal force matrix* $\{F\}$. Using Eq. (17.8.5), multiply the premultiplier matrix $[SA^T]$ by the postmultiplier matrix $\{X\}$, as shown in Fig. 18.2.3.

(k) *Obtain the final end moments.* The total moment F^* on the end of a member is the sum of the F-value from the matrix operations and the fixed-end moment F_0. In matrix notation,

$$\{F^*\} = \{F_0\} + \{F\}$$

For this example, using the F_0-values given in Fig. 18.2.1d,

$$\begin{array}{c}\{F^*\} = \\ \text{(ft-kips)}\end{array}\quad \begin{array}{c|c} 1 & 0 \\ \hline 2 & 0 \\ \hline 3 & -83.33 \\ \hline 4 & +83.33 \\ \hline 5 & 0 \\ \hline 6 & 0 \\ \hline 7 & -130.21 \\ \hline 8 & +130.21 \end{array}\ +\ \begin{array}{c} +12.01 \\ \hline +24.01 \\ \hline +28.07 \\ \hline +30.93 \\ \hline +22.52 \\ \hline +45.04 \\ \hline -29.09 \\ \hline -85.21 \end{array}\ =\ \begin{array}{c} +12.01 \\ \hline +24.01 \\ \hline -55.26 \\ \hline +114.26 \\ \hline +22.52 \\ \hline +45.04 \\ \hline -159.30 \\ \hline +45.00 \end{array}$$

$$\{X\} = $$

X \ LC	1
1	$+\dfrac{84.04}{EI}$
2	$+\dfrac{112.61}{EI}$
3	$-\dfrac{558.87}{EI}$

$$[SA^T] = $$

F \ X	1	2	3
1	$\dfrac{2EI}{14}$		
2	$\dfrac{4EI}{14}$		
3	$\dfrac{4EI}{20}$	$\dfrac{2EI}{20}$	
4	$\dfrac{2EI}{20}$	$\dfrac{4EI}{20}$	
5		$\dfrac{2EI}{10}$	
6		$\dfrac{4EI}{10}$	
7		$\dfrac{4EI}{25}$	$\dfrac{2EI}{25}$
8		$\dfrac{2EI}{25}$	$\dfrac{4EI}{25}$

$$\{F\} = $$

F \ X	1
1	$+12.01$
2	$+24.01$
3	$+28.07$
4	$+30.93$
5	$+22.52$
6	$+45.04$
7	-29.09
8	-85.21

Figure 18.2.3 Computation of internal force matrix $\{F\}$.

The solution agrees with those obtained in Examples 11.7.1 and 11.8.1.

(l) *Output checks.* Checks of statics and compatibility of deformations are always required to verify the correctness of the solution.

18.3 STATICS CHECKS

The statics checks are the same as for continuous beams; that is, verify that the clockwise moments on a joint are balanced by the counterclockwise moments. As always is the case for statically indeterminate structures, the check of statics is necessary but not sufficient to ensure correctness of the results.

Example 18.3.1

Illustrate the statics checks for the rigid frame analyzed in Example 18.2.1.

Solution. The results of the analysis obtained in Example 18.2.1 are presented in Fig. 18.3.1. The equilibrium of the actual condition is examined. At joints B, C, and D the sum of clockwise moments does indeed balance the counterclockwise moments.

Note that the shears on frame members cause axial forces in the adjacent members. This does not happen for the continuous beam but is always true when the members of a structure are joined at other than 180°. Further, even under gravity loading there are horizontal reactions at F, G, and D of the rigid frame, whereas a

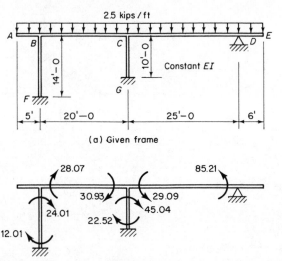

(a) Given frame

(b) Equilibrium under external joint forces causing actual joint displacements

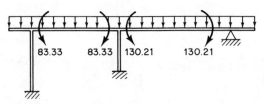

(c) Fixed condition (i.e., all joint displacements zero)

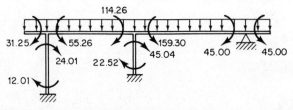

(d) Actual condition; sum of (b) and (c) above. (In addition the statically determinate cantilever moments are shown.)

Figure 18.3.1 Final results and statics checks; Examples 18.2.1 and 18.3.1.

continuous beam has only vertical reactions. The computation of these reactions and the presentation of results appear in Example 11.7.2, Fig. 11.7.2.

18.4 DEFORMATION CHECKS

These checks are made by determining the slopes to the elastic curve at each end of each member to verify that they are identical to the values of X from the matrix solution. (In rigid frames with sidesway, there would be one or more joint deflections; these deflections must also be computed.) There are two deformation checks per member.

Example 18.4.1

Illustrate the deformation checks for the rigid frame analyzed in Example 18.2.1.

Solution

(a) *Member FB.* Draw the free-body diagram and the M/EI diagrams for the component loading, as in Fig. 18.4.1a. Using the moment area method,

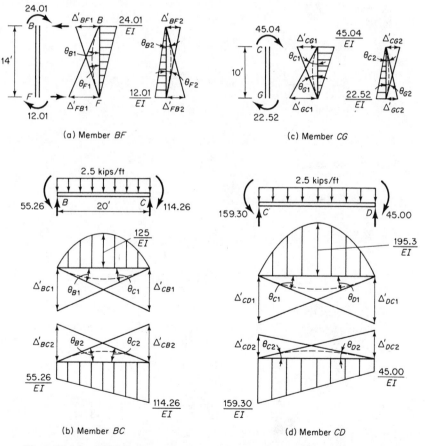

Figure 18.4.1 M/EI diagrams for deformation checks on rigid frame; Examples 18.2.1 and 18.4.1.

$$\theta_{B1} = \frac{\Delta'_{FB1}}{14} = \frac{1}{14EI}\left[\frac{1}{2}(24.01)(14)\left(\frac{28}{3}\right)\right] = \frac{112.05}{EI}$$
(clockwise)

$$\theta_{B2} = \frac{\Delta'_{FB2}}{14} = \frac{1}{14EI}\left[\frac{1}{2}(12.01)(14)\left(\frac{14}{3}\right)\right] = \frac{28.02}{EI}$$
(counter-
clockwise)

$$\theta_B = \frac{112.05 - 28.02}{EI} = \frac{84.03}{EI} \quad \text{(clockwise)}$$

This checks with X_1.

$$\theta_{F1} = \frac{\Delta'_{BF1}}{14} = \frac{1}{14EI}\left[\frac{1}{2}(24.01)(14)\left(\frac{14}{3}\right)\right] = \frac{56.02}{EI}$$
(counter-
clockwise)

$$\theta_{F2} = \frac{\Delta'_{BF2}}{14} = \frac{1}{14EI}\left[\frac{1}{2}(12.01)(14)\left(\frac{28}{3}\right)\right] = \frac{56.05}{EI}$$
(clockwise)

$$\theta_F = \frac{56.05}{EI} - \frac{56.02}{EI} = 0 \quad \text{(checks for fixed end)}$$

(b) *Member BC.* The M/EI diagrams of the component loadings are shown in Fig. 18.4.1b.

$$\theta_{B1} = \frac{1}{20EI}\left[\frac{2}{3}(125)(20)(10)\right] = \frac{833.33}{EI}$$
(clockwise)

$$\theta_{B2} = \frac{1}{20EI}\left[\frac{1}{2}(55.26)(20)\left(\frac{40}{3}\right) + \frac{1}{2}(114.26)(20)\left(\frac{20}{3}\right)\right] = \frac{749.26}{EI}$$
(counter-
clockwise)

$$\theta_B = \frac{833.33 - 749.26}{EI} = \frac{84.07}{EI} \quad \text{(clockwise)}$$

This checks with X_1.

$$\theta_{C1} = \frac{1}{20EI}\left[\frac{2}{3}(125)(20)(10)\right] = \frac{833.33}{EI}$$
(counter-
clockwise)

$$\theta_{C2} = \frac{1}{20EI}\left[\frac{1}{2}(55.26)(20)\left(\frac{20}{3}\right) + \frac{1}{2}(114.26)(20)\left(\frac{40}{3}\right)\right] = \frac{945.93}{EI}$$
(clockwise)

$$\theta_C = \frac{945.93 - 833.33}{EI} = \frac{112.60}{EI} \quad \text{(clockwise)}$$

This checks with X_2.

(c) *Member GC.* The M/EI diagrams of the component loadings are shown in Fig. 18.4.1c.

$$\theta_{C1} = \frac{1}{10EI}\left[\frac{1}{2}(45.04)(10)\left(\frac{20}{3}\right)\right] = \frac{150.13}{EI}$$
(clockwise)

$$\theta_{C2} = \frac{1}{10EI}\left[\frac{1}{2}(22.52)(10)\left(\frac{10}{3}\right)\right] = \frac{37.53}{EI}$$
(counter-
clockwise)

$$\theta_C = \frac{150.13 - 37.53}{EI} = \frac{112.60}{EI} \quad \text{(clockwise)}$$

This checks with X_2.

$$\theta_{G1} = \frac{1}{10EI}\left[\frac{1}{2}(45.04)(10)\left(\frac{10}{3}\right)\right] = \frac{75.07}{EI}$$
(counter-
clockwise)

$$\theta_{G2} = \frac{1}{10EI}\left[\frac{1}{2}(22.52)(10)\left(\frac{20}{3}\right)\right] = \frac{75.07}{EI}$$
(clockwise)

$$\theta_G = \frac{75.07 - 75.07}{EI} = 0 \quad \text{(checks for fixed end)}$$

(d) *Member CD*. The M/EI diagrams of the component loadings are shown in Fig. 18.4.1d.

$$\theta_{C1} = \frac{1}{25EI}\left[\frac{2}{3}(195.31)(25)(12.5)\right] = \frac{1627.58}{EI}$$
(clockwise)

$$\theta_{C2} = \frac{1}{25EI}\left[\frac{1}{2}(159.30)(25)\left(\frac{50}{3}\right) + \frac{1}{2}(45)(25)\left(\frac{25}{3}\right)\right] = \frac{1515.00}{EI}$$
(counter-
clockwise)

$$\theta_C = \frac{1627.58 - 1515.00}{EI} = \frac{112.58}{EI} \quad \text{(clockwise)}$$

This checks with X_2.

$$\theta_{D1} = \frac{1}{25EI}\left[\frac{2}{3}(195.31)(25)(12.5)\right] = \frac{1627.58}{EI}$$
(counter-
clockwise)

$$\theta_{D2} = \frac{1}{25EI}\left[\frac{1}{2}(159.30)(25)\left(\frac{25}{3}\right) + \frac{1}{2}(45)(25)\left(\frac{50}{3}\right)\right] = \frac{1038.75}{EI}$$
(clockwise)

$$\theta_D = \frac{1038.75 - 1627.58}{EI} = \frac{-588.83}{EI} \quad \text{(counterclockwise)}$$

This checks with X_3.

PROBLEMS

18.1–18.8. For any problem assigned, draw the P-X and F-e diagrams using the minimum number of degrees of freedom. Then obtain numerically the statics matrix $[A]$, the deformation matrix $[B]$, the member stiffness matrix $[S]$, the $[SA^T]$ matrix, the $[ASA^T]$ matrix, the $[ASA^T]^{-1}$ matrix, and the final answers in the form of the $\{X\}$ matrix and the $\{F^*\}$ matrix. Finally, show the statics and deformation checks.

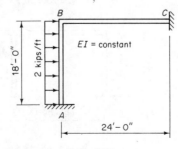

Prob. 18.1

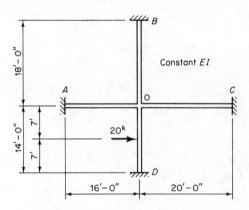

Prob. 18.2

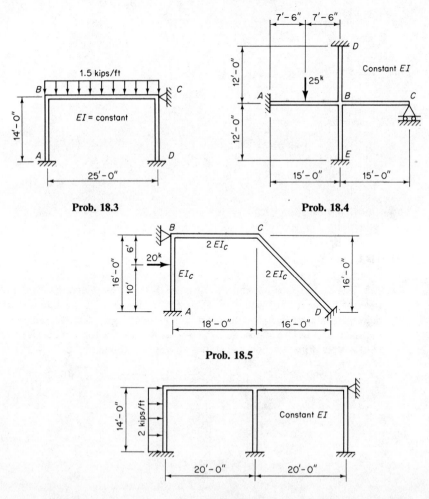

Prob. 18.3

Prob. 18.4

Prob. 18.5

Prob. 18.6

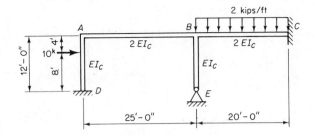

Prob. 18.7

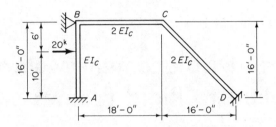

Wang & Salmon Prob. 18.5

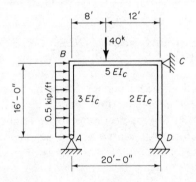

Prob. 18.8

[19]

Matrix Displacement Method Applied to Rigid Frames With Sidesway

---

19.1 DEGREE OF FREEDOM VS. DEGREE OF STATICAL INDETERMINACY

The degree of statical indeterminacy for a rigid frame was defined in Section 11.2; in accordance with the force method concept it equals the number of restraints (either reactions or internal forces) necessary to be removed to make a "basic" statically determinate structure. Alternatively, the degree of statical indeterminacy can be determined by use of the concept of *degree of freedom*, as shown in the following paragraphs.

Just as discussed for the beam in Section 17.1, the frame, whether statically determinate or statically indeterminate, may be separated into a finite number of "members" or "elements," at points called "joints" or "nodes." The "members" may be entire spans (or columns), or they may be segments of spans (or columns). The difference between a beam and a frame is that "members" are not oriented at 180° to one another in a frame, nor are there necessarily only two "members" intersecting at a "joint". In other words, the rigid frame is a more general structure, with the continuous beam being a special case.

When treating a structure according to the displacement method, the joint displacements are the primary unknowns. The joint displacements, including rotation and linear movement, define the deformed shape of the structure. Each joint displacement used in the solution is said to be a *degree of freedom*. When the minimum number of degrees of freedom is used, there may or may not be any joints having linear displacement. A continuous beam without internal hinges will have only rotational degrees of freedom when the minimum number of such degrees of freedom necessary for solution is used. A rigid frame without sidesway, as treated in Chapter 18, will likewise have only rotational degrees of freedom when the minimum number for solution is used.

The actual degree of freedom for a rigid frame depends on the number of members and joints taken in the analysis procedure because the effect of adding a joint at any point along a straight member is to create for solution one additional member. For instance, even though the two rigid frames of Fig. 19.1.1 are identical, one may be treated in analysis as having four joints and three members, while the other is treated in analysis as having six joints and five members.

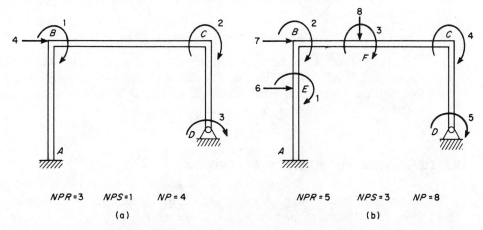

NPR = 3   NPS = 1   NP = 4   NPR = 5   NPS = 3   NP = 8

(a)   (b)

Figure 19.1.1 Degree of freedom for a rigid frame.

Pertinent to this chapter, however, is the fact that joints E and F in Fig. 19.1.1b are optional with the analyst, so that linear displacements (translations) numbered 6 and 8 occur only with joints E and F. The horizontal displacement at joint B (numbered 4 in Fig. 19.1.1a and 7 in Fig. 19.1.1b) must be used in the solution. In other words, the structure is said to have one minimum degree of freedom in *sidesway*.

It is a simple matter to observe the number of unknown joint rotations NPR, but in assigning the number of independent unknown joint deflections (sidesways) NPS, it is important to keep in mind the two previously stated fundamental assumptions: that axial deformation is ignored, and that transverse end displacements do not affect the member length. Thus, in Fig. 19.1.1a the only unknown joint deflection is the horizontal deflection of either joint B or C; in Fig. 19.1.1b the three unknown joint deflections are the horizontal deflection of E, the vertical deflection of F, and the horizontal deflection of any one of joints B, F, or C.

The degree of freedom NP of a rigid frame is dependent on both the number of unknown joint rotations and the number of unknown independent joint deflections; or

$$NP = NPR + NPS \qquad (19.1.1)$$

When axial deformation is neglected in the analysis, the axial forces in the members of a rigid frame are dependent on the member end moments and can be computed from the laws of statics. This will require that just a sufficient number of equations of equilibrium be available to determine these unknown axial forces.

As discussed in Section 11.2, when axial deformation is considered, the statics of each member requires the axial force to be an independent unknown as well, in addition to the two end moments on the member. The adaptation of the solution procedure to include axial deformation appears in Chapter 20.

For the common elementary analysis neglecting axial deformation, the number of unknown internal forces for each member is two (the moments at each end), in which case the total number of internal forces NF equals twice the number of "members" NM. Then, according to Eq. (11.2.6), the degree NI of statical indeterminacy is

$$NI = NF - NP = 2(NM) - (NPR - NPS) \qquad [11.2.6]$$

The degree of statical indeterminacy is an inherent property of the structure and cannot be affected by the analyst's choice of members or joints. The degree of freedom has a minimum value but above that value it depends on the number of members and joints used by the analyst in the solution.

19.2 DEFINITION OF STATICS MATRIX [A]

As discussed in Sections 15.1, 16.2, and 17.2, the statics matrix [A] expresses the joint forces $\{P\}$ in terms of the member end moments $\{F\}$. Keep in mind that any external loading may be converted into statically equivalent joint forces, and therefore a joint force P, causing a displacement X, may be considered as existing for each degree of freedom. Thus, each degree of freedom is identified by a rotational or translational arrow denoting both a force P and a joint displacement X. Referring to Fig. 19.2.1a, the P-X diagram is presented. The internal forces (end moments) F and member end rotations e are designated in Fig. 19.2.1b.

The free-body diagrams of the members and joints of the structure are shown in Fig. 19.2.1c. The equations of statics are obtained from the equilibrium of the joints,

$$\left. \begin{aligned} P_1 &= F_1 \\ P_2 &= F_2 + F_3 \\ P_3 &= F_4 + F_5 \\ P_4 &= F_6 \\ P_5 &= -\frac{F_1 + F_2}{L} - \frac{F_5 + F_6}{L} \end{aligned} \right\} \qquad (19.2.1)$$

The equation for P_5 is the one that arises from the "sidesway" degree of freedom. The shears on member AB cause axial force on member BC, which gives the force $(F_1 + F_2)/L$ on joint C. The shears on member CD give the force $(F_5 + F_6)/L$ on joint C.

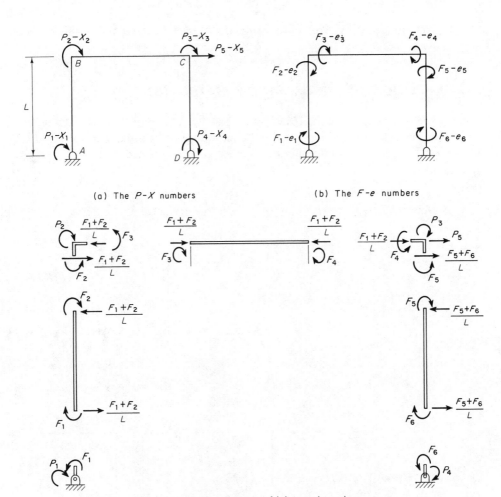

(a) The P-X numbers

(b) The F-e numbers

(c) Free-body diagrams of joints and members

Figure 19.2.1 Data for statics matrix $[A]$.

In matrix notation, the statics matrix $[A]$ is

$$[A]_{5 \times 6} =$$

P \\ F	1	2	3	4	5	6
1	$+1.0$					
2		$+1.0$	$+1.0$			
3				$+1.0$	$+1.0$	
4						$+1.0$
5	$-\dfrac{1}{L}$	$-\dfrac{1}{L}$			$-\dfrac{1}{L}$	$-\dfrac{1}{L}$

$$(19.2.2)$$

Note that for this two-hinged frame, the degree of statical indeterminacy is one (i.e., $NF - NP$).

19.3 DEFINITION OF DEFORMATION MATRIX $[B]$

The deformation matrix $[B]$ expresses the member end rotations $\{e\}$ in terms of joint displacements $\{X\}$. By observing the joint displacement diagrams of Fig. 19.3.1c and g, the following relationship between e and X is obtained:

$$e_1 = X_1 - \frac{X_5}{L} \qquad (19.3.1a)$$

Note that the clockwise rotation e *from* the straight line connecting the ends of the

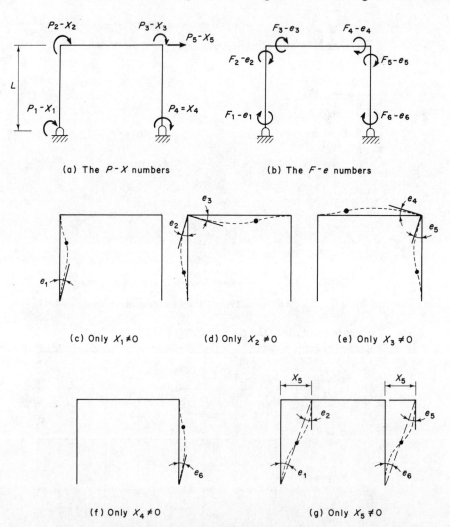

(a) The P-X numbers

(b) The F-e numbers

(c) Only $X_1 \neq 0$

(d) Only $X_2 \neq 0$

(e) Only $X_3 \neq 0$

(f) Only $X_4 \neq 0$

(g) Only $X_5 \neq 0$

Figure 19.3.1 Data for deformation matrix $[B]$.

member *to* the tangent to the elastic curve at the member end gives X_1 causing positive e but X_5 causing negative e. The rotation e_1 (Fig. 19.3.1g) equals X_5/L (tangent of angle equals angle in first-order analysis) and is a counterclockwise rotation from the sloped straight line to the vertical tangent to the elastic curve. The other equations are as follows:

$$
\left.
\begin{aligned}
e_2 &= X_2 - \frac{X_5}{L} \\[6pt]
e_3 &= X_2 \\[6pt]
e_4 &= X_3 \\[6pt]
e_5 &= X_3 - \frac{X_5}{L} \\[6pt]
e_6 &= X_4 - \frac{X_5}{L}
\end{aligned}
\right\} \qquad (19.3.1b)
$$

The deformation matrix $[B]$ consists of the coefficients of the X's in Eqs. (19.3.1), written as follows:

$$[B]_{6\times5} =$$

e \ X	1	2	3	4	5
1	+1.0				$-\dfrac{1}{L}$
2		+1.0			$-\dfrac{1}{L}$
3		+1.0			
4			+1.0		
5			+1.0		$-\dfrac{1}{L}$
6				+1.0	$-\dfrac{1}{L}$

$$(19.3.3)$$

The $[B]$ matrix is equal to the transpose of the $[A]$ matrix, by the principle of virtual work, discussed in Sections 16.4 and 17.4. Although not needed as an independent input matrix, it should be separately determined to check on the $[A]$ matrix. The proof for this transposition relationship, because of its importance, will be performed once more in the following section.

19.4 TRANSPOSITION RELATIONSHIP BETWEEN STATICS MATRIX [A] AND DEFORMATION MATRIX [B]

The transposition relationship between the statics matrix $[A]$ and the deformation matrix $[B]$ always holds true in first-order analysis; the proofs referring to trusses and continuous beams have been made in Section 16.4 and 17.4. It may be desirable at this point to present the necessary steps for making a general proof without reference to any specific type of structure. The steps for proving that A_{ij} equals B_{ji} are:

1. Draw the free-body diagrams of the members and joints representing the jth column of the $[A]$ matrix, which is said to be the jth equilibrium state.

2. Make note of the rigid-body movement (rotation and translation) of each joint and the member-end deformation (translational and rotational) at each member end representing the ith column of the $[B]$ matrix, which is said to be the ith compatible state.

3. Obtain the sum of the products of the external forces in step 1 and the corresponding displacement in step 2. Since the forces and the displacements, although multiplied, are in fact unrelated, their products "virtually" appear as "work," hence the name external virtual work.

4. Obtain the sum of the products of the member-end forces in step 1 and the corresponding member-end deformations in step 2; this sum is the internal virtual work.

5. Equate the external virtual work to the internal virtual work and the relationship $A_{ij} = B_{ji}$ is proven.

Referring to the rigid frame described in Figs. 19.2.1 and 19.3.1, the proof for $A_{23} = B_{32}$ can be made by applying the principle of virtual work to Fig. 19.4.1a and b. Thus,

$$(A_{23}F_3)(X_2) = (F_3)(B_{32}X_2)$$

from which

$$A_{23} = B_{32}$$

Note that all axial and shear forces do no virtual work because there is no linear movement of member ends in Fig. 19.4.1b. Similarly, the proof for $A_{52} = B_{25}$ can be made by applying the principle of virtual work to Fig. 19.4.2a and b. Thus,

$$(A_{52}F_2)(X_5) = (F_2)(B_{25}X_5)$$

from which

$$A_{52} = B_{25}$$

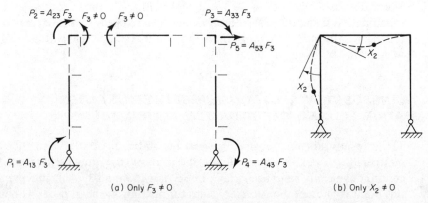

(a) Only $F_3 \neq 0$ (b) Only $X_2 \neq 0$

Figure 19.4.1 Proof for $A_{23} = B_{32}$.

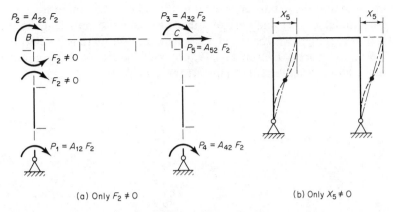

Figure 19.4.2 Proof for $A_{52} = B_{25}$.

In this case the internal virtual work done by the pair of axial forces acting on member BC is zero because there is no axial deformation (i.e., axial deformation is ignored). Note also that $B_{25}X_5$ is clockwise end rotation by definition, so that the expression $(F_2)(B_{25}X_5)$ represents a positive quantity.

19.5 MEMBER STIFFNESS MATRIX [S]

Exactly as described in Section 17.5 for beams, the member stiffness matrix consists of the coefficients of the end rotations e used to obtain the end moments F; that is, it expresses the relationship $\{F\} = [S]\{e\}$. The relationship is shown by Eq. (17.5.5). As long as the effect of axial deformation is not considered (i.e., the common situation is ordinary frames), the matrix $[S]$ is the same for a beam element and a frame element; the orientation of the element does not affect the member stiffness matrix.

19.6 GLOBAL STIFFNESS MATRIX [K]

The global stiffness matrix $[K]$ is the relationship between the external forces P and the joint displacements X. This was treated in detail in Section 17.6 for beams. The relationship is

$$\{P\} = [K]\{X\} \tag{19.6.1}$$

The matrix $[K]$ may be established by columns by letting one displacement equal unity simultaneously with all other displacements equal to zero. This was described in detail in Section 16.6 and 17.6. The derivation of the local stiffness matrix of a "typical beam element" (see Fig. 17.6.4) applies to the element of a rigid frame. In fact, the rigid frame element is the one where one usually expects the translational degrees of freedom ($P_3 - X_3$ and $P_4 - X_4$ of Fig. 16.6.4) to occur.

    If one is to establish directly the $[K]$ matrix for a frame such as that in Fig. 19.2.1a, columns one through four will be obtained essentially as shown for a beam in Fig. 17.6.1. The fifth column, involving the sidesway degree of freedom, is more involved and the details are shown in Fig. 19.6.1. To obtain the elements of the fifth column, let $X_5 = 1$ and keep all others X's equal to zero. The resulting elastic curve is shown in Fig. 19.6.1b. The end moments and shears thus developed on the members AB and CD are shown in Fig. 19.6.1c, and the free-body diagrams of the joints in Fig. 19.6.1d. Equilibrium of joints A through D gives

$$P_1 = -\frac{6EI_1}{L_1^2}$$

$$P_2 = -\frac{6EI_1}{L_1^2}$$

$$P_3 = -\frac{6EI_3}{L_1^2}$$

$$P_4 = -\frac{6EI_3}{L_1^2}$$

$$P_5 = +\frac{12EI_1}{L_1^3} + \frac{12EI_3}{L_1^3}$$

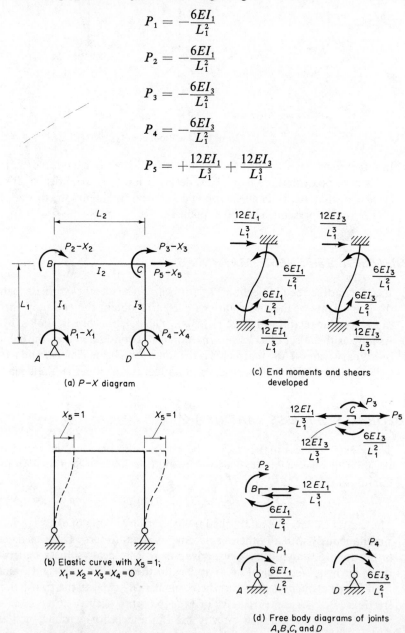

(a) $P-X$ diagram

(b) Elastic curve with $X_5 = 1$; $X_1 = X_2 = X_3 = X_4 = 0$

(c) End moments and shears developed

(d) Free body diagrams of joints $A, B, C,$ and D

Figure 19.6.1 Data for column five of the global stiffness matrix $[K]$.

The complete global stiffness matrix $[K]$ is as follows:

$$[K] =$$

P \ X	1	2	3	4	5
1	$\dfrac{4EI_1}{L_1}$	$\dfrac{2EI_1}{L_1}$	0	0	$-\dfrac{6EI_1}{L_1^2}$
2	$\dfrac{2EI_1}{L_1}$	$\dfrac{4EI_1}{L_1} + \dfrac{4EI_2}{L_2}$	$\dfrac{2EI_2}{L_2}$	0	$-\dfrac{6EI_1}{L_1^2}$
3	0	$\dfrac{2EI_2}{L_2}$	$\dfrac{4EI_2}{L_2} + \dfrac{4EI_3}{L_1}$	$\dfrac{2EI_3}{L_1}$	$-\dfrac{6EI_3}{L_1^2}$
4	0	0	$\dfrac{2EI_3}{L_1}$	$\dfrac{4EI_3}{L_1}$	$-\dfrac{6EI_3}{L_1^2}$
5	$-\dfrac{6EI_1}{L_1^2}$	$-\dfrac{6EI_1}{L_1^2}$	$-\dfrac{6EI_3}{L_1^2}$	$-\dfrac{6EI_3}{L_1^2}$	$\dfrac{12E}{L_1^3}(I_1 + I_3)$

Note that the end moments generated at the top and bottom of members AB and CD, when X_1 through X_4 act alone, cause shears to develop at the top and bottom of those members. The free-body diagram of joint C will thus have a horizontal force acting, giving a value for P_5 for each of X_1 through X_4 acting alone. Examination of the $[K]$ matrix shows it to be a square symmetric matrix, as it must be.

19.7 DIRECT METHOD OF OBTAINING GLOBAL STIFFNESS MATRIX

As discussed in Section 17.6, the *local* stiffness matrix $[K]$ may be established for each member to be used in the analysis. Considering only *plane* frames, there is a maximum of four degrees of freedom for any member, as shown in Fig. 17.6.4a. The local stiffness is established for each member according to the scheme shown by Eq. (17.6.9). Then for the global matrix the common matrix elements are added together.

For example, member AB will have a local stiffness matrix as follows:

$$[K] \text{ (for } AB\text{)} =$$

X	1	2	4	$P - X$, Fig. 17.6.4
P	1	2	5	$P - X$, Fig. 19.6.1
1	$+\dfrac{4EI_1}{L_1}$	$+\dfrac{2EI_1}{L_1}$	$-\dfrac{6EI_1}{L_1^2}$	
2	$+\dfrac{2EI_1}{L_1}$	$+\dfrac{4EI_1}{L_1}$	$-\dfrac{6EI_1}{L_1^2}$	
5	$-\dfrac{6EI_1}{L_1^2}$	$-\dfrac{6EI_1}{L_1^2}$	$+\dfrac{12EI_1}{L_1^3}$	

Next, consider member BC, which has a local stiffness matrix as follows:

X	1	2	$P - X$, Fig. 17.6.4
P	2	3	$P - X$, Fig. 19.6.1
2	$\dfrac{4EI_2}{L_2}$	$\dfrac{2EI_2}{L_2}$	
3	$\dfrac{2EI_2}{L_2}$	$\dfrac{4EI_2}{L_2}$	

$[K]$ (for BC) $=$

Finally, write the local stiffness matrix for member CD, as follows:

X	2	1	4	$P - X$, Fig. 17.6.4
P	3	4	5	$P - X$, Fig. 19.6.1
3	$+\dfrac{4EI_3}{L_1}$	$+\dfrac{2EI_3}{L_1}$	$-\dfrac{6EI_3}{L_1^2}$	
4	$+\dfrac{2EI_3}{L_1}$	$+\dfrac{4EI_3}{L_1}$	$-\dfrac{6EI_3}{L_1^2}$	
5	$-\dfrac{6EI_3}{L_1^2}$	$-\dfrac{6EI_3}{L_1^2}$	$+\dfrac{12EI_3}{L_1^3}$	

$[K]$ (for CD) $=$

Now to compile the global stiffness matrix, add the components from the three separate local stiffness matrices, according to the $P - X$ numbering from Fig. 19.6.1. The complete $[K]$ has already been shown in Section 19.6.

19.8 EXTERNAL FORCE MATRIX $\{P\}$

As discussed in Section 17.7, rarely are the real loads applied directly along the chosen degrees of freedom. Ordinarily, the loads occur at other locations and are transmitted to the joints by the members. This load transmission may be thought of as the sum of two parts: (1) the restraining forces acting on the member ends necessary to keep the joint displacements X equal to zero, plus (2) the active joint forces (opposite to the restraining forces) necessary to cause the true joint displacements. It was this same concept that has been used in the slope deflection method (see Fig. 9.2.3).

When the analysis is imagined to begin with zero displacements, it means there must be restraining forces acting at the member ends to hold the displacement at zero. For example, the structure of Fig. 19.8.1a will have its X_1 through X_5 equal to zero only if the restraining forces shown in Fig. 19.8.1c are acting on the member ends. Comparing then the force acting on each joint with the general joint force P (Fig. 19.8.1b) to be used in the analysis, the active joint forces are:

$$P_1 = +\frac{wL_1^2}{12} \qquad \text{(joint } A\text{)}$$

$$P_2 = +\frac{Wa_1b_1^2}{L_2^2} - \frac{wL_1^2}{12} \qquad \text{(joint } B\text{)}$$

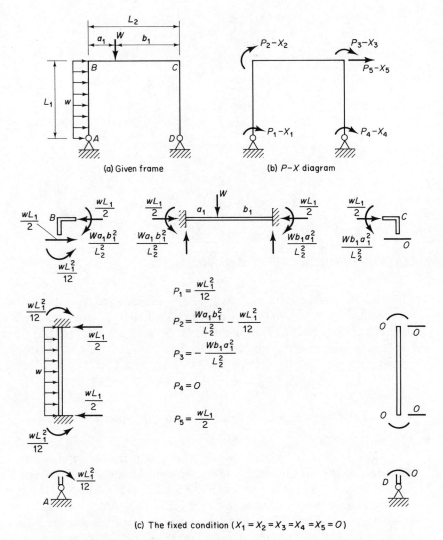

(a) Given frame

(b) P-X diagram

$$P_1 = \frac{wL_1^2}{12}$$

$$P_2 = \frac{Wa_1b_1^2}{L_2^2} - \frac{wL_1^2}{12}$$

$$P_3 = -\frac{Wb_1a_1^2}{L_2^2}$$

$$P_4 = 0$$

$$P_5 = \frac{wL_1}{2}$$

(c) The fixed condition ($X_1 = X_2 = X_3 = X_4 = X_5 = 0$)

Figure 19.8.1 Equilibrium in the fixed (i.e., zero displacement) condition to determine the external force matrix $\{P\}$.

$$P_3 = -\frac{Wb_1a_1^2}{L_2^2} \qquad \text{(joint } C\text{)}$$

$$P_4 = 0 \qquad \text{(joint } D\text{)}$$

$$P_5 = +\frac{wL_1}{2} \qquad \text{(joint } C\text{)}$$

In matrix format the equivalent joint forces become the external force matrix $\{P\}$, as follows:

$$\{P\} = \begin{array}{|c|c|} \hline \diagbox{P}{LC} & 1 \\ \hline 1 & +\dfrac{w_1 L_1^2}{12} \\ \hline 2 & \dfrac{W a_1 b_1^2}{L_2^2} - \dfrac{w L_1^2}{12} \\ \hline 3 & -\dfrac{W b_1 a_1^2}{L_2^2} \\ \hline 4 & 0 \\ \hline 5 & +\dfrac{w L_1}{2} \\ \hline \end{array}$$

19.9 MATRIX DISPLACEMENT METHOD

The matrix displacement method for frames is identical in mathematical operations and sequence of steps to that described in Section 16.7 for trusses and in Section 17.8 for continuous beams. The equations to be used are Eqs. (17.8.1) to (17.8.6). The following two examples illustrate the details of the method as applied to frames having sidesway.

Example 19.9.1

Analyze the frame of Fig. 19.9.1a by the matrix displacement method.

Solution

(a) *Draw the P–X and F–e diagrams.* This structure is the one used as the basis for discussing general concepts in the earlier sections of this chapter. The P–X and F–e diagrams are Fig. 19.2.1a and b. As always these diagrams are dependent on the structure only and are independent of the loading. The number of degrees of freedom NP is the minimum value of five necessary to solve the problem. Each of the four joints can rotate and joints B and C move to the side—both by the same amount unless axial deformation of member BC is considered.

(b) *Establish the statics matrix* $[A]$. This was shown in Eq. (19.2.2), and the free-body diagrams from which the matrix is obtained are shown in Fig. 19.2.1. For this numerical example, $L = 16$ ft.

(c) *Establish the deformation matrix* $[B]$. This was shown in Eq. (19.3.3), and the explanatory data appear in Fig. 19.3.1 For this numerical example, $L = 16$ ft. After establishing matrix $[B]$ independently from matrix $[A]$, check that one is the transpose of the other.

(d) *Establish the member stiffness matrix* $[S]$. This is the pair of equations involving EI/L, Eqs. (17.5.4), that relate the end moments F to the end rotations e. For this example, referring to Fig. 19.9.1 for the relative I values, the matrix $[S]$ is

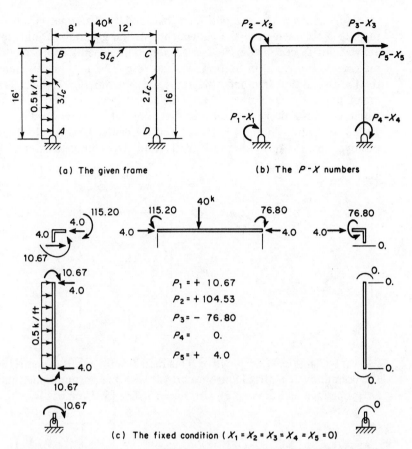

(a) The given frame

(b) The P-X numbers

$P_1 = + 10.67$
$P_2 = + 104.53$
$P_3 = - 76.80$
$P_4 = 0.$
$P_5 = + 4.0$

(c) The fixed condition ($X_1 = X_2 = X_3 = X_4 = X_5 = 0$)

Figure 19.9.1 Given data, P–X diagram, and data for external force matrix $\{P\}$ for Example 19.9.1.

$[S]_{6 \times 6} = EI_c$

F \ e	1	2	3	4	5	6
1	$\frac{3}{4}$	$\frac{3}{8}$				
2	$\frac{3}{8}$	$\frac{3}{4}$				
3			1	$\frac{1}{2}$		
4			$\frac{1}{2}$	1		
5					$\frac{1}{2}$	$\frac{1}{4}$
6					$\frac{1}{4}$	$\frac{1}{2}$

(e) *Establish the external force matrix* $\{P\}$. This matrix, also called the load matrix, is obtained from the equivalency of the generalized joint forces P with the active joint forces in the fixed condition. The general approach to obtaining this matrix was presented in Section 19.8 (see Fig. 19.8.1). For this numerical example, the fixed-condition free-body diagrams are given in Fig. 19.9.1c. The matrix values are obtained, for example, by taking the actual forces on joint B and comparing them with the forces P in the P–X diagram. At joint B, the net moment is $115.20 - 10.67 = 104.53$ clockwise. Thus, P_2 is 104.53 and is positive because clockwise is the chosen positive direction of P_2 on the P–X diagram. The complete external force matrix is

$$\{P\}_{5\times 1} = \begin{array}{c|c} \diagdown\!\!\!\begin{array}{c}LC\\[-4pt]P\end{array} & 1 \\ \hline 1 & +10.67 \\ \hline 2 & +104.53 \\ \hline 3 & -76.80 \\ \hline 4 & 0. \\ \hline 5 & +4.0 \end{array}$$

(f) *Obtain the* $[SA^T]$ *matrix.* This matrix is the result of matrix multiplication of the premultiplier matrix $[S]$ from part (d) with the postmultiplier matrix $[B]$ (that is, $[A^T]$) obtained as Eq. (19.3.3) with $L = 16$. The $[SA^T]$ matrix is

$$[SA^T]_{6\times 3} = EI_c \begin{array}{c|c|c|c|c|c} \diagdown\!\!\!\begin{array}{c}X\\[-4pt]F\end{array} & 1 & 2 & 3 & 4 & 5 \\ \hline 1 & +\dfrac{3}{4} & +\dfrac{3}{8} & & & -\dfrac{9}{128} \\ \hline 2 & +\dfrac{3}{8} & +\dfrac{3}{4} & & & -\dfrac{9}{128} \\ \hline 3 & & +1 & +\dfrac{1}{2} & & \\ \hline 4 & & +\dfrac{1}{2} & +1 & & \\ \hline 5 & & & +\dfrac{1}{2} & +\dfrac{1}{4} & -\dfrac{3}{64} \\ \hline 6 & & & +\dfrac{1}{4} & +\dfrac{1}{2} & -\dfrac{3}{64} \end{array}$$

(g) *Establish the global stiffness matrix* $[K]$. Matrix multiplication of the premultiplier matrix $[A]$ obtained as Eq. (19.2.2) with $L = 16$, with the postmultiplier matrix $[SA^T]$ from part (f), gives

$$[K] = [ASA^T]_{5\times5} = EI_c$$

X / P	1	2	3	4	5
1	$+\dfrac{3}{4}$	$+\dfrac{3}{8}$	0	0	$-\dfrac{9}{128}$
2	$+\dfrac{3}{8}$	$+\dfrac{7}{4}$	$+\dfrac{1}{2}$	0	$-\dfrac{9}{128}$
3	0	$+\dfrac{1}{2}$	$+\dfrac{3}{2}$	$+\dfrac{1}{4}$	$-\dfrac{3}{64}$
4	0	0	$+\dfrac{1}{4}$	$+\dfrac{1}{2}$	$-\dfrac{3}{64}$
5	$-\dfrac{9}{128}$	$-\dfrac{9}{128}$	$-\dfrac{3}{64}$	$-\dfrac{3}{64}$	$+\dfrac{15}{1024}$

The matrix $[K]$ could also have been established directly according to the procedure described in Section 19.7.

(h) *Obtain the inverse of the matrix* $[K]$. Using Eq. (17.8.4),

$$\{X\} = [ASA^T]^{-1}\{P\} = [K]^{-1}\{P\}$$

the $[K]$ matrix must be inverted. By Gauss–Jordan elimination as discussed in Section 13.4,

$$[K]^{-1} = \frac{1}{EI_c}$$

X / P	1	2	3	4	5
1	$+4.087722$	$+0.228071$	$+0.438597$	$+2.649126$	$+30.596521$
2	$+0.228071$	$+0.859649$	-0.192982	$+0.754387$	$+7.017549$
3	$+0.438597$	-0.192982	$+0.859649$	-0.087719	$+3.649125$
4	$+2.649126$	$+0.754387$	-0.087719	$+5.070179$	$+32.280737$
5	$+30.596521$	$+7.017549$	$+3.649125$	$+32.280737$	$+363.78982$

The inversion of a 5 by 5 matrix requires a considerable amount of work unless a computer is available. In general, the solution by the matrix displacement method is practical only when the degree of freedom is relatively small.

(i) *Obtain the joint displacement matrix* $\{X\}$. Using Eqs. (17.6.7),

$$\{X\} = [K]^{-1}\{P\}$$

which gives

$$\{X\} = \frac{1}{EI_c}\begin{Bmatrix} +156.16 \\ +135.18 \\ -66.92 \\ +242.98 \\ +2234.92 \end{Bmatrix}$$

(j) *Obtain the internal force matrix* $\{F\}$. Using Eq. (17.8.5), multiply the premultiplier matrix $[SA^T]$ by the postmultiplier matrix $\{X\}$ to give

$$\{F\} = \begin{array}{|c|} \hline +10.67 \\ \hline +2.81 \\ \hline +101.73 \\ \hline +0.67 \\ \hline -77.48 \\ \hline 0.00 \\ \hline \end{array}$$

These internal forces $\{F\}$ which cause the displacements $\{X\}$, must be added to the internal forces $\{F_0\}$ which cause zero displacements in the fixed condition.

(k) *Obtain the final end moments.* The total moment F^* on the end of a member is the sum of the F-value from the matrix operations and the fixed-end moment F_0. In matrix notation,

$$\{F^*\} = \{F_0\} + \{F\} = \{F_0\} + [SA^T]\{X\}$$

$$\{F^*\} = \begin{array}{|c|} \hline -10.67 \\ \hline +10.67 \\ \hline -115.20 \\ \hline +76.80 \\ \hline 0.00 \\ \hline 0.00 \\ \hline \end{array} + \begin{array}{|c|} \hline +10.67 \\ \hline +2.81 \\ \hline +101.83 \\ \hline +0.67 \\ \hline -77.48 \\ \hline 0.00 \\ \hline \end{array} + \begin{array}{|c|} \hline 0.00 \\ \hline +13.48 \\ \hline -13.47 \\ \hline +77.47 \\ \hline -77.48 \\ \hline 0.00 \\ \hline \end{array}$$

The slight discrepancy between the pairs of numerical values of $F_2^* - F_3^*$ or $F_4^* - F_5^*$ is due to the fact that between four and five significant figures have been used in the hand computations of this problem.

(l) *Check.* Checks of statics and compatibility of deformations are always necessary to verify the correctness of the answer. These checks were discussed in Sections 17.9, 17.10, 18.3, and 18.4, and are not repeated here.

Example 19.9.2

Analyze the frame of Examples 11.7.2. and 11.9.1 (shown again in Fig. 19.9.2a) by the matrix displacement method.

Solution

(a) *Draw the P–X and F–e diagrams.* The structure has rotational degrees of freedom at joints B, C, and D. In addition, the horizontal member $ABCDE$ can displace with a horizontal deflection because of the rollers at D. Using a minimum number of degrees of freedom, the cantilevers, being statically determinate, will not be used. Thus, as shown in Fig. 19.9.2b and c, $NP = 4$ and $NF = 8$.

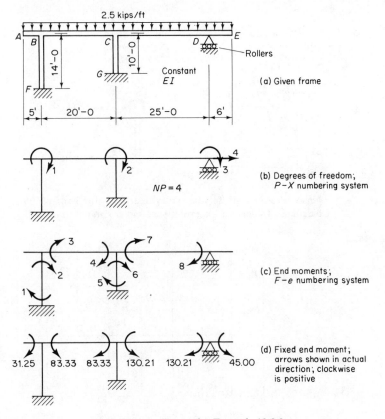

Figure 19.9.2 Frame for Example 19.9.2.

(b) *Establish the statics matrix* [A]. The statics matrix is obtained from the equilibrium of the joints when NP external forces are acting on the joints simultaneously with the NF internal forces (end moments). The free-body diagrams of joints B, C, and D are given in Fig. 19.9.3, where the vertical forces have not been shown since they do not affect the statics matrix. From the statics of each joint,

$$P_1 = F_2 + F_3$$
$$P_2 = F_4 + F_6 + F_7$$
$$P_3 = F_8$$
$$P_4 = -\frac{F_1 + F_2}{14} - \frac{F_5 + F_6}{10}$$

In matrix form, the statics matrix [A] is

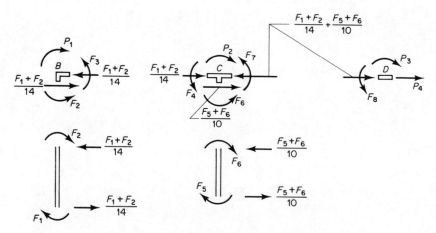

Figure 19.9.3 Data for the statics matrix $[A]$ for Example 19.9.2. Free-body diagrams of joints showing rotational and horizontal forces acting.

	F	1	2	3	4	5	6	7	8
$[A] =$	P								
	1		$+1.0$	$+1.0$					
	2				$+1.0$		$+1.0$	$+1.0$	
	3								$+1.0$
	4	$-\dfrac{1}{14}$	$-\dfrac{1}{14}$			$-\dfrac{1}{10}$	$-\dfrac{1}{10}$		

(c) *Establish the deformation matrix* $[B]$. In accordance with the equation

$$\{e\} = [B]\{X\}$$

the matrix $[B]$ may be established by columns, successively examining the effects of X_1 through X_4 acting alone with the other X's equal to zero. The effects of X_1 through X_3 were shown in Fig. 18.2.2b. The effect of the sidesway degree of freedom X_4 is shown in Fig. 19.9.4. The $[B]$, or $[A^T]$, matrix is as follows:

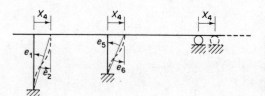

Figure 19.9.4 Compatibility of internal deformations with joint displacement $X_4 = 1$. (For the effects of X_1 through X_3 acting alone, see Fig. 18.2.2b.)

$$[B] = [A^T] =$$

e \ X	1	2	3	4
1				$-\dfrac{1}{14}$
2	+1.0			$-\dfrac{1}{14}$
3	+1.0			
4		+1.0		
5				$-\dfrac{1}{10}$
6		+1.0		$-\dfrac{1}{10}$
7		+1.0		
8			+1.0	

As usual after establishing the $[B]$ matrix independently from the $[A]$ matrix, verify that one is the transpose of the other.

(d) *Establish the member stiffness matrix* $[S]$. The member stiffness matrix arises from the relationship

$$\{F\} = [S]\{e\}$$

For this example, this matrix is identical to that for Example 18.2.1(d). The change from a hinged support at D in that example to a roller support at D in the present example does not affect member stiffness.

(e) *Establish the external force matrix* $\{P\}$. This load matrix, obtained from the equivalency of the generalized joint forces P with the active joint forces in the fixed condition, would differ from what was used in Example 18.2.1 only in the force P_4, which did not exist in that example. The net horizontal force acting on joint D when the structure is in the fixed condition will involve shears acting on members FB and CG in the fixed condition. Since there are no horizontal loads on those members, the fixed-condition shears are zero in this case, making the net horizontal force on D equal to zero, which means that P_4 is zero. Thus, the matrix $\{P\}$ is

$$\{P\}_{4 \times 1} =$$

P \ LC	1
1	+52.08
2	+46.88
3	−85.21
4	0.

LC = loading condition

(f) *Obtain the* $[SA^T]$ *matrix.* The matrix multiplication of the premultiplier matrix $[S]$ from Example 18.2.2(d) with the postmultiplier matrix $[B]$ of part (c) gives

$$[SA^T] =$$

F \ X	1	2	3	4
1	$\dfrac{2EI}{14}$			$-\dfrac{6EI}{196}$
2	$\dfrac{4EI}{14}$			$-\dfrac{6EI}{196}$
3	$\dfrac{4EI}{20}$	$\dfrac{2EI}{20}$		
4	$\dfrac{2EI}{20}$	$\dfrac{4EI}{20}$		
5		$\dfrac{2EI}{10}$		$-\dfrac{6EI}{100}$
6		$\dfrac{4EI}{10}$		$-\dfrac{6EI}{100}$
7		$\dfrac{4EI}{25}$	$\dfrac{2EI}{25}$	
8		$\dfrac{2EI}{25}$	$\dfrac{4EI}{25}$	

(g) *Obtain the global stiffness matrix* $[K]$. Matrix multiplication of the premultiplier matrix $[A]$ from part (b) above with the postmultiplier matrix $[SA^T]$ from part (f) gives

$$[K] =$$

P \ X	1	2	3	4
1	$\dfrac{17EI}{35}$	$\dfrac{2EI}{20}$	0	$-\dfrac{6EI}{196}$
2	$\dfrac{2EI}{20}$	$\dfrac{19EI}{25}$	$\dfrac{2EI}{25}$	$-\dfrac{6EI}{100}$
3	0	$\dfrac{2EI}{25}$	$\dfrac{4EI}{25}$	0
4	$-\dfrac{6EI}{196}$	$-\dfrac{6EI}{100}$	0	$+\dfrac{702EI}{42{,}875}$

(h) *Obtain the inverse of the matrix* $[K]$. Using Eq. (17.8.4),

$$\{X\} = [K]^{-1}\{P\}$$

	P X	1	2	3	4
$[K]^{-1} = \dfrac{1}{EI}$	1	$+2.335448$	$+0.056874$	-0.028437	$+4.574907$
	2	$+0.056874$	$+2.000872$	-1.000436	$+7.438591$
	3	-0.028437	-1.000436	$+6.750218$	-3.719295
	4	$+4.574907$	$+7.438591$	-3.719295	$+96.887935$

(i) *Obtain the joint displacement matrix* $\{X\}$. Using Eq. (17.6.7),

$$\{X\} = [K]^{-1}\{P\}$$

which gives

$$\{X\} = \frac{1}{EI} \begin{vmatrix} +126.720 \\ +182.010 \\ -623.568 \\ +903.903 \end{vmatrix}$$

(j) *Obtain the internal force matrix* $\{F\}$. Using Eq. (17.8.5),

$$\{F\} = [SA^T]\{X\}$$

which gives

$$\begin{matrix} \{F\} \\ \text{(ft-kips)} \end{matrix} = \begin{vmatrix} -9.568 \\ +8.535 \\ +43.545 \\ +49.074 \\ -17.832 \\ +18.570 \\ -20.764 \\ -85.210 \end{vmatrix}$$

(k) *Obtain the final end moments.* The total moment F^* on the end of a member is the sum of the F-value from the matrix operations and the fixed-end moment F_0. In matrix notation,

$$\{F^*\} = \{F_0\} + \{F\}$$

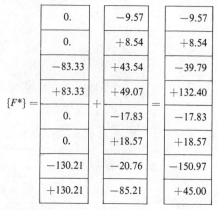

$$\{F^*\} = \begin{array}{c} 0. \\ 0. \\ -83.33 \\ +83.33 \\ 0. \\ 0. \\ -130.21 \\ +130.21 \end{array} \;+\; \begin{array}{c} -9.57 \\ +8.54 \\ +43.54 \\ +49.07 \\ -17.83 \\ +18.57 \\ -20.76 \\ -85.21 \end{array} \;=\; \begin{array}{c} -9.57 \\ +8.54 \\ -39.79 \\ +132.40 \\ -17.83 \\ +18.57 \\ -150.97 \\ +45.00 \end{array}$$

The solution is shown in Fig. 19.9.5.

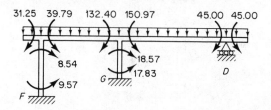

Figure 19.9.5 Solution for Example 19.9.2.

(l) *Check*. Checks of statics and compatibility of deformations are always necessary to verify the correctness of the answer. The checks were discussed in Sections 17.9, 17.10, 18.3, and 18.4, and will be discussed further with regard to frames in Sections 19.10 and 19.11. The answers above are, however, in agreement with the solution obtained in Example 11.9.1.

19.10 STATICS CHECKS

The statics checks are equal to the number of degrees of freedom; that is, see that rotational equilibrium is satisfied at each joint, and that the horizontal reaction at the roller is zero. From Fig. 19.9.4 for the frame of Example 19.9.2, it is seen that rotational equilibrium is satisfied at joints B, C, and D. For horizontal force equilibrium, the reaction at F is

$$H_F = \frac{9.57 - 8.54}{14} = 0.0736 \quad \text{(to the left)}$$

and that at G is

$$H_G = \frac{18.57 - 17.83}{10} = 0.0740 \quad \text{(to the right)}$$

Even though H_F and H_G should exactly balance one another, the calculation above shows that because of the loss of one significant figure in computing the numerical difference of the end moments, one would accept 0.074 for H_F and 0.074 for H_G.

19.11 DEFORMATION CHECKS

As discussed in Sections 16.9, 17.10, and 18.4, these checks are made by determining the slopes to the elastic curve at both ends of each member to verify them to be identical to the values of X from the matrix solution. When there is a joint deflection as specifically dealt with in this chapter, that deflection must also be computed to verify the matrix solution deflection.

Example 19.1.1

Illustrate the deformation checks for the rigid frame analyzed in Example 19.9.2.

Solution

(a) *Member FB.* Draw the free-body diagram of the member and the M/EI diagrams for the component loading, as shown in Fig. 19.11.1a. Using the moment area method,

$$\Delta_B = X_4 = \text{moment of } M/EI \text{ area between } F \text{ and } B \text{ about } B$$

$$= \frac{1}{EI}\left[\frac{1}{2}(8.54)(14)\left(\frac{14}{3}\right) + \frac{1}{2}(9.57)(14)\left(\frac{28}{3}\right)\right]$$

$$= \frac{904.21}{EI} \approx \left[X_4 = +\frac{903.90}{EI}\right]$$

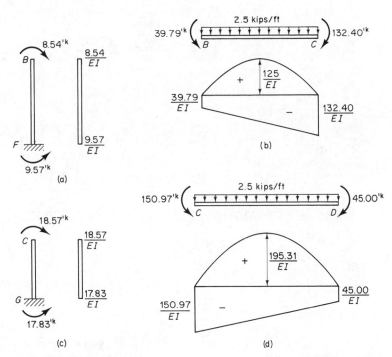

Figure 19.11.1 M/EI diagrams for deformation checks on rigid frame; Examples 19.9.2 and 19.11.1.

$\theta_B = X_1 = M/EI$ area between F and B
(clockwise)

$$= \frac{1}{EI}\left[\frac{1}{2}(8.54)(14) + \frac{1}{2}(9.57)(14)\right] = \frac{+126.84}{EI} \quad \text{(checks with } +126.720/EI)$$

(b) *Member GC.* The M/EI diagram is given in Fig. 19.11.1c.

$\Delta_C = X_4 = $ moment of M/EI between G and C about C

$$= \frac{1}{EI}\left[\frac{1}{2}(18.57)(10)\left(\frac{10}{3}\right) + \frac{1}{2}(17.83)(10)\left(\frac{20}{3}\right)\right]$$

$$= \frac{903.83}{EI} \approx \left[X_4 = +\frac{903.90}{EI}\right]$$

$\theta_C = X_2 = M/EI$ area between G and C
(clockwise)

$$= \frac{1}{EI}\left[\frac{1}{2}(18.57)(10) + \frac{1}{2}(17.83)(10)\right] = \frac{+182.00}{EI} \quad \text{(checks with } +182.010/EI)$$

(c) *Member BC.* The M/EI diagram is given in Fig. 19.11.1b.

$$\theta_{B1} = \frac{1}{20EI}\left[\frac{2}{3}(125)(20)(10)\right] = \frac{833.33}{EI}$$

(clockwise)

$$\theta_{B2} = \frac{1}{20EI}\left[\frac{1}{2}(39.79)(20)\left(\frac{40}{3}\right) + \frac{1}{2}(132.40)(20)\left(\frac{20}{3}\right)\right] = \frac{706.60}{EI}$$

(counter-clockwise)

$$\theta_B = X_1 = \frac{833.33 - 706.60}{EI} = \frac{+126.73}{EI} \quad \text{(checks with } +126.720/EI)$$

$$\theta_{C1} = \frac{1}{20EI}\left[\frac{2}{3}(125)(20)(10)\right] = \frac{833.33}{EI}$$

(counter-clockwise)

$$\theta_{C2} = \frac{1}{20EI}\left[\frac{1}{2}(39.79)(20)\left(\frac{20}{3}\right) + \frac{1}{2}(132.40)(20)\left(\frac{40}{3}\right)\right] = \frac{1015.30}{EI}$$

(clockwise)

$$\theta_C = X_2 = \frac{1015.30 - 833.33}{EI} = \frac{+181.97}{EI} \quad \text{(checks with } +182.010/EI)$$

(d) *Member CD.* The M/EI diagram is given in Fig. 19.11.1d.

$$\theta_{C1} = \frac{1}{25EI}\left[\frac{2}{3}(195.31)(25)(12.5)\right] = \frac{1627.58}{EI}$$

(clockwise)

$$\theta_{C2} = \frac{1}{25EI}\left[\frac{1}{2}(150.97)(25)\left(\frac{50}{3}\right) + \frac{1}{2}(45.00)(25)\left(\frac{25}{3}\right)\right] = \frac{1445.58}{EI}$$

(counter-clockwise)

$$\theta_C = X_2 = \frac{1627.58 - 1445.58}{EI} = \frac{+182.00}{EI} \quad \text{(checks with } +182.010/EI)$$

$$\theta_{D1} = \frac{1}{25EI}\left[\frac{2}{3}(195.31)(25)(12.5)\right] = \frac{1627.58}{EI}$$

(counter-clockwise)

$$\theta_{D2} = \frac{1}{25EI}\left[\frac{1}{2}(150.97)(25)\left(\frac{25}{3}\right) + \frac{1}{2}(45.00)(25)\left(\frac{50}{3}\right)\right] = \frac{1004.04}{EI}$$

(clockwise)

$$\theta_D = X_3 = \frac{1004.04 - 1627.58}{EI} = \frac{-623.54}{EI} \quad \text{(checks with } -623.568/EI)$$

19.12 USE OF SHORT SEGMENTS

As illustrated in Examples 17.8.2 and 17.11.1 and discussed in Section 17.11, a structure may be divided into as many elements as desired over and above the minimum for solution. There are no additional concepts required to use additional "joints" at any location in a rigid frame. Each additional "joint" used in the analysis of a plane frame where axial deformation is neglected will add two degrees of freedom (one rotational and one translational) and two internal forces (end moments). Unless there is need for a graphical output for the elastic curve on the computer, usually no advantage accrues by using more "joints" than the minimum because the size of the global stiffness matrix $[K]$ to be inverted equals the number of degrees of freedom used in the solution.

PROBLEMS

19.1–19.7. For any problem assigned, draw the P-X and F-e diagrams using the minimum number of degrees of freedom. Then obtain numerically the statics matrix $[A]$, the deformation matrix $[B]$, the member stiffness matrix $[S]$, the $[SA^T]$ matrix, the global stiffness matrix $[K] = [ASA^T]$, the $[K]^{-1}$ matrix, and the final answers in the form of the $\{X\}$ matrix and the $\{F^*\}$ matrix. Show the statics and deformation checks.

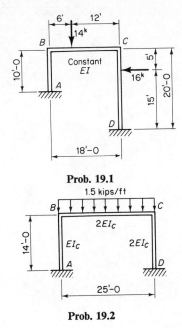

Prob. 19.1

Prob. 19.2

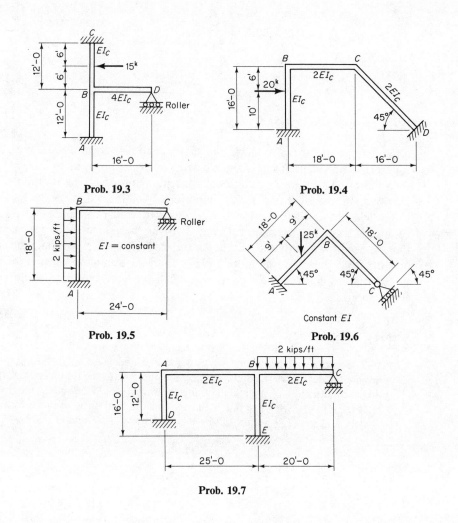

Prob. 19.3

Prob. 19.4

Prob. 19.5

Prob. 19.6

Prob. 19.7

[20]

Matrix Displacement Method Applied to Rigid Frames Containing Truss-Type Members

---

20.1 GENERAL

Following the treatment of truss members (two-force members) in Chapter 16 and flexural members disregarding axial deformation in Chapters 17 to 19, the next logical step is to combine the two types of members into the same structure. Such a structure has sometimes been called a *composite structure*. A composite structure as herein defined should not be confused with the more common use of the word "composite" to mean the interaction between two different materials.

The reader may recall that a truss member, as treated in Chapter 16, has axial force only as the internal force; thus, there will be one unknown internal force per member. For a flexural member in continuous beams or rigid frames with axial deformation neglected, the two end moments are treated as the unknown internal forces; thus there will be two unknown internal forces per member. The flexural members of Chapters 17 to 19 may or may not be subject to axial force in addition, but they are considered rigid with regard to axial deformation; that is, axial deformation is considered zero.

In reality, truss members are joined together by welded or bolted connections which induce some bending moment into the members. The true action of a truss, however, is primarily to cause axial forces in the members. Although some bending moment may also arise, its magnitude is usually small compared to the axial force, and thus the bending moment is neglected in any primary analysis of a truss. On the other hand, the primary action in rigid frames is bending and the axial deformation is frequently small relative to the transverse deflection due to bending moment. Thus even though axial force should be used in the design of the member, its deformation is not used in the conditions for compatibility of deformations. Tall building frames may have axial forces large enough to require consideration of axial deformation in the vertical members, particularly in the lower stories.

In the real design situation, the structural engineer must decide on the proper assumptions. Axial deformation may be important in some or all members of a structure, and flexural deformation may be important in some or all members of a structure. If the element has only unknown axial force without bending, its axial deformation should be considered and the element is treated as a *truss element* described in Chapter 16. If the element has two unknown end moments and zero axial force, it is called a *beam element*. If the element has two unknown end moments and an unknown axial force, the axial force is not an independent unknown when its effect on the change in length of that element is ignored. However, if the axial force in an element in bending is taken to cause axial deformation, this element must be treated as a *combined truss and beam element*; that is, there are three unknown internal forces for this element. In this chapter, structures containing truss elements, beam elements (with or without axial force), and combined truss and beam elements are treated.

20.2 DEGREE OF FREEDOM VS. DEGREE OF STATICAL INDETERMINACY

An explanation of degree of freedom and degree of statical indeterminacy for trusses, continuous beams, and rigid frames was given in Chapters 16 to 19. When more than one type of element is present in a single structure, degrees of freedom are assigned in accordance with the manner that deformation is assumed to occur. For example, the structure in Fig. 20.2.1, known as a *king-post truss*, contains member *ABC*, which is primarily a flexural member but which also serves as the top chord of the truss, thus

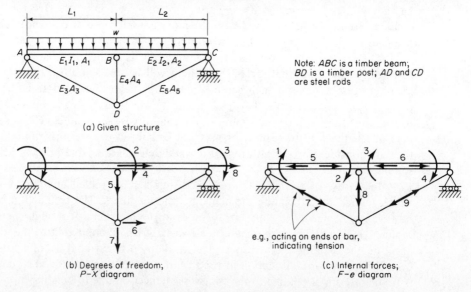

Note: *ABC* is a timber beam; *BD* is a timber post; *AD* and *CD* are steel rods

(a) Given structure

(b) Degrees of freedom; *P–X* diagram

e.g., acting on ends of bar, indicating tension

(c) Internal forces; *F–e* diagram

Figure 20.2.1 King-post truss containing flexural and truss members.

taking axial compression. The truss members *AD*, *BD*, and *CD* are subject only to axial force. For the flexural member, there are rotational degrees of freedom 1 to 3, and vertical translational degrees of freedom 5 at joint *B*. For the truss, there are translational displacement degrees of freedom 4 through 8. The unknown internal forces are shown in Fig. 20.2.1c, where end moments 1 through 4 and internal axial forces 5 through 9 are identified. Since the unknown internal forces *NF* exceed the number of degrees of freedom *NP* by one, the structure is statically indeterminate to the first degree. If the axial deformation in the timber beam *ABC* is to be neglected, the deflections along the degrees of freedom 4 and 8 are zero and the internal forces 5 and 6 will not be independent unknowns. Similarly, if the axial deformation in the timber post *BD* is to be neglected, the vertical deflections 5 and 7 should be of equal amount, requiring one less degree of freedom in the analysis, but the internal force 8 will not be an independent unknown. Therefore, the degree of statical indeterminacy of the king-post truss is 1 regardless of the method of analysis. From a classical analysis point of view, a truss *ABCD* would be statically determinate; likewise, a simply supported beam *AC* is statically determinate. When the beam and the truss are coupled, a compatibility of deformation requirement would be that the vertical deflection of the beam and of the truss at joint *B* must be identical. When one deformation requirement, in addition to the laws of statics, permits the analysis to be made, the structure is statically indeterminate to the first degree.

A second structure of the type to be treated in this chapter is the rigid frame with the tie rod shown in Fig. 20.2.2. In this case, the analyst recognizes that the

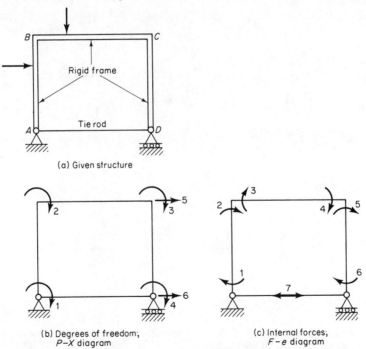

(a) Given structure

(b) Degrees of freedom;
P–X diagram

(c) Internal forces;
F–e diagram

Figure 20.2.2 Frame containing one member requiring consideration of axial deformation.

deformation of the tie rod will be large compared to the axial deformation of members *AB*, *BC*, or *CD*. Also the frame members have flexural stiffness, whereas the tie rod has negligible flexural stiffness. Thus, the degrees of freedom and internal forces used for the analysis follow assumptions reasonably conforming with actual behavior. The unknown internal forces are taken as the end moments F_1 through F_6 to recognize flexural stiffness for members *AB*, *BC*, and *CD*, and the axial force F_7 to recognize axial stiffness for the tie rod. The degrees of freedom are 1 through 4 for rotations, 5 for sidesway translation of beam *BC*, and 6 for the horizontal deflection at the roller support *D* resulting from the elongation of the tie rod. In this case, since $NF - NP = 1$, the structure is statically indeterminate to the first degree. If the tie had not been used, the degree of freedom *NP* would still have been 6 since joint *D* can move horizontally. However, the number *NF* of unknown internal forces would have been six and the structure would have been statically determinate.

20.3 DEFINITION OF STATICS MATRIX [A]

As discussed in Sections 15.1, 16.2, 17.2, and 19.2, the statics matrix [*A*] expresses the joint forces {*P*} in terms of the internal forces {*F*}, including member end moments and axial forces. For the structure of Fig. 20.2.1, the free-body diagrams are shown in Fig. 20.3.1. The equations of statics are

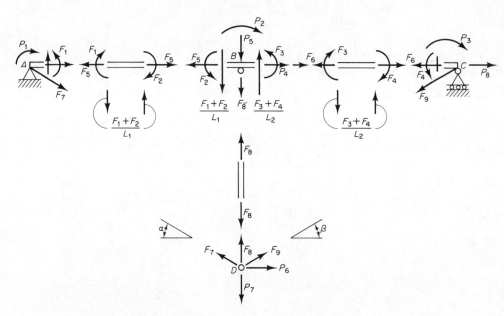

Figure 20.3.1 Free-body diagrams of joints and members; data for the statics matrix [*A*] of the structure of Fig. 20.2.1.

$$P_1 = F_1$$
$$P_2 = F_2 + F_3$$
$$P_3 = F_4$$
$$P_4 = F_5 - F_6$$
$$P_5 = -\frac{F_1 + F_2}{L_1} + \frac{F_3 + F_4}{L_2} - F_8$$
$$P_6 = F_7 \cos \alpha - F_9 \cos \beta$$
$$P_7 = F_7 \sin \alpha + F_8 + F_9 \sin \beta$$
$$P_8 = F_6 + F_9 \cos \beta$$

which in matrix form may be written

$$\{P\} = [A]\{F\}$$

where

$$[A] =$$

F\P	1	2	3	4	5	6	7	8	9
1	+1.0								
2		+1.0	+1.0						
3				+1.0					
4					+1.0	−1.0			
5	$-\dfrac{1}{L_1}$	$-\dfrac{1}{L_1}$	$+\dfrac{1}{L_2}$	$+\dfrac{1}{L_2}$				−1.0	
6							+cos α		−cos β
7							+sin α	+1.0	+sin β
8						+1.0			+cos β

20.4 DEFINITION OF DEFORMATION MATRIX [B]

The deformation matrix [B] expresses the internal deformations $\{e\}$, that is, member end rotations or member elongations, in terms of joint displacements $\{X\}$. To establish the matrix [B] by columns, let a particular joint displacement X equal unity while holding the other X's equal to zero. The resulting internal deformations are the column values for the [B] matrix.

For example, to obtain the deformation matrix for the structure of Fig. 20.2.1, refer to Fig. 20.4.1, where each of the eight X's is applied to the structure while the other seven X's are zero. The data in each of the eight parts of Fig. 20.4.1 provide the values for the eight columns of the deformation matrix [B]. Application of X_1 alone

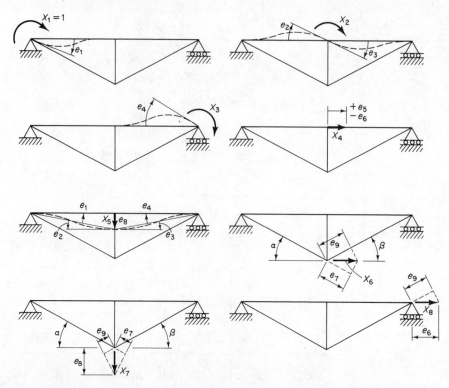

Figure 20.4.1 Effect on internal deformations e of X_1 through X_8 applied individually with all other X's equal to zero.

gives

$$e_1 = X_1$$

Application of X_5 alone gives

$$e_1 = -\frac{X_5}{L_1}$$

$$e_2 = -\frac{X_5}{L_1}$$

$$e_3 = +\frac{X_5}{L_2}$$

$$e_4 = +\frac{X_5}{L_2}$$

$$e_8 = -X_5$$

Application of X_7 alone gives

$$e_7 = +X_7 \sin \alpha$$

$$e_8 = +X_7$$

$$e_9 = +X_7 \sin \beta$$

Similarly, the coefficients for the other columns of the deformation matrix [B] may be obtained. The complete matrix [B] is as follows:

$$[B] =$$

X \ e	1	2	3	4	5	6	7	8
1	+1.0				$-\dfrac{1}{L_1}$			
2		+1.0			$-\dfrac{1}{L_1}$			
3		+1.0			$+\dfrac{1}{L_2}$			
4			+1.0		$+\dfrac{1}{L_2}$			
5				+1.0				
6				−1.0				+1.0
7						$+\cos\alpha$	$+\sin\alpha$	
8					−1.0		+1.0	
9						$-\cos\beta$	$+\sin\beta$	$+\cos\beta$

As usual, the [B] matrix is verified to be the transpose of the [A] matrix.

20.5 MEMBER STIFFNESS MATRIX [S]

For the structure containing flexural members as well as truss members, the member stiffness matrix is a combination of that used in Section 16.5 for trusses and that used in Section 17.5 for flexural members. The general equation

$$\{F\} = [S]\{e\}$$

is valid whether flexural stiffness or axial stiffness is involved.

For flexural members the internal forces F are in pairs, so the member stiffness equations are in pairs. For members in which axial deformation is to be considered, the axial stiffness EA/L is involved; and since there is one unknown axial force per member, the member stiffness equations include only one per member.

For the structure of Fig. 20.2.1, the member stiffness equations are as follows:

$$F_1 = \frac{4E_1I_1}{L_1}e_1 + \frac{2E_1I_1}{L_1}e_2$$

$$F_2 = \frac{2E_1I_1}{L_1}e_1 + \frac{4E_1I_1}{L_1}e_2$$

$$F_3 = \frac{4E_2I_2}{L_2}e_3 + \frac{2E_2I_2}{L_2}e_4$$

$$F_4 = \frac{2E_2I_2}{L_2}e_3 + \frac{4E_2I_2}{L_2}e_4$$

$$F_5 = \frac{E_1A_1}{L_1}e_5$$

$$F_6 = \frac{E_2A_2}{L_2}e_6$$

$$F_7 = \frac{E_3A_3}{L_3}e_7$$

$$F_8 = \frac{E_4A_4}{L_4}e_8$$

$$F_9 = \frac{E_5A_5}{L_5}e_9$$

In matrix form the member stiffness matrix $[S]$ becomes

$[S] =$

$_F\diagdown^e$	1	2	3	4	5	6	7	8	9
1	$\frac{4E_1I_1}{L_1}$	$\frac{2E_1I_1}{L_1}$							
2	$\frac{2E_1I_1}{L_1}$	$\frac{4E_1I_1}{L_1}$							
3			$\frac{4E_2I_2}{L_2}$	$\frac{2E_2I_2}{L_2}$					
4			$\frac{2E_2I_2}{L_2}$	$\frac{4E_2I_2}{L_2}$					
5					$\frac{E_1A_1}{L_1}$				
6						$\frac{E_2A_2}{L_2}$			
7							$\frac{E_3A_3}{L_3}$		
8								$\frac{E_4A_4}{L_4}$	
9									$\frac{E_5A_5}{L_5}$

20.6 GLOBAL STIFFNESS MATRIX [K]

The global stiffness matrix $[K]$ is the relationship between the external forces P and the joint displacements X. As discussed in Sections 16.6, 17.6, and 19.5, the relationship is

$$\{P\} = [K]\{X\}$$

Since $[K] = [ASB]$, the matrix $[K]$ obtained by matrix multiplication is shown in Table 20.6.1.

To illustrate the direct determination of the matrix $[K]$, the data for obtaining column five of the matrix $[K]$ are shown in Fig. 20.6.1. Each joint displacement in turn is considered to act alone, with the other displacements all held at zero. The free-body diagrams of the joints are drawn and the set of joint forces $\{P\}$ necessary for equilibrium gives the elements in that column of the $[K]$ matrix. The complete $[K]$ matrix is given in Table 20.6.1. As always, the $[K]$ matrix is seen to be a symmetric matrix.

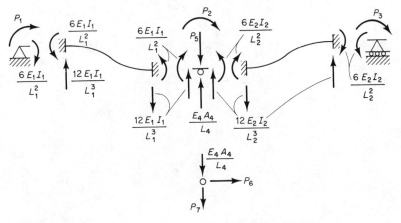

Figure 20.6.1   Data for column five of the global stiffness matrix $[K]$; structure of Fig. 20.2.1.

20.7 EXTERNAL FORCE MATRIX $\{P\}$

As discussed in Sections 17.7 and 19.7, the actual loading, which rarely occurs only at joints, is replaced by equivalent joint forces. These joint forces may be considered as consisting of two parts: (1) the forces acting on the member ends necessary to hold all joint displacements X equal to zero; and (2) the forces acting at the joints necessary to cause the true joint displacements.

To illustrate the concept for the structure containing both members having flexural stiffness and members having axial stiffness, refer again to the structure of Fig. 20.2.1. When the specific loading acts on the structure a set of fixed-end forces must act on the member ends to conform with the situation that all the joint displacements X are equal to zero. The forces shown on the free-body diagrams of the members in Fig. 20.7.1 are the fixed-end forces acting to keep all X's equal to zero. In terms of joint forces $\{P\}$, the action at each joint would be opposite to the direction of the fixed-end forces acting on the member ends. Thus, the equivalent joint forces are observed from Fig. 20.7.1 to be

TABLE 20.6.1 $[K]$ MATRIX FOR THE STRUCTURE OF FIG. 20.2.1

P\X	1	2	3	4	5	6	7	8
1	$\dfrac{4E_1I_1}{L_1}$	$\dfrac{2E_1I_1}{L_1}$	0	0	$-\dfrac{6E_1I_1}{L_1^2}$	0	0	0
2	$\dfrac{2E_1I_1}{L_1}$	$\dfrac{4E_1I_1}{L_1} + \dfrac{4E_2I_2}{L_2}$	$\dfrac{2E_2I_2}{L_2}$	0	$-\dfrac{6E_1I_1}{L_1^2} + \dfrac{6E_2I_2}{L_2^2}$	0	0	0
3	0	$\dfrac{2E_2I_2}{L_2}$	$\dfrac{4E_2I_2}{L_2}$	0	$+\dfrac{6E_2I_2}{L_2^2}$	0	0	0
4	0	0	0	$\dfrac{E_1A_1}{L_1} + \dfrac{E_2A_2}{L_2}$	0	0	0	$-\dfrac{E_2A_2}{L_2}$
5	$-\dfrac{6E_1I_1}{L_1^2}$	$-\dfrac{6E_1I_1}{L_1^2} + \dfrac{6E_2I_2}{L_2^2}$	$+\dfrac{6E_2I_2}{L_2^2}$	0	$\dfrac{12E_1I_1}{L_1^3} + \dfrac{12E_2I_2}{L_2^3} + \dfrac{E_4A_4}{L_4}$	0	$-\dfrac{E_4A_4}{L_4}$	0
6	0	0	0	0	0	$\dfrac{E_3A_3}{L_3}\cos^2\alpha + \dfrac{E_5A_5}{L_5}\cos^2\beta$	$\dfrac{E_3A_3}{L_3}\sin\alpha\cos\alpha - \dfrac{E_5A_5}{L_5}\cos\beta\sin\beta$	$-\dfrac{E_5A_5}{L_5}\cos^2\beta$
7	0	0	0	0	$-\dfrac{E_4A_4}{L_4}$	$\dfrac{E_3A_3}{L_3}\cos\alpha\sin\alpha - \dfrac{E_5A_5}{L_5}\cos\beta\sin\beta$	$\dfrac{E_3A_3}{L_3}\sin^2\alpha + \dfrac{E_5A_5}{L_5}\sin^2\beta + \dfrac{E_4A_4}{L_4}$	$+\dfrac{E_5A_5}{L_5}\sin\beta\cos\beta$
8	0	0	0	$-\dfrac{E_2A_2}{L_2}$	0	$-\dfrac{E_5A_5}{L_5}\cos^2\beta$	$\dfrac{E_5A_5}{L_5}\sin\beta\cos\beta$	$+\dfrac{E_2A_2}{L_2} + \dfrac{E_5A_5}{L_5}\cos^2\beta$

$$P_1 = +\frac{wL_1^2}{12} \qquad \text{(joint } A\text{)}$$

$$P_2 = \frac{w}{12}(L_2^2 - L_1^2) \quad \text{(joint } B\text{)}$$

$$P_3 = -\frac{wL_2^2}{12} \qquad \text{(joint } C\text{)}$$

$$P_4 = 0 \qquad\qquad \text{(joint } B\text{)}$$

$$P_5 = \frac{w}{2}(L_1 + L_2) \quad \text{(joint } B\text{)}$$

$$P_6 = 0$$

$$P_7 = 0$$

$$P_8 = 0$$

In matrix format the equivalent joint forces become the external force matrix $\{P\}$, as follows:

$$\{P\} =$$

LC / P	1
1	$+\frac{wL_1^2}{12}$
2	$+\frac{w}{12}(L_2^2 - L_1^2)$
3	$-\frac{wL_2^2}{12}$
4	0
5	$+\frac{w}{2}(L_1 + L_2)$
6	0
7	0
8	0

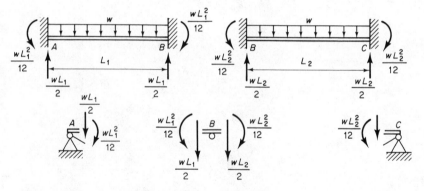

Figure 20.7.1 Fixed condition (X_1 through $X_8 = 0$) for the structure of Fig. 20.2.1.

20.8 MATRIX DISPLACEMENT METHOD

The mathematical equations for the matrix displacement method, as given by Eqs. (17.8.1) to (17.8.6), are still valid. It makes no difference whether (1) only axial stiffness is involved as for a truss described in Chapter 16; (2) only flexural stiffness is involved as for a beam in Chapter 17 and for a rigid frame with axial deformation ignored in Chapters 18 and 19; or (3) both axial stiffness and flexural stiffness are involved as for a rigid frame containing truss-type members described in this chapter.

The following examples illustrate the application of the matrix displacement method to structures containing more than one type of the following three types of elements: (1) truss element with axial stiffness only, (2) beam element with finite value of flexural stiffness but with infinitely large axial stiffness, and (3) combined truss and beam elements with both axial stiffness and flexural stiffness.

Example 20.8.1

Analyze the structure of Fig. 20.8.1 by the matrix displacement method.

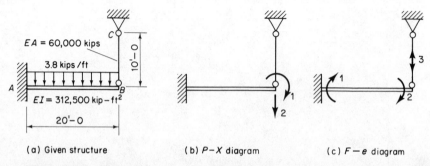

(a) Given structure   (b) P–X diagram   (c) F–e diagram

Figure 20.8.1 Given structure, degrees of freedom, and internal forces for Example 20.8.1.

Solution

(a) *Draw the P-X and F-e diagrams.* The analyst must first make the observation as to whether axial stiffness, or flexural stiffness, or both, are to be involved for each of the members of the structure. In this case, the member AB is a flexural member fixed at A and supported by member BC at end B. Member BC is subject only to axial load. Thus, member AB will be treated as a beam element and member BC as a truss element. Next, the degrees of freedom may be identified. The beam may have rotation and deflection at joint B, identifying P_1-X_1 and P_2-X_2. Member BC has axial deformation, thus permitting vertical deflection at B, which has already been taken care of by P_2-X_2. The P-X diagram is shown in Fig. 20.8.1b. The unknown internal forces are the end moments F_1 and F_2 on the beam element and F_3 on the truss element. The F-e diagram is given in Fig. 20.8.1c. Since $NF - NP = 3 - 2 = 1$, the structure is statically indeterminate to the first degree.

(b) *Establish the statics matrix* $[A]$. The free-body diagrams of Fig. 20.8.2a, par-

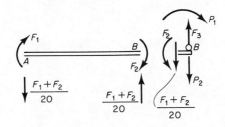

(a) Data for statics matrix $[A]$.

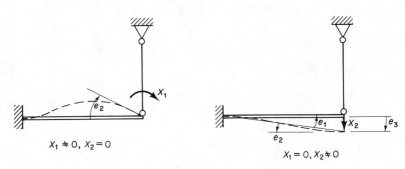

(b) Data for deformation matrix $[B]$.

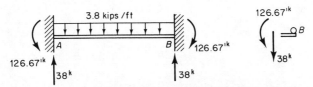

(c) Fixed condition; data for external force matrix $[P]$.

Figure 20.8.2 Data for Example 20.8.1.

ticularly the free-body diagram of joint B, provide the data for the statics matrix $[A]$. The statics equations are

$$P_1 = F_2$$

$$P_2 = -\frac{F_1 + F_2}{20} + F_3$$

or, in matrix form, according to

$$\{P\} = [A]\{F\}$$

$[A] =$	P \ F	1	2	3
	1	0	$+1.0$	0
	2	$-\dfrac{1}{20}$	$-\dfrac{1}{20}$	$+1.0$

(c) *Establish the deformation matrix* $[B]$. The displacement diagrams for the deformation matrix are given in Fig. 20.8.2b. Each displacement X is taken to act alone and the corresponding internal deformations comprise the column elements of the deformation matrix $[B]$, which is given as follows:

$[B] =$	e \ X	1	2
	1	0	$-\dfrac{1}{20}$
	2	$+1.0$	$-\dfrac{1}{20}$
	3	0	$+1.0$

Verify that the deformation matrix $[B]$ is the transpose of the statics matrix $[A]$.

(d) *Establish the member stiffness matrix* $[S]$. Use Eqs. (17.5.4) for members considered as having flexural stiffness, and Eq. (16.5.2) for members considered as having axial stiffness. For member AB,

$$\frac{4EI}{L} = \frac{4(312,500)}{20} = 62,500 \text{ ft-kips}$$

For member BC,

$$\frac{EA}{L} = \frac{60,000}{10} = 6000 \text{ kips/ft}$$

Thus, the member stiffness matrix $[S]$ becomes

F \ e	1	2	3
$[S] =$ 1	62,500	31,250	
2	31,250	62,500	
3			6000

(e) *Establish the external force matrix* $\{P\}$. The fixed (zero displacements) condition is shown in Fig. 20.8.2c, where the forces acting on joint B (opposite to the fixed-end forces acting on member ends) give the following external force matrix:

P \ LC	1
$\{P\} =$ 1	-126.67
2	$+38.0$

(f) *Obtain the* $[SA^T]$ *matrix.* For later use in Eq. (17.8.5),

$$\{F\} = [SA^T]\{X\} \qquad [17.8.5]$$

the matrix multiplication gives

F \ X	1	2
$[SA^T] =$ 1	$+31,250$	-4687.5
2	$+62,500$	-4687.5
3	0	$+6000$

(g) *Obtain the global stiffness matrix.* $[K]$. Matrix multiplication of the premultiplier matrix $[A]$ with the postmultiplier matrix $[SA^T]$ gives

P \ X	1	2
$[K] =$ 1	$+62,500$	-4687.5
2	-4687.5	$+6468.75$

(h) *Obtain the inverse of the matrix* $[K]$. For use in Eq. (17.8.4),

$$\{X\} = [K]^{-1}\{P\} \qquad [17.8.4]$$

the $[K]$ matrix must be inverted. After inversion the $[K]^{-1}$ matrix is

	P X	1	2
$[K]^{-1} =$	1	$+0.16919541$	$+0.12260537$
	2	$+0.12260537$	$+1.6347382$

($\times 10^{-4}$ applied to the first row as shown)

(i) *Obtain the joint displacement matrix* $\{X\}$. *Using Eq. (17.8.4),*

	$\begin{array}{c}LC\\ X\end{array}$	1
$\{X\} =$	1	-0.00167730 rad
	2	$+0.00465896$ ft

(j) *Obtain the internal force matrix* $\{F\}$. *Using Eq. (17.8.5),*

$$\{F\} = [SA^T]\{X\} \qquad\qquad [17.8.5]$$

gives

	$\begin{array}{c}LC\\ F\end{array}$	1
$\{F\} =$	1	-74.25 ft-kips
	2	-126.67 ft-kips
	3	$+27.95$ kips

These internal forces $\{F\}$, which cause the displacements $\{X\}$, must be added to the internal forces (fixed-end moments) in the fixed condition where all displacements are zero.

(k) *Obtain the final internal forces.*

$$\{F^*\} = \{F_0\} + \{F\}$$

$\{F^*\} =$	-126.67	$+$	-74.25	$=$	-200.92
	$+126.67$		-126.67		$0.$
	$0.$		$+27.95$		$+27.95$

(l) *Checks.* Checks of statics and compatibility of deformations are always necessary to verify the correctness of the solution. The checks are discussed in Sections 20.9 and 20.10 and are ommitted here. The logic of the result should be examined first. Is the answer obtained possible? Does it make sense? For this example, if the rod

BC were not there, the member *AB* would be a cantilever beam where the end moment at *A* would be $3.8(20)^2/2 = 760$ ft-kips. The hanger *BC* takes up a tensile force, reduces deflection, and decreases the end moment at *A* to 201 ft-kips. Note that this small deflection at *B* (0.056 in.) increased the moment at *A* from 190 ft-kips for $X_2 = 0$ (a propped cantilever) to 201 with $X_2 = 0.056$ in.

Example 20.8.2

Analyze the structure of Fig. 20.8.3 by the matrix displacement method, considering all members to have axial stiffness and in addition members *AB* and *BC* to have flexural stiffness.

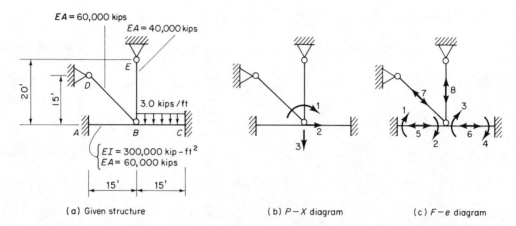

(a) Given structure   (b) *P−X* diagram   (c) *F−e* diagram

Figure 20.8.3 Given structure, degrees of freedom, and internal forces for Example 20.8.2.

Solution

(a) *Draw the P-X and F-e diagrams.* Referring to Fig. 20.8.3b, there will be a rotational displacement at joint *B* of the flexural member, and there will be both horizontal and vertical displacements of joint *B*. Thus, there are three degrees of freedom. For internal forces (Fig. 20.8.3c) there are end moments on members *AB* and *BC* which have flexural stiffness, and there are independently unknown axial forces in all members because all members are assumed to have finite values of axial stiffness. Since NF − NP = 5, the structure is statically indeterminate to the fifth degree.

(b) *Establish the statics matrix* [*A*]. The free-body diagrams are shown in Fig. 20.8.4. The equilbrium of the forces on joint *B* provides the statics equations, as follows:

$$P_1 = F_2 + F_3$$

$$P_2 = +F_5 - F_6 + \frac{\sqrt{2}}{2} F_7$$

$$P_3 = -\frac{F_1 + F_2}{15} + \frac{F_3 + F_4}{15} + \frac{\sqrt{2}}{2} F_7 + F_8$$

In matrix form the statics matrix is

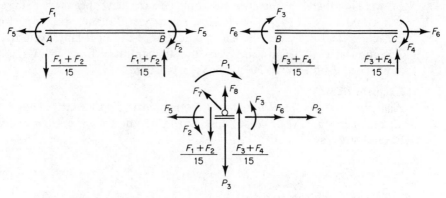

Figure 20.8.4 Free-body diagrams for the statics matrix $[A]$; Example 20.8.2.

$$[A] = $$

P \ F	1	2	3	4	5	6	7	8
1		+1.0	+1.0					
2					+1.0	−1.0	$+\frac{\sqrt{2}}{2}$	
3	$-\frac{1}{15}$	$-\frac{1}{15}$	$+\frac{1}{15}$	$+\frac{1}{15}$			$+\frac{\sqrt{2}}{2}$	+1.0

(c) *Establish the deformation matrix* $[B]$. The displacement diagrams for the deformation matrix $[B]$ are given in Fig. 20.8.5. The deformation matrix $[B]$ established by columns is as follows:

$$[B] = $$

e \ X	1	2	3
1			$-\frac{1}{15}$
2	+1.0		$-\frac{1}{15}$
3	+1.0		$+\frac{1}{15}$
4			$+\frac{1}{15}$
5		+1.0	
6		−1.0	
7		$+\frac{\sqrt{2}}{2}$	$+\frac{\sqrt{2}}{2}$
8			+1.0

Matrix Displacement Method Applied to Rigid Frames   Chap. 20

(f) *Obtain the [SA^T] matrix.* The matrix multiplication gives

$$[SA^T] =$$

F \ X	1	2	3
1	+40,000		−8000
2	+80,000		−8000
3	+80,000		+8000
4	+40,000		+8000
5		+4000	
6		−4000	
7		+2000	+2000
8			+2000

(g) *Obtain the global stiffness matrix [K].* Matrix multiplication of the premultiplier matrix [A] and the postmultiplier matrix [SA^T] gives

$$[K] =$$

P \ X	1	2	3
1	+160,000	0	0
2	0	+9414.2136	+1000$\sqrt{2}$
3	0	+1000$\sqrt{2}$	+5547.5469

(h) *Obtain the inverse of the matrix [K].* For use in Eq. (17.8.4),

$$\{X\} = [K]^{-1}\{P\} \qquad\qquad [17.8.4]$$

the [K] matrix must be inverted. After inversion the [K]$^{-1}$ matrix is

$$[K]^{-1} =$$

X \ P	1	2	3	
1	+0.006250	0	0	
2	0	+0.110453	−0.028157	× 10^{-3}
3	0	−0.028157	+0.187437	

(i) *Obtain the joint displacement matrix $\{X\}$. Use Eq. (17.8.4),*

$$\{X\} = \quad \begin{array}{|c|c|c|}
\hline
\diagbox{X}{LC} & 1 \\
\hline
1 & +0.00035156 \text{ rad} \\
\hline
2 & -0.00063354 \text{ ft} \\
\hline
3 & +0.00421735 \text{ ft} \\
\hline
\end{array}$$

(j) *Obtain the internal force matrix $\{F\}$. Using Eq. (17.8.5),*

$$\{F\} = [SA^T]\{X\} \qquad\qquad [17.8.5]$$

gives

$$\{F\} = \quad \begin{array}{|c|c|}
\hline
\diagbox{F}{LC} & 1 \\
\hline
1 & -19.676 \\
\hline
2 & -5.614 \\
\hline
3 & +61.8636 \\
\hline
4 & +47.8012 \\
\hline
5 & -2.53416 \\
\hline
6 & +2.53416 \\
\hline
7 & +7.16762 \\
\hline
8 & +8.4347 \\
\hline
\end{array}$$

(k) *Obtain the final internal forces.*

$$\{F^*\} = \{F_0\} + \{F\}$$

$$\{F^*\} = \begin{array}{|c|}
\hline
0. \\
\hline
0. \\
\hline
-56.25 \\
\hline
+56.25 \\
\hline
0. \\
\hline
0. \\
\hline
0. \\
\hline
0. \\
\hline
\end{array} + \begin{array}{|c|}
\hline
-19.68 \\
\hline
-5.61 \\
\hline
+61.86 \\
\hline
+47.80 \\
\hline
-2.53 \\
\hline
+2.53 \\
\hline
+7.17 \\
\hline
+8.43 \\
\hline
\end{array} = \begin{array}{|c|}
\hline
-19.68 \\
\hline
-5.61 \\
\hline
+5.61 \\
\hline
+104.05 \\
\hline
-2.53 \\
\hline
+2.53 \\
\hline
+7.17 \\
\hline
+8.43 \\
\hline
\end{array}$$

The internal forces are shown on an answer diagram in Fig. 20.8.7.

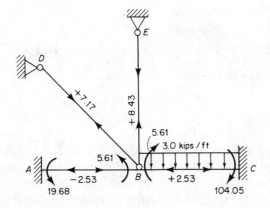

Figure 20.8.7 Answer diagram for Example 20.8.2. Moments in ft-kips and axial forces in kips.

(l) *Check.* Checks of statics and compatibility of deformations are always necessary to verify correctness of solution. The checks are discussed in Sections 20.9 and 20.10, where the checks for this example are shown.

20.9 STATICS CHECKS

The statics checks are equal to the number of degrees of freedom. For the frame of Example 20.8.2, the complete statics check is at joint B. Figure 20.9.1 shows the free-body diagrams of members AB and BC and joint B. For horizontal equilibrium on joint B,

$$\Sigma F_x = 0$$

$$2.53 + 2.53 - 5.07 = -0.01 \approx 0 \quad \text{(check)}$$

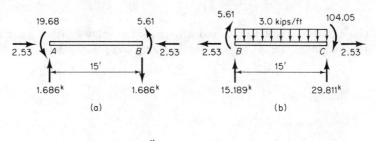

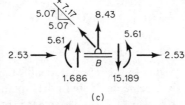

Figure 20.9.1 Free-body diagrams for statics checks for the structure of Example 20.8.2.

For vertical force equilibrium,

$$\Sigma F_y = 0$$

$$5.07 + 8.43 + 1.686 - 15.189 = -0.003 \approx 0 \quad \text{(check)}$$

Moment equilibrium is satisfied, as seen by inspection.

20.10 DEFORMATION CHECKS

As continually emphasized in Sections 16.9, 17.10, 18.4, and 19.11, making deformation checks is an essential part of analyzing statically indeterminate structures. The checks consist of determining the slopes to the elastic curve at both ends of each member to verify the values of rotational displacement X from the matrix displacement method solution. In addition, the translational displacements of joints must be computed from internal deformations to verify the matrix solution result. The number of deformation checks is equal to the number of independent internal forces as used in the $\{F\}$ matrix.

Example 20.10.1

Illustrate the deformation checks for the rigid frame analyzed in Example 20.8.2.

Solution

(a) *Member EB.* The elongation of this member is

$$\text{Elongation} = e_8 = \frac{F_8 L_8}{(EA)_8} = \frac{8.43(20)}{40,000} = 0.0042150 \text{ ft}$$

This agrees favorably with computed $X_3 = +0.004217$.

(b) *Member BD.* The elongation of this member is

$$\text{Elongation} = \frac{F_7 L_7}{(EA)_7} = \frac{7.17(15\sqrt{2})}{60,000} = +0.002535 \text{ ft}$$

This elongation is to be checked with the value obtained from using the joint displacement equation, Eq. (3.4.1), where X_1 through X_4 are as defined in Fig. 3.4.1,

$$e = (X_3 - X_1) \cos \alpha + (X_4 - X_2) \sin \alpha$$

For this member,

$$\text{Elongation} = e_8 = (0 - X_2)\left(-\frac{\sqrt{2}}{2}\right) + (0 + X_3)\left(+\frac{\sqrt{2}}{2}\right)$$

$$= (0 + 0.00063354)\left(-\frac{\sqrt{2}}{2}\right) + (0 + 0.00421735)\left(+\frac{\sqrt{2}}{2}\right)$$

$$= +0.002534 \text{ ft}$$

Thus, the elongation due to the internal axial force agrees with the elongation due to the external joint displacements. Note in the equation above that X_2 is positive to the right, whereas X_3 is positive for the downward direction; but the joint displacement equation, Eq. (3.4.1), is based on X's to the right and upward as positive. In fact, it is easier to find the elongation of member BD due to horizontal and vertical deflections of joint B by inspecting Fig. 20.8.7, or

$$\text{Elongation} = e_8 = 0.00421735 \left(\frac{\sqrt{2}}{2}\right) - 0.00063354 \left(\frac{\sqrt{2}}{2}\right)$$

$$= +0.002534 \text{ ft}$$

because the downward deflection of joint B makes the bar longer and the deflection to the left of joint B makes the bar shorter.

(c) *Member AB*. The elongation of this member is

$$\text{Elongation} = e_5 = \frac{F_5 L_5}{(EA)_5} = \frac{(-2.53)(15)}{60,000} = -0.0006325 \text{ ft}$$

which agrees with the computed X_2.

Next, the slope at B and the deflection at B must be computed in accordance with the principles of flexural deformation. Using the moment area method, the M/EI diagrams are required; these are given in Fig. 20.10.1.

First, compute the slope at B by the moment area method; clockwise θ_B is equal to the absolute value of the negative M/EI area between A and B (see Fig. 20.10.1), since θ_A equals zero.

$$\theta_B = \frac{1}{EI}\left[\frac{1}{2}(19.68)(15) - \frac{1}{2}(5.61)(15)\right] = \frac{105.5}{EI} = 0.3518 \times 10^{-3} \text{ rad} \quad \text{(clockwise)}$$

This checks with the computed X_1.

Next compute the vertical deflection at B by the moment area method; thus,

$$\Delta_B = \Delta'_{BA} = \frac{1}{EI}\left[\frac{1}{2}(19.68)(15)(10) - \frac{1}{2}(5.61)(15)(5)\right] = \frac{1265.6}{EI}$$

$$= 4.218 \times 10^{-3} \text{ ft} \quad \text{(downward)}$$

This checks with the computed X_3.

(d) *Member BC*. The elongation of this member is

$$\text{Elongation} = e_6 = \frac{F_6 e_6}{(EA)_6} = \frac{(+2.53)(15)}{60,000} = +0.0006325 \text{ ft}$$

This agrees with the computed X_2 of 0.00063354 ft to the left. The small discrepancy is due to the smaller number of significant figures used in the internal force.

Applying the moment area method to the component M/EI areas for member BC in Fig. 20.10.1 gives

$$\theta_B = \theta_C + \frac{1}{EI}\left[\frac{2}{3}(84.375)(15) - \frac{1}{2}(104.05)(15) + \frac{1}{2}(5.61)(15)\right]$$

$$= 0 + \frac{105.45}{EI} = 0.3515 \times 10^{-3} \text{ rad} \quad \text{(clockwise)}$$

This checks with the computed X_1.

$$\Delta_B = \frac{1}{EI}\left[\frac{1}{2}(104.05)(15)(10) - \frac{1}{2}(5.61)(15)(5) - \frac{2}{3}(84.375)(15)(7.5)\right]$$

$$= \frac{1265.2}{EI} = 0.004218 \text{ ft} \quad \text{(downward)}$$

This checks with the computed X_3.

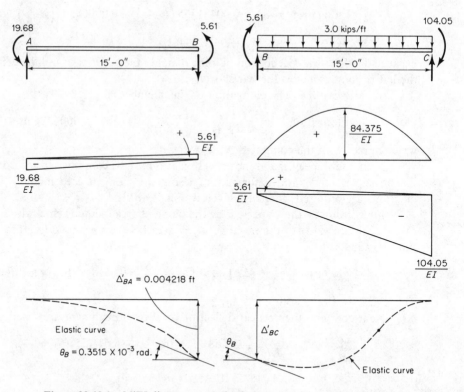

Figure 20.10.1 M/EI diagrams and elastic curve used in making deformation checks for Example 20.10.1 (structure of Fig. 20.8.7).

(e) *Discussion.* In general, there should be NF deformation checks, where NF is the number of independent internal forces used in the solution. For a truss element, there is one deformation check; for a beam element, two; and for a combined truss and beam element, three. In the deformation check for a truss element, the elongation due to axial force is checked against the elongation due to the horizontal and vertical displacements of the ends of the member. For a member in bending, the end rotations computed by treating the member as a simple beam are checked against the end rotations due to the rotations and transverse deflections at the ends of the member. The former values are due to loads applied on the member itself as well as due to the end moments (F^*-values). The latter values are measured from the new member axis direction (different from the original member axis direction if the transverse deflections of the member ends are not equal) to the tangents to the elastic curve as given by the joint rotations. Although the procedure described above is the most general one for an irregular-shaped structure with most joints displaced in translation, the deformation checks for bending in members AB and BC in the present problem are simplified by the fixed ends at A and C.

PROBLEMS

20.1. Verify the values in columns 6, 7, and 8 of Table 20.6.1.

20.2–20.6. Analyze the structure of the accompanying figure using the matrix displacement method. Consider axial deformation only in those members for which a value of *EA* is given.

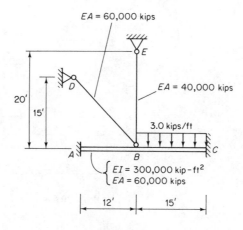

Prob. 20.2

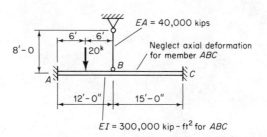

Prob. 20.3

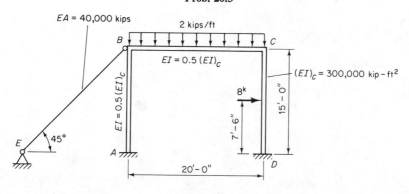

Prob. 20.4

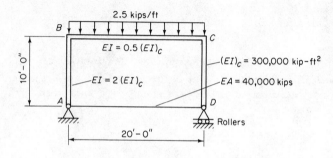

Prob. 20.5

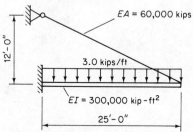

Prob. 20.6

Index

Bending moment (*Cont.*)
  minimum values, 163, 168, 173
  relationships, 94, 112
  sign convention, 90
Bendixen, Axel, 288, 302
Benscoter, Stanley U., 438
Bernoulli, Johann, 63
Bertot, 250, 280
Betti, E., 221, 236, 438
Biggs, John M., 33
Bow, Robert, 48, 51
Bowman, Harry Lake, 405
Bow's notation, 48, 50
Bryan, C. W., 392, 404
Buckling, 26
Buettner, Donald R., 280

Cable-supported roof, photo, 8
Cantilever method, 398, 401
Carry-over factor, 309
Castigliano, Carlo A., 223, 224, 235, 236, 275, 438
Castigliano theorem, part I, 223, 275, 276, 434, 435, 436
  formula, 225
  stated, 224
  using maximum displacement method, 434
Castigliano theorem, part II, 227, 276, 420, 427, 433
  formula, 228
  stated, 227
  using matrix force method, 427
  using redundant forces, 228
Centroids of areas, table of, 133
"Checkerboard" loading, 383
Checks:
  deformation, 474, 499, 507, 525, 553, 580
  statics, 473, 499, 505, 523, 552, 579
Chen, Pei-Ping, 438
Clapeyron, E. P. B., 2, 33, 243, 250, 254, 280
Classical methods, 32
Clough, Ray W., 34, 405, 438, 439
Coefficients, moment, 390
Compatibility conditions, 207, 355, 462
  rigid frames, 350
Compatible displacement, 64
Compatible state, 65, 115
Composite structure, 557
Computer methods, 32
Conjugate beam, 144
  alternate theorems, 145
  comparison with other methods, 152
  method, 144-152, 356
  theorems, 144
  typical situations, 146
Consistent deformation method, 209, 211
  applied to beams, 243
  applied to rigid frames, 346, 352
  support reaction as redundant, 211, 243

truss member as redundant, 215
Continuous beam, 242 (*see also* Beams)
  approximate analysis, 389
Criteria for maximum, 173, 175
  bending moment due to concentrated loads, 173
Cross, Hardy, 305, 324

Dead load, 9
Deflection:
  beams, 110
  conjugate beam method for, 144
  differential equation relationships, 112
  moment area method for, 130, 247, 342
  reciprocal, 127
  rigid frames, 342
  trusses, 62, 447
Deformation (*see* Deflection)
Deformation checks:
  matrix method, 474, 499, 507, 525, 580
Deformation matrix, 446, 456, 457, 464, 484, 495, 501, 511
  beams, 484, 495, 501
  rigid frames, 518, 534, 548, 553
    with truss-type members, 561, 570, 574
  truss, 456, 461, 464
Degree of freedom, 208, 454, 480, 530, 558
  rigid frame with truss-type members, 558
  sidesway, 333, 370, 376, 530
Degree of statical indeterminacy, 207-209, 454, 480, 530, 558
Denke, P. H., 438, 439
Design:
  definition of, 110
  relationship to slope and deflection of beams, 110
Determinate (*see* Statically determinate)
Diagram:
  answer for truss forces, 41
  free-body (*see* Free-body diagram)
Displacement (*see* Deflection)
Displacement method, 29, 31, 62, 211, 218, 287, 454, 480, 493
  concept applied to beams, 287
  concept applied to frames, 331
  defined, 29, 31
  matrix:
    for beams, 480, 493
    for rigid frames with sidesway, 530
    for rigid frames without sidesway, 516
    for trusses, 429, 454
Distribution factor, 310
Double integration method, 112
Duhem, P., 80

Elastic curve:
  defined, 22, 110
  differential equation for, 111